# EcoDesign for Circular Value Creation: Volume I

Shinichi Fukushige • Tomomi Nonaka
Hideki Kobayashi • Chiharu Tokoro
Eiji Yamasue
Editors

# EcoDesign for Circular Value Creation: Volume I

Springer

*Editors*
Shinichi Fukushige
School of Creative Science and Engineering
Waseda University
Tokyo, Japan

Tomomi Nonaka
School of Creative Science and Engineering
Waseda University
Tokyo, Japan

Hideki Kobayashi
Graduate School of Engineering
Osaka University
Suita, Osaka, Japan

Chiharu Tokoro
School of Creative Science and Engineering
Waseda University
Tokyo, Japan

Eiji Yamasue
Department of Mechanical Engineering
Ritsumeikan University
Shiga, Japan

ISBN 978-981-97-9067-8 ISBN 978-981-97-9068-5 (eBook)
https://doi.org/10.1007/978-981-97-9068-5



This Springer imprint is published by the registered company Springer Nature Singapore Pte Ltd.
The registered company address is: 152 Beach Road, #21-01/04 Gateway East, Singapore 189721, Singapore

# Preface

Circular economy aims at transforming the linear model of take-make-waste production in industrial economy toward a model that is restorative. Circularity is increasingly recognized as a source of value in the manufacturing industry over the last decades. EcoDesign has remained a core concept for circular value creation as a key technology of constructing sustainable socio-technical systems, including product service system, product lifecycle system, transportation system, energy system, and production system. To minimize their resource consumption and environmental impact while providing sufficient functionality to the society, the role of EcoDesign will continue to expand in the coming decades.

This book collates 72 papers out of 200 papers presented at EcoDesign2023—the 13th International Symposium on Environmentally Conscious Design and Inverse Manufacturing, which was held online from November 29 to December 1, 2023. All the 72 papers were peer-reviewed by the EcoDesign 2023 Executive Committee. EcoDesign 2023 provided the excellent platform to share the state-of-the-art research and practices in the field of EcoDesign. A total of 334 researchers and practitioners from 23 countries participated in EcoDesign 2023.

The book consists of two volumes, i.e., the first volume focuses on "Products, Services, and Business Models" and the second volume focuses on "Social Perspectives and Sustainability Assessment." Reflecting the expansion of the symposium scope, the book chapters cover broad areas—product and service design, business models and policies, circular production and life cycle management, digital technologies, sustainable manufacturing, user behavior and health, sustainable consumption and production, ecodesign of social infrastructure, sustainability education, sustainability indicators, and energy system design. We believe that the methods, tools, and practices described in the chapters are useful for readers to facilitate value creation for sustainability.

Last but not least, we would like to express our sincere appreciation to all the contributors, supporters, and participants of EcoDesign 2023. This book cannot be published without the help of the executive committee members who cooperated in the peer review of the papers.

Tokyo, Japan Shinichi Fukushige
Tokyo, Japan Tomomi Nonaka
Osaka, Japan Hideki Kobayashi
Tokyo, Japan Chiharu Tokoro
Shiga, Japan Eiji Yamasue

# Contents

**Part I Product & Service Design**

**Proposal and Performance Evaluation of a Fuel Consumption System for H-Bikes, Considering the Shared-Cycle-Use Phase** . . . . . . . . 3
Ryuta Nagado, Satoshi Sakai, Tomoya Ezawa, Koya Honda, Masami Sumita, Noboru Katayama, Shan Miao, and Kiyoshi Dowaki

**A Discussion of the Thermal Management System for Metal Hydride Cartridges for H-Bike** . . . . . . . . 25
Satoshi Sakai, Ryuta Nagado, Tomoya Ezawa, Koya Honda, Noboru Katayama, Shan Miao, and Kiyoshi Dowaki

**Replacement of Manipulator Modules Based on Workload Using Part Agents** . . . . . . . . 43
Arata Hori and Hiroyuki Hiraoka

**Development of a Knowledge Repository Model for Conflict Management in Life Cycle Design Collaborations** . . . . . . . . 57
Jing Tao, Yingyu Chen, Rujing Wu, Suiran Yu, and Haiwei Zhou

**Development of Internal Audit Process to Obtain EPD Certifications in Response to Ecodesign Regulations** . . . . . . . . 69
Yong-jin Lee, Han-bi Kim, Ki-woong Jung, and Noh-hyun Lim

**Verification of Operational Effect of Solar-Powered System for Mobile Off-Grid Architecture** . . . . . . . . 79
Satoko Nasu and Shota Tajima

**Urban Conscious Product Design Integrating a Life Cycle Perspective** . . . . . . . . 97
Sina Rudolf, Marius Dellbrügge, Max Juraschek, Mark Mennenga, Uwe Clausen, and Christoph Herrmann

**Part II Sustainable Design and User Behavior**

**Designing Labels with a Lasting Impact: How Eco-labels Influence Purchase Behavior and Lifetime Extension of Clothing** . . . . . . . . . . . . . 113
Marie Das, Gabriela Dias, Leene Remeysen, Els Du Bois, and Ingrid Moons

**An Analysis of a Circular Economy Strategy Combining Green Packaging Reuse Design and Sustainable Jewelry Development** . . . . . . . . . . . . . . . . . . . . . . . . . . . . . . . . . . . . . . 131
Yu-Chen Huang and Tzu-Yun Chen

**Exploring Suitable Growing Parameters of Mycelium Prototyping for Art Installations** . . . . . . . . . . . . . . . . . . . . . . . . . . . . . 147
Tarteel Abdelsadek, Kazutoshi Tsuda, and Tomohiro Inoue

**The Role of Shame and Guilt in Designing Anti-littering Interventions** . . . . . . . . . . . . . . . . . . . . . . . . . . . . . . . . . . . . . . 161
June Kyong Trondsen, Casper Boks, Wanjun Chu, Babak Nemat, and Renee Wever

**Enhancing Sustainable Behavior in Waste Management: A Social Perspective** . . . . . . . . . . . . . . . . . . . . . . . . . . . . . . . . . 177
Siti Khatijah Zamhari, Ahmad Fariz Mohamed, and Azahan Awang

**Factors Influencing Smart Technologies Social Acceptance: A Mini Review Study** . . . . . . . . . . . . . . . . . . . . . . . . . . . . . . . . 195
Andante Hadi Pandyaswargo, Meilinda Fitriani Nur Maghfiroh, Tanika Dewi Sofianti, and Hiroshi Onoda

**Part III Business Models & Policies**

**Plastic Packaging Policy in Malaysia: An Analysis Towards the Implementation of Extended Producer Responsibility (EPR)** . . . . . . . . . 215
Khalilulnisha Abu Bakar, Ahmad Fariz Mohamed, and Yusuke Kishita

**Second Life of Electric Vehicle Lithium-Ion Batteries from a Sustainable Business Model Perspective** . . . . . . . . . . . . . . . . . . . . . . . 229
Saad Ahmed, Elli Verhulst, and Casper Boks

**European Union Policy Developments Related to Ecodesign, Green Claims, and Eco-labeling: Implications for Voluntary Circular Labeling Initiatives** . . . . . . . . . . . . . . . . . . . . . . . . . . . . . . . . . . 245
Karolina Vikingsson, Carl Dalhammar, and Leonidas Milios

**Factors Affecting Local Digital Transformation Toward the Sustainable Local Development** . . . . . . . . . . . . . . . . . . . . . . . . . . 261
Takaharu Takikawa and Yuna Seo

**Analysis of the Effects of Adopting "Imaginary Future Generations" on the Design of Technology Development and Business Proposal: Case Study of Workshop at a Plating Processing Company** . . 277
Ken Fujita, Tetsusei Kurashiki, Keishiro Hara, Junji Ikeda, and Takashi Nakamura

**Exploring Participatory Approaches for Facilitating ESG Initiatives in Organizations** . . . . . . . . . . . . . . . . . . . . . . . . . . . . . . . . . . . 293
Kamila Kunrath and Devarajan Ramanujan

**Impact Assessment of Low-Carbon Electricity Transition by Applying Input–Output Analysis: A Case Study of the Impacts on Natural Gas Power Plants in Taiwan** . . . . . . . . . . . . . . . . . . 313
Ping-Hung Chen and Yuh-Ming Lee

**Part IV Circular Production and Management**

**Circular Strategies in Southeast Asia: Cambodian Customer Perception of Remanufactured Motorcycle Parts** . . . . . . . . . . . . . . . . . . 335
R. Casper, J. Vogt Duberg, S. Seang, J. Kurilova-Palisaitiene, E. Sundin, M. Lindahl, and R. Banomyong

**Information Needs for Establishing Circularity-Focused Collaborations Among Product Manufacturers and Recyclers** . . . . . . . . 355
Giovanni Formentini, Rembrandt H. E. M. Koppelaar, Deborah Leone, Siro Dell'Ambrogio, Alessandro Fontana, Fabiana Guggiari, Tina Frangez, Slavko Dvoršak, Janez Matavš, Kristina Obu Vazner, Lucía Herreras Martínez, Eniko Hajosi, and Devarajan Ramanujan

**Thermal-Based Allocation Method for Multiple Co-product Resource Consumption in Plastic Recycling: A Case Study in Malaysia** . . . . . . . . 371
Ming Hui Tan, Meng Soon Chiong, Invidiadoloria Faian Ferolin, Yoon-Young Chun, Kenichiro Tsukahara, and Kiyotaka Tahara

**Material Flow Analysis for the Biomass Waste Management of Palm Oil Mills with Regard to Sustainable Palm Oil Production** . . . . 383
Noor Azimah Darus, Ahmad Fariz Mohamed, and Nor Diana Mohd Idris

**Readiness Status of Smart Waste Collection and Processing Technologies for Plastic Waste Recycling** . . . . . . . . . . . . . . . . . . . . . . . 393
Chaoxia Shan, Andante Hadi Pandyaswargo, and Hiroshi Onoda

**Research Framework for Urban Industrial Symbiosis** . . . . . . . . . . . . . . 407
Philipp Grimmel, Katja Knecht, Grace Abou Jaoude, Mark Mennenga, Olaf Mumm, Christoph Herrmann, and Vanessa Miriam Carlow

**Methodology to Model the Environmental Impact of the Overall Maintenance of Aircraft Engines** . . . . . . . . . . . . . . . . . . . . . . . 427
Michel Nocture, Maud Lemagnen, Bertrand Laratte, and Sarah Atroun

**Part V Digital Technologies for Sustainability**

**Digital Twins for Real-Time Life Cycle Assessment of Products** . . . . . . . 447
Theresa Riedelsheimer, Anne Seegrün, and Kai Lindow

**Toward a Framework for Sustainable Cross-Border Electronic Commerce** . . . . . . . . . . . . . . . . . . . . . . . . . . . . . . . . . . . . . . . . . . . 465
Jiahe Chen and Yi-Chen Lan

**How Should AI Technique Support Wastewater Treatment and Sludge Management? A Case Study from Mongolia** . . . . . . . . . . . . 479
Richao Cong, Tumurtogtokh Oyunchimeg, Atsushi Fujiyama, and Toru Matsumoto

**Development of an Ideal Training Dataset for Visual Analysis-Based Waste Sorting Robot: An Experiment with Mixed-Construction Waste** . . . . . . . . . . . . . . . . . . . . . . . . . . . . 495
Tianhao Cheng, Hao Hu, Andante Hadi Pandyaswargo, and Hiroshi Onoda

**Exploring the Current Applications and Potential of Extended Reality for Environmental Sustainability in Manufacturing** . . . . . . . . . . 515
Huizhong Cao, Henrik Söderlund, Mélanie Despeisse, and Björn Johansson

**Part VI Sustainable Manufacturing**

**Exploring Factory Data for Resource Efficiency Assessment: A Case Study at a Truck Manufacturing Company** . . . . . . . . . . . . . . . . 535
Qi Fang, Mélanie Despeisse, Ebru Turanoglu Bekar, Lena Moestam, Helena Söderberg, Dennis Andersson, and Björn Johansson

**An Analysis Method Considering Manufacturing Skills and Technologies Toward Slow Manufacturing** . . . . . . . . . . . . . . . . . . . 547
Yuta Irino, Hidenori Murata, Hideki Kobayashi, and Giampaolo Campana

**Reconfigurable Tooling for Circular Manufacturing: A Review** . . . . . . . 563
Cherukupally Shivaprasad, S. Suryakumar, Carla Susana A. Assuad, Kristian Martinsen, Sverre Gulbrandsen-Dahl, and N. Venkata Reddy

**Sustainable Disjoining of Aluminium Multi-material Joints** . . . . . . . . . . 581
Siri Marthe Arbo, Sverre Gulbrandsen-Dahl, Kristian Martinsen,
Mitsutaka Matsumoto, and S. Surya Kumar

**Applying Principles of Biological Transformation for Product Development and Manufacturing** . . . . . . . . . . . . . . . . . . . . . . . . . . . . . . 599
Magda Berkhahn, Ina Peters, and Rainer Stark

# Part I
# Product & Service Design

# Proposal and Performance Evaluation of a Fuel Consumption System for H-Bikes, Considering the Shared-Cycle-Use Phase

**Ryuta Nagado, Satoshi Sakai, Tomoya Ezawa, Koya Honda, Masami Sumita, Noboru Katayama, Shan Miao, and Kiyoshi Dowaki**

**Abstract** Fuel cells (FCs) are power-supply devices that efficiently convert input fuel from chemical to electrical energy. Polymer electrolyte membrane FCs (PEMFCs) are used for co-generation systems and FC vehicles because they can be used at room temperature. In the future, PEMFCs will be used in small-scale applications such as mobile batteries and electrically power-assisted bicycles (E-bikes). This study focuses on the use of FC electrically power-assisted bicycles (H-bike). The use of FCs in mobility applications requires an additional power source. Li-ion batteries are the most commonly used chemical rechargeable batteries. However, they gradually deteriorate through repeated charging and discharging. In addition, the high frequency of battery replacement due to degradation is a critical factor increasing their environmental impacts. Their increased environmental impacts during production and use are expected to contribute to higher product prices in the future.

This study focuses on degradation and fuel consumption, which are operational issues for H-bikes equipped with FC systems and metal hydride tanks during shared-used urban cycling. Among the multiple degradation factors, we focus on the degradation caused by the operating conditions and changes in the equipment configuration. First, the operating conditions are set up using various driving data obtained from actual measurements, and the degradation is evaluated by simulation using actual and experimental data. Thereafter, the environmental impact is determined based on an LCA. Finally, referring to the international standards for

R. Nagado (✉) · S. Sakai · T. Ezawa · S. Miao · K. Dowaki
Department of Industrial and Systems Engineering, Graduate School of Science and Technology, Tokyo University of Science, Chiba, Japan
e-mail: 7422536@ed.tus.ac.jp

K. Honda · N. Katayama
Department of Electrical Engineering, Graduate of Science and Technology, Tokyo University of Science, Chiba, Japan

M. Sumita
Department of Electrical Engineering, Faculty of Science and Technology, Tokyo University of Science, Chiba, Japan

 
S. Fukushige et al. (eds.), *EcoDesign for Circular Value Creation: Volume I*,
https://doi.org/10.1007/978-981-97-9068-5_1

environmental impact indicators, a comparison with existing E-bikes is performed to quantitatively demonstrate the superiority of introducing H-bikes.

**Keywords** Fuel cell · Li-ion battery · Degradation · H-bike · Efficiency · Fuel consumption · Life cycle assessment

## 1 Introduction

Fuel cells (FCs) have become increasingly popular in recent years owing to their ability to reduce environmental impacts. FCs are power-supply devices that efficiently convert input fuel from chemical to electrical energy. The most popular FCs include polymer electrolyte membrane FCs (PEMFCs), solid oxide FCs, molten carbonate FCs, and phosphoric acid FCs. PEM5FCs are used in cogeneration systems and FC vehicles (FCVs) because they can be used at room temperature, and they are widely used in large-scale applications [1]. In the future, PEMFCs will be used in small-scale applications, such as mobile batteries and electrically powered bicycles (E-bikes). E-bikes have attracted considerable attention as a means of transportation to achieve last-mile connectivity [2]. There is considerable interest in the use of shared bikes in urban transportation [2]. This study focuses on the use of FC electrically power-assisted bicycles (H-bikes), E-bikes with FCs, and metal hydride (MH) tanks during shared cycling [3]. Compared to E-bikes, the environmental impact of H-bikes is approximately 10% lower in the context of global warming. In addition [4], H-bikes with MHs have been reported to have approximately 20 kg-$CO_2$ eq. lower carbon dioxide emissions compared to E-bikes.

The use of FCs in mobility applications requires additional power sources. As FCs are not responsive to instantaneous loads, additional power sources are required for support. Examples of devices that serve as power sources include Li-ion supercapacitors. FCVs are currently widely used in Li-ion batteries. The H-bike system proposed in this study also utilizes Li-ion batteries, which are the most commonly used rechargeable chemical batteries. Li-ion batteries have high energy densities and can be rapidly charged and discharged. However, they gradually deteriorate with repeated charging and discharging. In addition, their environmental impact during manufacturing is a major concern, and the increased frequency of battery replacement owing to degradation is a critical factor in exacerbating the environmental impact.

The environmental impacts of manufacturing are partly related to the structure of PEMFCs and the presence of hydrogen. PEMFCs use Pt/C with platinum and carbon black as catalysts. The environmental impacts of Pt catalysts are considerable, and these catalysts are degraded during power generation. The life cycle assessment (LCA) method recommended by the International Electrotechnical Commission (IEC) requires FCs to be replaced when their performance decreases by 10% from their initial status [5].

Hydrogen, a widely used fuel, is classified into three types based on the fuel and process used to produce it: gray hydrogen, blue hydrogen, and green hydrogen. Gray hydrogen is produced from fossil fuels and emits carbon dioxide, and blue hydrogen is also produced from fossil fuels. The carbon dioxide emissions can be reduced by capturing or storing the carbon dioxide emitted during production. Green hydrogen is produced from renewable energy sources. Our research group developed bio-hydrogen manufactured from biomass feedstock. This study explores the operation of H-bikes that use bio-hydrogen.

Biomass is a renewable energy source with a low environmental impact due to its carbon neutrality. However, its energy density is approximately 50% lower than that of fossil fuels [6]. Regarding the production and supply of hydrogen from biomass, bio-hydrogen cannot be supplied over a wide scale, which limits its use. Therefore, FCs with high energy-conversion efficiencies using bio-hydrogen are necessary for efficient operation [7]. Because FCs generate electricity through electrochemical reactions, they are not significantly affected by scale or the operating temperature. Therefore, they can be used in various applications and for products consumed on a small scale.

The low volumetric energy density of hydrogen has created a need for efficient transportation and storage methods. Currently, liquefaction and high-pressure conversion are the main transportation and storage methods. However, liquefaction and gasification require large amounts of energy and have high environmental impacts [8]. Therefore, we believe that local use with small-scale distributed power sources is suitable for minimizing the energy loss and environmental impact due to transportation and storage. We are considering the use of bio-hydrogen for shared cycling as a regional target. According to the literature [9], the shared-cycle system is used in approximately 2300 cities worldwide and 225 cities in Japan. The literature [10] also indicates that the market is expected to increase to approximately 14 trillion yen.

This study considered the use of bio-hydrogen by switching from Li-ion batteries to FCs because the use of biomass, which is carbon neutral, may contribute more to the environment than the use of conventional electricity derived from fossil fuels.

Because blue and gray hydrogen are produced from fossil fuels, there are concerns regarding the increased environmental impact of carbon dioxide emissions during production and the short life of fossil fuels. Conversely, green hydrogen does not emit carbon dioxide or cause resource depletion during production. However, because it uses renewable energy, it is highly dependent on the environment and may not be able to supply sufficient hydrogen to meet demands. Our research group is also investigating and developing bio-hydrogen from domestic sewage sludge. This approach has the advantage of stably providing hydrogen compared with green hydrogen, which uses photovoltaics and wind. In this study, we considered the use of H-bikes that make use of bio-hydrogen.

The increased environmental impact during production and use is expected to contribute to higher product prices in the future. The literature [11] states that, to use FC technology sustainably and comprehensively and incorporate it into environmentally friendly mobility, it is necessary to analyze the optimization potential of FC based on its life cycle. Therefore, products with very high power performance or longer-than-usual product lives may increase environmental impacts and hinder their widespread use. Thus, when designing a product, an optimal design of power performance and product life, considering the entire life cycle, is required to reduce the environmental impact and product diffusion. In addition, because the environmental load during the usage phase of a product's life cycle depends on its efficiency, it is important to propose and design a system that is maximally efficient.

This study focuses on the degradation of FC and Li-ion batteries, fuel consumption, and system efficiency. These aspects comprise the operational issues faced by urban shared-cycle H-bikes equipped with FCs and MH tanks. Previous studies [3, 4] have demonstrated the potential to reduce the environmental impact of H-bikes. However, these studies only evaluated H-bikes in specific driving environments and did not use actual driving data. However, the proposed system is simplified and does not consider detailed energy losses [12]. In this study, we propose an approach that considers the energy loss due to the converter and power supply to the fan, which exhausts the heat from the FC. The efficiency of the final system as a whole is then determined. Subsequently, among the multiple degradation factors, we focused on the degradation caused by operating conditions and equipment configuration changes. First, the operating conditions were set using various operation data obtained from actual measurements, and a simulation was used to evaluate the degradation based on the actual and experimental data. Thereafter, we estimated the environmental impact based on the LCA. The technical specifications (TS 62282-9-101) for FC-combined heat and power systems have been published by the IEC [13]. Finally, we quantitatively demonstrated the advantages of introducing H-bikes, by comparing them with existing electric bicycles in terms of international standards for environmental impact indicators.

## 2 System Overview and Simulation Condition

### *2.1 System Overview*

This study proposes a system configuration that allows actual riding based on the system configuration described in reference [12] (Fig. 1). It considers the power supply to the FC fan and the detailed energy losses through the DC/DC converter. In addition, it exhibits the weight restrictions outlined in [3]. The system can accommodate equipment weighing up to 8 kg, excluding the weight of the **Li-ion** batteries. This study assumed the inclusion of MH tanks in an H-bike. The number of MH tanks that could be installed was determined based on the weight range specified in the literature [14]. The MH tanks were considered to be a combination of four tanks, as reported in the literature [15], forming a single cartridge. The number of cartridges installed in this study was determined as the maximum number of cartridges that did not exceed the allowable weight. The weights of each piece of equipment comprising the H-bike are listed in Table 1.

MH tanks undergo endothermic reactions to release hydrogen, making it necessary to provide constant heat to their environment. This study attempted to solve this problem by providing an MH tank with exhaust heat from an FC.

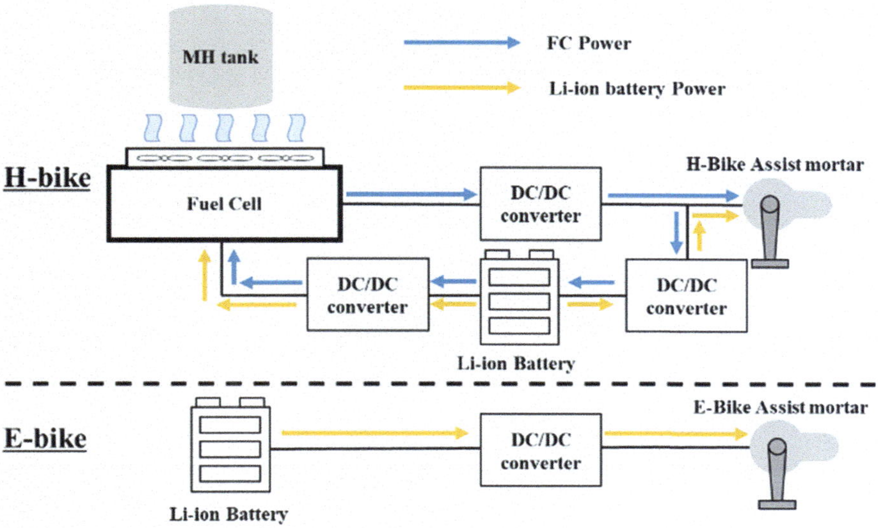


**Fig. 1** System configuration diagram of E-bike and H-bike

**Table 1** Weight of each device

| Device | Value | Unit |
|---|---|---|
| FC Stack | 2790 | g |
| Controller | 400 | g |
| DC/DC Converter | 824 | g |
| MH tank | 2040 | g |

## *2.2 Power Demand Data for Simulation*

The power demand data for the E- and H-bikes used in the simulation were obtained by running two different routes, as shown in Fig. 2a and b.

The obtained power demand data are shown in Fig. 3a and b. In this study, these data were used to validate multiple scenarios and evaluate the degradation of FC and Li-ion batteries, along with their associated environmental impacts.

# 3 Simulation Model and Life Cycle Assessment

## *3.1 Li-Ion Battery Degradation Model*

In this study, Li-ion batteries were used as the driving battery of the E-bike and the auxiliary battery of the H-bike. Cylindrical Li-ion batteries (18,650 type) were used.

**Fig. 2** Map of travel test roads used for the measurements

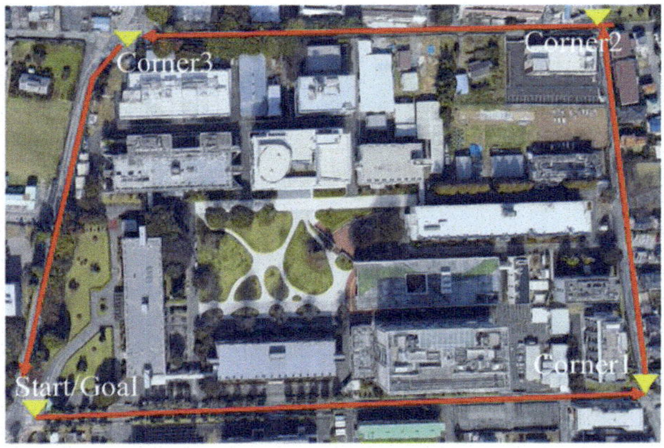


**(a)** Route #1.

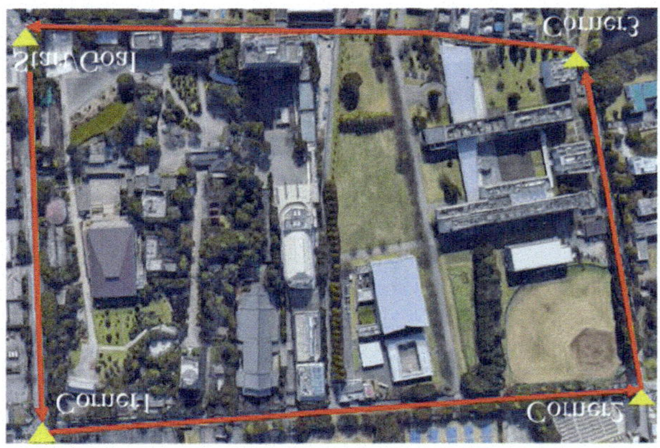

**(b)** Route #2.

This type of battery is commonly installed on E-bikes [16]. We considered the deterioration of Li-ion based on the charged/discharged current throughput per unit time [17].

$$\Delta Q_b = B(c)\exp\left(\frac{-E_a(c)}{RT}\right)A(c)^z, \tag{1}$$

where $\Delta Q_b$ [%] is the percentage capacity loss, $c$ is the charging/discharging rate ($C$-rate), and variable $A$ [Ah] is the charged/discharged current throughput per unit time depending on the $C$-rate. $B(c)$ is the pre-exponential factor as a function of the $C$ rate and is calculated using the following Eq. (15):

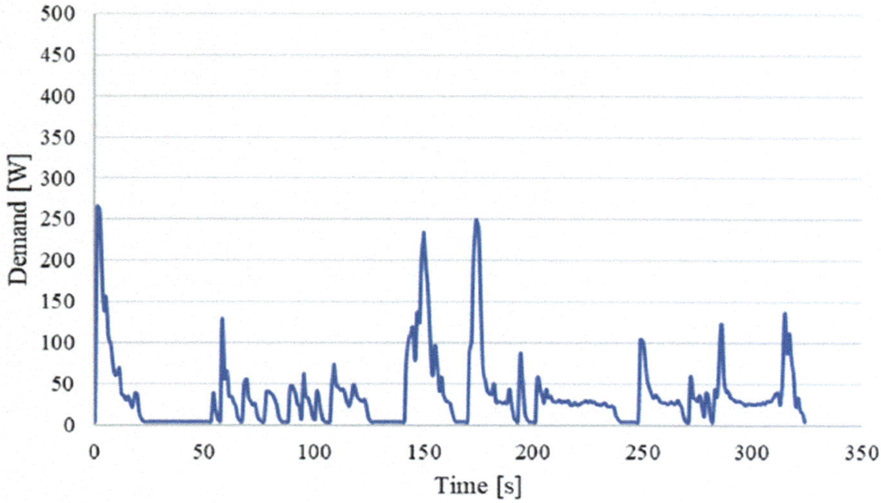


**(a)** Route #1.

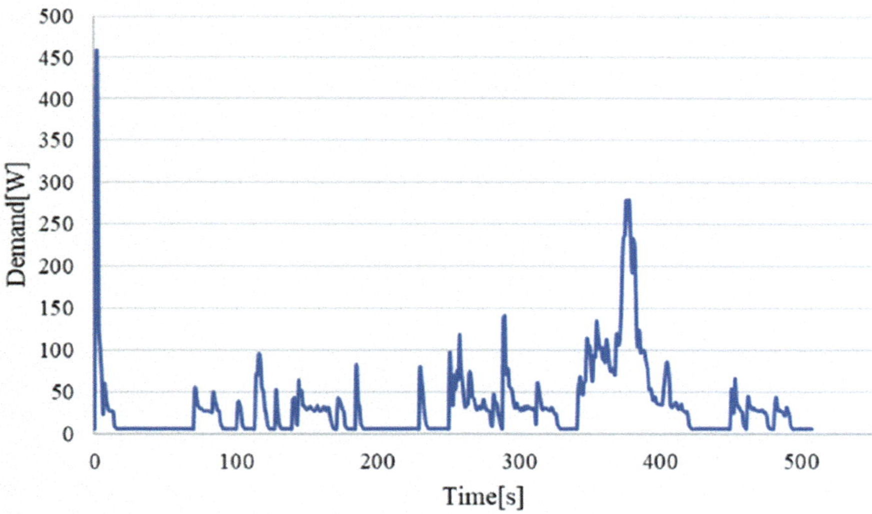


**(b)** Route #2.

**Fig. 3** Power demand in each route

$$B(c) = \left(a_1 T^2 + a_2 T + a_3\right) exp[(a_4 \mathrm{T} + a_5)(c)] \tag{2}$$

Parameters $a_1$, $a_2$, $a_3$, $a_4$, and $a_5$ are coefficients derived from the experimental results [16]. These parameters are listed in Table 2.

**Table 2** Parameters

| Parameter | Value | Unit | Reference |
|---|---|---|---|
| a_1 | 8.61E-6 | 1/Ah-K2 | [16] |
| a_2 | −5.13E-3 | 1/Ah-K | [16] |
| a_3 | 7.63E-1 | 1/Ah | [16] |
| a_4 | −6.73E-3 | 1/K-(C-rate) | [16] |
| a_5 | 2.53 | 1/(C-rate) | [16] |

The activation energy, $E_a$, and the power-law factor, $z$, are given as shown in Eq. (3) [17]:

$$E_{a(c)} = 31700 - 370.3c, \quad z = 0.55 \tag{3}$$

In addition, the total charged/discharged current throughput per unit time, $A_{\text{total}}$, and the corresponding number of cycles (N) until EOL are calculated as follows [17]:

$$A_{\text{total}}(c, T_c) = \left[ \frac{\Delta Q_b}{B(c) \exp\left( \frac{-E_a(c)}{RT} \right)} \right]^{\frac{1}{z}} \tag{4}$$

$$N(c, T_c) = \frac{A_{\text{total}}(c, T_c)}{C_{\text{bat}}} \tag{5}$$

$C_{\text{bat}}$ is the nominal capacity of the battery in Ah. The remaining battery capacity can be calculated using Eq. (6) [17].

$$Q(t) = \frac{|I(t)|}{2N(c, T_c) C_{\text{bat}}} \tag{6}$$

The E-bike state of charge was set as 0–100%, and that of the H-bike was set at 25–85%. The battery was replaced with a new battery when the state of health dropped to 80%.

## 3.2 *PEMFC Degradation Model*

In this study, the PEMFC was installed on an H-bike. The detailed specifications of the PEMFC are listed in Table 3.

The FC output differed from that reported in the literature [13]. The efficiency of the DC-to-DC converter was set to 85%. The efficiency of the DC-to-DC converter was experimentally determined. The FCs operated at a constant output. This was because the FCs could not follow the load and required a stable supply of waste heat to the MH tank. In this study, we focused on the degradation of PEMFCs owing to

**Table 3** Specifications of PEMFC

| | Value | Unit |
|---|---|---|
| Temperature | 25 | °C |
| Rated power | 300 | W |
| Low output-power range | 64 | W |
| High output-power range | 253 | W |

**Table 4** Parameters

| Parameter | Value | Unit | Reference |
|---|---|---|---|
| $D_m$ | $1.5 \times 10^2$ | mol/L | [18] |
| $c_{Pt}$ | $1.0 \times 10^{-7}$ | mol/L | [18] |
| $\alpha_m$ | 1 | – | [18] |

the elution of Pt particles, which is a multiple degradation factor. The degradation simulation was based on an electrochemical reaction Eq. (18).

$$Pt \leftrightarrow Pt^{2+} + 2e^- \tag{7}$$

$$Pt + H_2O \leftrightarrow PtO + 2H^+ + 2e^- \tag{8}$$

$$PtO + 2H^+ \leftrightarrow Pt^{2+} + H_2O \tag{9}$$

The FC degradation due to the elution of Pt particles was evaluated using the reduction in the electrochemical surface area (ECSA) and I–V characteristics.

For the ECSA degradation, we focused on individual Pt particles. The reaction related to the dissolution of each Pt particle $n_{i,\ j}$ is represented by the following Eq. (10) [18].

$$n_{i,j} = n_{i,j-1} + \Delta t\left(-v_1\left(r_{i,j}\right) - v_2\left(r_{i,j}\right) + v_{or}\left(r_{i,j}\right)\right) \tag{10}$$

$\Delta t$ is a time step of 1 second, and $v_1$ and $v_2$ indicate the reaction rates in Eqs. (7) to (9). $r_{i,\ j}$ [nm] is the radius of each Pt particle. The change in the Pt particle size due to Pt dissolution can be calculated using Ostwald's equation, as shown in Eq. (11) [18].

$$v_{\mathrm{or}}\left(r_{i,j}\right) = 4\pi D_m c_{Pt} \alpha_m \left(\frac{r_{i,j}}{\overline{r_j}} - 1\right) \tag{11}$$

$v_{\mathrm{or}}(r_{i,\ j})$ and $\overline{r_j}$ are the reaction rates of Ostwald ripening and the average radius of the Pt particles, respectively. The remaining values are constants. These parameters are summarized in Table 4 [18].

ECSA was determined by calculating the geometric surface areas (GSA). This is expressed as Eq. (12) [19]:

$$\text{GSA}(t) = 2\pi \sum_{i=1, j=1}^{N, M} \text{Num}_{i,j} n_{i,j}{}^{2} \tag{12}$$

$M$, $N$, and $\text{Num}_{i,j}$ are the number of particle size groups, the thickness of the entire cathode catalyst, and the number of particles, respectively [19]. The degradation of the ECSA [ECSA($t$)] due to temporal changes is determined using GSA($t$) and can be calculated using the following Eq. (13) [19]:

$$\frac{\text{ECSA}(t)}{\text{GSA}(t)} = \frac{\text{ECSA}(0)}{\text{GSA}(0)} \tag{13}$$

ECSA(0) is the initial value, which was set to 112 $m^2/g$. The cell voltage, V, used for evaluating the I–V characteristics was calculated using Nernst's equation, membrane resistance, and cathode overpotential. The calculation is described in Eq. (14) and [20].

$$V = V_0 - \frac{\text{RT}}{2F} \ln \frac{\text{p}_{\text{H}_2} \left(\text{p}_{\text{O}_2}\right)^{0.5}}{\text{p}_{\text{H}_2\text{O}}} - R_{\text{mem}} I - \eta_c \tag{14}$$

$V_0$ [$V$] is the open-circuit voltage of the FC. $p_{H_2}$ [Pa], $p_{O_2}$ [Pa], and $p_{H_2O}$ [Pa] are the pressures of $H_2$, $O_2$, and $H_2O$, respectively. $R_{mem}$ [ohm], $I$ [A/cm$^2$], and $\eta_c$ [V] denote the membrane resistance, current, and cathode overpotential, respectively, and F is Faraday's constant (96,485 C/mol).

## 3.3 *Efficiency Model*

The efficiency of the proposed H-bike system was calculated as follows:

$$\eta_{\text{H-bike}}(t) = \frac{E_{\text{output}}}{E_{\text{input}}(t)} \tag{15}$$

$\eta_{H-bike}$ [%] indicates the system efficiency of the H-bike; $E_{input}$ [J] and $E_{output}$ [J] indicate the amount of energy input and the amount of energy finally consumed, respectively.

## 3.4 Scenarios

Based on the H-bike characteristics, a combination of the FC scale and a supplementary Li-ion battery is discussed. For the two routes (#1 and #2), we assumed certain scenarios for the four other capabilities of Li-ion batteries (Table 5).

The internal structure of the batteries used in existing E-bikes consists of four modules connected in parallel and seven modules connected in series. Four Li-ion batteries with different capacities were assumed by reducing the number of modules connected in parallel to one set. The FC generated constant electricity in two output bands: 64 W on a flat road and 253 W on a sloping road.

The performance conditions of the H-bike were as follows: operating duration of 30 min per day, frequency of operation of six times per day, and lifetime of 8 years. These data were based on a demo test performed in Sendai City [21] (Table 6).

**Table 5** Operation scenarios

| Scenario | Route | Battery storage [Wh] | FC-power assist [W] |
|---|---|---|---|
| A | #1 | 306 | Assisted by a Li-ion battery |
| B | #2 | | |
| C | #1 | 306 | 64 |
| D | #2 | | |
| E | #1 | 267.75 | |
| F | #2 | | |
| G | #1 | 178.5 | |
| H | #2 | | |
| I | #1 | 89.25 | |
| J | #2 | | |
| K | #1 | 306 | 253 |
| L | #2 | | |
| M | #1 | 267.75 | |
| N | #2 | | |
| O | #1 | 178.5 | |
| P | #2 | | |
| Q | #1 | 89.25 | |
| R | #2 | | |

**Table 6** Frequency of use

| | | |
|---|---|---|
| Operating duration | 30 | min/day |
| Frequency of operation | 6 | times/day |
| Life year | 8 | years |

## *3.5 Life Cycle Assessment (LCA)*

A comparative evaluation of the environmental burden between E-bikes and H-bikes was conducted using LCA. In this study, the driving conditions and mileage were the same for the E-bikes and H-bikes. The driving conditions were based on the driving routes and electricity demand data presented in this article. The driving distance was assumed to be approximately 5 km, based on the results of a previous study [3], the demonstration experiment [21], which showed that each use lasted approximately 30 minutes, and the time required to measure electricity demand data in this study. The frequency of E-bike and H-bike service was assumed to be six times per day, based on the results of a previous study [3] and the demonstration experiment [21].

The performance guarantee of the H-bike was determined according to the conditions obtained from the experiments in a previous study [3]. Our research team has investigated this in an earlier study [3]. It was found that the power demand did not change depending on the weight of the E-bike load for a loading weight of up to 8 kg, except for that of the conventional Li-ion battery. Based on this study, actions due to changes in loading weight would be required from 11 kg. These actions include replacing the battery with the FC with the storage tanks, a process unaffected by the energy economy. Therefore, these E-bikes are treated as conventional E-bikes and are not affected by loading weight. In addition, we assumed that the energy and environmental impact of the replacement is negligible.

In the LCA, we focused on the environmental impact indicators of global warming potential (GWP) and abiotic depletion potential (ADP). We compared the use of FCs, which are carbon-neutral due to biomass utilization, with that of Li-ion batteries, which are fossil-fuel-derived, and we considered the GWP.

On the other hand, it is known that electricity-assisted bicycles are expected to become more widespread. Therefore, we can expect an accompanying increase in the consumption of the rare metals used in the manufacture of devices such as Li-ion batteries and FCs, which are the power sources of these bicycles. In addition, we believe it will be important to consider the impact of these rare metals on resource depletion; thus, we have considered ADP. The GWP and ADP were determined with reference to guidelines established by IEC TC105 [8]. Data for these indicators were obtained from the CML database [22, 23]. SimaPro v. 9.2.0.1 was used to construct the life cycle model. The functional unit was defined as one unit of bicycle operation per 8 years [24].

The scope of the assessment encompassed the production-to-use phases of both E-bikes and H-bikes. The detailed system boundaries are shown in Fig. 4. The inventory data for the PEMFC stack are listed in Table 7 [25]. For the converter, inventory data obtained from an electric vehicle converter, as indicated in the literature [26], were used, and its size was adjusted based on Eq. (16) using the 0.6 scaling rule on a weight basis to match the size of the converter in this study [28].

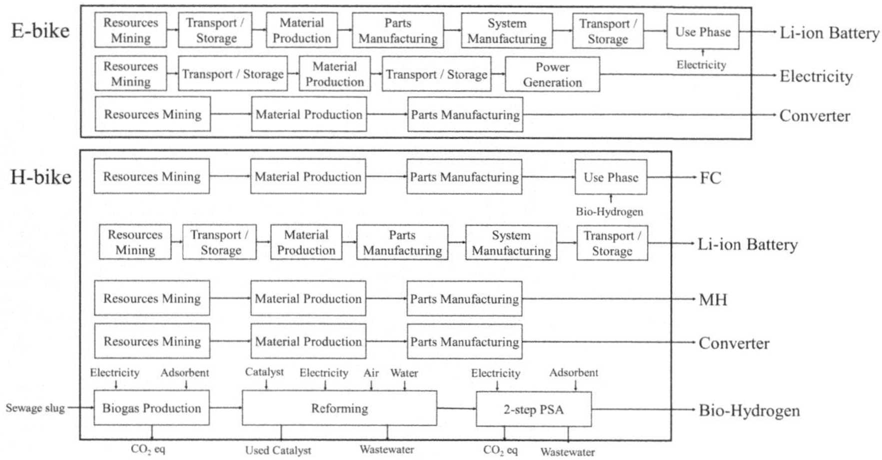


**Fig. 4** System boundary

$$\mathbf{Inv_{conv}} = \mathbf{Inv_{basic,conv}} \left( \frac{\mathbf{Output_{conv}}}{\mathbf{Output_{basic,conv}}} \right)^{\frac{2}{3}} \tag{16}$$

where $\mathbf{Inv_{conv}}$ and $\mathbf{Inv_{basic,conv}}$ are the inventory data of the target converter and the converter data reported in the literature [29], respectively. $\mathbf{Output_{conv}}$ indicates the weight of the target converter. In this study, we incorporated time by conducting an LCA that considers the replacement of Li-ion batteries with PEMFC. Notably, these battery devices experience performance degradation over time. The hydrogen used in the H-bike was assumed to be bio-hydrogen synthesized from sewage sludge. GWP and ADP were also evaluated for bio-hydrogen production.

The GWP and ADP in the LCA can be calculated using Eqs. (17) and (18) [5, 29]:

$$\mathbf{GWP}_{T,i} = \frac{\int_0^T \boldsymbol{a_i} \times \boldsymbol{c_i(t)dt}}{\int_0^T \boldsymbol{a_{CO2}} \times \boldsymbol{c_{CO2}(t)dt}} \tag{17}$$

$$\mathbf{ADP_{total}} = \sum_k \mathbf{ADP}_k \times \boldsymbol{m_k} \tag{18}$$

where $\boldsymbol{a_i}$ is the infrared absorption coefficient per unit of substance $i$, $c_i(t)$ is the decay of substance $i$ after emission over time $t$, and $T$ is time. Because the GWP varies depending on the lifetime of greenhouse gases, the values may differ when considering the time horizon. However, in this study, a commonly used value for a 100-year time horizon was employed. $\mathbf{ADP}_k$ represents the nonrenewable resource depletion potential per unit of resource $k$, and $\boldsymbol{m_k}$ represents the quantity of resource k.

**Table 7** Inventory data (FC stacks, Li-ion batteries, and others)

| PEMFC stack [25] | Input (kg/W) | |
|---|---|---|
| | NAFION | $7.0 \times 10^{-5}$ |
| | Platinum | $7.5 \times 10^{-8}$ |
| | Graphite | $4.5 \times 10^{-3}$ |
| | Polyvinylidene chloride (PVDC) | $1.1 \times 10^{-3}$ |
| | Aluminum | $3.0 \times 10^{-4}$ |
| | Chromium steel | $1.0 \times 10^{-4}$ |
| | Glass fires | $1.0 \times 10^{-4}$ |
| | Carbon black | $8.0 \times 10^{-8}$ |
| BoP | Steel product | $3.7 \times 10^{-3}$ |
| | High-density polyethylene (HDPE) granulate | $1.5 \times 10^{-3}$ |
| | Chromium steel | $1.1 \times 10^{-3}$ |
| | Cast iron component | $8.0 \times 10^{-4}$ |
| | Aluminum | $7.5 \times 10^{-4}$ |
| | Polypropylenes (PP) granulate | $2.5 \times 10^{-4}$ |
| Converter [26] | Input (g/converter) | |
| | Steel | 96.2 |
| | Iron | 343.7 |
| | Aluminum | 842.0 |
| | Copper | 123.7 |
| Li-ion battery | Input (kg/kWh) | |
| | Battery, Li-ion, rechargeable, prismatic {GLO}\| market for \| APOS, U | |
| MH | Input (g/MH tank) | |
| | Mischmetal {GLO}\| production, primary \| APOS, U | |
| Hydrogen | Input ($m^3$/s) | |
| | Bio $H_2$ from sewage sludge [27] | |
| Electricity | Input (kWh) | |
| | Electricity, medium voltage {JP}\| market for \| APOS, U | |

# 4 Results and Discussion

## *4.1 Performance Drops of PEMFC*

Figures 5 and 6 present the results of the PEMFC degradation simulations. The numerical results presented herein were evaluated based on a study by Tochigi et al. [30], which considered an FC power generation capacity of 300 W.

The Fig. 5 plots show the degradation of the ECSA. The degradation was the most severe for scenarios C to J, where the FC generated power at an output of 64 W. In contrast, the degradation of the ECSA in scenarios K to R, where the FC generated power at an output of 253 W, was relatively mild compared to those of scenarios

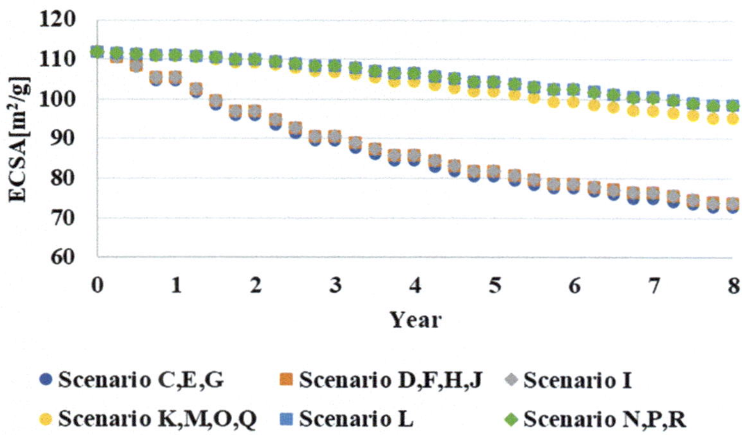


**Fig. 5** Comparison of ECSA

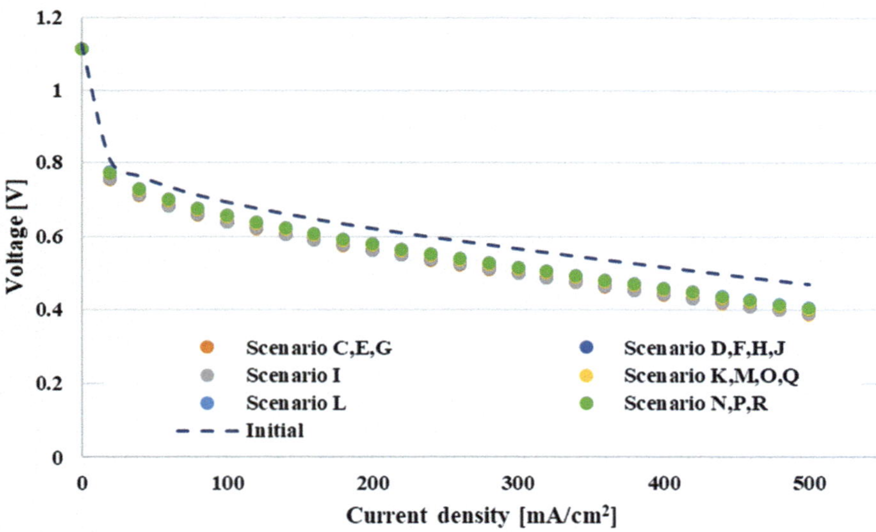


**Fig. 6** Comparison of I–V characteristics

C to J. This can be attributed to the longer operational time of the FC in scenarios C to J and the lower output range of the FC in comparison to scenarios K to R.

Figure 6 shows the initial I–V characteristics of the FC and the I–V characteristics at the EOL stage [30]. No differences in I–V characteristics due to the different scenarios were observed.

## 5 Fuel Consumption and System Efficiency

Figures 7 and 8 show the fuel consumption in each scenario and the system efficiency of the H-bike.

Figure 7 illustrates the fuel efficiency for each scenario. Scenarios of K to R, where the FC operated at an output of 253 W, exhibited better fuel efficiency. Scenarios O and L exhibited the best fuel efficiencies for each driving route. Scenario O achieved the highest fuel efficiency, 8.9 km/L, on Route #1, whereas scenario of L achieved the best fuel efficiency, 11.2 km/L, on Route #2. These results show that driving with an FC output of 253 W decreases the fuel consumption by H-bikes. Additionally, the impact of different battery capacities on fuel efficiency was not significant, as revealed by the results of this study.

Figure 8 shows the annual variations in the H-bike efficiency. Focus was placed on the system efficiency of the H-bike for scenarios K to P, which exhibited the best fuel efficiencies. The efficiency was calculated for each year, considering the degradation of the components, according to Eq. (14). Notably, scenario K exhibited the smallest variation in efficiency. However, scenarios in Route #2 exhibited considerable fluctuations in efficiency.

Moreover, we calculated the total efficiency from bio-hydrogen production to H-bike production and electricity production to E-bike production. The efficiency of bio-hydrogen generation was determined using the values calculated and those reported in the literature [31]. The system efficiency of the H-bike was determined based on the median value. The results are summarized in Table 8.

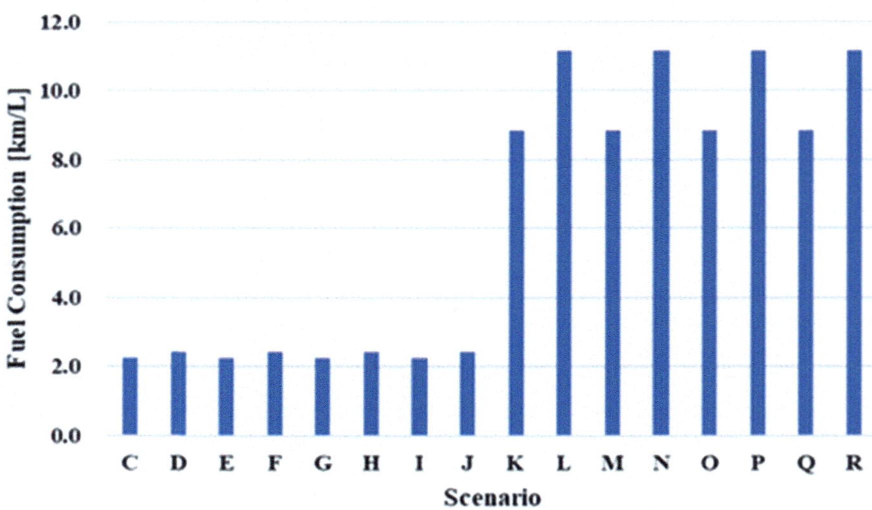


**Fig. 7** Fuel consumptions in scenarios C to R

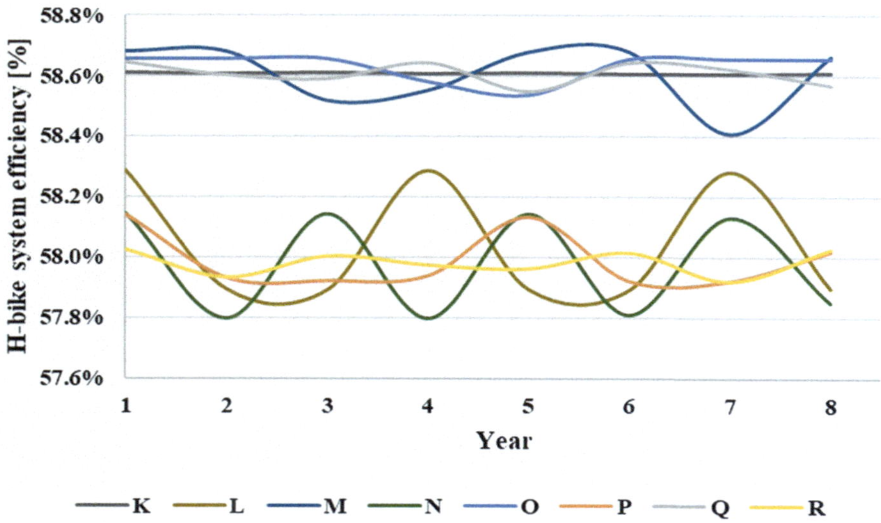


**Fig. 8** H-bike system efficiencies in scenarios K to R

**Table 8** Comparison of total efficiency of E-bike and H-bike

| | E-bike | H-bike |
|---|---|---|
| Bio-hydrogen generation efficiency [31] | – | 43.7% |
| Power generation efficiency [32] | 50.4% | – |
| System efficiency | 76.7% | 58.3% |
| Total efficiency | 38.7% | 25.5% |

## 6 LCA Results

Figures 9 and 10 show the LCA results.

Figure 9 shows the GWP results for each scenario. Scenarios C–J exhibited higher hydrogen consumption compared to scenarios K to R, contributing more to the overall environmental impact. In contrast, scenarios K to R exhibited lower hydrogen consumption but higher frequencies of Li-ion battery replacement. However, their overall environmental impact was lower than that of scenarios C–J. Additionally, their environmental impacts were lower than that of conventional E-bikes. The scenario with the smallest GWP value was scenario N, with a value of 82.9 kg-$CO_2$ eq. However, all H-bike scenarios had an environmental impact owing to the MH tank. In this study, we assumed that the maximum number of MH tanks that could be installed within the weight limit of 8 kg would be installed. Therefore, when considering usage in a bike-sharing system, more MH tanks than necessary should be installed. Reducing the number of tanks to the minimum required reduces the environmental impact of H-bikes.

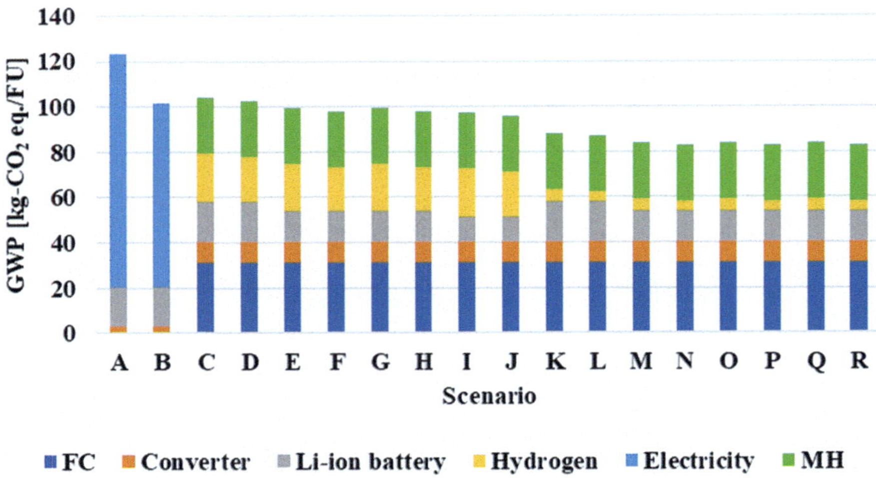


**Fig. 9** Results of GWP

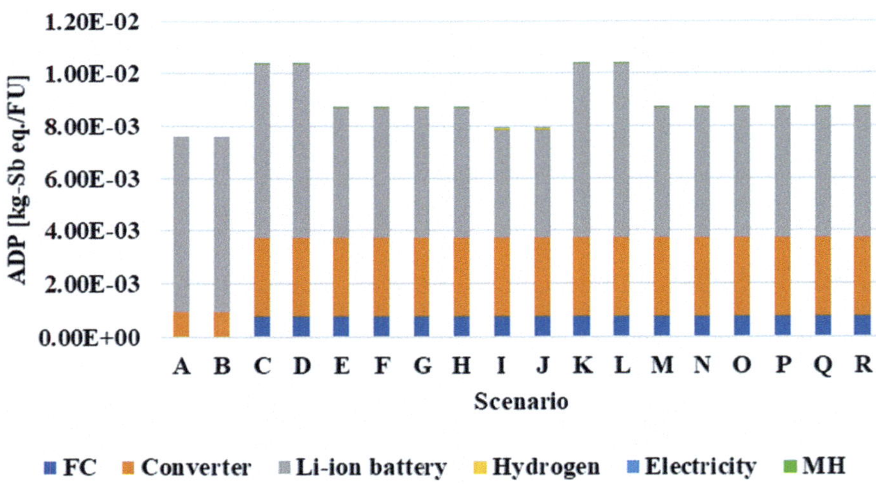


**Fig. 10** Results of ADP

Figure 10 shows the ADP results for each scenario. In all scenarios, the environmental impact was higher than that of conventional E-bikes. This can be attributed to the impact of the ecological burdens of converters and Li-ion batteries. For example, the converters accounted for 30% of the overall environmental impact of the H-bike. Moreover, Li-ion batteries had the most significant environmental impact. Scenarios C, D, K, and L adopted the largest-capacity Li-ion battery, resulting in a larger eco-burden even with fewer replacements. In contrast, scenarios I and J had the lowest environmental impacts among the H-bike scenarios, with both exhibiting a value of $7.89 \times 10^{-3}$ kg-Sb eq. This suggests that their environmental impact would be reduced even with a smaller Li-ion battery capacity and a higher frequency of battery replacement.

## 7 Conclusion

In this study, we focused on the degradation, fuel efficiency, and system efficiency of FC and Li-ion batteries in an H-bike, considering its use in an urban bike-sharing system. FC degradation was evaluated based on the ECSA and I–V characteristics. ECSA degradation was associated with a performance drop in FC power. Consequently, the degradation was more severe at a lower power output of 64 W and less significant at a higher power output of 253 W.

Regarding the system and fuel efficiencies, the system efficiency was determined to be 58.3%, and the overall efficiency, considering the bio-hydrogen generation efficiency, was 25.5%. The fuel efficiency indicated that the scenarios with an FC power output of 253 W performed better. Furthermore, the battery capacity did not significantly affect the fuel efficiency.

Regarding the LCA results in the scenarios, the GWP had the lowest environmental impact, 82.9 kg-$CO_2$ eq. In contrast, the ADP was $7.89 \times 10^{-3}$ kg-Sb eq.

## References


1. Zhang H, Shen PK (2012) Recent development of polymer electrolyte membranes for fuel cells. Am Chem Soc 112:2780–2832
2. Raposo M, Silva C (2022) City-level E-bike sharing system impact on final energy consumption and GHG emissions. Energies 15:6725
3. Hosobuchi E, Misaki C, Katayama N, Dowaki K (2021) Impact of increased tire pressure on fuel consumption and environment for fuel-cell-assisted shared bicycles. J Jpn Inst Energy 100: 116–121
4. Hara D, Misaki C, Sugihara H, Kako S, Katayama N, Dowaki K (2021) Exergy and environmental analysis of a bio-hydrogen supply chain using data envelope analysis. In: Kishita Y, Matsumoto M, Inoue M, Fukushige S (eds) EcoDesign and sustainability II. Springer Nature, Singapore, pp 525–540
5. International Electrotechnical Commission (2020) Evaluation methodology for the environmental performance of fuel cell power systems based on life cycle thinking–streamlined lifecycle considered environmental performance characterization of stationary fuel cell combined heat and power systems for residential applications. In: IEC/TS 62282-9-101. IEC, Geneva, Switzerland
6. Magdziarz A, Wilk M (2013) Thermogravimetric study of biomass, sewage sludge, and coal combustion. Energy Convers Manag 75:425–430
7. Ellis MW, Von Spakovsky MR, Nelson DJ (2001) Fuel cell systems: efficient, flexible energy conversion for the 21st century. Proc IEEE 89:1808–1818

8. Hara D, Misaki C, Sugihara H, Kako S, Katayama N, Dowaki K (2021) Exergy and environmental analysis of a bio-hydrogen supply chain using data envelope analysis, vol 2021. Springer, pp 525–540
9. Current status and issues regarding bicycle sharing | MLIT; https://www.mlit.go.jp/road/ir/ir-council/sharecycle/pdf01/03.pdf. Accessed September 8, 2023
10. Sharing economy related survey 2021 survey results | ICR; https://sharing-economy.jp/ja/wp-content/uploads/2022/01/1d6acc7e6a69d1938f054c88778ba43b.pdf. Accessed September 8, 2023
11. Rückert P, Khabipova A, Tracht K (2023) Development of a sustainability strategy for fuel cells using life cycle analysis and expert interviews. In: Development of a sustainability strategy, pp 892–900
12. Nagado R, Sakai S, Ezawa T, Shimogawa J, Katayama N, Miao S, Kiyoshi Dowaki K (2022) Dynamic LCA of fuel cell-assisted bike in consideration of Pt catalyst degradation in various operating conditions. In: IOP Conf. Series: Earth and Environmental Science, p 1187
13. Fuel cell technologies - Part 9–101: Evaluation methodology for the environmental performance of fuel cell power systems based on life cycle thinking - Streamlined life-cycle considered environmental performance characterization of stationary fuel cell combined heat and power systems for residential applications; https://standards.iteh.ai/catalog/standards/iec/567cf021-cbd3-4627-ac27-bf219cf2e922/iec-ts-62282-9-101-2020. Accessed September 4, 2023
14. Hara D, Miao S, Shimogawa J, Katayama N, Dowaki K (2021) A suitable design of metal hydride cartridge for a fuel cell assisted bicycle in consideration of heat transfer performance. IOP Conf Series Earth Environ Sci 1034:012049
15. Shimogawa J, Hara D, Miao S, Katayama N, Dowaki K (2022) Design and temperature analysis of a metal hydride cartridge using exhaust heat of a fuel cell for electric-assisted bicycles. J Jpn Inst Energy 101:152–161
16. Wang J, Purewal J, Liu P, Hicks-Garner J, Soukazian S, Sherman E, Sorenson A, Luan V, Tataria H, Verbrugge MW (2014) Degradation of li ion batteries employing graphite negatives and nickelecobaltemanganese oxide þ spinel manganese oxide positives: part 1, aging mechanisms and life estimation. J Power Sources 269:937–948
17. Wang Y, Moura SJ, Advani SG, Prasad AK (2019) Power management system for a fuel cell/battery hybrid vehicle incorporating fuel cell and battery degradation. Int J Hydrogen Energy:448479–448492
18. Ao Y, Chen K, Laghrouche S, Depernet D (2021) Proton exchange membrane fuel cell degradation model based on catalyst transformation theory. Fuel Cells 21:254–268
19. Zheng Z, Yang F, Lin C, Zhu F, Shen S, Wei G, Zhang J (2020) Design of gradient cathode catalyst layer (CCL) structure for mitigating Pt degradation in proton exchange membrane fuel cells (PEMFCs) using mathematical method. J Power Sources 451:227729
20. Al-Baghdadi MAS (2008) Three-dimensional computational fluid dynamics model of a tubular-shaped PEM fuel cell. Renew Energy 33:1334–1345
21. We will inform you of the results of the "Sendai Community Cycle Social Experiment;" https://www.city.sendai.jp/jitensha/kurashi/machi/kotsu/jitensha/torikumi/documents/komisaikekka_1_1.pdf. Accessed: June 27, 2022
22. CML - Department of Industrial Ecology; https://www.universiteitleiden.nl/en/research/researchoutput/science/cml-ia-characterisation-factors#getting-and-using-the-database. Accessed May 29, 2021
23. Aitor PA, Cristina R, Andreas C (2019) Impact assessment methods in life cycle assessment and their impact categories
24. Huang Y, Jiang L, Chen H, Dave K, Parry T (2022) Comparative life cycle assessment of electric bikes for commuting in the UK. Transp Res D 105:103213
25. Chen ZS, Lam JSL (2022) Life cycle assessment of diesel and hydrogen power systems in tugboats. Transp Res D 103:103192

26. Koroma MS (2018) Estimating the potential life cycle environmental impacts of current and future electric passenger cars. TRITA-ABE-MBT-18334
27. Kako S, Seo Y, Dowaki K, Kameyama M (2019) A proposal and design of H2 production system using sewage sludge in residential sector. Proc Conf Biomass Sci:137–138
28. Chung CA, Yang SW, Yang CY, Hsu CW, Chiu PY (2013) Experimental study on the hydrogen charge and discharge rates of metal hydride tanks using heat pipes to enhance heat transfer. Appl Energy 103:581–587
29. Kendall A (2012) Time-adjusted global warming potentials for LCA and carbon footprints. Int J Life Cycle Assess 17:1042–1049
30. Tochigi S, Dowaki K (2023) Environmental impact assessment of PEM fuel cell combined heat and power generation system for residential application considering cathode catalyst layer degradation. Energies 16(4):1985
31. Hamazaki M, Torii K, Shan M, Kameyama M, Dowaki K (2022) Analyzation of the temperature distribution in the pyrolizer by using CFD modeling to discuss the rate of H2 and tar production in the indirect pyrolysis plant. In: 10th Asian conference biomass science. November 25, 2022, Online
32. Tan J, Xie S, Wencong W, Qin P, Ouyang T (2022) Evaluating and optimizing the cold energy efficiency of power generation and wastewater treatment in LNG-fired power plant based on data-driven approach. J Clean Prod 334:130149

# A Discussion of the Thermal Management System for Metal Hydride Cartridges for H-Bike

**Satoshi Sakai, Ryuta Nagado, Tomoya Ezawa, Koya Honda, Noboru Katayama, Shan Miao, and Kiyoshi Dowaki**

**Abstract** As one of the means to realize a sustainable society, fuel cell applications using hydrogen as an alternative fuel are being studied. Hydrogen can be synthesized from biomass feedstock, a renewable resource. For instance, fuel cells efficiently convert sewage sludge and wood chips into electrical energy. Two main types of fuel hydrogen storage technologies exist high pressure and liquefaction. However, these methods require additional measures regarding the energy necessary for compression or liquefaction and the system's safety.

On the other hand, hydrogen storage using metal hydride is being introduced because it can operate safely under a relatively average temperature and pressure atmosphere, depending on the type of alloy. Other features include high volumetric energy density and weight density, suitable for systems requiring no weight constraints and space-saving. In metal hydride, hydrogen storage, and release are caused by reversible equilibrium reactions. The hydrogen release reaction is endothermic and requires additional heat to maintain a sufficient hydrogen release rate. However, due to the low thermal conductivity of the hydrogen storage alloy, the amount of heat transferred from the surrounding environment to the hydrogen storage alloy tank is usually insufficient to match the amount of heat required by the reaction rate corresponding to the hydrogen release rate needed for the fuel cell stack. Therefore, the temperature inside the tank decreases due to the endothermic heat of the reaction heat. As a result, the equilibrium reaction stops because the hydrogen supply rate to the fuel cell drops, which may cause the fuel cell to stop. One way to solve this problem is to use an external heat source to make the hydrogen storage container hot. Therefore, as a solution to improve the system's performance, using the exhaust heat dissipated through the fuel cell cooling system is considered one of the attractive means that does not require additional systems.

S. Sakai (✉) · R. Nagado · T. Ezawa · S. Miao · K. Dowaki
Department of Industrial and Systems Engineering, Graduate School of Science and Technology, Tokyo University of Science, Chiba, Japan
e-mail: dowaki@rs.tus.ac.jp

K. Honda · N. Katayama
Department of Electrical Engineering, Graduate of Science and Technology, Tokyo University of Science, Chiba, Japan

 
S. Fukushige et al. (eds.), *EcoDesign for Circular Value Creation: Volume I*,
https://doi.org/10.1007/978-981-97-9068-5_2

This research group focuses on the fact that no weight limit and space is saved that can take advantage of the advantages of the metal hydride and is researching the thermal management of a small-scale system that combines a storage alloy tank mounted on the assisted fuel cell bike (H-bike) that runs on a fuel cell including the configuration and the geometric design. In this study, we experimented with a small, air-cooled fuel cell (300 W) combined with multiple small self-made tanks (diameter 30 mm, length 30 mm). In addition, we examined the thermal characteristics of the cover around the tank as thermal management to improve emission performance. At this time, the conditions necessary for maintaining the tank's internal temperature were estimated, and the conditions required for mounting on H-bike were examined.

**Keywords** Metal hydride · Fuel cell · Thermal management

## 1 Introduction

In recent years, fuel cells (FC) that can use hydrogen as an alternative fuel have attracted attention for realizing a sustainable society. Polymer electrolyte membrane FCs (PEMFC) have advantages over other FC, such as high energy efficiency and low operating temperatures [1–3]. Hydrogen is the fuel for FC and can be synthesized from biomass materials such as sewage sludge. Hydrogen is a category of green energy with no direct $CO_2$ emissions. Additionally, FC can efficiently generate electricity using green hydrogen.

However, the storage technology is important for the use of hydrogen. Currently, two mainstream methods are used: high-pressure compression and liquefaction. However, these methods increase the overall cost and volume of the systems as they require additional energy and equipment for safe operation [4]. Metal hydride (MH) is an alternative higher hydrogen storage method that has a higher capacity per unit volume, safety, and reliability than compression and liquefaction [4]. Owing to these advantages, there are several suitable ways to use MH. However, the volume weight density causes operational difficulties because of the considerable mobility required to prioritize fuel consumption [5]. Thus, issues related to the application performance of MH need to be resolved.

In combining the battery of FC with the MH storage, the weight demerit would be a certain problem in performance. However, if the surface resistance against the road is reduced, the fuel consumption would be little affected. In addition, the energy density per weight would be more advantageous than the battery. The alternative with the applications is possible compared to the conventional Li-ion battery. In this study, we discuss the operating conditions of MH batteries in H-bikes.

Among the various types of MH, alloys with compositions AB2 and AB5, known as low-temperature MHs, absorb and release hydrogen under relatively high-temperature and high-pressure atmospheres through a reversible equilibrium reaction [6]. The reaction heat is important in this equilibrium reaction. An exothermic reaction occurs during hydrogen absorption, which is determined by the relationship between the equilibrium pressure and pressure around the MH, which is determined by the type of alloy, amount of hydrogen stored per weight, and temperature. The equilibrium pressures are shown in the hydrogen absorption and desorption isotherms of the MH material.

An issue arises from the variation in the equilibrium pressure owing to the changes in the MH temperature resulting from the heat released during the reaction. During the storage process, if the equilibrium pressure increases owing to an increase in the temperature of the MH bed caused by the exothermic reaction, the equilibrium pressure becomes higher than the pressure around the MH, terminating the storage process. Similarly, in the release process, if the equilibrium pressure drops as the temperature of the MH bed decreases owing to the endothermic reaction, the equilibrium pressure becomes lower than the pressure around the MH, and the release process stops. This makes release over an extended period difficult and reduces the discharge-to-storage volume ratio, affecting operational performance.

The thermal management of MH has been proposed to solve these issues. Many previous studies have confirmed that using phase change materials, heat pipes, and other heat exchangers can improve performance by efficiently delivering heat to the MH floor [7, 8]. Another heat management method uses systems that dispose of the heat generated during FC operation. Reza et al. developed a system combining an air-cooled PEMFC with a rated power of 2.5 kW scale and 6100 g MH × 9 tanks capable of storing 900 L of hydrogen that operated continuously for 5 h under conditions of 500 W FC power [9]. However, limited research has been conducted on medium- and small-scale fuel cells [9, 10]. The smaller the PEMFC, the less heat is available from the PEMFC and the more difficult it is to manage the heat.

Therefore, this study investigated a heat management system that combines a small FC and an MH tank. The system was mounted on an H-bike, which was chosen for two reasons: first, the MH can be loaded in a space-saving manner; second, the weight-limit disadvantage of the MH, can be mitigated to some extent. This is because previous research by this team showed that the required power output of an electric bicycle does not change when carrying a weight of up to 8 kg [11].

This study combined a small air-cooled FC (300 W) with several small homemade tanks (30 mm in diameter and 30 mm in length) to study the thermal management for bicycle operation. Moreover, four tanks were connected and considered as one cartridge because the smaller size is thought to make it easier to manage the heat in each tank and because dividing the tank into sections increases the surface area and facilitates thermal management. The cartridge contained approximately 380 g of $LaNi_5$, which can be released in an atmosphere of relatively average temperature and pressure and corresponds to approximately 50 L of hydrogen storage. For thermal management, we considered the hydrogen release flow rate as an internal tank factor and the effect of the FC exhaust heat and the cover surrounding the cartridges as external tank factors. The reasons for considering these factors are as follows.

1. The hydrogen release flow rate affects the release reaction rate inside the tank. This implies that the amount of heat absorbed changes, leading to a change in the amount of heat required for heat management.
2. The amount of heat released from the FC varies depending on the power band of the FC, which is necessary when considering the amount of heat that can be supplied.
3. Regarding the cover, more efficient use of waste heat from the FC would make it easier to deliver heat.

This study was conducted using actual measurements from experiments and calculations to examine the conditions under which an H-bike can run. The temperature conditions necessary to maintain the operation were considered. In addition, we compared the environmental performance of this system with that of existing Li-ion batteries, considering weight limitations and other factors, to demonstrate the significance of introducing this system.

# 2 Methodology

## *2.1 Outline of FC Performance Test*

An experiment was conducted to verify the operating time of the actual combination of the PEMFC and MH tank. The FC used in this experiment was a Horizon PEMFC (H-300), air-cooled and rated at 300 W. The detailed specifications and system schematics are presented in Table 1 and Fig. 1.

The power load device was connected to the PEMFC side to control the output. In addition, a pressure-reducing valve was placed between the MH tank and the PEMFC so that the required pressure of the FC was reached. A flow meter was placed between the tank and pressure-reducing valve to measure the flow rate; the PEMFC outlet side was opened to atmospheric pressure and sucked into the draft chamber.

**Table 1** PEMFC details

| | |
|---|---|
| Type of FC | Polymer electrolyte membrane FC |
| Number of cells | 60 stacks |
| Rated Power | 36 V at 8.3 A |
| External temperature range [°C] | 5–30 |
| Inlet $H_2$ Pressure [kPa] | 4.5–5.5 |
| Cooling method | Air (integrated cooling fan) |
| Stack weight [g] (incl. Fan and casing) | 2790 (±50) |
| Controller weight [g] | 400 (±30) |
| Dimension [cm] | 11.8 × 26.2 × 9.4 |

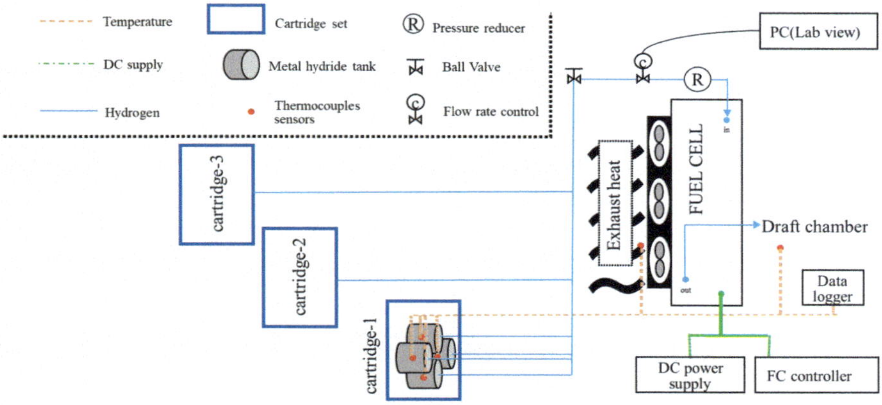


**Fig. 1** Schematic diagram of the FC apparatus

The tanks supplying hydrogen to the PEMFC were made of aluminum to efficiently capture heat from the outside. They used a sintered filter at the outlet to prevent the $LaNi_5$ from leaking into the flow path. One tank with an outer diameter of 30 mm and height of 30 mm contained an average of 95 g of hydrogen. This scale is intended to facilitate thermal management [12]. $LaNi_5$ can store up to 1.4% of its weight in hydrogen, implying that a single tank can contain approximately 14 liters of hydrogen. Thermocouples were used to measure the temperature at the center of the tank, and pressure gauges were placed near the joints of each cartridge. The cartridges were prepared at room temperature so that they were not exposed to the exhaust heat from the PEMFC.

In addition, a fixing device was fabricated, and two cartridges were lined up parallel to the exhaust air. The cartridges were lined up two by two (Fig. 2).

## 2.2 *Experimental Conditions*

The PEMFC-generated power was 65 W, which corresponds to the required power output on a flat road with a converter efficiency of 85% [13].

The measurement was ended when the pressure on the tank side fell below the required pressure of the FC (below 4.5–5.5 kPa) or when the discharge flow rate was less than the flow rate needed to generate 65 W. In addition, three sets of three cartridges were used, ranging from 1 to 3 sets of the number of cartridges to be connected. After each experiment was completed, hydrogen was absorbed at approximately 0.3 MPa for the next experiment, and hydrogen equal to the released flow rate was stored before the next experiment under the following conditions.

(a) PEMFC device

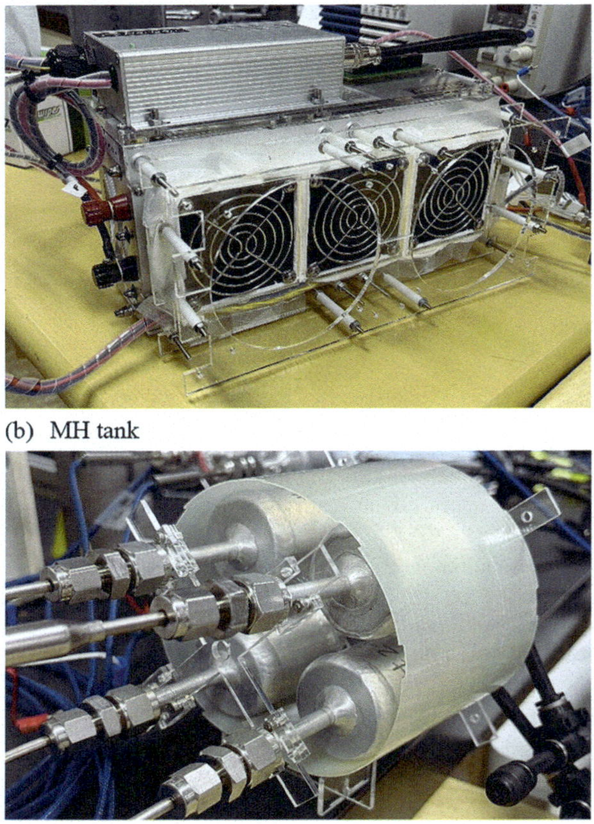

(b) MH tank

**Fig. 2** Configuration of PEMFC and MH tank

## 2.3 *Tank Thermal Management*

In a system combining a PEMFC and MH tank, the output of the FC side can be stabilized by loading an auxiliary battery into a system in which the required output fluctuates depending on the operation, such as an H-bike. In the case of stable operation of the FC side output, the temperature around the central portion of the MH tank can be stabilized and no temperature drop occurs. This can be considered a steady state in a circular cylinder with internal heat absorption. If the distance from the surface of the cylinder in the central direction is $r$ [m], the following equation holds:

$$\Delta T_{\mathrm{tank}} = \frac{Q_{\mathrm{tank}}}{4\lambda_{\mathrm{tank}}}\left(R^2 - r^2\right) \quad (1)$$

**Table 2** Coefficient details of Eqs. (1)–(3)

| | |
|---|---|
| Tank radius $R$ [m] | $1.5 \times 10^{-3}$ |
| Tank volume $V$ [m$^3$] | $2.1 \times 10^{-5}$ |
| Porosity $\varepsilon$ | 0.5 |
| $LaNi_5$ thermal conductivity $\lambda_{MH}$ [W/(m· K)] | 1.0 |
| $H_2$ thermal conductivity $\lambda_{Hydrogen}$ [W/(m· K)] | 0.1672 |
| Discharged mass flow rate m [kg/s] | $8.033 \times 10^{-5}$ |
| Heat of formation $\Delta H$ [J/kg] | $-1.539 \times 10^{-7}$ |
| Number of cartridges | 1, 2, 3, 4, 5 |

where $Q_{tank}$[W/m$^3$], $\lambda_{tank}$[W/(m · K)], and $R$ [m] are the heat absorption per unit volume, thermal conductivity, and radius of the cylinder of MH tank. Because the temperature difference between the center and wall can be calculated by setting $r = 0$ [m], Eq. (1) is used with $r = 0$.

The amount of heat absorbed inside $Q_{tank}$ per tank was determined using the following equation:

$$Q_{tank} = m\Delta H \frac{1}{4Vn} \tag{2}$$

where $m$ [kg/s], $\Delta H$ [J/kg], $V$ [m$^3$], and $n$ are the flow rate of the hydrogen released, heat produced in the reaction, tank volume, and number of cartridges, respectively. In addition, $\lambda_{tank}$ can be obtained using the following equation because the alloy in the tank exists with hydrogen as a porous material with porosity $\varepsilon$.

$$\lambda_{tank} = (1 - \varepsilon)\lambda_{MH} + \varepsilon\lambda_{hydrogen} \tag{3}$$

where $\lambda_{MH}$ [W/(m · K)] and $\lambda_{hydrogen}$ [W/(m · K)]are the thermal conductivities of $LaNi_5$ and hydrogen, respectively. The required temperature difference outside the tank was examined using Eqs. (1) and (2).

The constants in Eqs. (1) and (2) were set as listed in Table 2. The number of tanks, n, was assigned to the number of cartridges in one to five sets and was calculated to examine the change with the number of cartridges.

## 2.4 Minimum Inner Temperature of MH Tank

Considering a stable release, the PEMFC operation may become impossible when the equilibrium pressure of the release falls below the required pressure of PEMFC. The equilibrium pressure in $LaNi_5$ can be classified into three states depending on the amount of hydrogen stored, and formulas exist for each state (0 or less than 0.3 wt. %, 0.3 or less than 1.1 wt.%, and more than 1.1 wt.%).

**Table 3** Coefficients of Eqs. (4) and (5)

| | |
|---|---|
| Adsorption plateau pressure coefficient $A_d$ | 18.3 |
| Plateau pressure coefficient $B$ [K] | 3704.6 |
| alpha phase coefficient a | 0.0819 |
| Storage rate at 0.3 wt. $H/M|_{0.3wt\%}$ | 0.23008 |
| Reference pressure $p_{ref}$ [Pa] | 1000 |
| PEMFC required pressure $p_{FCreq}$ [Pa] | 4500 |

The following equation calculates the temperature below the PEMFC, which requires a pressure of 0.3 wt. %, which is the boundary of the storage volume at which the equilibrium pressure drops sharply in release and release becomes difficult [14].

$$p_{eq.d} = \exp\left( A_d - \frac{B}{T_{tank\_min} + a \cdot H/M|_{0.3wt\%}} \right) \cdot p_{ref} \tag{4}$$

$$p_{eq.d} \geq p_{FCreq} \tag{5}$$

where $p_{eq.\ d}$ [Pa], $A_d$, $B$[K], $T_{tank\ \_\ min}$[K], $a$, $p_{ref}$ [Pa], and $p_{FCreq}$[Pa] are the release equilibrium pressure, adsorption plateau pressure coefficient, plateau pressure coefficient, minimum internal tank maintenance temperature, alpha-phase factor, $H/M|_{0.3\ wt.\ \%}$ of the hydrogen stored percentage at 0.3 wt.% storage, the reference pressure, and the required pressure of FC.

The values of the constants in Eqs. (4) and (5) are listed in Table 3.

## 2.5 Exhaust Heat Through PEMFC

In the case where the exhaust heat of the PEMFC is provided with a cover against the card carriage and the temperature of the exhaust gas from the PEMFC is higher than the atmospheric temperature, the heat balance is as follows when Fig. 3 is considered as one system and Eq. (6) is valid.

$$\Delta Q_{exhaust} = -(4Q_{tank} + Q_{out}) \tag{6}$$

$Q_{tank}$ [J] was considered in Eq. (2) and because the four tanks were regarded as a one-card carriage, four times the amount of heat absorption was considered. In addition, $Q_{out}$ [J] was calculated as follows:

$$Q_{out} = U_{cover} \cdot (T_{in} - T_{air}) \cdot S_{cover} \tag{7}$$

where $T_{in}$ and $T_{air}$[K] are the temperatures of the intake gas and the gas outside the cover, respectively; $S_{cover}$[m$^2$] and $U_{cover}$[W/(m$^2$ · K)] are the surface area and thermal transmittance of the cover, respectively, based on the cover thickness $d$ [m] and the thermal conductivity of $\lambda_{cover}$ [W/(m · K)].

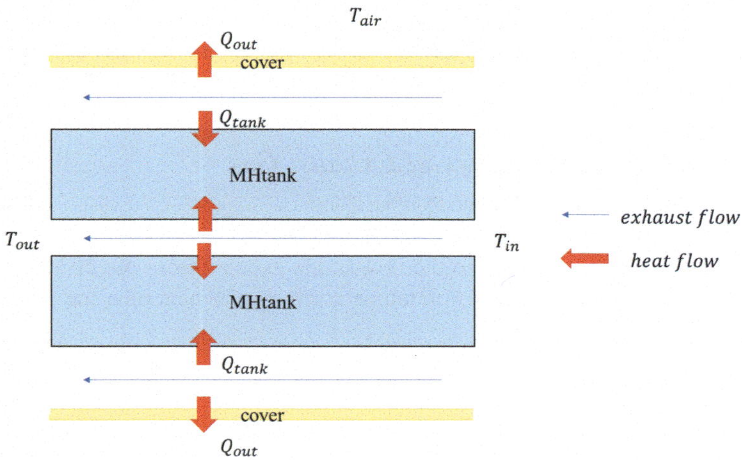


**Fig. 3** Heat balance in the system

**Table 4** Coefficients of Eqs. (6)–(9)

| | |
|---|---|
| Cover surface area $S_{cover}$ [m$^2$] | $9.4 \times 10^{-3}$ |
| Cover Thickness $d$ [m] | $5 \times 10^{-4}$ |
| Cover Thermal Conductivity $\lambda_{cover}$ [W/(m$^2$·$K$)] | 55 |
| PEMFC exhaust airflow $V_{exhaust}$ [m$^3$/h] | 62 |
| Inlet air temperature $T_{in}$ [°C] | 25.9 |
| Ambient temperature $T_{air}$ [°C] | 23.2 |

$$U_{cover} = 1/(d/\lambda_{cover}) \tag{8}$$

In addition, since $\Delta Q_{exhaust}$ [J] can be viewed as the amount of sensible heat lost in the airflow in the system, the following equation holds:

$$\Delta Q_{exhaust} = \rho_{air} \cdot V_{exhaust} \cdot (T_{in} - T_{out}) \tag{9}$$

where $\rho_{air}$[kg/m$^3$], $V_{exhaust}$[m$^3$/s], and $T_{out}$ [K] are the air density, exhaust gas volume, and outlet temperature from the PEMFC device.

These equations describe the temperature drop when passing through the system. In addition, because it can be assumed that these systems are lined up consecutively and that the cartridges are lined up in a connected manner, the minimum exhaust temperature conditions emitted by the PEMFC when mounted on the H-bike were calculated.

In this research group, a cover made of CFRP was created as a cover, focusing on the lightweight aspect (Fig. 2b); its thermal conductivity and thickness are listed in Table 4. The exhaust air volume of the PEMFC was measured using an anemometer. The temperature of the exhaust heat from the PEMFC was measured through a mockup test to determine the extent of the temperature increase above the outside air temperature [15]. In this calculation, $T_{in}$ and $T_{air}$ were calculated considering 25.9 °C near the PEMFC exhaust fan and 23.2 °C at room temperature at 65 W operation.

### 2.6 Minimum Temperature of Exhaust Gas

Next, from Eqs. (1), (3), and (8), the minimum maintained temperature of the tank and the temperature drop owing to the flow in the exhaust from the PEMFC can be calculated to determine the minimum temperature requirement near the air inlet:

$$T_{in\_min} = T_{tank\_min} + \Delta T_{tank} + \sum_{k=1}^{n} \Delta T_{cartridge_k} \quad (10)$$

where $\Delta T_{tank}$ and $T_{tank\_min}$ [K] are the temperatures based on Eqs. (1) and (3), respectively; and $\Delta T_{cartridge_k}$ [K] is the temperature difference of the k-th cartridge.

### 2.7 Life Cycle $CO_2$ ($LCCO_2$)

Next, we compared the environmental aspects in terms of equivalent $CO_2$ emissions using the LCA method. The comparison targets were an existing Li-ion battery and the proposed system, combining cartridges and FC. Background data were obtained from Ecoinvent 3 and evaluated using SimaPro v. 9.2.0.1, using the CML model as the environmental impact assessment method.

The system boundaries are shown in Fig. 4. The boundaries included the MH, PEMFC, and Li-ion battery manufacturing phases and the use phase of hydrogen for FC or electricity for Li-ion batteries from the production to charging stages.

The Li-ion battery to be compared was the Yamaha X0T-82110-20, which has been used in existing e-bikes [13]. The weight and battery capacity of the E-bike were 2.2 kg and 306 Wh, respectively. Note that electricity was supplied from a conventional grid in Japan and hydrogen of sewage sludge origin was used as fuel.

The functional unit is the total power output per fill for each system. Inventory data based on the Ecoinvent 3 database are listed in Table 5.

## 3 Results

### 3.1 FC Operation Tests

The temperature, pressure, and flow rate results for varying numbers of connected cartridges are shown in Figs. 5, 6, and 7. All experiments were terminated when the pressure was below the PEMFC pressure requirement. The release times were

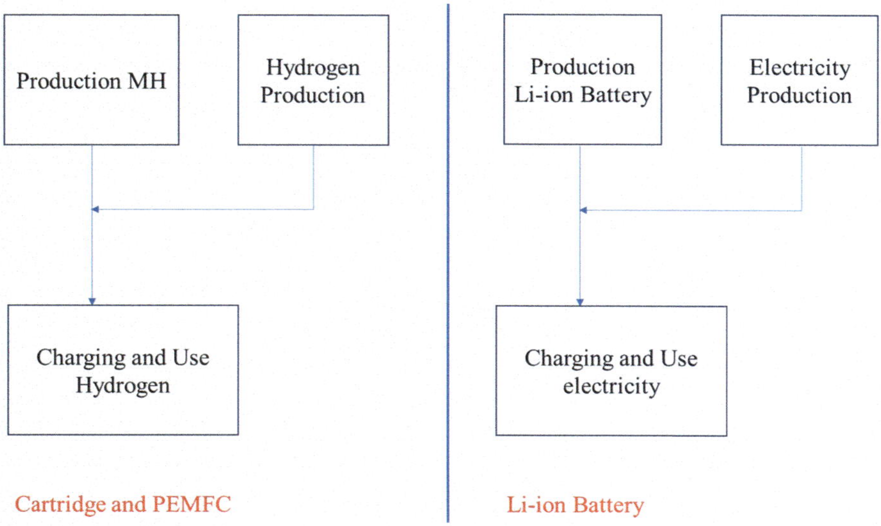


**Fig. 4** System boundary

**Table 5** Inventory data

| Li-ion battery | Contents | Input (one Li-ion battery) |
|---|---|---|
| | Li-ion battery production | 13.64 [kg $CO_2$ eq] |
| Electricity | | Input (one Li-ion battery) |
| | Conventional electricity of Japan | $2.06 \times 10^{-2}$ [kg $CO_2$ eq] |
| MH | | Input (one cartridge) |
| | Mischmetal Production [9] | 2.07 [kg $CO_2$ eq] |
| Hydrogen | | Input (one cartridge) |
| | Bio $H_2$ from Sewage sludge at Home | $2.9 \times 10^{-2}$ [kg $CO_2$ eq] |

approximately 100, 255, and 350 s for one, two, and three connected cartridges, respectively. Note that the required hydrogen flow rate of the PEMFC for 65 W power generation was approximately 900 sccm on average. At the end of the experiment, the tank temperatures were 19.47 °C for one connected cartridge, 19.74 °C for two cartridges, and 23.06 °C for three cartridges.

## 3.2 *Tank Thermal Management*

The results are shown in Fig. 8. It showed that the temperature difference required to reach a steady state was 23.41, 11.71, 7.80, 5.85 and 4.68 °C when there were one, two, three, four, and five cartridges. Consequently, as the number of tanks increased, the amount of heat required per tank decreased and the required temperature difference decreased.

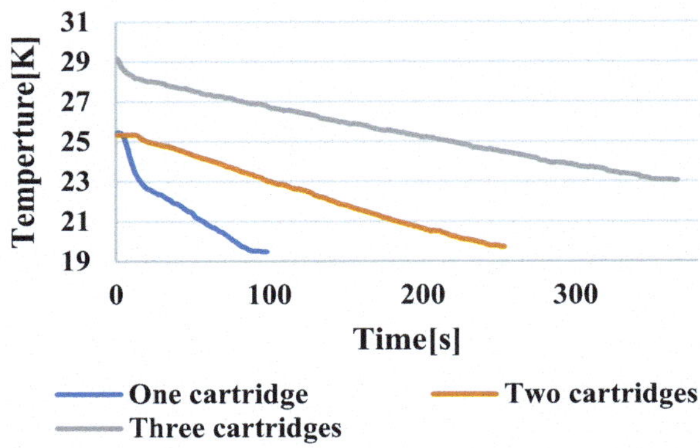


**Fig. 5** Temperature transition during the discharge test

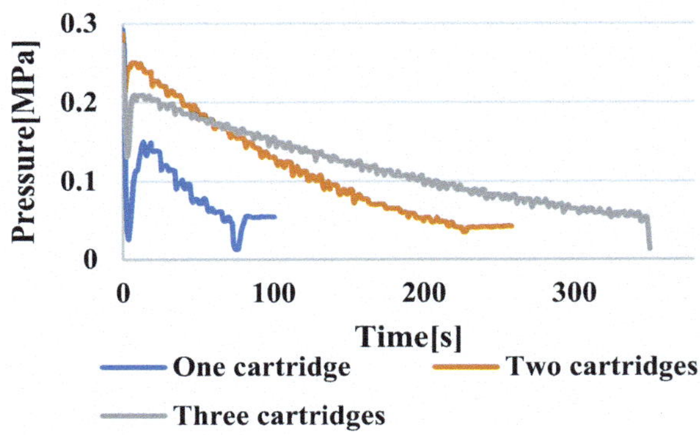


**Fig. 6** Pressure transition during the discharge test

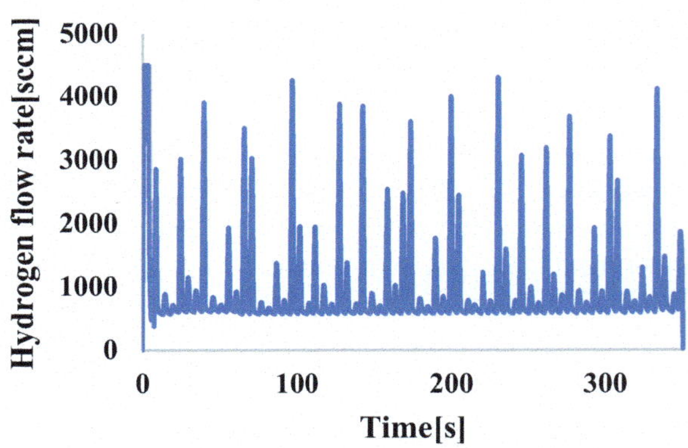


**Fig. 7** Flow rate transition during the discharge test

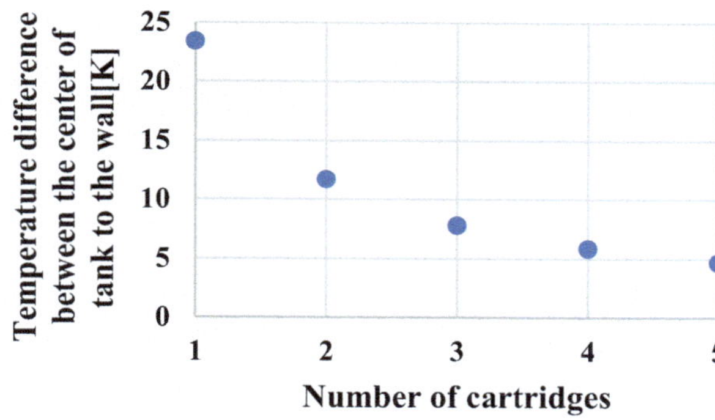


**Fig. 8** Temperature difference from the center of the MH tank to the wall

**Table 6** Temperature drops in exhaust air due to system

| Number of cartridges | $\sum \Delta T_{cartridge\_k}$ [K] |
|---|---|
| 1 | 0.96 |
| 2 | 0.99 |
| 3 | 1.0 |
| 4 | 1.0 |
| 5 | 1.1 |

**Table 7** Minimum temperature requirement of the air inlet

| Number of cartridges | $T_{in\_min}$ [°C] |
|---|---|
| 1 | 31 |
| 2 | 19 |
| 3 | 15.2 |
| 4 | 13.3 |
| 5 | 12.2 |

Based on these calculations, the minimum temperature $T_{tank\_min}$ at the center of the tank was 6.29 °C. In the following, we assumed that the minimum temperature requirement should not be below this value:

Moreover, the PEMFC exhaust air temperature decreased as it passed through the entire system, which varied with the number of connected cartridges calculated from Eqs. (6)–(9), as listed in Table 6.

Thus, the minimum temperature of the exhaust air near the fan of the PEMFC, which is required as a condition for stable output in the operation of the H-bike assuming a flat road, was calculated for each number of cartridges connected using Eq. (10) (Table 7). The operating temperature condition of an atmosphere above 30 ° C is not recommended for FC use (Table 1). Therefore, it is realistic to operate these systems using two or more cartridge combinations.

**Table 8** Total weight (PEMFC + Controller + Cartridge) of each system

| Number of cartridges | System weight [kg] |
|---|---|
| 1 | 3.9 |
| 2 | 4.6 |
| 3 | 5.2 |
| 4 | 5.9 |
| 5 | 6.6 |

## 3.3 Weight Balance

Next, we confirmed that the total weight of the FC battery satisfied the specific weight limitations. The weight per tank with MH enclosed was 170 g. The cartridge, including the four tanks, weighed 680 g. The total weight, including the PEMFC, controller, and required number of cartridges, with different numbers of connected cartridges, are listed in Table 8. The weight of the cover of one card carriage was approximately 3 g, which was negligible considering the entire system. The results showed that none of the systems exceeded the 8 kg weight limit for H-bikes. The results of this study can be used as a reference for future research. However, it should be noted that converters and other equipment must also be loaded in actual operation.

## 3.4 Energy Density and LCCO$_2$

Next, we compared the Li-ion batteries in terms of energy weight density, which is said to be superior to other batteries [16].

The Li-ion battery was based on conventional specifications and the cartridge side was calculated from the amount of stored hydrogen and weight of the cartridge (Eqs. (11) and (12)).

$$d_{\mathrm{Li_{ion}}} = V_{\mathrm{Li_{ion}}} \times \frac{A_{\mathrm{Li_{ion}}}}{M_{\mathrm{Li_{ion}}}} \tag{11}$$

$$d_{\mathrm{MH}} = V_{\mathrm{H_2}} \times Q_{\mathrm{H_2}} / M_{\mathrm{tank}} \tag{12}$$

where $d_{\mathrm{Li_{ion}}}$ [Wh/kg] and $d_{\mathrm{MH}}$ [Wh/kg] are the energy densities; $V_{\mathrm{Li_{ion}}}$ [V], $A_{\mathrm{Li_{ion}}}$ [Ah], and $M_{\mathrm{Li_{ion}}}$ [kg] are the battery-rated voltage, battery-rated capacity, and battery weight; $V_{\mathrm{H_2}}$ [L], $Q_{\mathrm{H_2}}$ [J/L], and $M_{\mathrm{tank}}$ [kg] are the storage volume of hydrogen, volumetric energy density of hydrogen, and MH tank weight, respectively. The energy densities of the batteries based on these equations are shown in Fig. 9.

Figure 10 shows the results of life cycle $CO_2$ (LC$CO_2$) analysis varying the numbers of MH tanks. The systems with 1–5 cartridges emitted lower LC$CO_2$ emissions than the Li-ion battery.

It can be seen from Figs. 9 and 10 that the developed system is superior in terms of environmental impact and battery performance. Thus, from the viewpoint of weight balance, optimal conditions for obtaining benefits can be determined.

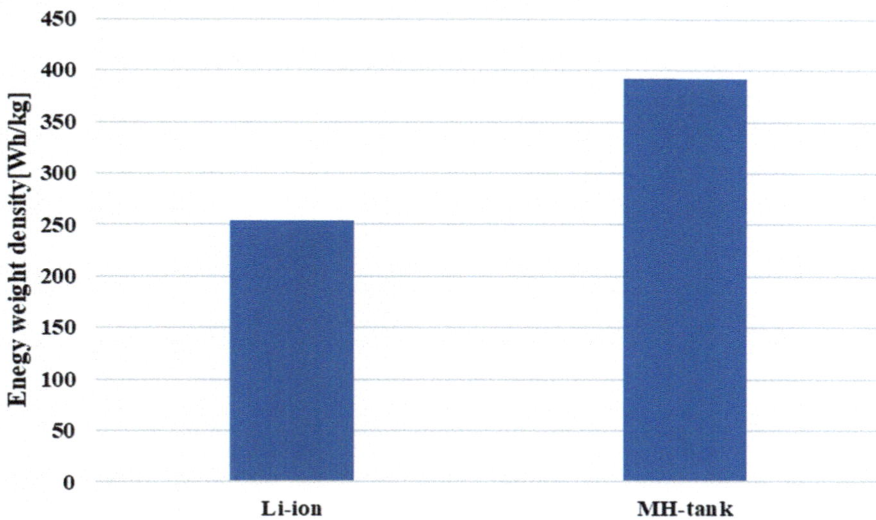


**Fig. 9** Result of energy density

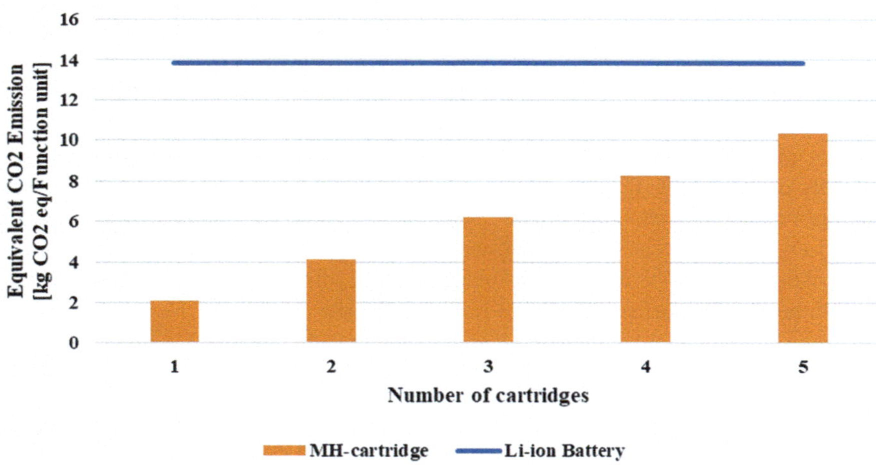


**Fig. 10** $LCCO_2$ results

## 4 Discussion

The results in Table 7 suggest that even with multiple cartridges installed, it is difficult to provide a sufficient temperature around MH tanks in winter from the exhaust heat alone, and it may be challenging to maintain the center temperature of the tank in a steady state. Therefore, it is necessary to study the amount of heat required to maintain the center temperature of the tank in a non-steady state. In

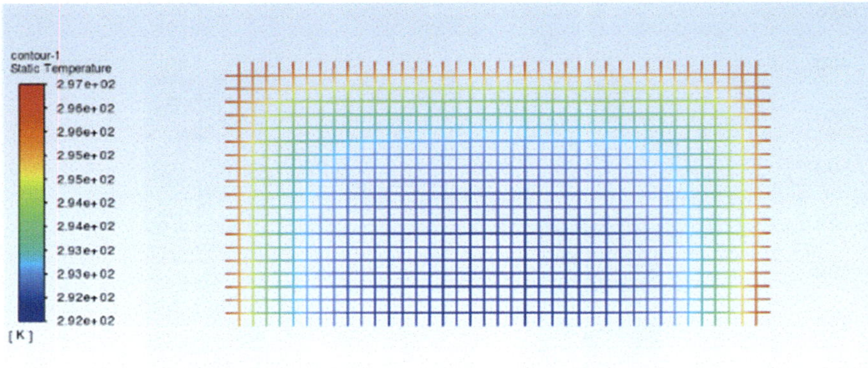


**Fig. 11** Temperature distribution in MH tank

addition, it is known that the auxiliary battery needed to operate the H-bike under steady-state conditions has a significant environmental impact when combined with the system; therefore, a study on a non-steady-state system with a smaller-scale auxiliary battery combination needs to be conducted [13].

During the release of hydrogen in a tank containing $LaNi_5$, an endothermic reaction due to the release generates an internal temperature distribution owing to the low thermal conductivity of $LaNi_5$. The temperature distribution was reproduced using transient analysis software [12, 14, 17] (Fig. 11):

Considering the temperature gradient from the wall to the center of the tank, the distance of 11.75 mm from the center is the center of the rise in the tank we are currently dealing with. Therefore, we considered thermal management based on the assumption that all reactions in the release occur from the center of the tank to 11.75 mm from the center. The following equation is valid according to Fourier's law because it can be assumed that the heat flux is sufficient to compensate for the reaction heat at the surface where the reaction occurs:

$$Q_{\text{tank}} = \frac{-2\pi\lambda_{\text{tank}} L \Delta T_{\text{tank}}}{\ln\left(\frac{r_{\text{tank}}}{r_{\text{area}}}\right)} \tag{13}$$

where $Q_{\text{tank}}$[W], $\lambda_{\text{tank}}$[W/(m · K)], $L$[m], $\Delta T_{\text{tank}}$[K], $r_{\text{tank}}$ [m], and $r_{\text{area}}$[m] are the reaction heat in the tank (Eq. 2), thermal conductivity (Eq. 3), height of the tank, temperature difference between the inner and outer of the tank, radius of the tank, and reacting position.

Using the values in Table 9, Eqs. (3) and (13) can be converted into a proportionality equation between the heat of the reaction and the temperature difference. This can be considered as a condition for not causing a temperature drop according to the reaction rate per MH tank.

$$Q_{\text{tank}} = 4.6 \times 10^{-2} \times \Delta T_{\text{tank}} \tag{14}$$

**Table 9** Coefficients of Eq. (13)

| | |
|---|---|
| Height of tank $L$[m] | $3.0 \times 10^{-3}$ |
| Radius of tank $r_{tank}$[m] | $1.5 \times 10^{-3}$ |
| Radius of reacting area $r_{area}$[m] | $1.17 \times 10^{-3}$ |

Based on Eq. (14), the relationship between the hydrogen flow rate discharged from one cartridge and the required temperature difference is proportional.

From the results, considering the atmospheric conditions, weight balance, and demand, we need to determine suitable thermal conditions to supply hydrogen fuel to FC. Thus, investigations of the cartridge insulator and control of the flow rate from the MH tank, considering the total weight balance, will be executed in future studies.

We have some issues which can achieve the practical level. The main problem is to control the flow rate of $H_2$ from the tank to meet the bike assist demand.

First, the flow rate might be inadequate for the travel distance and/or the duration, even if the stored $H_2$ in the tank is sufficient. In addition, due to the less capacity to follow the demand of the FC, the supplementary battery is necessary. This means the battery's scale and performance supplied from fossil electricity would be affected. Consequently, the eco benefit might not be obtained sufficiently.

Second, in our plan, the cartridges are assumed to be set up corresponding to the travel distance. However, deciding on the appropriate number of cartridges would be challenging for more users with different travel conditions because the uncertain aspects of space and/or duration are not considered. For several operating conditions (e.g., road, using duration, atmospheric climatic conditions, the loading weight, etc.), the flow rate test must be investigated by varying the number of cartridges in the future.

## 5 Summary

This study investigates thermal management in a PEMFC and MH tank system. We conducted an actual operation experiment by combining an MH tank ($LaNi_5$) and PEMFC-rated output of 300 W.

Our study examined the required temperature difference when the PEMFC generated electricity at the rated output of the auxiliary battery, assuming that the PEMFC was installed on an H-bike. Moreover, the minimum temperature required to maintain the pressure of the PEMFC below the required pressure was verified and the differences were presented depending on the number of cartridges installed in the system.

The results showed that the minimum required exhaust temperature of the PEMFC was 15.22, 13.29, and 12.15 °C when there were three, four, and five cartridges.

Moreover, regardless of the number of cartridges installed in the system (1–5), the PEMFC emitted lower $LCCO_2$ emissions than the Li-ion battery. This implies that the alternative FC battery provides environmental benefits in comparison with the Li-ion battery.

## References

1. Islam R, Shabani B, Andrews J, Rosengarten G (2017) Experimental investigation of using ZnO nanofluids as coolants in a PEM fuel cell. Int J Hydrog Energy 42:19272–19286
2. Shabani B, Omrani R, Andrews J (2017) Energy security and sustainability for road transport sector: the role of hydrogen fuel cell technology
3. Shabani B, Andrews J (2011) An experimental investigation of a PEM fuel cell to supply both heat and power in a solar-hydrogen RAPS system. Int J Hydrog Energy 36:5442–5452
4. Vijayakumar A, Kannan R, Ramesh S, Vijayakumar M, Raghavendran PS, Siva Ramkumar M, Anbarasu P, Sundramurthy VP (2022) Review on li-ion battery vs Nickel metal hydride battery in EV. Adv Mater Sci Eng 2022, 7910072. 7 pages
5. Rowsell JL, Yaghi OM (2005) Strategies for hydrogen storage in metaleorganic frameworks. Angew Chem Int Ed 44:4670–4679
6. Boris PT, Pavel VF, Alexey AV, Mikhail SB, Yustinas YS, Aleksey MK, Volodymyr AY, Stanford C, Sivakumar P, Mykhaylo VL (2021) Metal hydride hydrogen storage and compression systems for energy storage technologies. Int J Hydrog Energy 46:13647–13657
7. Alqahtani T, Mellouli S, Bamasag A, Askri F, Phelan PE (2020) Thermal performance analysis of a metal hydride. Int J Hydrog Energy 45:23076–23092
8. Chung CA, Yang S-W, Yang C-Y, Hsu C-W, Chiu P-Y (2013) Experimental study on the hydrogen charge and discharge rates of metal hydride tanks using heat. Appl Energy 103:581–587
9. Omrani R, Nguyen HQ, Shabani B (2020) Thermal coupling of an open-cathode proton exchange membrane fuel cell with metal hydride canisters: an experimental study. Int J Hydrog Energy 45:28940–28950
10. Omrani R, Nguyen HQ, Shabani B (2019) Open-cathode PEMFC heat utilization to enhance hydrogen supply rate of metal hydride canisters. Energy Procedia 160:542–549
11. Hosobuchi E, Misaki C, Katayama N, Dowaki K (2021) Impact of increased tire pressure on fuel consumption and environment for fuel-cell-assisted shared bicycles. J Jpn Inst Energy 100: 116–121
12. Hara D, Miao S, Shimogawa J, Katayama N, Dowaki K (2021) A suitable design of metal hydride cartridge for a fuel cell assisted bicycle in consideration of heat transfer performance. IOP Conf Series Earth Environ Sci 1034:012049
13. Nagado R, Sakai S, Ezawa T, Shimogawa J, Katayama N, Miao S, Dowaki K (2022) Dynamic LCA of fuel cell-assisted bike in consideration of Pt catalyst degradation in various operating conditions. IOP Conf Series Earth Environ Sci 1187
14. Chung CA, Chen Y-Z, Chen Y-P, Chang M-S (2015) CFD investigation on performance enhancement of metal hydride hydrogen storage vessels using heat pipes. Appl Therm Eng 91:434–446
15. Shimogawa J, Hara D, Miao S, Katayama N, Dowaki K (2022) Design and temperature analysis of a metal hydride cartridge using exhaust heat of a fuel cell for electric-assisted bicycles. J Japan Inst Energy 101:152–161
16. Quinn JB, Waldmann T, Richter K, Kasper M, Wohlfahrt-Mehrens M (2018) Energy density of cylindrical li-ion cells: a comparison of commercial 18650 to the 21700 cells. J Electrochem Soc 165(14):A3284–A3291
17. Hara D, Misaki C, Katayama N, Dowaki K (2021) Computational fluid dynamics simulation and energy consumption analysis of metal hydrides in the hydrogen charging process. J Jpn Inst Energy 100:294–300

# Replacement of Manipulator Modules Based on Workload Using Part Agents

**Arata Hori and Hiroyuki Hiraoka**

**Abstract** It is important to promote the reuse of products in order to create a circulating society. For this purpose, the management of products and parts is necessary not only at their manufacturing stage but also throughout their entire life cycle, including use, collection, and disposal. To achieve this, we are developing a part agent system. The part agent system assigns a dedicated network agent to each product and part. The agent moves on the network, following the corresponding product or part in the real world. It retains and collects information on the product or part, including the lifecycle, manufacturing information, current status, and operation history. Based on the information, the part agent advises the user in making decisions regarding the maintenance of the product or part.

In this study, we focus on the workload of the module as a value related to the deterioration and environmental load of the module. The part agent monitors the workload of the corresponding module, determines the appropriate replacement of the module based on communication and negotiation with other part agents, and generates advice to the user on the replacement. In this paper, we describe a scheme applied to a modularized manipulator, propose a method for a part agent to replace modules based on the joint workload of the manipulator, show some early simulation results, and discuss the issues with the method.

**Keywords** Part agents · Workload · Reuse · Replacement of modules

A. Hori
Graduate School of Engineering, Chuo University, Tokyo, Japan

H. Hiraoka (✉)
Faculty of Science and Engineering, Chuo University, Tokyo, Japan
e-mail: hiraoka@mech.chuo-u.ac.jp

 
S. Fukushige et al. (eds.), *EcoDesign for Circular Value Creation: Volume I*,
https://doi.org/10.1007/978-981-97-9068-5_3

## 1 Introduction

In recent years, energy resources on the earth have been depleted, so there is an urgent need to shift from a consumption society with mass production, mass consumption, and mass disposal to a circulating society [1]. The reuse of parts and products, which can reduce energy consumption and waste generation, is regarded as an important measure to this end. To achieve efficient reuse of parts, it is necessary to manage the life cycle of individual parts. For that purpose, we are developing a part agent system that manages the status of parts and supports the maintenance actions of users [2].

A part agent system manages the lifecycle of parts by making network agents follow the actual parts. A network agent is a software module that autonomously works according to the situation while moving around the network. A part agent corresponding to each individual part manages all information about the part throughout its lifecycle, predicts its possible states in the near future, and generates appropriate advice on part maintenance [3, 4].

The replacement of parts is a crucial activity for promoting the reuse of parts. An appropriate decision is required for when and which part is replaced with which part. The decision is made based not only on the deterioration of the part, but also on various factors such as its lifecycle stage, its environmental burden, the current market situation and the user's preferences. We are developing a system to support the replacement of parts using the part agent system.

We chose a manipulator that consists of multiple modules consisting of links and joints as the target product of this research. A part agent is assigned to each module to manage its life cycle and replace it.

In this study, we use the workload performed by a module as a key value for the decision on its replacement. The value of workload relates both to the deterioration of the module and to the environmental load it produces. In our simulation, the workload of the joints of a modularized manipulator is estimated based on the motion of the manipulator. We regard this joint workload of a module as the workload of the module.

Researchers in robotics studied the efficiency of the workload; Carnier et al. proposed the optimization of energy consumption of the walking behavior of two-leg biped models [5], and Li et al. proposed constraining a link of the manipulator to improve accuracy of the trajectory and minimize energy consumption [6]. For manufacturing systems, Suwa et al. proposed an empirical model to predict the power consumption of machine tools [7]. The literature concerns mainly the efficiency and reduction of energy consumption in the behavior and operations of machines.

In this study, we focus on the workload and power consumption of the replacement of parts. The parts are replaced for various reasons, for example, when their deterioration level is insufficient for the assigned work, when the user prefers a new part, or when their environmental load becomes too large. As stated above, the workload of modules of the manipulator is a critical value that can be applied to many of these purposes. In order to reduce the workload, we adopted a strategy [8] that equalizes the workload of the modules. If the same products are used for a certain period of time by a group of users, each of whom requires a different amount of workload, the number of products that exhaust their lives in the period can be reduced by exchanging the products among users to equalize the workload of the products.

In this paper, our proposal for the replacement of modules based on their workload is described to equalize the workload of the modules, along with the preliminary results of the simulated replacement. After this introduction, our scheme to replace modules of a manipulator using a part agent system and its simulation are described in Sect. 2. Module-based manipulator we are developing, and the estimation of its workload based on the simulated behavior of the manipulator are explained in Sect. 3. The preliminary simulation result is shown in Sect. 4. After the future issues are discussed in Sect. 5, we conclude the paper in Sect. 6.

## 2 Workload-Based Replacement of Modules

### *2.1 A Scheme for the Replacement of Modules*

Figure 1 shows the proposed scheme for the replacement of parts based on the part agent system. The colored block at the bottom of the figure depicts the modularized manipulator to be maintained. The control system controls the manipulator to do a designated task. A part agent is assigned to each module and obtains information about the motion of the corresponding module from the control system. It estimates the accumulated workload of the module based on the obtained information and the

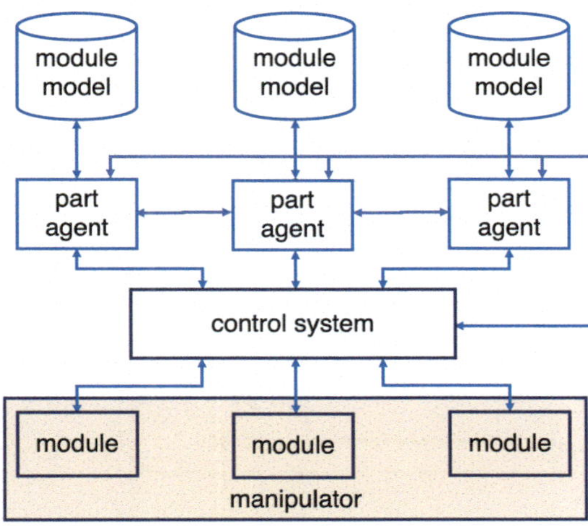


**Fig. 1** A scheme for the replacement of manipulator modules

state of the module contained in the module model. The estimation is performed with the collaboration of the modules. If a part agent considers that a replacement is necessary for its part, it sends a request for the replacement to other part agents. The user is advised to replace the part when the replacement is decided by negotiation among part agents. We are developing the module replacement system based on this scenario.

## 2.2 Concept of Part Agent

In the part agent system, an agent dedicated to each product or part collects information such as the state of the product or part, manufacturing information, and usage history, and presents it so that a user can make an accurate judgment on the maintenance of the product or part. Figure 2 shows a conceptual scheme of the part agents. With radio frequency identification (RFID) tags [9] attached to each part, a part agent on the network follows its corresponding part that moves among the production factory, repair factory, consumer, retail store, etc. The agent communicates with functions within the network to obtain the necessary information to manage its corresponding part such as product design information. The agent then devises the most appropriate maintenance action for the corresponding parts at that time and proposes it to the user. We assume that a widespread network is available for developing the part agent system, via which activities of all product life cycle stages are connected and are accessible by the users and part agents from anywhere.

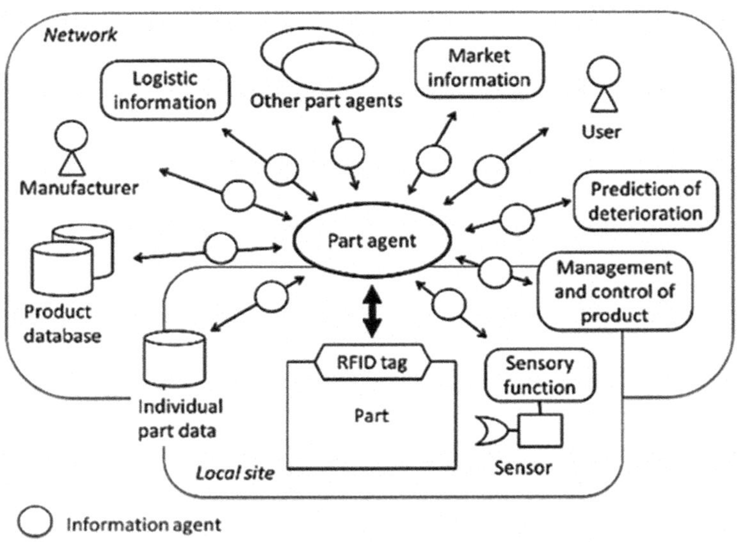


**Fig. 2** Conceptual scheme of part agent

We attach an RFID tag to a part to associate the part with its corresponding part agent. This enables the part agent to follow the part that changes its position in the real world.

### *2.3 Simulation of the Replacement System*

Prior to implementing the module replacement system, we developed a simulation of the replacement system to check its feasibility. In the simulation, the control system works virtually, and the behavior of the part agents and modules is generated. Negotiation among the modules is not implemented, and the replacement decision is made based on the predetermined workload threshold. Details of the simulation are described in the next section.

## 3 Module Replacement of Manipulator

### *3.1 Modularized Manipulator*

The target is a small manipulator, as shown in Fig. 3, that has three links whose dimensions and weights are shown in Table 1. The manipulator, whose kinematic

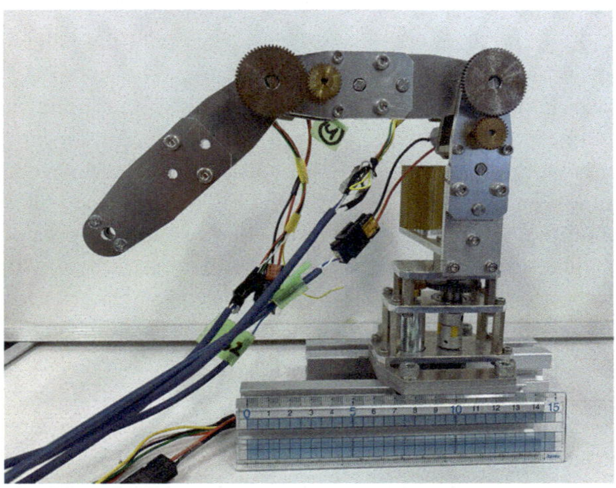

**Fig. 3** Modularized manipulator

**Table 1** Characteristics of the links of the manipulator

| Link | Base link | First link | Second link |
|---|---|---|---|
| Length [m] | 0.135 | 0.07 | 0.07 |
| Width [m] | 0.03 | 0.03 | 0.03 |
| Weight [kg] | 1.215 | 0.63 | 0.63 |

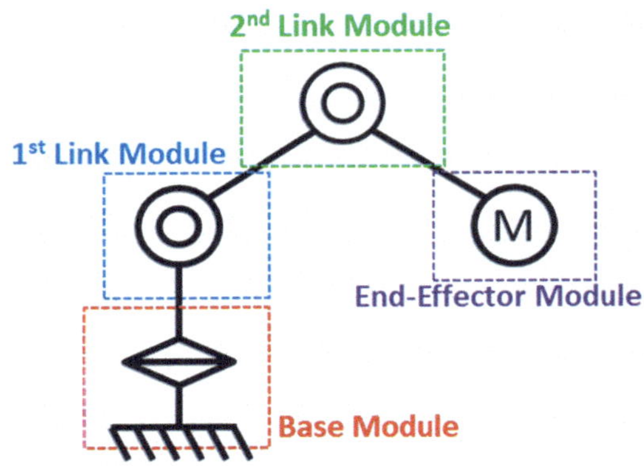


**Fig. 4** Kinematic structure of the manipulator [10]

structure is depicted in Fig. 4, consists of four modules. Each module, except the end-effector module, has a joint and two half-links and is designed for ease of replacement [10]. We assume that each module is managed by a part agent and is replaced and reused based on its estimated workload.

## 3.2 *Behavior of Modules*

In the simulation, the behavior of the modules is estimated by the following calculation.

Figure 5 depicts the simulation process of the behavior of modules based on the assembly structure of the manipulator. The assembly model that represents the manipulator consists of multiple models of the modules. A module model contains geometric and kinematic information as well as information for dynamics. In the simulation, calculating the forward kinematics and dynamics of the module determines its behavior. The calculation is performed in series from one module to another along the kinematic structure of the manipulator.

The forward kinematics that calculates the posture of the manipulator is performed by following its kinematic links. The transformation matrix representing the position and orientation of a module is calculated by multiplying the transformation matrix representing the position and orientation of the former module, by the displacement between two joints of the link, and the joint motion between two links connected. Thus, the calculation is performed from one module to another via the structure of the manipulator, as shown in Fig. 5.

The following dynamics Eqs. (1) through (4) calculate the force and torque necessary to evaluate the workload of the joint of a module [4]. In the equation, $\boldsymbol{C}$ and $\boldsymbol{r}$ denote the transformation and vector of the displacement, respectively. $\boldsymbol{F}$ and $\boldsymbol{M}$ denote the force and torque. $\boldsymbol{a}$, $\boldsymbol{g}$, $\boldsymbol{\omega}$, and $\dot{\boldsymbol{\omega}}$ denote acceleration, gravity acceleration, rotational velocity, and rotational acceleration, respectively. $m$ and $I$ denote the mass and inertial moment of the link. The suffixes $J$ and $L$ denote joint and link,

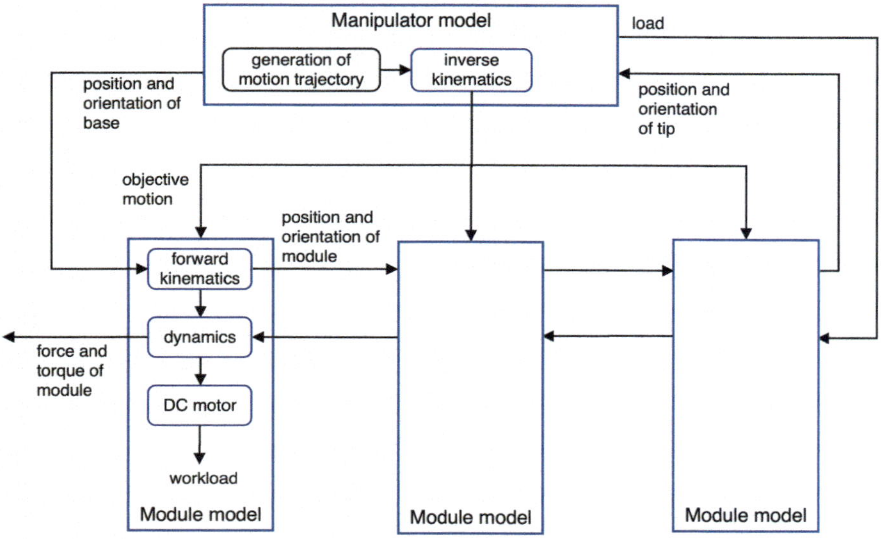


**Fig. 5** Estimation of module behavior

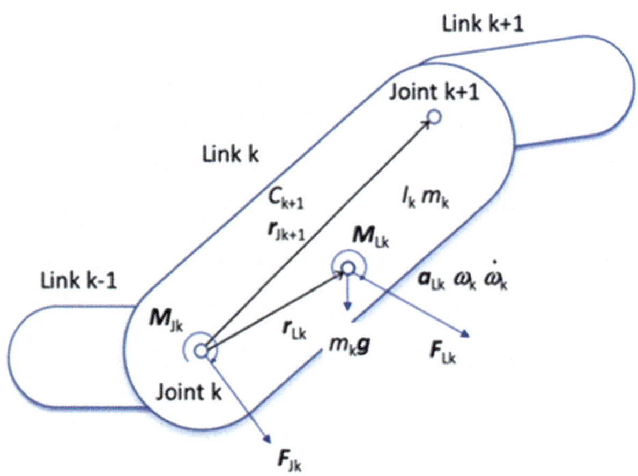


**Fig. 6** Symbols for dynamics of a link module [12]

respectively. $k$ represents the number of the link. Figure 6 shows the nomenclature for the symbols.

$$\boldsymbol{F}_{Lk} = m_k(\boldsymbol{a}_{lk} + \boldsymbol{g}) \tag{1}$$

$$\boldsymbol{M}_{Lk} = \boldsymbol{I}_k\dot{\boldsymbol{\omega}}_k + \boldsymbol{\omega}_k \times \boldsymbol{I}_k\boldsymbol{\omega}_k \tag{2}$$

$$\boldsymbol{F}_{Jk} = \boldsymbol{C}_{k+1}\boldsymbol{F}_{Jk+1} + \boldsymbol{F}_{Lk} \tag{3}$$

$$\begin{aligned} \boldsymbol{M}_{Jk} = \boldsymbol{C}_{k+1}\boldsymbol{M}_{Jk+1} + \boldsymbol{r}_{Jk+1} \times \boldsymbol{C}_{k+1}\boldsymbol{F}_{Jk+1} \\ + \boldsymbol{M}_{Lk} + \boldsymbol{r}_{Jk+1} \times \boldsymbol{F}_{Lk} \end{aligned} \tag{4}$$

Note that both the kinematic behavior and dynamic behavior of a module affect and are affected by its connected modules [11].

### 3.3 Estimation of the Workload

There are two aspects to the workload: one is the mechanical workload, and another is the consumed electric power. The former is calculated from the motion of the manipulator, i.e., the load and the trajectory, and it might affect the deterioration of the mechanical structure of the manipulator. The latter is calculated by estimating the electric energy consumed at the motors of the manipulator, and it might affect the electric power charge and the amount of greenhouse gas generated in producing it. The workload of the manipulator is generated by the DC motors of its joints of the modules. We estimate the consumed electric power using the following Eq. (5), which represents the conservation of energy of the motor.

$$\mathbf{V}\boldsymbol{I}_a = \boldsymbol{R}_a\boldsymbol{I}_a{}^2 + \boldsymbol{K}_E\boldsymbol{\Omega}\boldsymbol{I}_a \tag{5}$$

where $\mathbf{V}$ [V] and $\boldsymbol{I}_a$ [A] are the voltage and electric current which are imposed on the motor. $\boldsymbol{R}_a$ [Ω], $\boldsymbol{K}_E$ [V/m/s], and $\boldsymbol{\Omega}$ [rpm] are the resistance, back-emf constant, and rotation speed of the motor, respectively.

This equation shows that the supplied electric power on the left-hand side of the equation is consumed as Joule heat in the first item on the right hand and mechanical workload in the second item. The rotation speed of the motor and the current can be calculated using the angular velocity and the torque of the joint, which are derived from the trajectory and load. The Eqs. (6) and (7) calculate those values for estimating the joint workload of a module.

$$\boldsymbol{\Omega} = \boldsymbol{\omega} * \frac{\mathbf{60}}{\mathbf{2}\boldsymbol{\pi}} \tag{6}$$

$$\boldsymbol{I}_a = \frac{\boldsymbol{T}}{\boldsymbol{K}_T} \tag{7}$$

where $\boldsymbol{\omega}$ is angular velocity of the joint, $\boldsymbol{K}_T$ [Nm/A] is the torque constant of the motor, and $\boldsymbol{T}$ [Nm] is the required torque. We will investigate both the mechanical workload and the consumed electric power using the simulation and the experiments.

### 3.4 Decision of Replacement

In this simulation, the replacement is decided according to the predetermined threshold of the estimated workload, as described in Sect. 2.3. The value of the threshold is increased by the basic threshold value multiplied by the number of requests for replacements the module issued previously. Hence, the second request for replacement is issued based on a threshold twice that of the first request, and the third request is issued based on a threshold three times that of the first request. Note that the threshold of a module is not changed when the replacement is requested by another module.

When the workload of the joint of a module reaches the threshold thus determined, replacement of the module is requested, and the module is replaced with another module that has the joint with the lowest workload value.

## 4 Simulation of Module Replacement by the Predetermined Workload

### 4.1 Simulation of Replacement of the Modules

Two manipulators were used in the simulation. The tip load of one manipulator was set at 1 kg and the other at 0.01 kg. The manipulators were operated for half an hour, and the cumulative joint workload of each module during the period was estimated. In this simulation, the modules subject to replacement were the two link modules shown in Fig. 4. The threshold for the first replacement of modules was 100,000 [Nm]. This value was chosen to promote multiple replacements of modules deliberately during the experiment duration of the simulation, i.e., for half an hour. The simulation was performed according to the following procedure:

1. Set initial values for manipulators and set each module as a part agent.
2. Set the mechanism information, the tip load, and the trajectory of the manipulator.
3. Calculate the amount of workload from the required torque to move the manipulator.
4. Check the workload of the joint at each module once every 10 s while repeating the simulation for half an hour.
5. If a part agent notices that the total workload of the corresponding module reaches the threshold, the module is exchanged with another one.
6. Output the workload and motion of the manipulator.

## 4.2 Experimental Results on Workload Estimation

The results of the simulation were compared between the case when the modules were replaced and the case when they were not replaced. Figure 7 shows the variation in the workload over time at each module when replacement was performed, and Fig. 8 shows the workload without replacement. Here, we refer to the four link modules subject to replacement as modules A, B, C, and D. At the beginning of the simulation, modules A and B are set to the first and second links of manipulator 1, and modules C and D are set to the first and second links of manipulator 2, respectively.

In the case when the modules were replaced, smaller variations were observed in the workload of joints of modules. As every module experiences the similar workload, it is expected to equalize the lifetime of modules. Table 2 shows the power consumption and workload of the joint at each module of the manipulators. Most of the power of the motor is consumed as Joule heat, and only a small amount of the power is used for workload. The larger the load applied to the joint of the manipulator, the smaller the ratio of workload to power. We will conduct further experiments with different conditions.

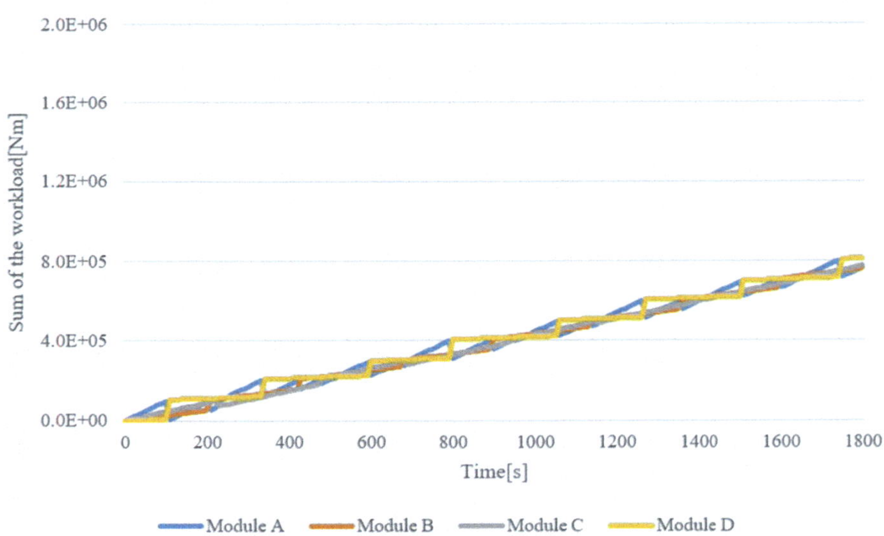


**Fig. 7** Workload transition with replacement

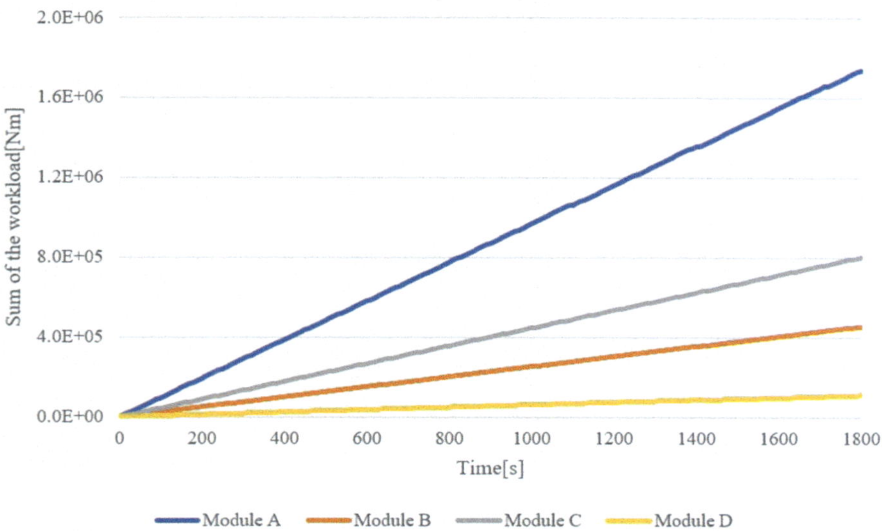


**Fig. 8** Workload transition without replacement

**Table 2** The power consumption and workload of the joint at each module of the manipulators

| Module | Module A | Module B | Module C | Module D |
|---|---|---|---|---|
| Power consumption [J] | 3.9E+10 | 7.9E+09 | 1.1E+09 | 6.9E+07 |
| Workload [J] | 1.7E+06 | 8.0E+05 | 4.6E+05 | 1.1E+05 |

# 5 Issues for Future Development

## *5.1 Implementation of the Replacement System*

Issues remain to implement the replacement system. We are developing a control system for the modularized manipulator shown in Fig. 3, with which part agents communicate and get necessary information to estimate the workload of modules. We will verify the estimated workload obtained by the simulation through an experiment. We will measure the electric power consumption of the manipulator and evaluate the validity of the simulation results.

The workload of modules is determined not only by the position of the joint in the kinematic structure but also by the trajectory of the manipulator. A combined planning of the replacement and the trajectory may be a future issue.

### *5.2 Negotiation Among Part Agents for the Replacement of Modules*

As each module has a different status, different deteriorations, and a different history of use, it may have different requirements for its usage. Part agents should recognize those requirements of the corresponding module and should request appropriate maintenance actions, including replacement. A negotiation method among the agents is necessary to coordinate those various requirements. For example, a part agent may request a replacement of the corresponding part because a notable deterioration of the part is detected, and another part agent may request a replacement of its part because the user prefers a newer part. If the negotiation among agents is successful, the former agent will replace its part with the part of the latter agent, and the latter agent will purchase a new part.

### *5.3 Combination of Task Scheduling, Module Replacements, and Planning of the Motion*

When a task is assigned to a group of manipulators in a factory, a schedule is planned for each manipulator to do a series of subtasks. Then, a sequence of operations is planned for each manipulator to perform the assigned subtasks. If necessary, modules are replaced with those with different specifications, such as length of links or performance. For efficient execution of a task for a group of manipulators with a lower environmental load, it is necessary to make task scheduling of manipulators, replacements of manipulator modules, and planning of operations for each manipulator in combination. As this issue is an optimization problem with multiple purposes, we will address it in the future.

Even if the plan is available, the progress of deterioration and the occurrence of failure are unpredictable. That is a reason why we are developing a part agent system that monitors the state of the corresponding module and provides advice on maintenance, including module replacements.

## 6 Conclusion

This paper describes our proposal for an effective module replacement based on the estimation of the power consumption and the workload of a simple manipulator. Early simulation results are shown, which present equalized workloads of modules. In the future, we will verify the power consumption for the trajectory of the manipulator.

## References

1. Umeda Y, Takata S, Kimura F, Tomiyama T, Sutherland JW, Kara S, Herrmann C, Duflou JR (2012) Toward integrated product and process life cycle planning—an environmental perspective. CIRP Ann 61:681–702
2. Yokoki Y, Hiraoka H (2017) Life cycle simulation of mechanical parts with part agent considering user behavior. Procedia CIRP 61:116–121
3. Nagahata T, Saito H, Hiraoka H (2020) Application of causality model to propose maintenance action of parts. In: EcoDesign and sustainability, vol I, pp 325–338
4. Tanaka T, Hiraoka H (2021) State prediction and parts replacement of a manipulator based on assembly model. In: EcoDesign, pp 491–498
5. Carnier RM, Fujimoto Y (2018) Comparison of energy consumption of an optimized gait cycle between human-like and bird-like leg models. In: IECON 2018 - 44th Annual Conference of the IEEE Industrial Electronics Society. Washington, DC, 21–23 October 2018
6. Li X, Shen K, Minami M (2017) Merits concerning energy-consumption and trajectory-tracking accuracy derived from elbow-bracing robot. J Adv Mech Des Syst Manuf 11(5)
7. Suwa H, Samukawa T (2022) Energy efficiency in machining systems based on power-low model for specific energy consumption. Procedia CIRP 107:931–936
8. Takata S (2013) Maintenance-centered circular manufacturing. Procedia CIRP 11:23–31
9. Bornello G (2005) RFID: tagging the world. Commun ACM 48(9):34–37
10. Fukazawa Y, Honda Y, Hiraoka H (2020) Part agents for exchanging modules of manipulators. In: EcoDesign and sustainability, vol I, pp 281–295
11. Luh JYS, Walker MW, Paul RPC (1980) On-line computational scheme for mechanical manipulators. J Dyn Syst Meas Control 102(2):69–76
12. Hiraoka H, Tanaka T, Sugahara T, Saito H (2021) Assembly model of mechanical products for their life cycle simulation. Procedia CIRP 98:553–558

# Development of a Knowledge Repository Model for Conflict Management in Life Cycle Design Collaborations

**Jing Tao, Yingyu Chen, Rujing Wu, Suiran Yu, and Haiwei Zhou**

**Abstract** The pursuit of sustainability is fundamentally impacting the way companies develop new products across the world. Life cycle design (LCD) is believed to be a promising approach to the realization of a sustainable production and consumption system. Within the methodology framework of LCD, comprehensive and systematic considerations of different aspects (e.g., product, process, service) of the product development are required, as well as the holistic optimization of the whole life cycle (i.e., close the loop). Thus, cross-organizational collaboration is believed to be essential, but challenging for LCD. Conflict in collaboration is a key factor hindering the effective implementation of product lifecycle design: the differences in corporate strategy, interest demands, and technological level among stakeholders involved in product life cycle design, as well as the special "competition-cooperation" relationship between them, result in many obstacles in LCD collaborations. The introduction of environmentally friendly product attributes can disrupt the optimized balancing of existing product system functionality, performance, and cost, leading to conflicts between new and existing attributes. This study aims to develop a knowledge repository for conflict management in LCD collaboration and to facilitate collaborative eco-design practices of complex mechatronic product (i.e., manufacturing equipment, construction machinery, vehicles). An ontology-based representation of conflict typology and conflict-resolving process in life cycle design collaborations are first proposed, which enable the structuring and standardization of related data, the storage of such data in a computer-processable manner, and

J. Tao (✉) · Y. Chen
School of Mechatronic Engineering and Automation, Shanghai University, Shanghai, P. R. China
e-mail: taojing1016@shu.edu.cn

R. Wu · S. Yu
School of Mechanical Engineering, Shanghai Jiao Tong University, Shanghai, P. R. China

H. Zhou
School of Mechatronic Engineering and Automation, Shanghai University, Shanghai, P. R. China

PISX Shanghai Pairui Information Technology Co Ltd, Shanghai, P. R. China

 
S. Fukushige et al. (eds.), *EcoDesign for Circular Value Creation: Volume I*,
https://doi.org/10.1007/978-981-97-9068-5_4

generation of conflict management case knowledge. Descriptive features of conflict management case are defined in accordance to the ontology model, based on which the similarity between case and case queries is calculated using the nearest neighbor method and case retrieval is realized based on the calculated similarity. With the proposed ontology model, a cloud-based and micro-service architectured conflict-managing knowledge repository system prototype is then developed, which provide the benefits of low-cost deployment, better accessibility and easy maintenance.

**Keywords** Life cycle design · Collaboration · Conflict management · Knowledge repository · Ontology

## 1 Introduction

Design has an important impact on product life cycle sustainability. Life cycle design (LCD) is an advanced design methodology based on the principle of improving the environmental, economic and social sustainability of products, and on the premise of meeting the needs of users and the requirements of social well-being [1, 2]. It is believed to be an inevitable choice to implement the green manufacturing strategy and realize the sustainable development of industry [3].

LCD is affecting the way companies engineer and production their products [4]. Within the methodology framework of LCD, comprehensive and systematic considerations of different aspects (e.g., product, process, service) of the product development are required, as well as the holistic optimization of the whole life cycle (i.e., close the loop). This also means that a single designer or design team may no longer be able to complete all the work of product development independently, but need to carry out cross-organizational collaborative design [5]. The most typical example is end-of-life cycle design through the coordination of parts and material suppliers, manufacturers, distributors, recycling service providers, over the selection of recyclable materials, design for disassembling, retired products collection, recycling process. Stakeholders along the product life cycle work together to achieve the recycling strategy through the necessary knowledge, technology trading.

Thus, cross-organizational collaboration is believed to be essential, but challenging for LCD. Cross-organizational collaboration in LCD may lead to increased risk of conflicts, as distributed multidisciplinary personals with diverse value and background ought to work concurrently to complete multidimensional design goals (e.g., environment-friendly, highly profitable, well performed, etc.) [6]. This naturally leads to the needs for methods and tools of conflict managing in LCD collaborations.

Thus, this study aims to develop a knowledge repository for conflict management in LCD collaboration and to facilitate collaborative eco-design practices of complex mechatronic product (i.e., manufacturing equipment, construction machinery, vehicles).

In the following sections, the characteristics of conflicts in LCD collaboration are first analyzed. Then, the ontology-based representation of conflict managing knowledge is proposed. With the proposed ontology model, Web service-based conflict managing knowledge repository system is developed to help LCD conflict resolution. The proposed knowledge repository system provides the benefits of low-cost deployment, better accessibility, and easy maintenance.

## 2 Characterizing Conflicts in LCD Collaboration

Conflict is a key issue that restricts the effective implementation of product life cycle design: The introduction of environmentally friendly attributes of products may break the balance between the functional performance and economics of the original product system, resulting in conflicts between new attributes and existing attributes [7]. The increase in collaborative activities among relevant parties in the life cycle design and the complexity of design may add to conflicts [6, 8].

In this study, conflicts in LCD collaboration are characterized as follows:

- Compliance conflict are those between design decisions and external, universal, and mandatory regulations such as local laws and regulations, international industry standards, and technical specifications. For example, conflicts arise between WEEE's requirements for scrap recycling management and RoHS's requirements for hazardous substance restriction management, and product engineering technical solutions such as material selection, manufacturing processes, and disposal methods.
- Interest conflicts between entities regarding the sharing of technological risks, research and development costs, and expected benefits from the sustainable products or technologies. However, different organizations have different needs for interests, some pursue higher economic benefits; Some want to establish a better corporate image through green transformation; Some want to upgrade technology to improve their market competitiveness. How to propose an unanimously accepted interest-allocation scheme in LCD remains an unsolved problem. A typical scenario is the mismatch of contributors and beneficiaries. For example, the cost and risk of eco-material technology are mainly borne by suppliers, but the benefits (such as remanufacturing profits, corporate image, and policy support) are mainly obtained by manufacturers and recycling service provider, which leads to conflicts between the allocation of R&D cost of eco-material research and the allocation of tangible and intangible benefits form eco-material (i.e., easy to recycle).
- Technical conflicts, including conflicts in design goals, design methods employed, and capabilities to fulfill certain engineering requirements. Goal conflicts in LCD, typically caused by inconsistent recognition of the importance of different attributes of product systems by collaborative entities in life cycle design. For example, conflicts between the processing convenience that manufacturers hope for and the ease of disassembly and replacement that recycling and repair service providers hope for; method conflict, which refers to inconsistencies

in technical specifications, principles, and methods used to solve specific design problems; capability conflict, which refers to a mismatch in the technical capabilities required to solve specific design problems.
- Information conflict refers to problems such as incompatible information forms, inconsistent information semantics, inaccurate information transmission, and untrue information in communication between collaborative entities.

LCD is a complex process, resulting in a variety of conflicts. At different stages of the life cycle, the types of conflicts manifests or users care about most can be different. For examples, in the stage of market demand analysis, the compliance conflict becomes the focus of consideration. In the stage of business model establishment, interest conflicts may manifest. When it comes to the stage of product design (either tangible or intangible), technical conflicts may dominate.

## 3 Ontology-Based LCD Conflict-Managing Knowledge Representation

As the extension of "case" [9], LCD conflict-managing knowledge refers to the expression of knowledge set with certain relevance to past experience, to solve a specific conflict in LCD collaboration. In this study, the LCD conflict managing knowledge shall include the crucial information about the whole process of resolving a particular conflict [10]. That is, LCD conflict-managing knowledge, in general, includes the description of conflict, the stakeholders involved, the object dealt with, the methods and process employed, the resource invested, results generated, and managing performance evaluated and should be presented in a form which is highly instructive and easy to understand.

Ontology can accurately express and describe knowledge in a certain field, which is the basis of human–computer interaction [11]. It provides a shared knowledge carrier with structural characteristics.

Thus, ontology method is introduced to conflict managing knowledge representation. An ontology can be viewed as a graph where nodes denote concepts and the edges between the nodes denote relations between the corresponding concepts. A concept is defined as a class in the ontology formalism. Each concept or thematic class is characterized by properties.

In this study, the LCD conflict managing knowledge ontology is proposed, as shown in Fig. 1. The following attributes characterize the relations between the concepts (or classes) in the ontology formalism:

- Involve, such as "ConflicInLCD" "Involve" "Stakeholder";
- InRegardTo, such as "Conflict" "InRegardTo" "Conflict Object";
- DealWith, such as, "Resolution" "DealWith" "ConflictInLCD";
- AppliedTo, such as "Resource" "AppliedTo" "ConflictResolution", "Evaluation" "Use" "Resource" "AppliedTo" "ConflictManageEval(uation)";

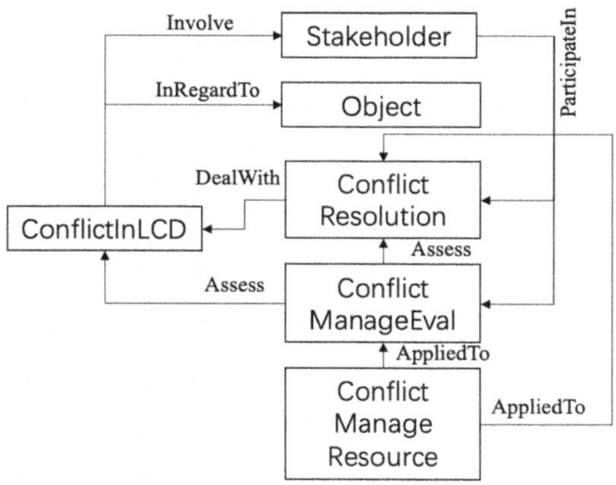


**Fig. 1** LCD conflict managing knowledge ontology—concepts and relations

- Assess, such as "ConflictManageEval(uation)" "Assess" "ConflictResolution";
- ParticipateIn, such as "Stakeholder" "ParticipateIn" "ConflictResolution", "Stakeholder" "ParticipateIn" "ConflictManageEval(uation)".

Data properties are used to define the attributes of the concepts. Table 1 shows the data properties of the LCD conflict managing knowledge classes. Then, based on the proposed ontology model, the conflict managing knowledge instantiation is realized by giving (numerical or linguistic) values for the data properties.

## 4 LCD Conflict Managing Knowledge Retrieval

When users encounter conflicts in LCD collaboration, they may turn to the managing knowledge of similar conflict for reference. In this study, the LCD conflict managing knowledge instance retrieval is realized as shown in Fig. 3. The conflict managing knowledge retrieval conditions are templated based on six conflict descriptive features defined according to the knowledge ontology model. Users may input queries by describing the conflict to be resolved with the features in natural languages.

As shown in Fig. 2, the relevance of knowledge instance is evaluated based on the similarity between the conflict to be resolved and those included in the knowledge instances. The Nearest Neighbor Algorithm is employed in this study for the similarity calculation. The specific similarity calculation is as follows:

$$\mathrm{Sim}(Xi, X'i) = \cos\theta = \frac{\sqrt{vxi \cdot vx'i}}{|\ vxi\ | \cdot |vx'i|} \tag{1}$$

**Table 1** LCD conflict managing knowledge ontology—data properties

| Concepts | Data properties | Value type | Value |
|---|---|---|---|
| ConflictInLCD | ConflictName | String | |
| | ConflictNumber | String | |
| | TimeofCreation | Date | |
| | ConflictProposer | String | |
| | ConflictSubject | String | |
| | ConflicImportance | Enum | Level 1; Level 2; Level 3 |
| | ConflictInLife CycleStage | Enum | ProjectApproval; SchemeDesign; DetailedDesign; Purchase; Manu-facture; Fabrication; Marketing; Logistics; Sales; Recycle; Remanufacture |
| | ConflictType | Enum | Compliance; RiskSharing; CostSharing; BenifitSharing; Goal; Method; Decision-making; Capabil-ity; Communication; RoleAssignment; |
| | Conflict Description | String | |
| Stakeholder | StakeholderName | String | |
| | Type | Enum | Manufacturer; Supplier; Distributor User; ServiceProvider; Recycler; Administrator; Consultor; Other; |
| | AgentName | String | |
| | AgentRole | String | Conflict Proposal; Conflict Resolution; Evaluation |
| Object | Type | Enum | ProdDesign; ProcDesign; ServiceDesign; Other |
| | Description | String | |
| | Product | String | |
| | DesignIndex | String | |
| | LifeCycle Stage | String | |
| Conflict Resolution | StepNumber | String | |
| | StepName | String | |
| | EventDiscript | String | |
| | EventType | Enum | Negotiation; Arbitration |
| | StrategyEmployed | String | |
| | ResolutionScheme | String | |
| | Efficacy | Enum | Remarkable; Common; Fail |
| | Comment | String | |

(continued)

**Table 1** (continued)

| Concepts | Data properties | Value type | Value |
|---|---|---|---|
| ConflictManageEvaluation | Type | Enum | ConflictResolution Difficult; ConflictResolution Effect |
| | Comment | String | |
| ConflictManageResource | Type | Enum | Tool; Knowledge; Service; Others |
| | Name | String | |
| | Domain | String | CAD Engineering Design; CAE Engineering Analysis; LCA&EnvironAss; Info&Knowledge Services; Experiment&Test; Others |

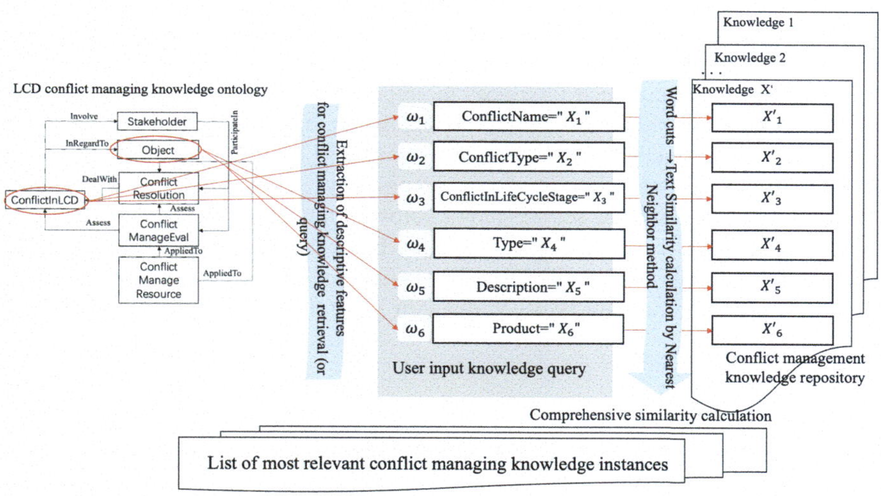


**Fig. 2** LCD conflict managing knowledge retrieval

$$\mathrm{Sim}(X, X') = \frac{\sum_{i=1}^{m} wi\,\mathrm{Sim}(Xi, X'i)}{\sum_{i=1}^{m} wi} \tag{2}$$

in which, $A$ represents the conflict to be resolved, and $A'$ represents the conflict already stored in the knowledge repository; $X$ is the description of $A$ by descriptive features, while $X'$ is the description of $A'$; $W$ represents the weight assignment. The calculated Sim ($X$, $X'$) is the comprehensive similarity between the new conflict and the conflict in the repository. The similarity Sim ($X_i$, $X_i'$) is calculated by the vector cosine law.

It should be noted that the six retrieval conditions selected in this paper are not the conditions that can fully describe the conflict managing knowledge, but the conditions that are most convenient for users to describe the conflict they encountered and to be resolved.

Among the proposed query conditions, three of them are about the conflict (i.e., name, type, related to what life cycle stage) and the rest are about the object the conflict in regard to. This is because, this study is particularly about conflict in LCD collaboration, and designers may need to know more about the conflicts in regard to certain type (or similar) "Product."

## 5 Development of a Web Service-Based LCD Conflict Managing Knowledge Repository

With the development of information network technology, Web service technology has been very mature. It has the advantages of easy development and deployment, easy maintenance, good accessibility, and convenient use.

Figure 3 shows functional interfaces the repository system, on which users can add, modify, delete and retrieve LCD conflict managing knowledge.

Figure 4 shows methods and framework employed to develop the Web service-based LCD conflict managing knowledge repository, which includes the browser-based user interface and the server side of software functionality and database.

The repository is developed by the Vue.js framework with UI plugins of IView/Element UI/Vue-Atlas for the user interface layer, the Springboot framework for the business logic layer, and the JDBS-Redis with MySQL for database access layer.

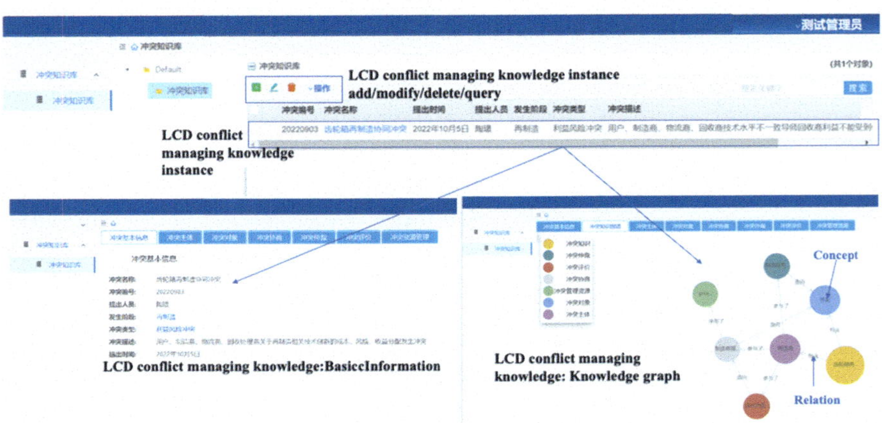


**Fig. 3** Web service-based LCD managing knowledge repository

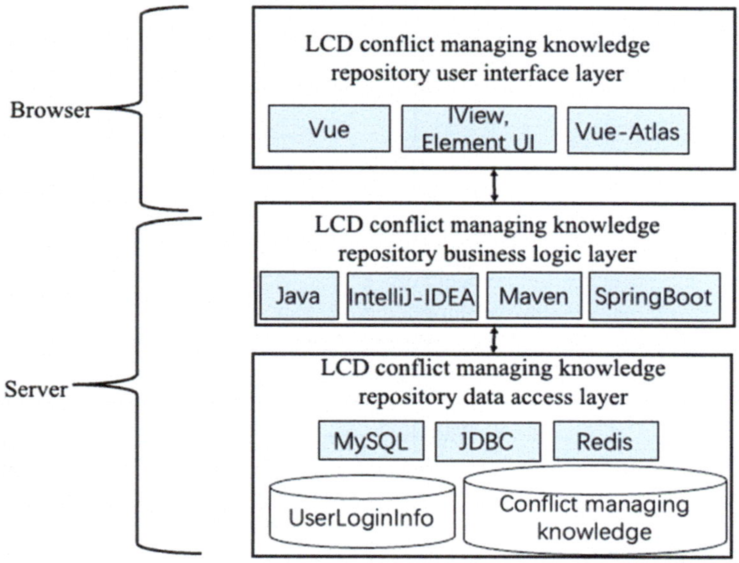


**Fig. 4** System architecture of cloud-based LCD managing knowledge repository

# 6 Appliciation of Conflict-Managing Knowledge Repository

The proposed LCD conflict managing knowledge repository is currently a functioning web service component of the eco-design service platform developed by PISX Shanghai Pairui Information Technology (http://internal.pisx.com:18081/login).

The proposed conflict managing knowledge repository enables the conflict managing knowledge capture and explication, sharing and reuse. The repository provides a convenient retrieval method, with which users can search for conflict managing knowledge for reference.

PLM/BPM system is commonly used for project management. In particular, the proposed repository has realized the interaction with certain PLM/BPM system. A "LCD Conflict Management" function module was developed in the PLM/BPM system. This module can export conflict managing data to the repository and generate LCD conflict knowledge instance in accordance to Table 1 (Fig. 5).

To implement the data migration (or knowledge capture and translation) from the "Conflict Management" of the PLM system to the proposed knowledge repository:

First, establish the mapping relationship between the conflict management data and that of the conflict knowledge model; develop the data interface to realize the data flow and call; program a parameter transfer function module to realize the migration of conflict-resolution knowledge data from PLM/BPM system to the knowledge repository.

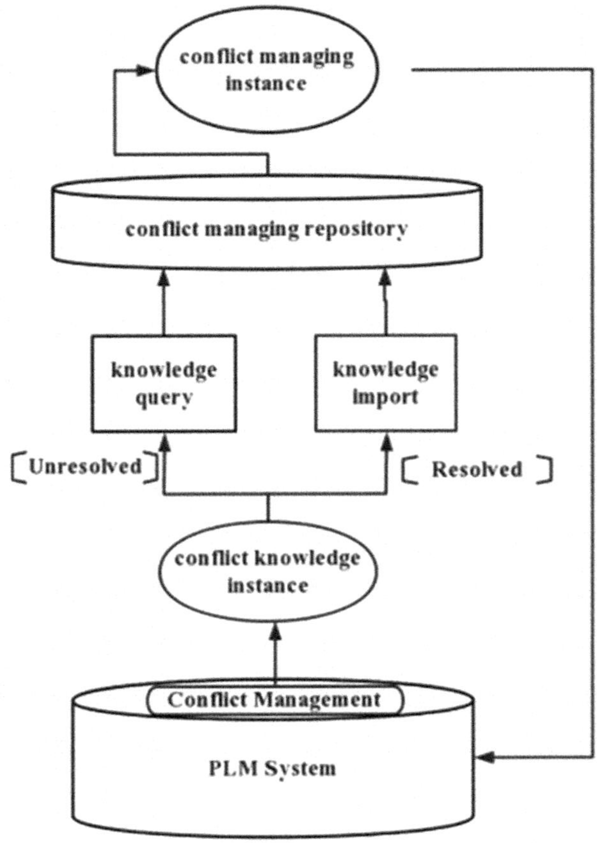


**Fig. 5** Integration LCD conflict managing knowledge repository with PLM/BPM

## 7 Conclusion

How to overcome the difficulties of decentralized design process, diverse design objectives, inconsistent design concepts and conflicting interest objectives in the process of collaborative LCD design is a bottleneck problem that needs to be solved.

This study aims to develop LCD conflict managing knowledge repository system to facilitate LCD practices. The contributions of this study may include (1) the formal LCD conflict managing knowledge ontology is proposed, which enables the structuring and standardization of related data and the storage of such data in a computer-processable manner and (2) a Web service-based LCD conflict managing knowledge library is developed, which enables the conflict managing knowledge capture and explication, sharing and reuse.

**Acknowldegment** This study is funded by National Key R&D program of China (2020YFB1711604).

## References


1. Alting L (1995) Life cycle engineering and design, CIRP annals - manufacturing technology. Bd 44S:569–580
2. Hauschild M, Herrman C, Kara S (2017) An integrated framework for life cycle engineering. Procedia CIRP 61:2–9
3. Chen W, Hauschild M, Huang B et al (2022) Life cycle engineering and sustainable manufacturing for net-zero targets and environmental sustainability. Resour Conserv Recycl 186:106480
4. Ramani K, Ramanujan D, Bernstein WZ et al (2010) Integrated sustainable life cycle design: a review. J Mech Des 132(9)
5. Riesener M, Kuhn M, Hellwig F et al (2023) Design for circularity—identification of fields of action for ecodesign for the circular economy. Procedia CIRP 116:137–142
6. Dekoninck EA, Domingo L, O'Hare JA et al (2016) Defining the challenges for ecodesign implementation in companies: development and consolidation of a framework. J Clean Prod 135:410–425
7. Perez-Gallardo RJ, Azzaro-Pantel C, Astier S (2018) Combining multi-objective optimization, principal component analysis and multiple criteria decision making for ecodesign of photovoltaic grid-connected systems. Sustain Energy Technol Assess 27:94–101
8. Bey N, Hauschild MZ, Mcaloone TC (2013) Drivers and barriers for implementation of environmental strategies in manufacturing companies. CIRP Ann Manuf Technol 62(1):43–46
9. Issa G, Shen S, Chew M (1994) Using analogical reasoning for mechanism design. IEEE Expert 9(3):60–69
10. Brambila-Macias SA, Sakao T (2020) Effective ecodesign implementation with the support of a lifecycle engineer. J Clean Prod 279:123520
11. Zou M, Basirati MR, Bauer H et al (2020) Facilitating consistency of business model and technical models in product-service-systems development: an ontology approach [C]. In: IFAC conference on manufacturing modelling management, and control

# Development of Internal Audit Process to Obtain EPD Certifications in Response to Ecodesign Regulations

**Yong-jin Lee, Han-bi Kim, Ki-woong Jung, and Noh-hyun Lim**

**Abstract** The EU commission has announced amendments to ecodesign from Directives to Regulations. This ecodesign regulation includes expanding the scope of targets from existing ErP (Energy-related Products) to physical products placed on the market, which adds requirements for product performance and information. In particular, among these requirements, information on environmental impacts is also required, and the LCA methodology is used to derive such environmental impact information, and EPD is a representative certification system. EPD is a global standard, an environmental label for Type III (ISO 14025), which can evaluate environmental performance across products based on LCA. These environmental labels are an effective way for corporations to provide clear and reliable information to consumers, and corporations need to establish various business systems and processes to respond quickly to international change. To this end, EPD-Norge and IGSC developed an EPD certified auditor program curriculum to train manpower, and established an "Internal Audit Process" through such trained internal auditing experts. Through the established "Internal Audit Process," the environmental performance of a corporation's products can be directly calculated, and its own internal auditing can also be participated before the certification auditing. Through this process, the product's EPD certification acquisition period can be shortened by 1 month, and EPD information that can be used internationally can be calculated and disclosed through internal calculations.

**Keywords** Ecodesign · Environmental product declaration · Internal audit process · Auditor

---

Y.-j. Lee · H.-b. Kim · K.-w. Jung · N.-h. Lim (✉)
Institute of Global Sustainability Certification (IGSC), Seoul, South Korea



S. Fukushige et al. (eds.), *EcoDesign for Circular Value Creation: Volume I*,
https://doi.org/10.1007/978-981-97-9068-5_5

## 1 Introduction

ESG is a nonfinancial evaluation method consisting of environmental (minimizing environmental impact), social (corporate social responsibility), and governance (financial performance), and is attracting worldwide attention as a measure to evaluate the value of a corporation. The performance of a corporation's qualitative/quantitative activities for each constituent item can be measured and evaluated as a corporate performance indicator.

The Ecodesign for Sustainable Products Regulation (ESPR) expands upon the existing Ecodesign Directive. Its overarching goal is to reduce the environmental impact of products throughout their lifecycle. By emphasizing circular economy principles, the ESPR aims to encourage the creation of products that are more durable and easier to repair, reuse, and recycle (Fig. 1). To reduce the environmental impact of products throughout their lifecycle, environmental impact can be derived through LCA's methodology, and EPD is a representative certification system that can derive environmental impact results.

EPD is a global standard, an environmental label for Type III (ISO 14025), that can evaluate environmental performance across products based on LCA. These environmental labels are an effective way for a corporation to provide clear and reliable information to consumers, and a corporation needs to establish various work systems and processes to respond quickly to international changes. A corporation builds various internal systems to respond quickly in line with global changes, and as part of that, they can build an "Internal Audit Process" to maximize work efficiency.

A new product certification process can be established through the "Internal Audit Process," which simplifies the certification process to efficiently calculate the environmental impact information of the product. A corporation will be able to establish an internal auditor's curriculum to establish an "Internal Audit Process," and through this, they will be able to train internal auditing experts. The trained internal auditing experts will be able to directly perform calculations to derive

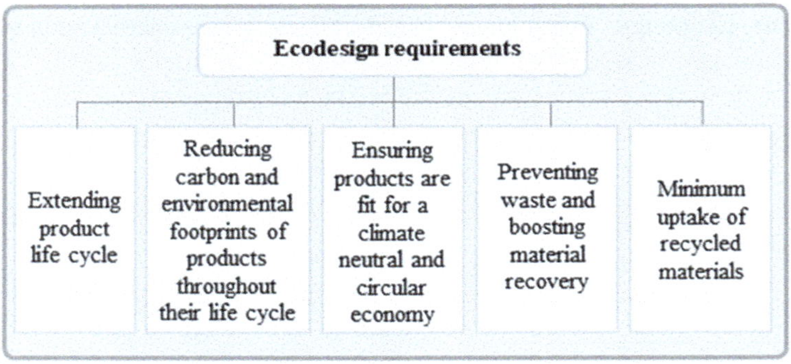


**Fig. 1** Factsheet on sustainable products initiative

environmental impact information on the product, and reflect these results directly in the certification review process. In addition, through this process, internal certification auditors directly or indirectly participate in the certification auditing process. Through this process, the period required for the EPD certification process can be significantly reduced.

In this study, we would like to review the work efficiency of a corporation for the "Internal Audit Process" established through internal auditing experts trained through the curriculum developed with EPD-Norge.

## 2 Description of Term

### *2.1 EPD-Norge*

EPD-Norge is a certification body with an environmental product reporting system that can quantitatively evaluate the environmental impact of the entire process from extracting raw materials to disposal of products. It is also a global certification body with a global network of partners studying various auditing and sustainable solutions, and is expanding its cooperation with various EPD certification bodies, including mutual recognition with certification bodies such as IBU (Germany), International EPD (Sweden), and EPD-Italy (Italy).

### *2.2 IGSC (Institute of Global Sustainability Certification)*

IGSC serves as EPD-Norge's Korea office and is a certification body that efficiently supports the acquisition of EPD certification by corporations that want to export products to various countries such as Europe, the United States, and Australia. In addition, we developed an EPD certification auditor training program using EPD-Norge to provide training for EPD certification, and we conducted certification auditor training in Oslo, Norway, and Seoul, Korea, respectively, to train a number of certification auditors [5].

### *2.3 EPD (Environmental Product Declaration)*

The EPD (Environmental Product Declaration) is a certification system that quantitatively displays the environmental impact on the entire process, such as extraction of raw materials, manufacturing, transporting, using, and disposal for products and services to enhance the environment of products and services [1].

EPD aims to potentially induce market-led continuous environmental improvement by easily and transparently disclosing accurate environmental impacts on products to support purchasing activities that take into account the consumer's environment.

EPD can provide environmental impact information for each category of influence such as ADP (Abiotic Depletion Potential), GWP (Global Warming Potential), ODP (Ozone Depletion Potential), POCP (Photochemical Oxidant Creation Potential), AP (Acidification Potential), EP (Eutrophication Potential), ETP (Eco-Toxicity Potential), HTP (Human Toxicity Potential) [2].

### *2.4 Internal Auditors*

An internal auditor refers to a person who acts as an internal certification auditor for a corporation's certification auditing. The internal auditor performs tasks such as document preparation, review, and self-internal auditing in relation to certification.

## 3 Internal Audit Process

### *3.1 Development of Internal Auditor Curriculum*

Programs of various curricula are required to train internal auditing experts. The Programs curriculum consists of auditing related General Programme Instructions (GPI), Product Category Rules (PCR), LCA Methodology, EPD Tools, report composition and auditing, certification-related standards, use of certification marks, auditor's roles, and auditing processes. Each curriculum organizes the contents of the education so that the subject can acquire knowledge of the certification audit. The curriculum of internal auditor programs is shown in Fig. 2.

### *3.2 Certification Audit for EPD*

The procedures for obtaining EPD certification can be largely divided into preparation for certification audit, document auditing, on-site auditing, submission of final reports, deliberation by the certification deliberation committee, and certification issued.

Among the certification auditing procedures, it takes the most time to prepare certification documents for certification auditing. In particular, LCA (Life Cycle Assessment) is required to apply for EPD certification, and the LCA performance period may take more than 6 months depending on system boundaries and the level of collected data. In addition, it takes about a month to prepare documents for application for certification based on the LCA results.

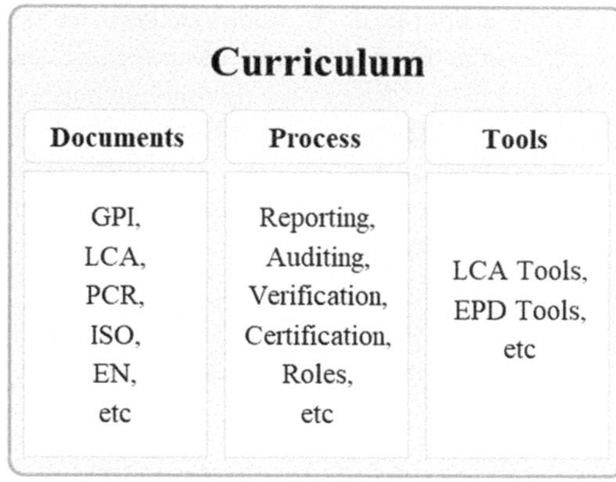


**Fig. 2** The curriculum of internal auditor programs

Once the documents for the certification application are prepared, they will apply to the EPD certification authority. When the EPD application is completed, the EPD certification auditor will be assigned after checking whether the submitted documents are omitted. The assigned EPD certification auditor will proceed with the document based on the submitted documents. When the auditing of all documents is completed, the on-site auditing will be conducted. In general, the certification audit takes 1–2 months, and 1–3 months may be added depending on the level of supplementation. When all the document review and on-site audit are completed, the final report on the certification audit will be submitted to the certification deliberation committee, which will be reviewed to decide whether to approve the certification. If there is no special reason for disqualification, the certification is approved, and the certificate is finally issued. As such, it takes a considerable amount of time depending on the various procedures for obtaining EPD certification, and it may take up to 1 year if applied for the first time. In addition, if an entity wants to obtain EPD certification for a variety of products, it will take a lot of time and money due to repeated certification procedures. Accordingly, corporations are looking for ways to save time and cost in the EPD certification process. Separately, as the number of certification applications for EPD certification increases, the certification body's certification audit process may be delayed.

In fact, as the number of applications for certification increases than the number of auditors of the certification body, the assignment of certification audit may be delayed, and it may also be delayed by the certification deliberation committee after the review is completed. Corporations establish various measures to shorten the time required to obtain certification, and generally form and operate a dedicated department for obtaining certification. Since professional work is carried out through this dedicated department, the time required in the process of issuing certification can be further reduced. In addition, the time incurred by repeated applications for certification may be further reduced. However, if the size of a corporation is large, a dedicated team is formed for each business area, and each dedicated team performs certification acquisition, so it is required to minimize the time and cost required for the entire corporation. In addition, security issues that occur during the product certification audit process may have a significant impact on the certification period.

Finally, in the face of global issues such as COVID-19, the auditing period can be affected even if there is a lack of certification auditors for auditing or if it is difficult to proceed with on-site auditing. The certification acquisition period may be delayed due to such various factors. Accordingly, corporations are looking for various ways to minimize delays.

## *3.3 Actual Period of Certification Audit*

After applying for EPD certification, the time it takes for a corporation to receive certification depends on the time it takes during the certification audit process. Certification audit generally takes 1.0 months for document review, 0.8 months for on-site audit, 0.5 months for final report on the summary of certification audit results, 1.2 months for certification and issuance review. The flow of the actual period of the certification audit is shown in Fig. 3.

This will take a total of 3.5 months for a corporation to apply for certification to be issued. However, additional time may be required depending on the level of document preparation related to certification required by the certification body to obtain certification. If supplementation is made at each stage of audit and approval, it may take a lot of time for corporations to receive certification because it takes additional time. In addition, as discussed above, the number of applications for certification to the certification authority is also a variable.

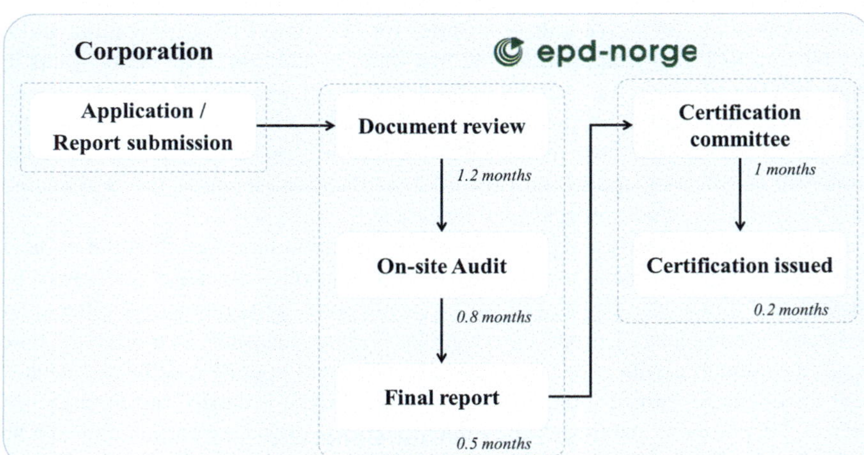


**Fig. 3** Actual period of the certification audit

### 3.4 Establishment of Internal Auditors Process

It is important for corporations to market their products at the right time and provide consumers with a variety of information to choose their products. In addition, environmental regulations are being strengthened around the world, so providing environmental information on products can have an important impact on consumers' choices. Therefore, since obtaining certification for products can act as an important factor for corporations, the time required to obtain certification is an important part. As a method of minimizing the time required for authentication, a method of minimizing the time required for the authentication auditing step may be selected.

One of these methods can minimize the auditing stage by training internal personnel of a corporation as internal auditors through the training of certification auditors of the certification body.

The certification body conducts internal auditor training in the same way as the certification auditor so that the internal auditor can play the role of the certification auditor. When the training is completed, the internal auditor prepares documents related to the corporation's certification audit in the same way as the certification auditor. In addition, to save the time required to prepare certificates, sufficient and appropriate documents can be prepared for the main audit of the product certification auditing process. When the document review and on-site audit are completed, the certification body submits the final audit results to the certification deliberation committee.

Documents submitted to the Certification deliberation committee may minimize the time required for the certification audit process by appropriately preparing all necessary matters during the certification audit process of the certification body.

However, sufficient education should be provided so that the internal auditor can replace the role of the certification auditor, and various information necessary for the certification audit process should be fully understood. In order to systematically manage certification-related tasks, an entity may introduce a dedicated department related to internal audit. Through the education and introduction of internal auditors, the time required for the certification audit process can be reduced by about 1.8 months from 3.5 months to 1.7 months. The flow of the actual period of the certification audit is shown in Fig. 4.

### 3.5 Requirements for Internal Auditor Program

If the internal auditor is replaced instead of the certification auditor during the certification audit process, at least two people are required to be secured equal to or more than the number of certification auditors required during the certification audit process. One of these internal auditors shall perform data collection and environmental impact's calculation, and one shall participate in the internal audit.

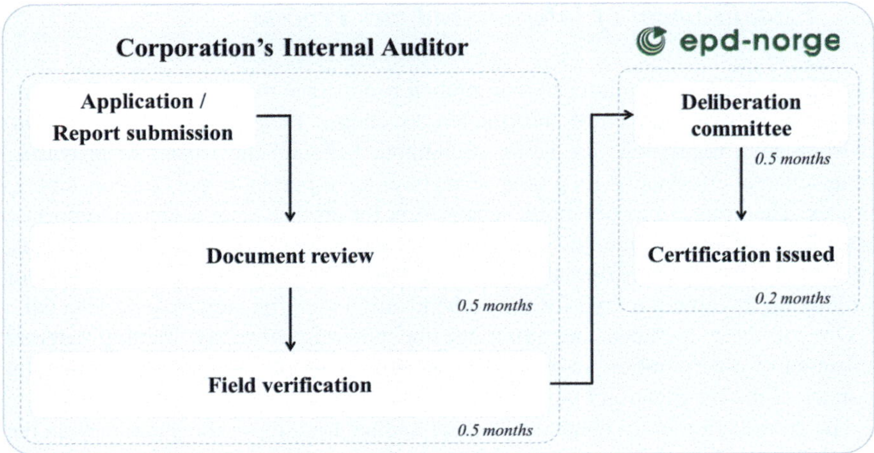


**Fig. 4** Certification Audit Process reflecting Internal Auditor Process

Establish internal systems and processes to enable environmental impact's calculations to:

- Create an internal data processing system for collecting data, management, and calculating environmental impact.
- Create a manual that establishes R&R by organization and defines the process for calculating environmental impact.

## 3.6 *Effect*

Through the introduction of an internal auditor, the time required to obtain certification for a corporation could be reduced by more than 1 month. In addition, internal auditor training enables corporations to meet international needs for LCA performance related to certification. In addition, internal auditors of a corporation may provide certification-related training to various stakeholders involved.

# 4 Activation and Plan

## 4.1 *Internal Auditor Training Program*

Through the internal auditor training program, a corporation's certification-related capabilities and the corporation's global capabilities can be strengthened.

First, an internal audit process for obtaining a corporation's certification can be established, and the corporation can build a system accordingly. The internal audit process can be used to systematically organize personnel and apply them to work from the certification preparation stage.

Second, corporations with overseas branches can establish procedures for obtaining certification for overseas branches, and personnel who have obtained qualifications as internal auditors can provide training to certification-related personnel. Through this corporations can strengthen their internal capabilities related to certification and efficiently perform certification-related tasks (Fig. 5).

Third, it is possible to provide training on certification-related processes to supply chains. In the case of supply chains, since linkage with corporation's business promotion is required, obtaining certification, and performing related tasks are also related to suppliers. Therefore, education related to certification of supply chains is required, and through this, not only corporations but also their capabilities can be strengthened.

## 4.2 *Utilizing the Internal Audit Process [5]*

It can be used in various ways through the internal audit process. Corporations that produce a variety of products can extend the internal audit process built for product certification to other products. Through this, internal audit procedures can be simplified, saving time and cost, and above all, strengthening internal capabilities. In addition, corporations that export products can respond to a variety of international regulations by leveraging various internal audit processes to strengthen their internal capabilities. In addition to strengthening the internal capabilities of corporations, the internal capabilities of suppliers who supply a variety of production parts can increase the global competitiveness between corporations.

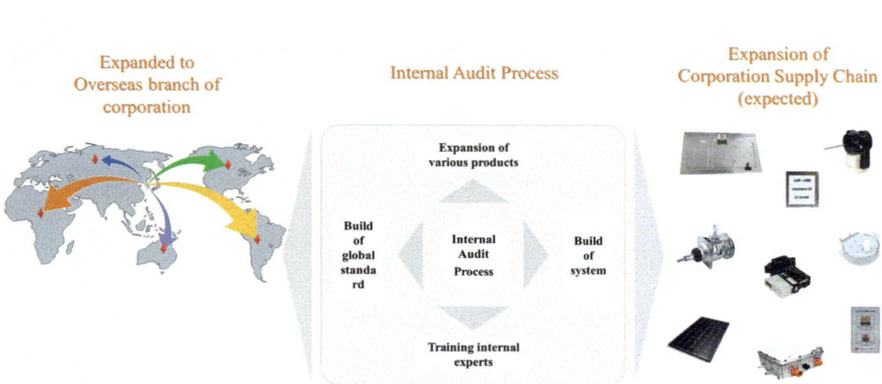


**Fig. 5** Utilizing the internal audit process

## 5 Summary

Establishing an internal audit process for obtaining the corporation's EPD certification has the following advantages:

- By establishing an internal audit process, it is possible to efficiently prepare documents related to certification audit.
- As document review and on-site audit are greatly simplified, deliberation for rapid certification issuance could proceed quickly.
- It is possible to shorten the time required for the audit process by about 50%.
- The process of certification is related to the products of all affiliates of a corporation.
- It is possible to strengthen a corporation's internal capabilities by training internal auditors.
- Various training programs related to certification could be conducted internally for domestic and foreign affiliates of a corporation.
- Systematic education can be conducted with corporate partners.

## References

1. General Programme Instructions for the International EPD® System
2. Korean Environmental Product Declaration
3. Regulation of the European Parliament and of the Council on Ecodesign for sustainable products, amending Regulation (EU)2019/1020 and repealing Directive 2009/125/EC
4. 2023 EPD Norway Verifier Training Course
5. Lim N-h (2017) Development of the internal audit process to obtain the carbon trust's carbon/water footprint certifications. Korean Soc Life Cycle Assess 1:55–60

# Verification of Operational Effect of Solar-Powered System for Mobile Off-Grid Architecture

**Satoko Nasu and Shota Tajima**

**Abstract** Mobile Off-Grid Architecture (MOA) has been developed with solar-powered systems as the self-electric power supply facility in desired places. For the effective utilization of MOA abilities, it is necessary to verify MOA movability and the advantages of solar system operation. The purpose of our research is to verify the effects of the method of movability usage of the MOA with its solar cell system contributing to energy efficiency. Our prototype MOA had been field-tested for its electrical power generation on the four equinoxes and solstices of the 24 solar terms in the solar year 2022–2023. We operated it manually, rotating to obtain the optimum direction for adjusting the sunlight depending on the sun's movement, with changes in the azimuth angle and elevation angle of each of the four seasons in the solar year. As a result of the rotation, the measurement power generation values for the wall solar cell at the summer solstice of 2022, the winter solstice of 2022, and the equinox of 2023 were approximately 1.9, 1.3, and 1.3 times the estimated values, respectively, compared to the scenario where the solar cell was fixed facing south without rotation. In conclusion, the movability usage of our MOA with a wall solar cell tilted at an angle of 90° through rotation is more effective than fixed south, especially in the summer solstice.

**Keywords** Mobile off-grid · Solar-powered system · Energy efficiency · Four seasons of the solar year

## 1 Introduction

Solar-powered systems with lightweight solar cells and high-performance storage batteries give mobile architecture a function for off-grid use as the self-electrical power supply in desired places. In the case of daily use, even in a nonelectric power-

S. Nasu (✉)
Independent Scholar (Former Chiba University), Chiba, Japan

S. Tajima
Graduate School of Global and Transdisciplinary Studies, Chiba University, Chiba, Japan

 
S. Fukushige et al. (eds.), *EcoDesign for Circular Value Creation: Volume I*,
https://doi.org/10.1007/978-981-97-9068-5_6

source area, and in the event of a natural disaster, the function of mobile off-grid architecture (MOA) is now available [1]. The role of MOA has started to gain attention, as various types of it have been developed by manufacturers and university projects around the world [2].

However, the mounted solar cells of MOAs have been mostly overlooked as an option to gain the advantage of mobility. MOA generally has several solar cell panels on a roof area in a horizontal or tilted state. Once MOA is fixed in place after moving to its destination, it cannot be moved much. Due to traffic law's height and weight restrictions, MOA cannot install heavy systems, and the area available for solar batteries is limited. Given such a limited installation area, MOA cannot have excessive solar battery panels and equipment.

For the most effective utilization of MOA's abilities, you must verify its movability and advantages within the operation of a solar-powered system. The ability to track the movement of sunlight with simultaneous changes in azimuth angle and elevation angle varies greatly depending on the season. MOA has a wheel device that can be rotated to the best azimuth and can be moved to the best destination for receiving sunlight.

There has been international and global research on the potential use of PV-powered vehicles for transport [3]. The purpose of our research is to verify the effects of MOA use on the on-site energy efficiency of an off-grid mobile system.

## 2 Method

### *2.1 MOA Prototype*

We have proposed and developed an MOA prototype for a sustainable social infrastructure called *EcoCabin*, which is a small-sized, lightweight vehicle installed with an ultra-lightweight, ultra-thin film-type solar cell system [4]. Its solar cells are set up on the roof and one side wall area, installed in a horizontal or vertical state [5], as shown in Fig. 1. There are two roots for the charger controller. The MOA's solar generation system is divided into two parts by an angle with different 90° light-receiving surfaces to avoid the influence of different voltages (refer to Figure 2 and Table 1).

### *2.2 Experimental Procedure*

We examined the effective operation of MOA with a rotation method that maximized sunlight reception. By using an operational method such as manual rotation on an axis, our MOA prototype can change its direction to the optimum direction to receive sunlight maximum. Detecting the relative direction relation with solar azimuth, our MOA prototype was field-tested, and experimental data was evaluated.

**Fig. 1** MOA prototype—EcoCabin

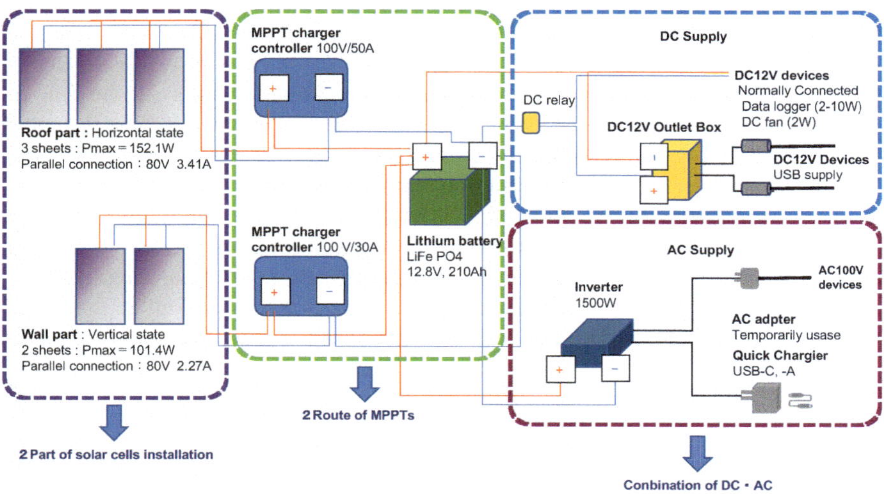


**Fig. 2** Solar cell system of the MOA

**Table 1** Specification of the MOA's solar-powered system

| Equipment | | Size/Capacity | Data |
|---|---|---|---|
| Amorphous solar cell | Ultra-lightweight/ ultra-thin film-type | W: 642 mm L: 1354 mm D: 2 mm | Voc = 80 V, Vpm = 58.7 V, Pmax = 50.7 W ipm = 1.137A |
| | Roof part 3 sheets parallel connection | 80 V 3.41A | 152.1 kW |
| | Wall part 2 sheets parallel connection | 80 V 2.27 A | 101.4 kW |
| Lithium battery | LiFe PO4 | 12.8 V 210 Ah | 2.68 kWh |

Three steps were used for the experiment with the rotation method, as follows:

1. Estimation of electrical power generation in the relative direction
2. Measuring electrical power generation data by field testing the rotation method
3. Comparison of estimated value with measured value

## 2.3 *Four Seasons in the Solar Year*

For the field test, we focused on specific days, such as "two equinoxes and two solstices in the solar year" generated on the earth by the yearly round and daily rotation of the earth. Three days out of the year were chosen: the one near the vernal or autumnal equinox, the summer solstice, and the winter solstice, all under fine weather conditions.

Early in the morning and evening, the sun is low in the East or West, and the vertical side wall solar cell of the MOA is most suitable for the light-receiving face in the vernal equinox, the summer solstice, and the autumnal equinox. At the time of the culmination of the sun, the horizontal roof solar cell of the MOA is the most suitable for the light-receiving face in the vernal equinox, the summer solstice, and the autumnal equinox. At the summer solstice, the rooftop solar cell receives the maximum sunlight, and the south-side wall cell receives the minimum. On the contrary, at the winter solstice, the south side wall cell receives the maximum, and the rooftop cell receives less than the south side wall cell, as shown in Figs. 3 and 4. The image of the rotation method for the MOA is summarized in Fig. 5.

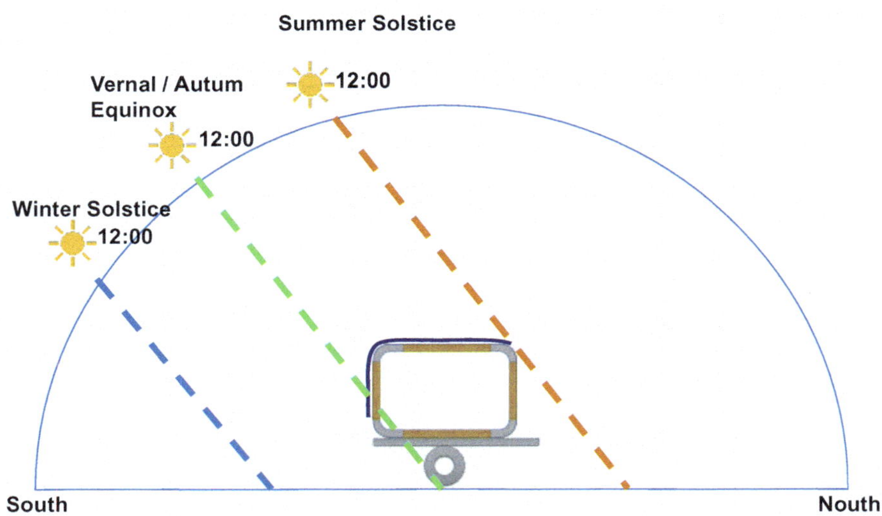


**Fig. 3** Solar elevation and azimuth angle with the MOA light-receiving face (cross-section South-North)

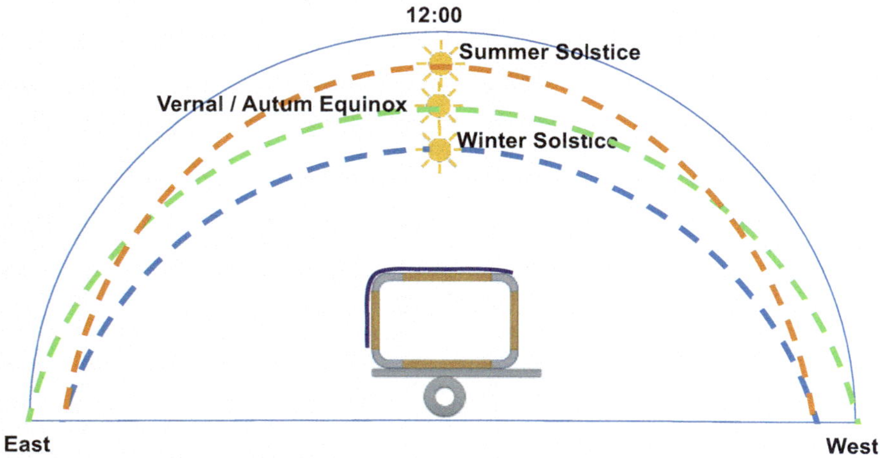


**Fig. 4** Solar elevation and azimuth angle with the MOA light-receiving face (cross-section East-West)

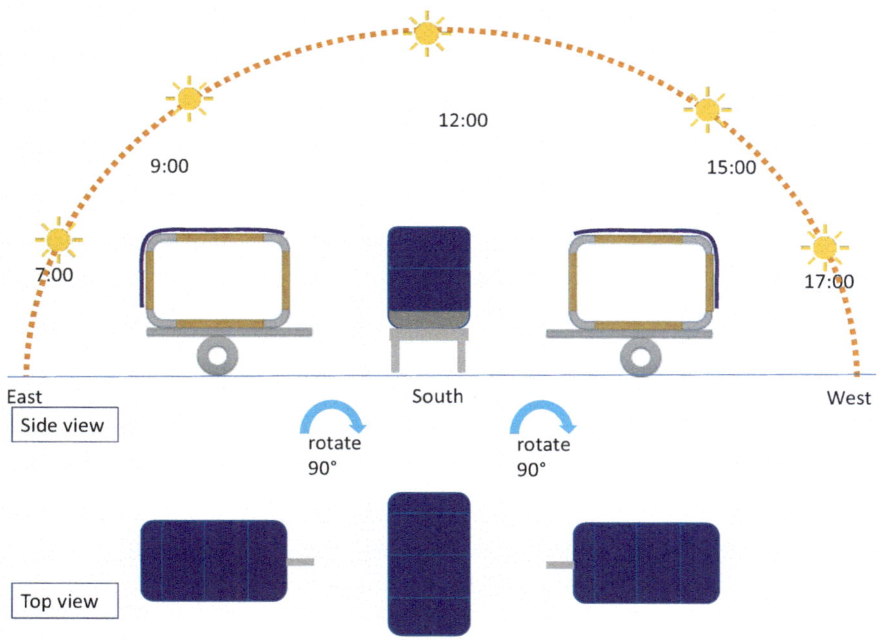


**Fig. 5** Rotation image of the MOA adjusting the sun's movement

## 2.4 Estimation

The prototype was planned to be field-tested to verify the differential in its power generation across the year, specifically at the vernal or autumnal equinox, the summer solstice, and the winter solstice from June 2022 to March 2023.

The location for our research and experimentation has been in Nagara Town Hall (lat.35.43° N, long.140.23° E), Chiba Prefecture. Before the field test, we obtained the insolation value of each day according to past meteorological information data from NEDO 7. Due to data from Nagara Town being unavailable, for convenience, the observation data of the Ushiku area near Nagara was used.

We chose days with nice weather close to each of the three seasonal turning points. By finding the maximum value of the insolation value of every hour, each relative solar azimuth of each hour at which the isolation value takes the maximum value for each angle of the light-receiving face was detected. The optimum direction of the MOA for each hour was determined from the maximum value of the insolation per hour.

These values allowed us to calculate the estimated value of the MOA's power generation per hour with relative direction and daily summation. Also, the rotation plan, based on tracking the optimum direction, determined that the MOA rotates on an axis from east to west through south, tracking sunlight.

## 2.5 Field Test

With the scheduled rotation plan for the MOA, our prototype MOA traveled as a towed vehicle to the experiment's location, the Nagara Town Hall parking area, on the fine, clear days of July 1, 2022, December 27, 2022, and March 15, 2023, for the summer solstice, the winter solstice, and the vernal equinox, respectively. Note that the autumnal equinox can be regarded as practically equivalent to the spring vernal equinox, so the field test on the autumnal equinox was omitted.

According to the planned rotation schedule, we operated the prototype MOA by hand, rotating to obtain the optimum direction to adjust the sunlight depending on its movement, with changes in azimuth angle and elevation angle for each of the three days, as shown in Figs. 3 and 4.

In its initial state, the MOA was facing east, then it began to be rotated on its axis manually; the point toward which one vertical side of the wall solar cell faces turns to the east (approximately 7:30–11:30 AM)—south (approximately 11:30 AM–1:00 PM)—west (approximately 1:00–5:00 PM) by angle 10–20° per hour as described in Fig. 5. If the MOA is not rotated to face east, then the side wall of the solar cell will not receive sunlight before noon, and the MOA will not be able to generate electricity from the wall solar cell. On the contrary, the point toward which the horizontal roof solar cell faces, when positioned vertically, does not track sunlight.

During the field test, the solar-powered system of the MOA measured the voltage and the current of each of the two roots of the charger controller and stored them with a data logger; these values were stored as CSV data every minute.

# 3 Results

## 3.1 Evaluation of the Measurement Value

Figures 6, 7, 8, 9, 10, 11, 12, 13, and 14. Show the difference in the MOA's electric power generation between the measured value and the estimated value on each of the three experimental days for the summer solstice, the winter solstice, and the vernal equinox, respectively.

The measured electric power generation of the roof solar cell, along with the estimated data, is shown in Figs. 6, 9, and 12. The results showed that the measured power generation of the roof solar cell waveform matched the estimated value. Note that the volume of the power generation on July 1, 2022 only, decreased by 70% because the roof solar cell system was affected by a failure of two out of three cells at this time due to unexpected gusty winds.

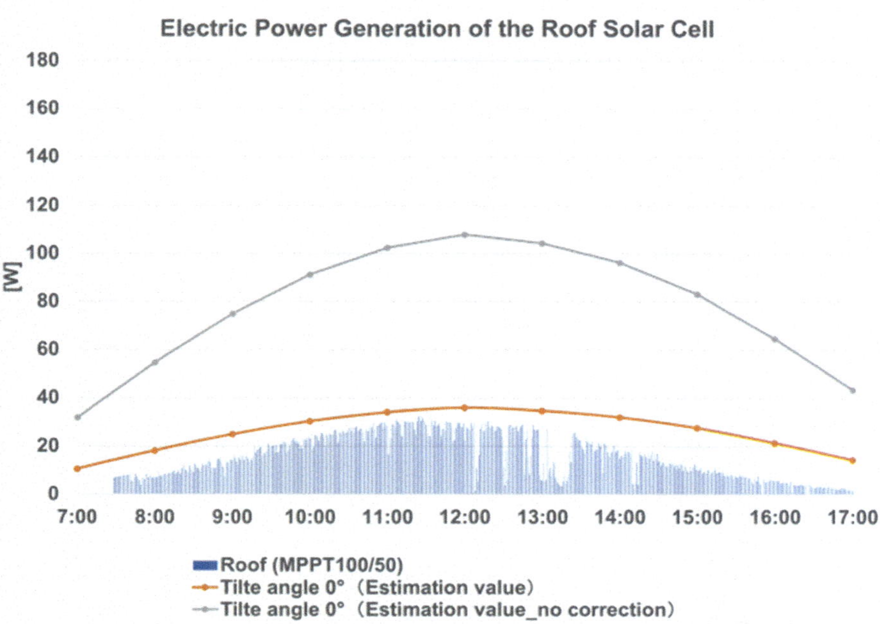


**Fig. 6** Electric power generation of the roof part (comparing measured value with estimated value) on the summer solstice, July 1, 2022

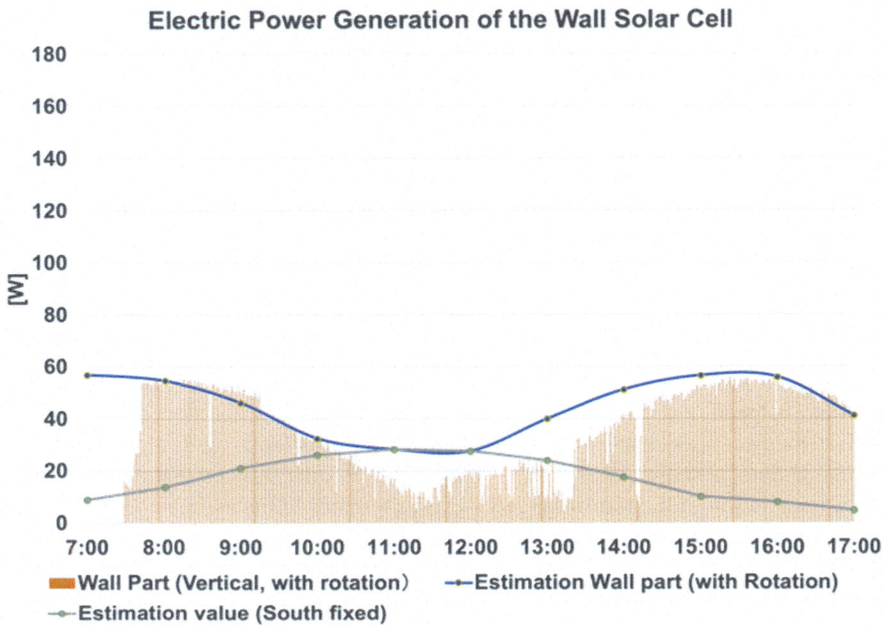


**Fig. 7** Electric power generation of the wall part (comparing measured value with estimated value) on the summer solstice, July 1, 2022

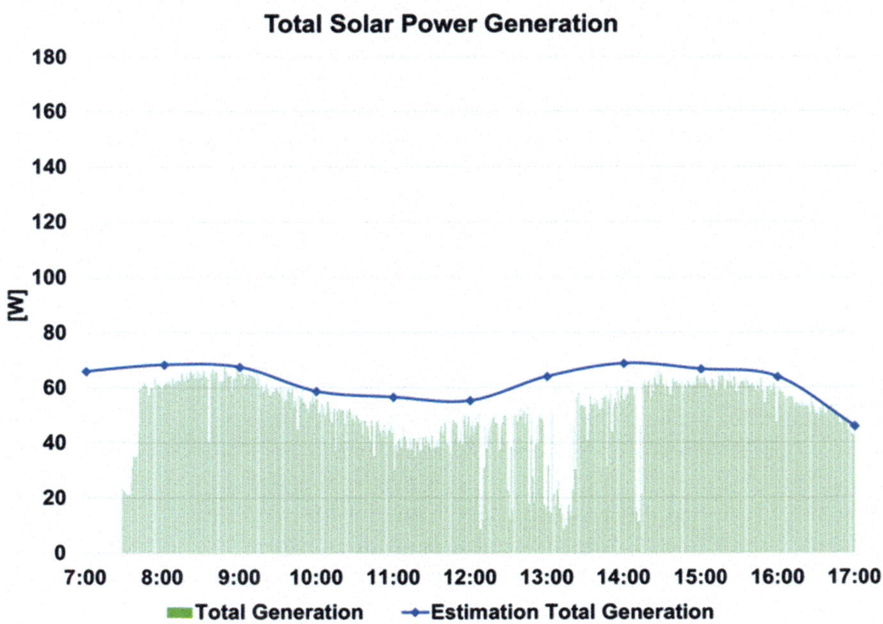


**Fig. 8** Electric power generation of all part(comparing measured value with estimated value) on the summer solstice, July 1, 2022

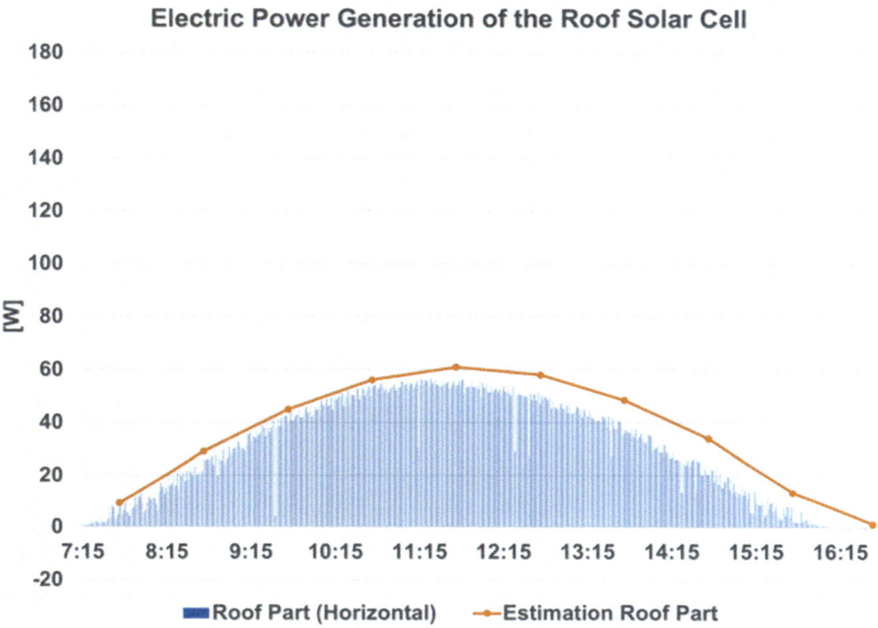


**Fig. 9** Electric power generation of the roof part (comparing measured value with estimated value) on the winter solstice, December 27, 2022

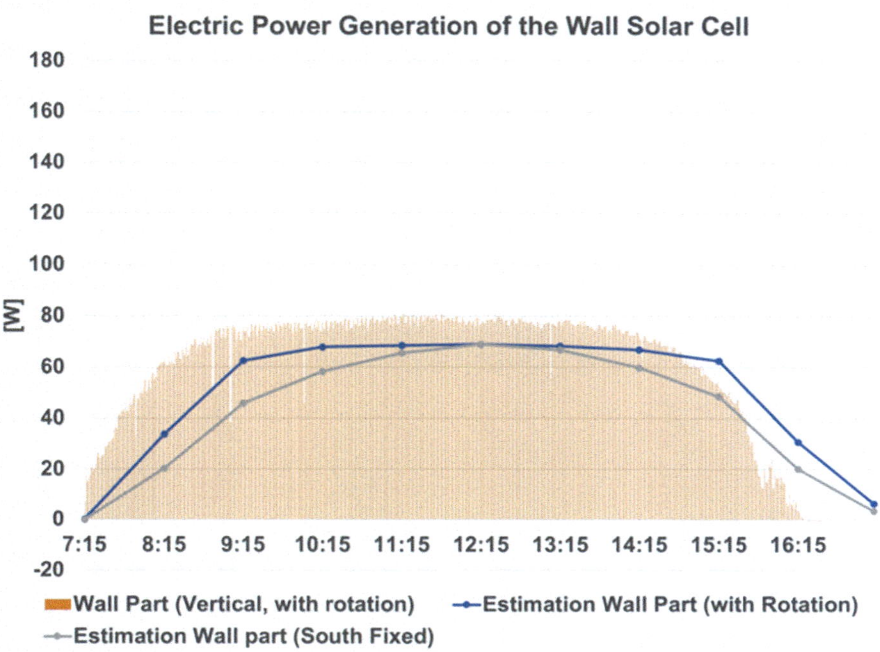


**Fig. 10** Electric power generation of the wall part (comparing measured value with estimated value) on the winter solstice, December 27, 2022

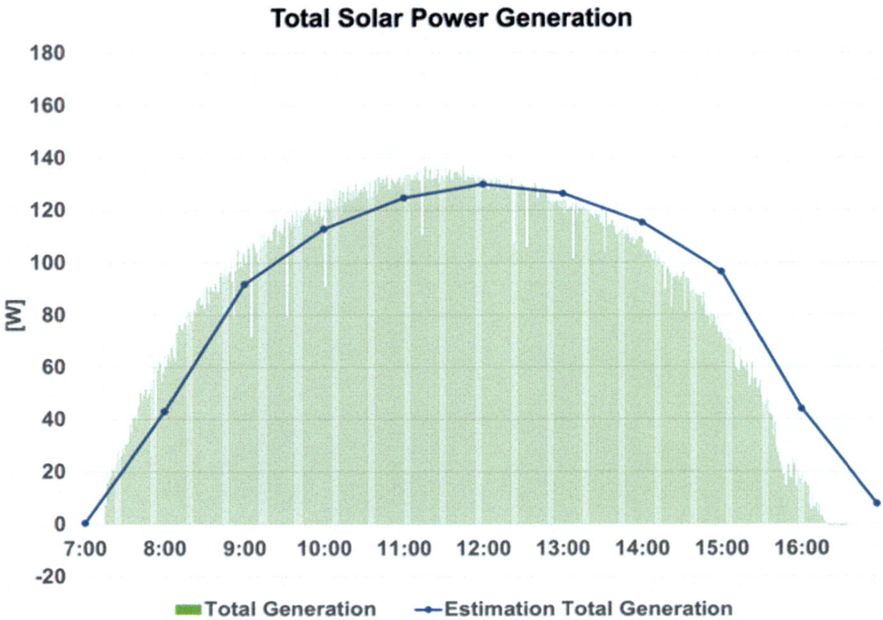


**Fig. 11** Electric power generation of all part (comparing measured value with estimated value) on the winter solstice, December 27, 2022

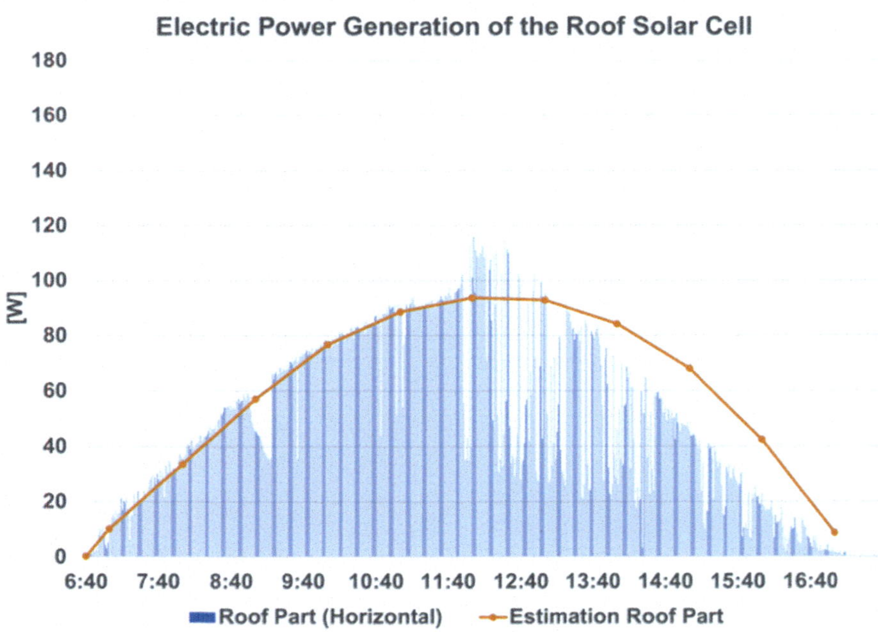


**Fig. 12** Electric power generation of the roof part (comparing measured value with estimated value) on the vernal equinox, March 15, 2023

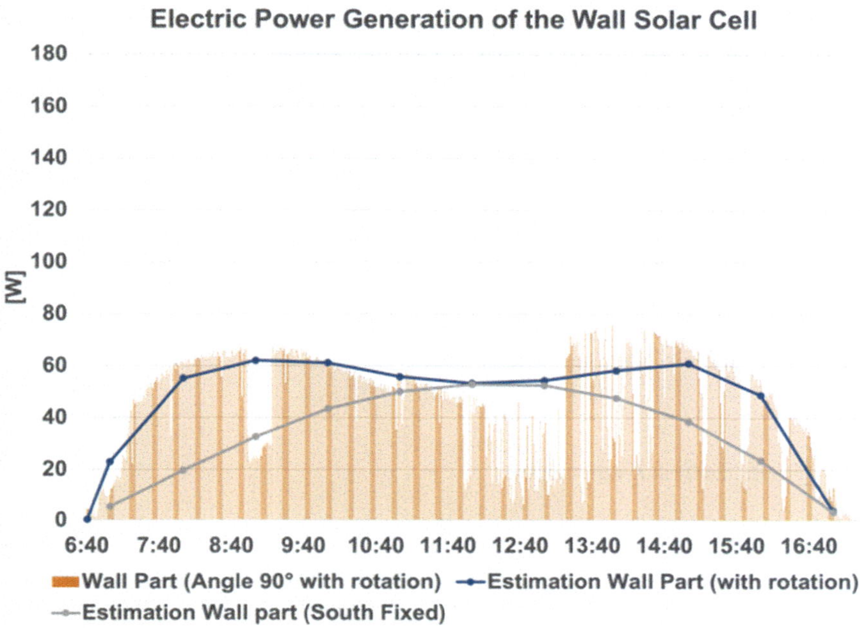


**Fig. 13** Electric power generation of the wall part (comparing measured value with estimated value) on the vernal equinox, March 15, 2023

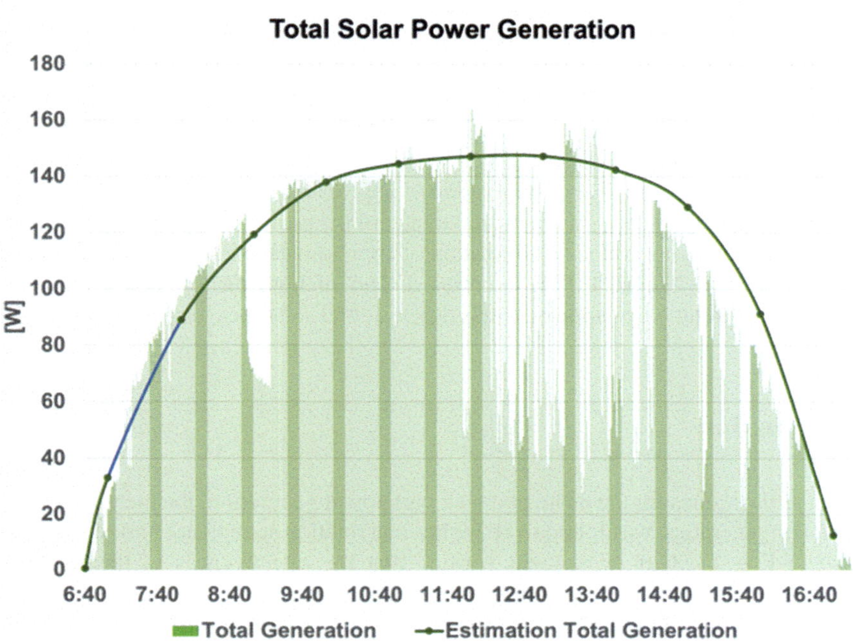


**Fig. 14** Electric power generation of all part (comparing measured value with estimated value) on the vernal equinox, March 15, 2023

The measured electric power generated by both the wall solar cell (with rotation) and the one that was fixed south (without rotation), along with the estimated data, are shown in Figs. 7, 10, and 13. The results of the test were that the measured generation power of the wall solar cell waveform matched the estimated one. In contrast, the estimated value of south fixed (no rotation) was less than the one with rotation, especially in the morning and evening. This effect was remarkable on the summer solstice.

The measured effect on electric power generation of rotation on the total solar cell, along with the estimation data, is shown in Figs. 8, 11, and 14. The results were that the measured power generation of all part waveform matched the estimated one. These waveforms were smooth and less precipitous—this tendency was remarkable on the vernal equinox and the winter solstice.

### *3.2 Quantitative Evaluation*

The total integrated values of the MOA's solar power generation are shown in Table 2. The results of the case where rotation occurred showed that the estimated amounts of total electric power generation from the wall solar cell, according to past data on the summer solstice of 2022, the winter solstice of 2022, and the vernal equinox of 2023, were 433.9, 536.7, 536.9 Wh /day, respectively. By using these results, the estimated values of the power generation of the wall solar cell, with and without rotation, were calculated on those days.

To verify the rotation effect, the rotation effective ratio is defined as the ratio of a measured value with rotation to an estimated value without rotation. As shown in Table 3, MEG_WR/EEG_WSF indicates a ratio of the measured power with rotation to the estimated power without rotation of approximately 1.9, 1.3, and 1.3 times, respectively. Also, EEG_WR/EEG_WSF indicates a ratio of the estimated value of rotation to the estimated value of no rotation of approximately 2.4, 1.2, and 1.5 times, respectively. MEG_T/EEG_TSF indicates a ratio of the measured total power value of rotation to the estimated value of no rotation of approximately 1.1, 1.1, and 1.0 times, respectively. EEG_TR/EEG_TSF indicates a ratio of both the estimated value of rotation and no rotation of approximately 1.5, 1.1, and 1.2 times, respectively.

## 4 Discussion

As well as the values for the estimated wall part power generation, the measured wall part power generation was found to be higher in a rotation state than in no rotation as shown in Fig. 16. Our experiment verified that the effectiveness of the rotation method was related to the side wall part.

In contrast, the value for the total power generation was found to be a little higher in the rotation state than with no rotation as shown in Fig. 17. This result is likely due to both some failure in the solar cell at the moment of the experiment and the

**Table 2** Quantitative evaluation of the MOA's solar power generation

| Target | Value [Wh/day] | Abbreviations | The Summer Solstice (July 1, 2022) | The Winter Solstice (December 27 2022) | The Vernal Equinox (March 15 2023) |
|---|---|---|---|---|---|
| Roof Part (Horizontal) | Measured value | MEG_R | 151.6 | 285.2 | 524.6 |
| | Estimated value | EEG_R | 284.8 | 356.1 | 656.7 |
| | Estimated value (no correction) | EEG_R no correction | 847.5 | – | – |
| Wall Part (Vertical rotation) | Measured value (with rotation) | MEG_WR | 341.5 | 586.5 | 477.3 |
| | Estimated value (with rotation) | EEG_WR | 433.9 | 536.7 | 536.9 |
| | Estimated value (south fixed, no rotation) | EEG_WSF | 181.1 | 457.5 | 369.3 |
| Total Generation | Measured value (with rotation) | MEG_TR | 493.1 | 871.7 | 1001.9 |
| | Estimated value (with rotation) | EEG_TR | 718.8 | 892.8 | 1193.6 |
| | Estimated value (south fixed, no rotation) | EEG_TSF | 465.9 | 813.6 | 1026 |
| | Estimated value (no correction) | EEG_TR no correction | 1281.4 | – | – |
| | Estimated value (south fixed, no correction) | EEG_TSF no correction | 1028.6 | – | – |

seasonal effect of changing the optimum light-receiving surface on the summer solstice, July 1, 2022. Upon noting failure, the measured value of the roof part generation and total generation also would be approximately three times as the estimation value shown in Table 3. As shown in Figs. 15, 16, and 17, the point EEG_TR (no correction) and EEG_TFS(no correction) indicate the potential power generation of 3 solar cells on the roof in a normal state. Consequently, the ideal point MEG_TR, the value of the perfect power generation of 3 cells, can be situated between 1,000 and 1,200 W.

During the specific period of the summer solstice, the sun's changing azimuth angle ranged widely from 270° to 90°. With the setting angle at south at 0°, a wall with a tilt angle of 90° can receive more sunlight both in the early morning and early evening, except at midday. In our experiment on July 1, 2022, the wall part and total generation of EEG_WR/EEG_WSF and EEG_TR/EEG_TSF ratios of the estimated values with rotation to those with no rotation are approximately 2.4 and 1.5, respectively. Therefore, the rotation effect could be attributed to the increase in total power generation on the summer solstice.

In contrast, during the winter solstice and the vernal equinox period, the changing azimuth angle of the sun is of a narrower range than the summer solstice, and the

**Table 3** The comparison of the measured value and the estimated value

| Target | Ratio | Abbreviations | The Summer Solstice (July 1, 2022) | The Winter Solstice (December 27, 2022) | The Vernal Equinox (March 15, 2023) |
|---|---|---|---|---|---|
| Wall Part (Vertical Rotation) | The ratio of measured value (with rotation) to estimated value (with rotation) | MEG_WR/ EEG_WR | 0.8 | 1.1 | 0.9 |
| | The ratio of the measured value (with rotation) to the estimated value (south fixed, no rotation) | MEG_WR/ EEG_WSF | 1.9 | 1.3 | 1.3 |
| | The ratio of estimated value (with rotation) to estimated value (south fixed, no rotation) | EEG_WR/ EEG_WSF | 2.4 | 1.2 | 1.5 |
| Total Generation | The ratio of measured value (with rotation) to estimated value (with rotation) | MEG_TR/ EEG_TR | 0.7 | 1.0 | 0.8 |
| | The ratio of the measured value (with rotation) to the estimated value (south fixed, no rotation) | MEG_T/ EEG_TSF | 1.1 | 1.1 | 1.0 |
| | The ratio of estimation value (with rotation) to estimated value (south fixed, no rotation) | EEG_TR/ EEG_TSF | 1.5 | 1.1 | 1.2 |
| | The ratio of measured value (with rotation) to estimated value (with rotation, no correction) | MEG_T/ EEG_TR | 0.4 | – | – |
| | The ratio of estimated value (with rotation, no correction) to estimated value (south fixed, no rotation) | EEG_TR/ EEG_TSF | 2.7 | – | – |
| | The ratio of measured value (with rotation) to estimated value (south fixed, no rotation, no correction) | MEG_T/ EEG_TSF | 0.5 | – | – |
| | The ratio of estimated value (south fixed, no rotation, no correction) to estimated value (south fixed, no rotation) | EEG_TSF/ EEG_TSF | 2.2 | – | – |

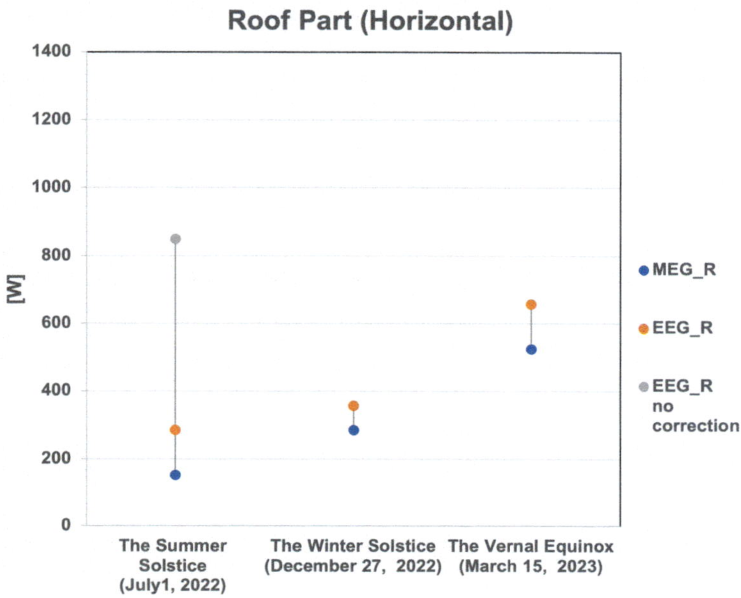


**Fig. 15** The differential of power generation (roof part)

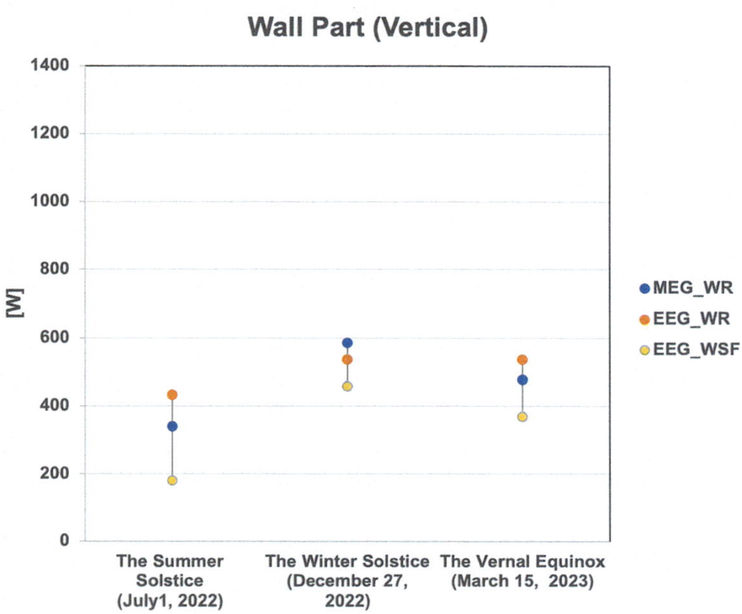


**Fig. 16** The differential of power generation (wall part)

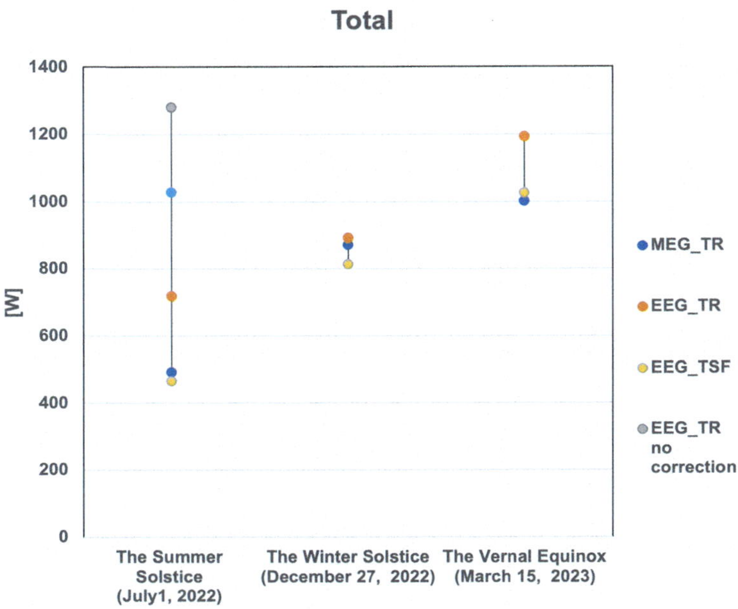


**Fig. 17** The differential of power generation (total)

elevation angle of the sun largely changes. Therefore, even with no rotation, the wall with a tilt angle of 90° has the advantage of receiving sunlight more than the roof area with a tilt angle of 0°. In the experiment of the vernal equinox on March 15, 2023, unfortunately, with some influence from the clouds at noon, the wall part and total generation of MEG_WR/EEG_WSF ratio would increase close to the EEG_TR/EEG_TSF ratio of 1.5 times.

It is unusual to have not only a rooftop solar panel but also a tilt angle of 90° wall as our MOA prototype. Usually, solar cell systems in an MOA will be equally treated as stationary solar panel types with no rotation. Moreover, some MOAs have expansion or portable type solar panels. However, the sun's movement problem in azimuth and elevation angle still exists where it largely changes depending on the four seasons.

Since MOA has mobile ability, our rotation method and system design can contribute to the effective utilization of MOA with solar energy in desired places. Our MOA is limited by its installation area and carrying capacity when using the rotation method, as a 1.5 times power generation increase in wall area has the same effect as the extension of approximately one solar cell sheet of 50 W.

However, as our rotation method is done by hand, there are some tasks that we must resolve, and further replication is required to confirm the efficient operation of MOA.

## 5 Conclusion

In this paper, we've introduced a simple method for the effective utilization of MOA ability in the manual rotation method, as well as verified our prototype MOA's solar cell system integrated as a two-angle solar cell system.

Our MOA was field-tested in three fine days in 2022–2023, close to the seasonal turning points of the twenty-four seasons in the solar year at Nagara town in Chiba.

The effective use of a technique with rotation for the optimum direction of the solar cell on the MOA by its mobility contributes to energy efficiency and makes up for the lack of installation area of its solar cell.

The integration of MOA's solar-powered system, like a solar-powered calculator and application example using MOA's power generation, needs to be addressed in the future.

Furthermore, MOAs such as EcoCabin also have a concept for regional revitalization effects in rural areas by electrifying with onsite energy for making "The Third Place" [6]. There is a possibility to provide a place to communicate and activate local mobile service in non-lectrified rural outdoor space through its mobility and providing a small but efficient source of energy.

**Acknowledgments** The authors thank F-WAVE Company Limited for providing their solar cells for EcoCabin and also thank Nagara Town Hall for the experiment support.

## References

1. Shota T, Satoko N (2019) Proposal for an off-grid Mobile architecture prototype for disaster relief. In: Proceedings of the 51st environmental design research association conference, p 189
2. Shota T, Satoko N (2020) Mobile off-grid architecture -a potential tool for the resilience of rural communities in the event of natural disasters. AGATHON- Int J Archit Art Des 08:118–127
3. The International Energy Agency (IEA) (2021) State-of-the-art and expected benefits of PV-powered vehicles. In: Report IEA-PVPS task 17 PV and transport
4. Satoko N, Shota T (2022) Development of an off-grid solar-powered system for mobile architecture. In: Proceeding of grand renewable energy 2022, pp 13–20
5. Shota T, Satoko N (2022) Mobile off-grid architecture prototype: a new application for regional revitalization. In: Proceeding of grand renewable energy 2022, pp 13–20
6. Ray O (2001) Celebrating the third place: inspiring stories about the "great good places" at the heart of our communities. Marlowe, New York
7. METPV-20, NEDO Solar Radiation Database, *NEDO*, Ver 3.0, https://appww2.infoc.nedo.go.jp/appww/index.html (2021).

# Urban Conscious Product Design Integrating a Life Cycle Perspective

**Sina Rudolf, Marius Dellbrügge, Max Juraschek, Mark Mennenga, Uwe Clausen, and Christoph Herrmann**

**Abstract** Cities are places of consumption for a variety of consumer products. At the same time, they are places of value creation and provide several potentials for manufacturing companies, as for instance proximities to their customers. With a production site located in urban environment it is possible to fulfill further demands of the urban system besides just providing goods to fulfil the needs of citizens. An urban factory may provide urban functions, such as education, innovation or waste disposal, which are generally provided by city authorities today. As "neighbors," factories can raise interest and fulfill needs of their urban surrounding actors along the whole life cycle of products manufactured in urban area. In order to make use of the outlined potentials, the product design must be adapted to the demands of the urban environment. For this purpose, it is necessary to identify not only the requirements of customers but also those that arise due to the proximity between the place of production and the place of consumption to fulfill further urban needs. This work aims at developing a design methodology that integrates requirements oriented at fulfilling urban functions into the product design along its life cycle. As a first step, urban functions and needs are identified. Then, the life cycle-oriented design methodology is developed to integrate the functional requirements into the design process. Within the methodology, strategies and design principles are defined for each product life cycle stage in order to fulfill urban functions. A case study is used to apply the methodology and derive specific design strategies as examples.

**Keywords** Urban conscious product design · Urban manufactured products · Urban factory

---

S. Rudolf (✉) · M. Juraschek · M. Mennenga · C. Herrmann
Sustainable Manufacturing and Life Cycle Engineering, Institute of Machine Tools and Production Technology, Technische Universität Braunschweig, Braunschweig, Germany
e-mail: s.rudolf@tu-braunschweig.de

M. Dellbrügge · U. Clausen
Institute of Transport Logistics, TU Dortmund University, Dortmund, Germany

 
S. Fukushige et al. (eds.), *EcoDesign for Circular Value Creation: Volume I*,
https://doi.org/10.1007/978-981-97-9068-5_7

# 1 Introduction

Cities are vital hubs for value creation and the production of goods [1]. Today, they account for more than half of the world's population and their importance is expected to rise further [2]. Cities are connected to higher affluence of their citizens, which makes them a hotspot for the consumption of products and services. Furthermore, they are expected to cater for all needs necessary for human well-being, including for instance the provision of adequate housing, work places, utilities, recreation, education, and mobility [3]. These core capabilities of a city can be summarized as urban functions [4], which are shown in Fig. 1.

However, the production, consumption, and end-of-life treatment of products are associated with resource exploitation and negative environmental impacts. It is therefore of importance to develop technologies and methods to lower, eliminate or positivize the effects of making, using and disposing goods for human needs. One strategy is the concept of sustainable urban production, in which factories are not seen as disturbances that need to be contained but rather as active elements of the urban environment that can exert positive impacts on their surroundings [5]. The interaction of a factory and its products with the urban environment can be conceptualized as a city–factory–product nexus that emphasizes the cross-influence of each element [6]. Previous studies have shown that urban and non-urban manufactured products can have a significantly different environmental performance over their life cycle [7].

The dual function of making and consuming in the city allows to establish interdependencies between the product design, manufacturing, use and recycling of products that have been manufactured in an urban environment. Following a strategy of effectiveness, products for consumption in urban areas can be purposefully designed to support the fulfillment of urban needs by contributing to the urban functions.

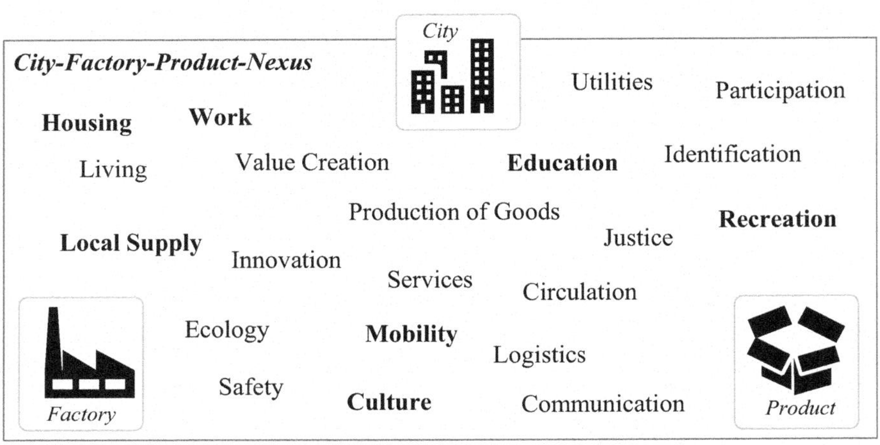


**Fig. 1** Selected urban functions in the city–factory–product nexus. (According to [4])

Consequently, a need arises to integrate a life cycle perspective with a functional perspective in the design of urban manufactured products in order to meet the urban needs and fulfill them along the whole life cycle. With an urban conscious product design (UCPD) approach the product design can integrate the demand of a city beyond the mere product functionality. This allows the consideration of potential impacts of products along their life cycle already in the design stage—from raw material extraction, to production, to usage until the end of life. Especially in an era, where there is a growing need for attaining the Sustainable Development Goals (SDGs), it becomes imperative to ensure that products not only meet sustainable practices but also contribute positively to sustainability.

By following a concept of UCPD, products can be created that not only meet the individual needs of consumers in urban environments but also offer added value along the life cycle of the product for the city and its inhabitants as urban stakeholders. This requires the development of new design strategies that specifically address both the product life cycle as well as urban functions. It leads to the research question of how urban functions can be integrated into the life cycle design of urban manufactured products to offer a positive contribution to urban stakeholders and their needs. To answer this research question, we first review existing approaches for the design of urban manufactured products. Based on this, we present an urban conscious product design concept. This approach is applied in a case study and the main findings are discussed.

## 2 Approaches for Life Cycle Design and Urban Production

A needs-oriented product design asks the question which kind of product a company produces instead of how a product is produced [8]. This type of design paradigm questions whether the product function justifies the environmental impact. The consideration of impacts and the resulting environmental and economic requirements on product design are issued in the research domain of ecodesign. As a systemic concept, ecodesign provides tools and methods striving for products with a low environmental impact throughout the complete life cycle [9]. Besides avoiding negative outcomes, ecodesign methods may lead to positive effects by improving the company's competitiveness as well as its image [10–12]. However, ecodesign may result in an increasing complexity of the product development [13]. There exists a large number of terms related to ecodesign that deal with similar concepts such as life cycle design (LCD), environmental design, or design for sustainability. Rousseaux et al. recognized more than 600 methods and tools in their review [14]. They vary with respect to their purpose, such as strategic tools, qualitative assessment tools, tools for the early stages of the product development process and many more [15].

It is not only the fulfillment of consumer needs that is crucial for a product manufactured in an urban area. In the city–factory–product nexus, the needs of further urban stakeholders, such as neighbors or city residents, must also be taken into account if the goal of fostering a positive relationship between the product and the urban environment is to be pursued. Our literature review shows that urban stakeholders and their needs are so far not explicitly addressed in product design approaches. Like many others, they are merely understood as external stakeholders in the design process [16]. However, few approaches already reflect the necessity to incorporate urban needs. For example, meeting the demand of urban stakeholders is addressed in the approach of urban services with which an urban factory provides urban functions (see Fig. 1) [4]. Urban services can be offered by the production site as factory-services or as a product-service system. With this service-based approach, a factory pursues a positive influence on its urban surroundings and its stakeholders. However, this does not yet consider the contribution that the product itself can already provide to the potential positive impact of the company. Looking from a system thinking perspective, another approach investigates the influence of product characteristics of urban manufactured products on the urban factory design. Triggered by the factory design, this in turn has an effect on the urban environment. This interrelationship was analyzed in a two-stage influence analysis [17]. With the information about the product characteristics with the main leverage on factory design, product engineering can shape the product design by addressing positive effects on the urban environment [17]. As products may cause positive as well as negative influence, there are different approaches investigating the impacts of products. One of them is the Positive Impact Product Engineering (PIPE) approach that is based on a four-phase model to capture product characteristics and their effects in the social, environmental, and economic dimensions of sustainability [18].

In the domain of urban production, there are a few approaches that already deal with ecodesign and LCD. Some of the ecodesign and LCD strategies can be found in a catalog of measures that was created as part of the Urban Factory project. An example is establishing a circular economy by making waste and by-products locally available or offering urban services [19]. Another life cycle-oriented approach in the context of urban factories focuses on industrial value creation in cities [20]. In a field study, the authors explore that in addition to the production stage (especially prefabrication but also assembly), sales and services connected to their products are offered on-site, and that the treatment of the end of life of a product is also covered.

In summary, research gaps exist with regard to the positive contribution of urban manufactured products to fulfill urban functions while at the same time reflecting on the impacts throughout the whole life cycle. Even though product development and design approaches in the literature consider potential influences of products, they do not yet take into account the demands of urban stakeholders to fulfill urban functions, nor do they explicitly focus on the positive effects that urban manufactured products can have on the city. We therefore propose a product design methodology, in which urban functions are integrated into the product design process in order to fulfill the needs of urban stakeholders and consider those for the whole life cycle of these products.

## 3 Urban Conscious Product Design

Understanding and positively shaping the relationship between an urban manufactured product and its urban surroundings requires a methodology that incorporates the fulfillment of the needs of a city and that integrates them into the life cycle-oriented product design.

The presented methodology seeks to improve decision support for product designers of urban manufactured products. The methodology is intended to be applied at the early stage of the design process, in which product designers determine the functional structure of the product and its basic mechanisms. As a result, applicable design principles and related design strategies can be derived for both consumers and industrial products that are manufactured in urban areas. Overall, the methodology aims at increasing sustainability in the three dimensions (environmental, economic, and social), in order to positively contribute to the city development and to a more sustainable usage of the product. The product design should enable not only a sustainable use but also a sustainable production as well as an environmentally friendly end-of-life treatment. This integrated life cycle perspective aims at addressing the complete product life cycle, closing the loop, making long-lasting, intensively used products and meeting the customer needs [15]. For this purpose, the targeted urban functions (see Fig. 1) join the essential requirements for meeting the needs of entities within a city [4]. In this way, the proposed approach builds on the conventional design processes from LCD and expands them to include the integration of urban functions in order to focus even more strongly on the needs of consumer and further urban stakeholders of an urban factory.

Figure 2 depicts the process which consists of four main steps embedded into the overall product design process. It considers an additional pathway in the conventional product development process to continuously improve the design according to the dynamic urban needs. The steps of the methodology are described in the following in more detail.

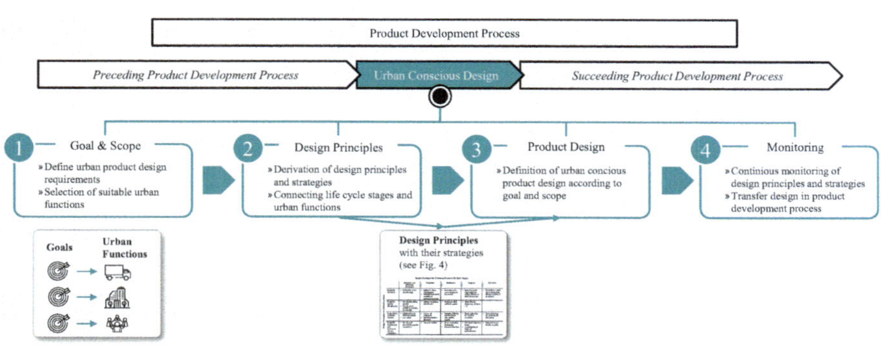


**Fig. 2** Steps of application of the proposed methodology for the life cycle design for urban manufactured products

## 3.1 Step 1—Goal Setting and Selection of Suitable Urban Functions

The product designer first needs to define goals for the product design as well as product requirements, functionalities and added value of the product with respect to their contribution to a positive relationship with the urban environment and its stakeholders. According to the goals and requirements, the product designer selects interesting, relevant or attractive urban functions that can be targeted by the product. With a prioritization of the urban functions, the decision on suitable design strategies is supported if several possible options will arise in the following steps. For this first step, the designer is provided with a list to document the main goals. Additionally, there is a compilation of targeted urban functions from which the most relevant and suitable urban functions for the company can be selected and prioritized, as indicated in Fig. 2.

## 3.2 Step 2—Derivation of Design Principles and Strategies

The main element of the methodology is the derivation of design principles that aim at realizing the urban functions, issued in Fig. 1, throughout the product life cycle. It establishes a connection to both, life cycle perspective and the fulfillment of urban functions in order to implement UCPD. A design matrix combining the product life cycle on the horizontal axis and the urban functions on the vertical axis supports this crucial step.

To address the urban functions, urban conscious design principles, such as *design for education* or *design for local product and service supply*, are derived. Additionally, the relevant life cycle stages for urban-manufactured products are considered: material extraction and processing, production, distribution, utilization and end of life. Since the intermediate logistics activities are particularly important in the city, the distribution stage is also explicitly listed as a single life cycle stage within the methodology. Distribution encompasses the entire process of making the product available to the consumer. In distinction, the urban function of *mobility* comprises a fulfillment of all accruing freight transport needs of the city residents in a way that is as compatible with the city as possible. It could be achieved by a reduction of freight transports and, as far as no reduction is possible, a sustainable and emission-reduced way of freight transport. Due to their density and diversity, urban areas introduce innovative local logistics concepts.

The intersections of the axes/dimensions form the cells of the matrix in which applicable, operationalized design strategies are entered. Accordingly, an overall design principle, such as *design for value creation* or *design for urban living*, consists of various design strategies, each of which addresses the product life cycle stage, as indicated in Fig. 3.

**Design Strategies for Product Life Cycle Design**

| | Extraction and processing of material | Production | Distribution | Use phase | End of life |
|---|---|---|---|---|---|
| Design for ***Urban Living*** | ▪ Use of household „waste" streams as secondary raw material | ▪ Matching the design of by-products to the needs of the neighborhood | ▪ Packaging and mobility solutions with minimal negative effects on neighborhood | ▪ Integration of upcycling and 2nd use options for urban reuse | ▪ Recovery of materials or components on-site |

**Fig. 3** Example for a design principle with respective design strategies for each product life cycle stage

The derivation of design principles and their corresponding strategies are oriented to the question how the product design can contribute to fulfill specific urban functions in each life cycle stage. Therefore, an analytical analysis of urban product design approaches, as well as an empirical analysis of existing examples and (best) practices, is conducted. From this research, a collection of general principles for all identified urban functions (see Fig. 1) has emerged. An example for the urban function of *housing* is given in Fig. 3, called *design for urban living*. For the material extraction stage, the operative design strategy could include the usage of "urban waste" streams from surrounding households as secondary raw material to be used for the urban manufactured product. The production could conceptualize also the design of unavoidable by-products that meet the needs of the surrounding neighborhood. With this strategy, the company can also increase their added value. In the distribution stage, packing and mobility concepts may be considered that have minimal negative effects on the neighborhood. For the usage stage, the design strategy may incorporate a solution for upcycling or a second use option in the urban area. At the end of life, a recovery of materials or components on-site of the factory could enable a circular and closed-loop secondary raw material supply of the urban manufactured product.

## 3.3 *Step 3—Definition of Specific Product Design*

The generically formulated design strategies proposed for the selected urban functions need to be reviewed with regard to their feasibility and suitability for the specific product. As a result, the user obtains appropriate strategies that need to subsequently adapt to the product. For this step, the user reviews and fills in the proposed matrix, leaving those generic strategies that have been described specifically enough for the product, and adapting those that still need specification. It is valuable to validate the derivation and selection of appropriate design strategies by a team of experts to obtain an interdisciplinary perspective on fulfilling urban functions. The input from a diverse team can bring different insights, expertise, and considerations together, helping to identify potential challenges, opportunities, and trade-offs. This collaborative approach enables a more holistic understanding of urban functions and ensures that the design strategies align with the desired goals and objectives. In the last step, a prioritization of the most relevant principles and strategies according to the goal and scope set in step 1 takes place in order to define the final product design.

### *3.4 Step 4—Continuous Monitoring and Product Development Process*

Using the matrix for the derivation of design strategies supports establishing variations in product design and serves as a starting point for further product development. The results from the design process are transferred into the proven steps of the product development process, such as proposed in the guideline "VDI 2221" [16]. It is important to note that the presented methodology is not intended to substitute the conventional product development process but rather to supplement it. As a tool it can be used in conjunction with other methods and techniques to support the design process and to make informed decisions. In addition, there is a continuous review of the design principles and strategies according to the dynamic urban environment, in order to integrate the current needs in the product design.

## 4 Application and Discussion

The proposed methodology is exemplarily applied to the case of a private stand-up electric scooter (e-scooter), manufactured in urban environment in direct proximity to the function of *living* and *housing*. An e-scooter generally consists of the following main components, such as frame, handlebars, deck, wheels, electric motor, battery, controller, and brakes.

The product design pursues two main goals for its e-scooter (*step 1*): fundamentally, the company follows a socioeconomic policy, so that the common good is highly valued. This is supported by the secondary objective of striving for sustainable development. According to these goals, the product designer, as the fictive user of the methodology, values the functions of *education*, *identification*, *participation*, and *logistics* as particularly relevant as they have the strongest impact on the *housing* and *living* function in direct neighborhood to the production site. *Education* is an essential function related to the common good of cities. The urban functions of *participation* and *identification* are considered as nearly equally important fields of action for the socioeconomic-driven company as they establish a personal relationship between the customer and the product. An e-scooter produced in the neighborhood or co-produced by the consumer can contribute to a durable consumer relationship and at the same time enhance the image of the city or neighborhood with a high-quality product and might lead to prolonging the usage time. The selection of the *logistics* function is based on its potential for conflict with city

**Design Strategies for E-Scooter Product Life Cycle Design**

| Design Principles for Urban Functions | Extraction and processing of material | Production | Distribution | Use phase | End of life |
|---|---|---|---|---|---|
| Design for ***Education*** | ▪ Information on the material origin | ▪ References about CO2 footprint<br>▪ Information on social condition of production processes | ▪ Instruction on the correct transport of the e-scooter | ▪ Instruction on the long-lasting and energy-efficient use of the product | ▪ Declaration / symbol for environmentally friendly recycling of the product |
| Design for regional ***Identification*** | ▪ Use of urban mining material<br>▪ Use of „waste" streams as secondary raw material | ▪ Marking the product manufacturer | ▪ Packing provided with local symbols | ▪ Urban lifestyle design (e.g. „made in ..") | |
| Design for city compatible ***Logistics*** | ▪ Minimized no. of different materials<br>▪ Low weight | ▪ Less no. of components<br>▪ Minimized product dimension | ▪ Packaging fulfilling requirements for urban mobility concepts | ▪ Sensor technology for condition assessment | ▪ Sensor technology for estimating the EoL timing |
| Design for ***Participation*** *and involvement of urban stakeholders* | ▪ Use of „own" (household) waste for own product | ▪ Value co-creation | ▪ 2nd use or upcycling of packaging<br>▪ Product as a package | ▪ DIY-repair supported design<br>▪ Customizability & degree of personalization | ▪ Instruction on eco-friendly recycling |

**Fig. 4** Design principles and their respective strategies for e-scooter product life cycle design

residents, as the mobility infrastructure such as streets may be heavily impacted by additional production caused traffic. Furthermore, the trade-off between the benefits of e-scooters as a flexible mobility solution and their space requirements and improper parking of the scooters should be considered.

For the design requirements, the exemplary product designer derives the following: (i) designed for private use, (ii) service-oriented design because most consumers live in the same city where the e-scooter is produced, (iii) design is aligned with the principles of circular economy, specifically to support R-strategies, (iv) sustainability-oriented product design, which presupposes a resource-saving product life cycle from extraction through production to the end of life.

In *steps 2 and 3*, the generic design strategies are assessed in terms of their applicability to the e-scooter. If they are principally applicable, the strategies are specified in more detail where needed. For the case of the e-scooter several design strategies are derived, both, specific as well as generic, and their respective urban conscious design strategies are formulated as shown in Fig. 4.

With regard to the function of *education*, the design principle is called *design for education*, mainly aiming at giving information to and sensitizing the consumer. Addressing the two life cycle stages of material extraction and production, the design strategies include, for example, labels on the product or information leaflets that include information about the origin of materials or the condition during the production stage regarding social or environmental values (e.g., $CO_2$ footprint). The design strategy for the distribution stage contains instructions for the way of transporting an e-scooter that is considered as a dangerous good due to the battery. The design of the e-scooter should raise the customer's awareness of handling the product during the use stage, for example by providing information on the proper usage or charging of the battery to ensure efficient use. Similarly, the design strategy for the end of life includes providing explanations about environmentally harmful recycling practices, e.g., by labels or (digital) customer information.

The general design principle for the urban function of *identification* is merged under *design for regional identification*. The principle aims at strengthening the relationship between citizens, the product, and the city, creating an identification with the place where the product is manufactured. For the stage of material extraction, this could be addressed by labelling where the material comes from or that primary material extracted from the own urban area should be used. Information on which people were involved in the manufacturing of the product could be included on an enclosed note, perhaps written personally. The packaging design (distribution stage) should contain local symbols or be produced by materials from the urban area. In the use stage, the label of "made in..." could be marked on the e-scooter.

The overall design principle for the function of *mobility* is considered as *design for city compatible logistics*. It contains a design that requires minimized weight, dimension, and use of material during the life cycle stages of material extraction and production as the logistics need to adapt to the mobility infrastructure of the city. The same requirements apply to the distribution stage, where the design mainly refers to the minimized design of the product packaging of the e-scooter, so that it can be distributed, e.g., by CEP (Courier, Express, and Parcel), cargo bike or parcel robots. Furthermore, a specific design option for the product is an installation of a sensor system for condition assessment of manufactured parts. The information about when the product lifetime ends can be used to initiate tailored urban logistics with regard to the end-of-life stage and, if necessary, a circular process.

The design principle that concentrates on the urban function of *participation* is called *design for participation and involvement of urban stakeholders*. With this principle, the urban actors, such as the customer or neighbor, will be involved in the activities along the life cycle of a product. Beginning with the material extraction, the strategies may incorporate "bring your own waste" for being used as secondary raw material for the product. In the case of the e-scooter, this might be plastics being melted for the handlebars. For the distribution stage, the packaging can be designed with a focus on a second life. The packaging could be in the form of a bag that can be used for further carriage of the e-scooter, for example in public transportation. A design that enables DIY-repair at the factory site meets the fulfillment of *participation* during the use stage.

In the last stage (*step 4*), the defined product design is transferred to the product development process that the company uses as standard. In addition, the product designers are required to check whether the framework conditions of the city and the needs of the urban stakeholders have changed as part of the continuous improvement process. If this is the case, it should be investigated to what extent additional services or design changes can still be incorporated.

## 4.1 Discussion

By applying the methodology to the case study of the e-scooter, we were able to demonstrate its general applicability. However, due to the generic approach, product details, production restrictions or legal conditions of a specific urban manufactured product were not taken into account. Thus, the derived strategies remain largely superficial within the presented case study.

We observed that some design principles and strategies can be adopted from existing approaches. From empirical investigations in practice within the Urban Factory II project, we observe that so far often no LCD-related approaches have been pursued and that urban functions have not been addressed either, although factory operators have shown strong interest in this. Therefore, this design methodology is intended to provide a comprehensive approach to enrich the approaches from LCD and customer-oriented design with the fulfilment of urban functions in order to support the positive impact and integration of the factory into its urban surroundings.

The methodology cannot be equally applied to all urban functions in each life cycle stage. We observed that not every urban function has a relevant purpose in every life cycle stage. Therefore, during the development of the strategies, a few gaps appeared in the matrix, or, in some cases, more strategies were developed than in other stages. This may be due to varying degrees of product-city relevance among different urban functions, such as value creation and the stage of material extraction. For the specific derivation of design strategies, depending on the product, material extraction may not take place within the city or prefabricated parts are acquired. However, considering that the product is produced and primarily sold in an urban area, a potential connection between the production, distribution, as well as the use stage and the city is very likely. In cases where the development of strategies poses difficulties, creativity-driven approaches that stimulate innovation in meeting the urban functions can support the process.

In the realm of strategies for the distribution stage, the focus lies primarily on packaging. Further logistic requirements need to be formulated in the future and specified for realization in collaboration with experts. A design that is compatible with urban characteristics may also be suitable for logistics in other environments, such as rural areas. However, some peculiarities are better suited for urban environments and only make economic sense there, such as parcel robots. This insight also applies to all design strategies: by fulfilling urban functions, design strategies are adapted to the context of the surrounding city. Of course, they can be equally useful and valuable in other areas, such as a rural production environment.

## 5 Conclusion and Further Research

In line with ecodesign principles, the product designer of urban manufactured products is provided with an interdisciplinary and life cycle-oriented design methodology that takes into account the complete life cycle, prioritizes urban product functionalities, and considers urban stakeholders needs [21]. The four-step methodology comprises an urban conscious product design (UCPD) approach which enables decision-making on life cycle-oriented design strategies and principles. By providing a design matrix, the product designer as user of the methodology is enabled to derive design principles and strategies that not only consider life cycle-oriented aspects but also fulfill urban needs, such as education.

The methodology provides an accessible and easy-to-use framework for decision-making and assists in aligning the product design with the desired positive urban outcomes and impacts towards urban stakeholders such as neighbors or city residents and consumers. Positive effects exist from the city and residents' point of view (e.g., by using UCPD and thus the fulfillment of urban functions) as well as from an ecological perspective.

The principles and strategies can be further developed and formulated holistically for all targeted urban functions. A validation phase is needed involving experts of various urban functions, such as experts from urban logistics as well as urban stakeholders (consumers, neighbors, etc.). In particular, the consumers determine whether the integration of urban functions makes the product more attractive and contributes to the purchasing decision. The principles should be expanded with best practices in order to create thought-provoking specific design strategies for the decision makers. Furthermore, it is necessary to apply the principles and strategies in the product development of urban manufacturing companies in order to further develop the methodology based on the emerging requirements.

For an advancement of the methodology, the approach of functional analysis should be strengthened in the future and quantitative methods such as life cycle assessment can be integrated to prioritize the trade-off between environmental impacts and the integration of design principles and strategies that meet urban needs in an early stage of the product development process [22]. The methodology should also be applicable to the continuous improvement and redesign of existing products in the future. Here, an approach could involve evaluating current design variations with a rating scale to indicate the potential for change. One possible method that could be used and further developed for this purpose is the Life Cycle Design Strategy Wheel [23].

**Acknowledgments** The research leading to the presented results has received funding from the German Federal Ministry for Economic Affairs and Climate Action (BMWK) for the research project "Urban Factory II" (Grant no. 03EN3042A).

## References

1. Jacobs J (1969) The economy of cities. Random House, New York
2. Khor N, Arimah B, Otieno Otieno R, van Oostrum M, Mutinda M, Martins Oginga J (2022) World cities report: envisaging the future of cities. United Nations Human Settlements Programme (UN-Habitat), Nairobi, Kenya
3. Reicher C (2014) Städtebauliches Entwerfen. Springer, Wiesbaden
4. Juraschek M (2022) Analysis and development of sustainable urban production systems. Springer, Cham

5. Juraschek M et al (2018) Urban factories and their potential contribution to the sustainable development of cities. Procedia CIRP 69:72–77
6. Herrmann C, Juraschek M, Kara S, Thiede S (2019) Urban factories: identifying products for production in cities. In: Hu AH, Matsumo M, Kuo TC, Smith S (eds) Technologies and eco-innovation towards sustainability I. Singapore, Springer, pp 185–198
7. Juraschek M, Becker M, Thiede S, Kara S, Herrmann C (2019) Life cycle assessment for the comparison of urban and non-urban produced products. Procedia CIRP 80:405–410
8. Heiskanen E, Jalas M (2003) Can services lead to radical eco-efficiency improvements? - a review of the debate and evidence. Corp Soc Responsib Environ Mgmt 10(4):186–198
9. UNEP, Cleaner production (1998) A guide to sources of information, 2nd edn. United Nations Environment Programme Industry and Environment, Paris
10. Del Segarra Ona MV, De Miguel MM, Paya Martinez A (2011) A review of the literature on eco-design in manufacturing industry: are the institutions focusing on the key aspects? RBIS 15(5):61–68
11. Ehrenfeld JR (1995) Design for environment: a new framework for strategic decisions. Environ Qual Manag 4(4):37–51
12. Rossi M, Germani M, Zamagni A (2016) Review of ecodesign methods and tools. Barriers and strategies for an effective implementation in industrial companies. J Clean Prod 129:361–373
13. Keoleian GA, Menerey D (1994) Sustainable development by design: review of life cycle design and related approaches. Air Waste 44(5):645–668
14. Rousseaux P et al (2017) "Eco-tool-seeker": a new and unique business guide for choosing ecodesign tools. J Clean Prod 151:546–577
15. Schäfer M, Löwer M (2021) Ecodesign - a review of reviews. Sustain For 13(1):315
16. VDI 2221 Blatt 1 (2019) Entwicklung technischer Produkte und Systeme (Design of technical products and systems): Modell der Produktentwicklung (Model of product design). Beuth Verlag, Berlin
17. Rudolf S et al (2023) Influencing factors of urban factories and their products for sustainable urban development. Procedia CIRP 116:167–172
18. Mörsdorf S, Vielhaber M (2023) Positive impact product engineering (PIPE) model - the way to net-positive sustainable products. Procedia CIRP 116:474–479
19. Bucherer M et al (2019) Maßnahmenkatalog zur Anpassung von urbanen Fabriken. Braunschweig
20. Juraschek M, Becht EJ, Büth L, Thiede S, Kara S, Herrmann C (2018) Life cycle oriented industrial value creation in cities. Procedia CIRP 69:94–99
21. Boks C, Stevels A (2007) Essential perspectives for design for environment. Experiences from the electronics industry. Int J Prod Res 45, 18–19:4021–4039
22. Moreno PR, Rohmer S, Ma H-W (2015) Analysis of potential relationships between functional analysis and life cycle assessment. Procedia CIRP 29:390–395
23. Brezet H, van Hemel C (1997) Ecodesign: a promising approach to sustainable production and consumption, 1st edn. UNEP, Paris

# Part II
# Sustainable Design and User Behavior

# Designing Labels with a Lasting Impact: How Eco-labels Influence Purchase Behavior and Lifetime Extension of Clothing

**Marie Das, Gabriela Dias, Leene Remeysen, Els Du Bois, and Ingrid Moons**

**Abstract** For consumers, it is not always clear how to be more sustainable. They find it difficult to buy more sustainable fashion and do not use their clothing long enough. Eco-labels have great potential to guide consumers into more sustainable behavior. The purpose of this paper is to understand how eco-label design and communication can be used to make more sustainable choices and extend the life of fashion products through repair, maintenance, and reuse. A survey ($n = 307$) was conducted to investigate the respondent's reaction to different label designs and their reaction to quality, maintenance, repair and second hand. We learn that labels have indeed a great potential in informing people about sustainability aspects (environmental, ethical, and local). When acquiring new clothing, a traceability tag is the most effective way to persuade the end-user, a label that should provide them mainly with information on materials and quality. For secondhand, an additional focus should lie on hygiene quality assurance. Next to that, labels for clothing in use should focus on care instructions (maintenance) and opportunities for qualitative repair (repair). This research provides us with more in-depth insights for design and communication opportunities. Future research is needed to understand consumers' quality perception, mediators and moderators in consumers' purchase reactions, how resale value is determined, where and how to introduce repair and how washing instructions should be re-designed.

**Keywords** Eco-label · Sustainable purchase behavior · Lifetime extension · Clothing · Fashion

---

M. Das (✉) · E. Du Bois
Faculty of Design Sciences, UAntwerpen, Antwerp, Belgium
e-mail: marie.das@uantwerpen.be

G. Dias
Faculty of Business and Economy, UAntwerpen, Antwerp, Belgium

School of Economics, Business, Accounting and Actuarial, University of Sao Paulo, Sao Paulo, Brazil

L. Remeysen · I. Moons
Faculty of Business and Economy, UAntwerpen, Antwerp, Belgium

 
S. Fukushige et al. (eds.), *EcoDesign for Circular Value Creation: Volume I*,
https://doi.org/10.1007/978-981-97-9068-5_8

# 1 Introduction

It has become common knowledge that the fashion industry is putting lots of pressure on our planet [1, 2] and on people [3, 4]. Fast fashion stimulates a high demand and therefore relies on large amounts of non-renewable resources and operates in a very linear way. Consequently, all these resources have a high probability (73%) of being landfilled or incinerated at the end of their very short life [5]. Moreover, the fashion industry is characterized by a complex, long, and fragmented supply chain [5, 6]. The rise of fast fashion has spurred overconsumption and an overall decrease in the number of times an item of clothing is worn [2]. Our current research takes a look at the extension of the user phase of garments and addresses the following question: What kind of claims can persuade consumers to make more sustainable fashion choices and what can eco-labels offer in this respect?

## *1.1 Sustainability and Fashion*

The sustainability issues within the fashion industry are multifaceted and relate to environmental (SDG14, SDG15, SDG13) as well as social (SDG1, SDG5) mismatches with the sustainable development goals (SDGs) [7, 8]. Consequently, to evolve toward a more sustainable fashion industry, ethical-social aspects must be faced [9, 10] as well as environmental related aspects [5, 11]. As for ethical social aspects, child labor, wages and working environments are aspects related to this concern [4] that gained importance after the collapse of the Rana Plaza in 2013 [12]. Striving toward a more circular system is high on the agenda to become more environmentally friendly. Within the "user phase" of this more circular system, the importance of product life extension is stressed by the Ellen MacArthur foundation [5] who visualized this in a butterfly diagram [13]. Important aspects that motivate consumers to keep their garments longer may be quality perception, maintenance, and repair [14]. One example is the evolution of the Patagonia brand from a brand that uses recycled materials, toward a brand that also offers garments of a high quality that people can maintain and repair and therefore use for a longer time [15]. Longevity can also be supported by new business models that trigger reuse, as for instance the creation of a secondhand market, which indeed seems to be a growing retailing model [16]. Apart from general secondhand stores (e.g., Kringwinkel.be) and sites (e.g., Vinted.com), brands also evolve toward this more circular system. An example is the Belgian fashion brand Essentiel Antwerp that started a secondhand platform "RE-ssentiel" [17]. Typical for the fashion industry is the main location of its production units in underdeveloped countries, which is associated with unsustainable working conditions, low quality, and unsustainable transportation. Therefore, a short supply chain that stresses a more local production may improve the sustainability of the industry [18]. Some fashion brands adhere to this sustainability trend (e.g., Studio AMA).

## 1.2 Communicating with Eco-labels

Although consumers in general are increasingly aware of the consequences of their choices [19] and look for more sustainable products and services [20, 21], it is not always clear what makes fashion sustainable and how they can make more sustainable fashion choices [22]. The communication about the sustainable impact of a garment is blurred [23]. A standard misconception hereby is that the reason for not buying sustainable fashion is the lack of information [24]. Indeed, there's a lot of information available on sustainable fashion. More information can lead to higher levels of knowledge but does not necessarily change behavior [25]. The problem is not the amount of information but the lack of clear and understandable information [26]. One way of quickly informing consumers on sustainable choices is through eco-labels. The goal of this medium is to evaluate, authenticate and standardize "green" claims. In this way, consumers can distinguish green products from non-green products [27]. Thus, an eco-label is a policy instrument that is intended to increase transparency in the market and that confirms that a product has been checked by an independent third party [28]. Consequently, eco-labels can play a role toward more clear and transparent communication between producers and consumers and may be able to persuade consumers to make a sustainable choice [29]. Apart from its benefits for consumers, they may be useful for designers as well. Clancy et al. [30] pointed to the importance of the upstream supply chain [30]. Harris et al. [31] posits that quality and lifetime extension is a challenge for the design and production process and that good eco-labels can be helpful from the beginning of the design process onward. They call for more insights that support the development of eco-labels [30]. Although Horne [32] made a useful classification of environmentally friendly labels, till today no coherent or ideal eco-label is put forward in the fashion industry. The existing labels are not well known, not always clear, and are not always trusted [33]. A lot of labels are only recently developed, and their design needs some improvement [34]. Moreover, research pleas for the use of only a few good labels. Indeed, the excessive growth of eco-labels can result in an overload of information, confusion, and distrust of these labels [35, 36]. According to White et al. [28], eco-labels should be eye-catching [37], easy to understand [38], and consistent across different product types [28]. Eco-labels are only effective if the consumers know what the labels stand for and see the difference between labeled and non-labeled products [39].

More recently, the initial call for transparency often leads to the call for traceability. It is not only about the third-party control that indicated sustainability through labeling but also about the idea that the consume can control the supply chain of its product by him/herself. Rossi and Rivetti [34] plea in this context for legislation about product information. With the evolution of the Internet of Things, traceability has become feasible and is already commonly applied in the food sector [34]. Nowadays, also for the fashion industry, traceability has become feasible and some companies have already applied it. Thinking Mu, a Spanish brand, discloses for each individual product supply chain information like how many people were involved in each step in the production process and how many kilometers the product traveled [40]. To reduce the negative effect of eco-label use, recent research analyzed the relationship between the provision of additional environmental cues with an eco-label [41]. The authors considered "traceability" as an additional sustainable cue and identified that its provision together with an eco-label positively influences the purchase intention of sustainable products. Also, the provision of both at the same time adds value to sustainable products and enhances the perception of product quality [42]. The goal of this research is to investigate if these statements are also true for the fashion industry and labeling of clothing.

Given the complexity of the eco-labels in the fashion industry, the current research aims to unravel the effects of different eco-labels on some aspects of consumer behavior hereby focusing on the core message that an eco-label communicates (environmentally friendly, social, local). Moreover, the research addresses the question of transparency and unravels whether tags that make all processes traceable have persuasive power for consumers' sustainable choices. This can be related to purchasing quality clothing that can be either new or secondhand, but also to help extend the products' lifetime through correct maintenance and repair.

## *1.3 Eco-labels and Sustainable Choices*

A study from 2023 has proved that there is a strong correlation between a company's green marketing communication and green purchasing behavior [43]. Eco-labels may play a role in this communication strategy. The emerging use of eco-labels has resulted in a growing segment of well-informed consumers [44]. But the success of such eco-labels is still ambivalent [29]. The current study looks at whether clear recognizable eco-labels that put forward one clear sustainability claim can be persuasive in the sense that they elicit the perception of sustainability and trust (PT), intention to buy (ITB), and willingness to pay more (WTP) for a sustainable product.

### Recognizability

Koszewska [33] reports on the awareness of some often-used eco-labels. Overall, the awareness is not overwhelming with EU eco-label, GOTS (Global Organic Textile Standard) and OEKO-tex as most well-known (all between 18% and 20%). Though the Fair-trade label is a more general social label, it is also applied to textile, and it is recognized by 10% of the respondents. As for recognizability, the current study addresses the question which message should be put forward in a clear and consistent way [33]. As mentioned earlier, sustainability can be associated with expected impact on the natural environment, on the social environment and specific for the fashion industry may be associated with production location. So, eco-labels may address three kinds of sustainable concerns. Therefore, in the current study three different labels were developed related to these concerns (natural environment, social environment, and location). The persuasive impact of each claim will be tested.

### Environmental Claim

Natural environment refers to the problems related to water and chemical use, carbon emissions, resource consumption, waste, etc. [2]. Eco-labels such as OEKO-TEX Standard 100 and GOTS report on this issue. The first label guarantees that cotton used for fabrics and products is free from toxic chemicals [45]. The GOTS-label promises the same but also ensures the cotton is grown and processed in fair and safe environments [46]. In our study the label with the environmental claim focuses on water use, carbon emissions, and material selection.

### Social Claim

This refers to the social problems the fashion industry faces. The Fair Wear Foundation is a nonprofit organization that works together with companies and fights for improvement of the labor conditions [47]. The Fair-Trade label is committed to fair international trade from developing countries to rich, Western countries [48]. In our study, the label with an ethical claim has a Fair Wear Foundation and Fairtrade label. Next to that it is emphasized that the garment is made in a fair and healthy environment.

### Local Claim

Buying locally is one of the tips the United Nations gives consumers who want to become more sustainable [49]. Products that are locally sourced and made, emit less carbon and are linked to good working conditions [50]. There are many campaigns that promote buying locally meaning choosing small, local companies over long chain companies [51]. In our study the local label shows that the garment is made and sold locally in Europe.

### Traceability Tag

Moreover, recent research and legislation plead for transparency and traceability [3]. Following the earlier research of Loureiro and Umberger [42], which is in favor of traceability tags [42], we compare each eco-label (environmental, social, local) with the same label but with an extra traceability tag.

## 1.4 *Three Consumer Reactions*

**Intention to Buy (ITB)**

Intention to buy is the probability that a consumer buys a product or service in a given period of time. It is an important predictor of actual buying behavior as it takes place just before the final action in the traditional conversion funnel [52]. Eco-labels may help green consumers in their decision making toward buying sustainable products [53]. Results from previous research indicate positive attitudes toward eco-labels, but they do not always influence consumers' purchase intention. This is partly due to a lack of understanding what they stand for [54]. Therefore, the current study takes the relationship between eco-labels and ITB as a research question.

**Perception of Sustainability and Trust (PT)**

Environmental product attributes are not easily verifiable by consumers which makes environmental friendliness a credence good [24, 55]. For a consumer it is hard to tell if the producer's claims are based on facts or falsehoods [56]. Eco-labels have the power to reveal environmental attributes but can also (unintentionally) greenwash the product [57]. Greenwashing is described as the act of misleading consumers by falsely informing the consumer on the environmental benefits of a product [58]. A report from 2010 reveals that 95% of green products in North America commit to at least one of the seven sins of greenwashing [24]. Misleading and vague information about a product's "positive impact" can be misinterpreted by consumers. Especially when they have no background information about the company [59]. Various investigations indicate that consumers lose confidence in the eco-labels or green information [60–63] or do not perceive them as sustainable, which can result in a decrease in purchase intention or willingness to pay for actual green products [64].

**Willingness to Pay More (WTP)**

As fast fashion relies on underpayment of people working in the industry, relying on mass production using unsustainable production processes and unsustainable materials that are more cost efficient, aiming for a more sustainable system, may lead to different value creation and to higher prices for the end users. So, can eco-labels help to convince consumers to pay more? The use of eco-labels already initiates the perception of higher prices [65]. Previous research indicated that consumers in developed countries are willing to pay higher prices for eco-labeled products and products from environmentally friendly and responsible companies [66, 67]. In a study among Slovenian consumers for instance, respondents were willing to pay 10% more for clothing with an eco-label [68].

### *1.5 Understanding the Meaning of Sustainable Choices for Consumers*

As mentioned earlier and as put forward by the Ellen McArthur Foundation [13] the extension of product lifetime by repair, maintenance and reuse is a crucial element in the attempt to reach more sustainability. Therefore, the current research describes some aspects of the current consumer behavior related to lifetime extension of garments [69, 70]. We try to grasp what they mainly understand by sustainability. We look at aspects that contribute to the way they perceive quality, maintenance, and repair. Moreover, motivators and barriers toward product life extension via other retail models (secondhand clothing) are explored. This more descriptive research provides us with insights for design and communication opportunities.

## 2 Research Aim and Research Questions

The current research investigates if eco-labels can persuade consumers toward more sustainable decisions when buying clothing. The main research question is:

Can eco-labels related to one of the three sustainable claims (environmental, social, and local) and a more neutral label influence consumers to choose for more sustainable clothing items? This main question is divided into sub-questions:

RQ1: What is the effect of adding a specific designed eco-label on ITB, PT, and WTP?

RQ2: What is the effect of a traceability tag on ITB, PT, and WTP?

RQ3: What is the combined effect of adding a traceability tag and a specific eco-label (environmental, social, and local) on ITB, PT, and WTP? (Fig. 1)

**Fig. 1** Example of presented eco-labels with social claim left with traceability (tag) and right without (no tag)

Moreover, the current research explores how labels can be better designed by describing what people understand and desire about aspects that may influence longevity of garments.

RQ4: What are consumers' views on lifetime extending aspects such as quality, maintenance repair, and secondhand offerings.

RQ5: How should labels be redesigned to optimally match product longevity?

# 3 Methodology

## 3.1 Sample

A convenience sample of 307 Flemish respondents fully completed the questionnaire. It consisted of respondents of different ages: 45% between 18 and 25 years old, 32% between 26 and 45 years old, and 23%, 45+. Most respondents are female (77%), and one person identified as non-binary. Most of the respondents work full-time (53%), followed by students (37%). The sample mainly consists of higher educated people (master's degree, 45%, followed by a professional bachelor's degree, 24%). It is relevant to describe our sample in terms of relevant ecological characteristics. Two scales were used for this purpose: the first one measuring ecologically conscious consumer behavior (ECCB) (eight items, 5-point Likert scale, $\alpha = 0.836$) [71] and the second one is the Brief Ecological Paradigm (BEP) scale (five items, 5-point Likert scale, $\alpha = 0.781$) [72]. Results indicate that the respondents in our sample are relatively high concerned with the environment (mean BEP = 4.1726, SD = 0.62059), and are acting more than averagely in an ecologically friendly way (mean ECCB = 3.421; SD = 0.72911).

## 3.2 Research Design

A between subject design was used in which each respondent was randomly allocated to one experimental condition (environmental, social, local, and neutral). The sample composition was kept invariant concerning "age" over the four conditions (for each condition: 18–25 = 45,3%, 26–45 = 31,9%, 45 + =22,8%). Respondents were presented with a scenario in which they were asked to imagine buying a T-shirt. For each condition a fictitious label was designed with a sustainability claim that corresponds with the condition as described under heading 1.3. (social, environmental, local, neutral). A within subject design tested the consumer reactions toward traceability tags (as for each condition the respondent got the label randomly with and without the tag). Qualtrics software was used to program the questionnaire around three main blocks: (1) consumer reaction toward different eco-labels, (2) questions regarding lifetime extension strategies, and (3) psychographics and socio-demographics.

### 3.3 Measures

First a spontaneous reaction about which of both labeled T-shirts one would choose was asked (tag, no tag). Then, three consumer reactions were asked for both stimuli: ITB ([73], $\alpha = 0{;}85$; mean = 37,215, SD = 0.71,408); PT ([34]; $\alpha = 0.890$, mean = 37,044, SD = 0.689); WTB (mean = 29,3806, SD = 13,916). Further, the recognizability and understanding of different eco-labels was tested by presenting well-known eco-labels. Next, respondents were asked what a sustainability label for clothing should stand for in the first place. They could choose between (1) minimal impact on the environment (environmental claim), (2) ethical aspects related to human welfare (ethical claim), (3) the distance the garment must travel (local claim), and (4) the lifespan of the garment (quality claim). Afterward, they were asked what different aspects of the chosen claim they value the most. Longevity aspects such as cues of quality, maintenance, and repair was evaluated using 5-point Likert scales, as well as by reaction to propositions about secondhand garments.

## 4 Results

To test the different effects of the different eco-claims, a between group comparison was done by running one way ANOVA's that test for significant differences on ITB, PT, WTP. To compare the effects of each of the eco-labels on the different consumer reactions (ITB, WTP, PT) a Bonferroni t-test was conducted. For the ITB, there is no statistically significant difference between the different eco-labels. (F (2,47) = 2,46, $p = 0.063$). For the WTP, there is a significant difference between the different eco-labels (F (2,47) = 9,82, $p = 0{,}001$). Bonferroni t-tests show that respondents want to pay significantly more for a T-shirt with an environmental ($p = 0{,}036$) an ethical ($p < 0{,}001$) or a local claim than for a T-shirt with a neutral claim. Even though we notice the absolute largest impact of the local label followed by the ethical label, there are no statistical differences between the different sustainability claims. For the PT, the ANOVA analysis shows a significant difference between label conditions (F (2,47) = 40,04, $p < 0{,}001$). The Bonferroni t-tests show a significant difference between the ethical claim and the neutral condition ($p < 0{,}001$), between the environmental claim and the no claim condition ($p < 0{,}001$) and between the local claim and the no claim condition ($p < 0{,}001$). There is no statistically difference between the different sustainability claims.

## 4.1 The Effect of a Traceability Tag

In a within group design, the respondents had to choose between T-shirts with (T) or without a trace tag (NT) (randomly presented). In general, when confronted with either sustainability claim, the T-shirt with the traceable tag was chosen by 78%. Over the four conditions, 80% choose the traceability label in the ethical, 80% in the environmental, 70% in the local and 77% in the control condition. This means that over all the possible conditions, traceability has a substantial effect. Moreover, the differences in the effects between NT and T are significant (all effects $p < 0{,}001$) as well as for ITB (NT: mean = 35,657; T: mean = 37,560); PT, (NT: mean = 34,748; T: mean = 39,340) as WTP (NT: mean = €28,0194; T: mean = €30,7319). In order to look at the combined effects of the tag and the different claims, paired sample t-tests were conducted to determine the differences in ITB, WTP and PT between labels with or without an additional traceability tag. We hereby consider all the sustainability claims together, and the four conditions (environmental claim, local claim, ethical claim, and neutral claim). In general, adding a trace tag to a sustainable label (general) results in a significant positive effect on ITB, WTP and PT. For both the environmental claim and the local claim, adding a traceable tag to environmental claims has a significant positive effect on WTP and PT ($p < 0.001$) and on ITB ($p < 0{,}01$). Adding a traceable tag on an ethical claim, the three reactions are positively influenced. (ITB, PT, WTP) ($p < 0{;}001$). Interesting is that in the control condition (without a sustainability claim) we notice significant effects of traceability on all the consumer reactions.

## 4.2 Important Aspects of an Eco-label

When asked about *knowledge* of different existing sustainability labels, we see that the Fairtrade label is the best-known. 50% recognize the label, and 28% also understand what it stands for. Remarkable is that the OEKO-TEX label and GOTS-label are not well known. Respectively 46% and 47% do not recognize the label. 30% and 32% recognize the label but do not know what it stands for. In relation to eco-labeling, the environmental claim is considered most important (50%), followed by ethical claim (37%), lifetime (11%), and distance (2%). Using a Likert scale, respondents indicated how much they valued different aspects of the chosen sustainability claim. Respondents who selected the environmental claim (50%) highly value (important to very important) all the mentioned aspects which are water (88%) and chemical use (90%), distance (79%), carbon emissions (92%), and transparency (94%). Respondents who selected the ethical claim (37%) highly value (important to very important) the reassurance of no child labor (99%), and good working conditions (99%). Transparency (75%) and fair wages (96%) are valued less but are still very important. Thirty-five (11%) and seven (2%) respondents selected the lifetime claim and the local claim as important. However, we cannot analyze these results due to the small subsamples.

## 4.3 *Aspects That Influence the Longevity*

**Perception of Lifetime Extension Aspects**

When respondents were asked ($n = 307$) what 3 aspects they consider when buying new clothes, we see that quality (28%, $n = 261$) and price (26%, $n = 242$) are mentioned most. Next, when asked how they recognize *quality* ($n = 307$), 27% indicate that they check the material composition and 20% touches the product as an indication for quality. Eighteen percent check the construction of the garment to evaluate its quality. One searches mainly for information on the *maintenance* of a garment ($n = 307$) by looking at the wash instructions (47%) followed by touching the garment (22%) and looking at the materials composition (15%). Only 14% look at the care label of the garment. Estimating *repairability* ($n = 307$) during the purchase phase is mostly done by looking at the construction of the garment (54%), followed by checking the materials (23%) and touching the garment (16%). Only 4% looks at label. When asked what the respondents do when a garment is damaged ($n = 307$), 29% have it repaired, 23% throw it into the clothing bin and only 19% repair it themselves. Sixteen percent donate the garment and 11% throw it away.

**Second Hand**

Thirty-eight percent of the respondents ($n = 307$) sometimes buy secondhand, but the majority (47%) seldom or never buy secondhand. The two main reasons ($n = 261$) are concerns about hygiene (27%) and quality (20%). Moreover, 18% complain about the shopping experience and 1% say that buying secondhand is bad for their appearance. Only 15% of the respondents ($n = 46$) indicate that they usually or always buy secondhand. Saving money and looking for unique pieces are equally most important (35%). Only 27% buy second hand for environmental considerations. secondhand clothes are bought in secondhand stores (35%) or specialized sites (35%).

# 5 Discussion and Future Research

We learned that labels have a great persuasive potential; however, they must be used in the correct way to achieve the right goal. Companies often assume that consumers attach great importance to labels and certificates such as GOTS and OEKO-TEX. Nevertheless, research shows that end-users do not know what these stands for [74]. This could be explained by the large difference between objective quality and subjective perceptions of quality. Objective quality is determined by companies and standards and subjective quality is how consumers perceive quality. Our research contributes to understanding what (subjective) information an eco-label should contain to stimulate sustainable consumer behavior. Based on our findings, we can confirm that consumers do not recognize the established eco-certificates and do not understand what they stand for. When asked what an eco-label should stand for, they associate it mostly with environmental considerations. This may be due to the stronger ties in consumers' brain between eco and environment than between eco and ethic or local.

## 5.1 The Effect of Eco-labels on Consumer Reactions (RQ1, 2, and 3)

Our results indicate that it does not matter which eco-claim the label communicates, they all increase PT and WTP toward the garment compared to a neutral claim. This is also confirmed by Roozen et al. [75]. The persuasive influence is not significant concerning the intention formation, which is in line with previous research [53, 54]. Labels with a local claim have the highest WTP. This may be due to embedded, almost stereotypical perception of "Asia" as a low-price product provider, due to bad work conditions. The estimated European wages may be associated with the estimation of a reference price for a garment. The motivation to support local industry may be another reason to value this proposition. Adding a traceability tag has an overall persuasive effect as well as an effect on all the consumer reactions (ITB, PT, and WTP). These results are in line with earlier research from Loureiro and Umberger [42] and Wu et al. [42, 76]. This additional positive effect of the traceability tag compared to non-traceability is significant for each of the eco-claims as well as for the neutral condition. Just providing consumers with the opportunity to check and control may stimulate their self-efficacy and self-control [77] and exert influence on sustainable behavioral reactions. This is in line with the stream of literature related to adoption processes [75, 78, 79]. The significant effect of adding a trace on the ethical claim is in line with recent fashion campaigns that stress the importance of transparency and the need of traceability [80]. This is mostly linked with the social problems in low-cost countries. The addition of traceability on an environmental or local claim is already more common in the food industry [81, 82].

## 5.2 Opportunities for Better Eco-labels That Influence Reuse and Lifetime Extension (RQ4 and 5)

To make eco-labels more relevant for the end user, our research provides us with some descriptive insights on consumer purchase decisions and sustainability perceptions that relate to longevity: quality, materials, and price. It was found that quality is assessed by looking at the material composition and sensing the garment like touching the fabric. This is confirmed by Aakko and Niinimäki [83], who explain that materials and construction are two factors that influence quality perception before, during, and after purchase and during usage of the product [83]. Connor-Crabb and Rigby [84] confirm that materials and the interaction with material (how it feels) influence the quality perception during the usage of clothing [84]. Especially in an online context it is important that labels mention the material but also that communication campaigns stress the touch and feel of high-quality materials and construction. Materials (composition and tactility) contribute to estimating the maintenance of a garment, which is, however, mainly done by looking at the wash instructions [85] that should be carefully provided. Damage to a garment due to incorrect washing can lead to lower quality perception and customer dissatisfaction [88]. Besides informing the consumer with good wash instructions, a new design of care instruction labeling can also mention other guidelines. Altering care instructions can nudge consumers into a more sustainable maintenance behavior. As washing, drying, ironing, etc., are all energy-intensive processes [87], a more sustainable way of caring for clothing is to wash them less frequently [88]. This can be done by using techniques such as airing clothing, using refreshment sprays, and spot treatments. The use of traceability can provide an opportunity to track stains and connect a fun meaning to them. For example, the wine stain on the dress is the result of bumping onto the wearer's future love of their life. It is proven that providing the history behind a garment helps form favorable attitudes and purchase intention [89]. This is especially important for secondhand clothing. This personalized traceability labeling can be a subject of further research. For repair, labels can also provide information on how to repair the garment if it gets damaged or where repair services are available. This can also be applied to nontextile products like bikes or electronics. The label can provide information on where to buy spare parts, how to repair common malfunctions or where to find a local repair-shop. On the other hand, repair services can use labels to show that the repair has met certain quality standards and therefore can guarantee quality. Again, in the secondhand environment, this is very important.

Especially for secondhand clothing, quality assessment is very important as quality concern is, next to perception of hygiene, one of the reasons to not buy (more) secondhand. Even for other products such as electronics it is important to highlight that the product is still functioning correctly. Considering the perception of hygiene, it can again be linked to the materials and how the product looks. The cleanliness of a product is determined by stains and smell. Body odors are considered gross and have resulted in washing our clothing more frequently [87]. Even if a garment is freshly washed but still has stains, it is considered not clean [90]. Water temperatures and used wash cycle also determine how clean the consumer perceives the garments [91]. Perception of contamination can alter how secondhand clothing is perceived [92]. The use of labeling can help with the quality perception of secondhand clothing, and traceability may be helpful. For example: a trace-code could allow the consumer to view when, where and how the specific item was washed. Follow-up research taking this as a subject is recommended.

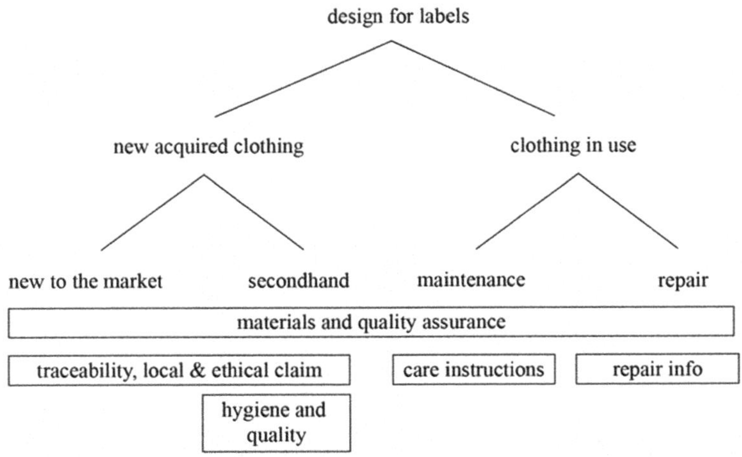


**Fig. 2** Design drivers for different categories of products

Combining these insights enables us to detail the *design drivers* for ideal label design, we can distinguish two categories, as visualized in Fig. 2: design drivers for labels for (newly) acquired clothing and for clothing in use. *Labels for new clothing* include both new clothing in the market and secondhand clothing. Overall, we noticed that these labels should focus on materials and quality assurance as it is the aspect consumers focus on during the entire use phase. When buying new clothing, we observed a stronger persuasive effect of ethical and local claims, as the ITB of labels with an environmental claim does not increase. Labels with a local claim have the highest WTP and labels with an ethical claim have the highest ITB, especially with an additional traceability tag. More specific, for secondhand clothing, focus should also lie on hygiene and quality of the garment. *Labels on in-usage clothing* should serve both for maintenance and repair. In the first case, labels should inform on optimal washing, drying, and other tasks to extend the product lifetime. This can be done by providing spot treatment tips, suggesting ventilation instead of washing, or informing on how to correctly wash a certain material. For repair, a label should provide information on how to easily repair the garment. This could be done by teaching the consumer or showing them where available repair shops are that guarantee quality. Lastly, in the survey we did not question end-of-life behavior. However, we assume that optimal behavior should also be suggested through the label. Further research is needed to detail how these design drivers can be realized and how consumers should be supported in further sustainability and lifetime extension activities.

## 6 Conclusion

This paper looks into the optimization of eco-labels and how they can be designed to increase sustainable purchase choices of clothing and into opportunities to use eco-labels for increasing the reuse of fashion items. When acquiring new clothing (new to the market or secondhand), we found that a traceability tag is the best way to persuade the consumer. Our research helps to detect on "what to trace" and which information can be followed up. We see that the willingness to pay more is the highest for labels with a local claim and the intention to buy is the highest for labels with an ethical claim. Labels for secondhand should focus on hygiene and quality assurance. Labels for clothing in use should contain information on how to correctly wash clothing (maintenance) and where or how clothing can be repaired, with the goal of safeguarding quality.

Aside from optimized redesign of clothing labels, future research is needed to understand (i) not only the objective quality but also the subjective quality perception of consumers, (ii) the resale value of clothes (including hygiene and quality), (iii) how washing instruction should be redesigned to wash smarter and less (as the washing process has the highest influence on quality), and (iv) how and where repair should be reintroduced to enable lifetime extension.

Further research may also elaborate on the pathways toward the persuasive effects of eco-labels on different consumer reactions by looking at mediators and moderators that provide us with more in depth insights. In addition, further research is also needed to systematically understand how consumers can be supported in all activities toward lifetime extension of clothes.

## References

1. McKinsey (2020) Fashion on climate: how the fashion industry can urgently act to Recude its greenhouse gas emissions
2. Niinimäki K, Peters G, Dahlbo H, Perry P, Rissanen T, Gwilt A (2020) The environmental price of fast fashion. Nat Rev Earth Environ 1:189–200
3. Simpliciano L, Galvin M, Barry C, Williot D (2022) Fashion transparency index
4. Mukherjee S (2015) Environmental and social impact of fashion: towards an eco-friendly, ethical fashion
5. Ellen Macarthur Foundation (2017) A new textiles economy: redesigning fashion's future
6. Obser S (2015) Transparency and traceability in the textile and clothing supply chain
7. Zijlstra Y (2023) Sustainable development goals voor de mode
8. United Nations (2023) The 17 goals
9. Amed I, Berg A, Balchandani A, Andersson J, Hedrich S, Young R, Beltrami M, Kim D, Rölkens F (2019) The State of Fashion 2019
10. Amed I, Berg A, Balchandani A, Hedrich S, Rölkens F, Young R, Poojara S (2018) The State of Fashion 2020
11. Goworek H, Fisher T, Cooper T, Woodward S, Hiller A (2012) The sustainable clothing market: an evaluation of potential strategies for UK retailers. Int J Retail Distrib Manag 40: 935–955

12. Clean Clothes Campain (2023) Rana Plaza
13. Ellen McArthur Foundation (2019) The butterfly diagram: visualising the circular economy
14. Haase LM, Lythje LS (2022) User strategies for prolonging product lifetimes: a new starting point for circular conceptual design. Sustain For 14:15133
15. Patagonia (2023) Repair portal
16. Ruben A, Dersarkissian S, Ellis S, Sierfert R, Sigenlaub Y, Cohen N, Fleming J (2023) The brand resale index
17. Essentiel Antwerp (2023) RE-SSENTIEL
18. Shen B (2014) Sustainable fashion supply chain: Lessons from H&M. Sustain For 6:6236–6249
19. Dhir A, Talwar S, Sadiq M, Sakashita M, Kaur P (2021) Green apparel buying behaviour: a stimulus–organism–behaviour–consequence (SOBC) perspective on sustainability-oriented consumption in Japan. Bus Strateg Environ 30:3589–3605
20. Musova Z, Musa H, Drugdova J, Lazaroiu G, Alayasa J (2021) Consumer attitudes towards new circular models in the fashion industry. J Competitiveness 13:111–128
21. Blenkhorn DL, MacKenzie HF (Herb) (2017) Categorizing corporate social responsibility (CSR) initiatives in B2B markets: the why, when and how. J Bus Ind Mark 32:1172–1181
22. Jung S, Jin B (2016) From quantity to quality: understanding slow fashion consumers for sustainability and consumer education. Int J Consum Stud 40:410–421
23. Kusá A, Urmínová M (2020) Communication as a part of identity of sustainable subjects in fashion. J Risk Financ Manag 13:305
24. Grolleau G, Ibanez L, Mzoughi N, Teisl M (2016) Helping eco-labels to fulfil their promises. Clim Pol 16:792–802
25. Abrahamse W, Steg L, Vlek C, Rothengatter T (2005) A review of intervention studies aimed at household energy conservation. J Environ Psychol 25:273–291
26. Das M, Herweyers L, Moons I, du Bois E (2021) Strategic design opportunities to increase sustainable fashion awareness and behaviour. In: Proceedings of the design society. Cambridge University Press, pp 2711–2720
27. Simi TB (2009) Eco-labels: trade barriers or trade facilitators?
28. White K, Habib R, Hardisty DJ (2019) How to SHIFT consumer Behaviors to be more sustainable: a literature review and guiding framework. J Mark 83:22–49
29. Rex E, Baumann H (2007) Beyond ecolabels: what green marketing can learn from conventional marketing. J Clean Prod 15:567–576
30. Clancy G, Fröling M, Peters G (2015) Ecolabels as drivers of clothing design. J Clean Prod 99: 345–353
31. Fiona, Harris H, Roby S, Dibb (2016) Sustainable clothing: challenges barriers and interventions for encouraging more sustainable consumer behaviour Abstract International Journal of Consumer Studies 40(3) 309-318 10.1111/ijcs.2016.40.issue-3 10.1111/ijcs.12257
32. Ralph E, Horne (2009) Limits to labels: The role of eco-labels in the assessment of product sustainability and routes to sustainable consumption Abstract International Journal of Consumer Studies 33(2) 175-182 10.1111/ijc.2009.33.issue-2 10.1111/j.1470-6431.2009.00752.x
33. Koszewska M (2011) Social and eco-labelling of textile and clothing goods as means of communication and product differentiation. Fibres Text East Eur 19:20–26
34. Rossi C, Rivetti F (2020) Assessing Young consumers' responses to sustainable labels: insights from a factorial experiment in Italy. Sustain For 12:10115
35. Brécard D (2014) Consumer confusion over the profusion of eco-labels: lessons from a double differentiation model. Resour Energy Econ 37:64–84
36. Lyon TP, Montgomery AW (2015) The means and end of greenwash. Organ Environ 28:223–249
37. Connell KYH (2010) Internal and external barriers to eco-conscious apparel acquisition. Int J Consum Stud 34:279–286
38. Tang E, Fryxell GE, Chow CSF (2004) Visual and verbal communication in the Design of eco-Label for green consumer products. J Int Consum Mark 16:85–105

39. Daugbjerg C, Smed S, Andersen LM, Schvartzman Y (2014) Improving eco-labelling as an environmental policy instrument: knowledge, trust and organic consumption. J Environ Policy Plann 16:559–575
40. Thinking Mu (2023) https://thinkingmu.com/en
41. Lee EJ, Bae J, Kim KH (2020) The effect of environmental cues on the purchase intention of sustainable products. J Bus Res 120:425–433
42. Loureiro ML, Umberger WJ (2007) A choice experiment model for beef: what US consumer responses tell US about relative preferences for food safety, country-of-origin labeling and traceability. Food Policy 32:496–514
43. Correia E, Sousa S, Viseu C, Larguinho M (2023) Analysing the influence of green marketing communication in consumers' green purchase behaviour. Int J Environ Res Public Health 20: 1356
44. Maniatis P (2016) Investigating factors influencing consumer decision-making while choosing green products. J Clean Prod 132:215–228
45. Oeko-tex (2021) What OEKO-TEX® Labels Mean and Why They Matter
46. Global Standard gGmbH (2021) GOTS – Philosophy
47. Fair Wear Foundation (2023) Get to know fair wear
48. Fair Trade (2023) Onze Basisprincipes
49. United Nations (2023) The Lazy Person's Guide to Saving the World
50. Henninger CE, Alevizou PJ, Oates CJ (2016) What is sustainable fashion? J Fash Mark Manag 20:400–416
51. McCaffrey SJ, Kurland NB (2015) Does "local" mean ethical? The U.S. "buy local" movement and CSR in SMEs. Organ Environ 28:286–306
52. Eagly AH, Chaiken S (1995) The psychology of attitudes. Psychol Mark 12:459–466
53. Peattie K, Belz F-M (2010) Sustainability marketing – an innovative conception of marketing. Mark Rev St. Gallen 27:8–15
54. D'Souza C, Taghian M, Lamb P (2006) An empirical study on the influence of environmental labels on consumers. Corp Commun Int J 11:162–173
55. Van Amstel M, De Brauw C, Driessen P, Glasbergen P (2007) The reliability of product-specific eco-labels as an agrobiodiversity management instrument. Biodivers Conserv 16:4109–4129
56. van Amstel M, Driessen P, Glasbergen P (2008) Eco-labeling and information asymmetry: a comparison of five eco-labels in The Netherlands. J Clean Prod 16:263–276
57. Brécard D (2017) Consumer misperception of eco-labels, green market structure and welfare. J Regul Econ 51:340–364
58. Lindwall C (2023) What is greenwashing?
59. Lyon TP, Maxwell JW (2011) Greenwash: corporate environmental disclosure under threat of audit. J Econ Manag Strateg 20:3–41
60. Erskine CC, Collins L (1997) Eco-labelling: success or failure? The Environmentalist 17:125–133
61. Peattie K (2001) Towards sustainability: the third age of green marketing. Mark Rev 2:129–146
62. de Boer J (2003) Sustainability labelling schemes: the logic of their claims and their functions for stakeholders. Bus Strateg Environ 12:254–264
63. Thøgersen J (2000) Psychological determinants of paying attention to eco-labels in purchase decisions: model development and multinational validation. J Consum Policy (Dordr) 23:285–313
64. Kilbourne WE (1995) Green advertising: salvation or oxymoron? J Advert 24:7–20
65. Pancer E, McShane L, Noseworthy TJ (2017) Isolated environmental cues and product efficacy penalties: the color green and eco-labels. J Bus Ethics 143:159–177
66. Wei S, Ang T, Jancenelle VE (2018) Willingness to pay more for green products: the interplay of consumer characteristics and customer participation. J Retail Consum Serv 45:230–238
67. Cohn & Wolfe (2010) 2010 ImagePower green brands survey

68. Žurga Z, Forte Tavčer P (2014) Apparel purchasing with consideration of eco-labels among Slovenian consumers. Fibres Text East Eur 22:20–27
69. Ackermann L, Mugge R, Schoormans J (2018) Consumers' perspective on product care: an exploratory study of motivators, ability factors, and triggers. J Clean Prod 183:380–391
70. Dissanayake DGK (2022) Enabling circular fashion through product life extension, pp 21–40
71. Roberts JA (1996) Green consumers in the 1990s: profile and implications for advertising. J Bus Res 36:217–231
72. López-Bonilla LM, López-Bonilla JM (2016) From the new environmental paradigm to the brief ecological paradigm: a revised scale in golf tourism. Anatolia 27:227–236
73. Assael H (1995) Consumer behavior and marketing action, 5th edn. South-Western College Publishing, Cincinnati
74. Weikerstorfer C (2021) Eco-labels in the fashion industry. Johannes Kepler Universität Linz
75. Roozen I, Raedts M, Meijburg L (2021) Do verbal and visual nudges influence consumers' choice for sustainable fashion? J Glob Fash Market 12:327–342
76. Wu X, Xiong J, Yan J, Wang Y (2021) Perceived quality of traceability information and its effect on purchase intention towards organic food. J Mark Manag 37:1267–1286
77. Michie S, van Stralen MM, West R (2011) The behaviour change wheel: a new method for characterising and designing behaviour change interventions. Implement Sci 6:42
78. Taylor S, Todd P (1995) Decomposition and crossover effects in the theory of planned behavior: a study of consumer adoption intentions. Int J Res Mark 12:137–155
79. Klöckner CA, Blöbaum A (2010) A comprehensive action determination model: toward a broader understanding of ecological behaviour using the example of travel mode choice. J Environ Psychol 30:574–586
80. Ditty S, Somers S, Seier B, Dewan K (2020) Why we still need a fashion revolution
81. Hoque MZ, Akhter N, Chowdhury MSR (2022) Consumers' preferences for the traceability information of seafood safety. Food Secur 11:1675
82. Nygaard A, Silkoset R (2022) Sustainable development and greenwashing: how blockchain technology information can empower green consumers. Bus Strategy Environ. https://doi.org/10.1002/bse.3338
83. Aakko M, Niinimäki K (2022) Quality matters: reviewing the connections between perceived quality and clothing use time. J Fash Mark Manag 26:107–125
84. Connor-Crabb A, Rigby ED (2019) Garment quality and sustainability: a user-based approach. Fash Pract 11:346–374
85. Intertek (2019) Care label recommendations: Weaving quality into your clothing, textiles and footwear
86. Coats (2023) Care labels – guide on care labelling systems – Apparel | Coats – Coats
87. Laitala K, Klepp IG, Boks C (2012) Changing laundry habits in Norway. Int J Consum Stud 36:228–237
88. WRAP (2017) Valuing our clothes: the cost of UK fashion
89. Kim I, Jung HJ, Lee Y (2021) Consumers' value and risk perceptions of circular fashion: comparison between secondhand, upcycled, and recycled clothing. Sustainability (Switzerland) 13:1–23
90. Godin L, Laakso S, Sahakian M (2020) Doing laundry in consumption corridors: wellbeing and everyday life, sustainability: science. Pract Policy 16:99–113
91. Lin J, Iyer M (2007) Cold or hot wash: technological choices, cultural change, and their impact on clothes-washing energy use in China. Energy Policy 35:3046–3052
92. da Costa Junior JC, da Silva Nascimento L, da Silva RVAF, de Barros Jeronimo T, Melo FVS, Primo MAM (2021) New clues on contamination and perceived quality of second hand clothing. Navus Revista de Gestão e Tecnologia 11:01–12

# An Analysis of a Circular Economy Strategy Combining Green Packaging Reuse Design and Sustainable Jewelry Development

**Yu-Chen Huang and Tzu-Yun Chen**

**Abstract** The fashion industry has set SDG 12, "Ensure Sustainable Consumption and Production Patterns," as a transformation goal, promoting the concept of "cradle-to-cradle design" by using urban mining to recycle precious metal materials and combining packaging design as a product to reduce environmental impact and create a sustainable and friendly circular economy model.

This study explores the possibility of developing sustainable jewelry through the combination of a green packaging reuse design and a circular economy. First, through literature review and expert interviews, the strategies for a green packaging reuse design and a circular economy model for sustainable jewelry will be established, including "packaging design and material selection," "establishing the concept of recycling and circular economy with consumers," and "achieving a sustainable economy." Then, a survey will be conducted among female consumers aged 19–30 in Taiwan to investigate their willingness to purchase sustainable jewelry. Through analyzing the needs, psychology, and behavior of the jewelry consumer group, a people-oriented green packaging reuse design, and circular economy strategy and design guidelines will be proposed.

**Keywords** Circular economy · Green packaging design · Sustainable jewelry · Urban mining · Fashion industry

## 1 Introduction

Since the beginning of industrialization, humans have consumed a lot of energy and resources in producing what we need and have mainly operated in a linear economic model. Raw materials continuously generate waste throughout mining and making goods and are eventually discarded directly without recovery—with most of the

Y.-C. Huang · T.-Y. Chen (✉)
Department of Crafts and Creative Design, National University of Kaohsiung, Kaohsiung, Taiwan
e-mail: a1091348@mail.nuk.edu.tw

 
S. Fukushige et al. (eds.), *EcoDesign for Circular Value Creation: Volume I*,
https://doi.org/10.1007/978-981-97-9068-5_9

energy and resources being used only once. To create a friendly environment, it is imperative to construct a green packaging reuse design strategy and a sustainable jewelry circular economy model.

The basic concept of a circular economy is the circulation of resources, i.e., resources can be reused and recycled in multiple life cycles to reduce waste emissions as much as possible. The core concept is maximizing resources' value while minimizing environmental impact. This study will use relevant concepts such as urban mining and green packaging utilization, combined with expert interviews and surveys, to propose a human-centered circular economy and design strategy for the green packaging reuse design combined with sustainable jewelry.

Urban mining, also known as raw material recovery, was introduced by Professor Nanjo of the Research Institute of Mineral Processing and Refining at Tohoku University in 1988 to replace the term "waste disposal." Urban mining is the process of actively recovering precious metals and raw materials, which can transform the damage caused by waste into secondary material sources through various methods, forming a circular economy [1]. In Asia, for example, Taiwan produces approximately 19.13 kg of electronic waste per person per year, second only to Hong Kong's 21.7 kg and Singapore's 19.95 kg [2]. The manufacturing industry's main output and export are items such as electronic components, semiconductors, and circuit boards (the "3C Products") [3]. However, the global recovery rate of electronic waste is only 17.4% [4]. If the electronic waste produced by the 3C Products can be transformed into urban mines, the extracted precious metals such as gold, silver, copper, nickel, and tin can be reused by the semiconductor industry, which can reduce production costs while making effective use of metal resources.

Electronic waste contains precious metals and has become a strategic resource that countries strive to possess. However, the precious metal extraction process used in the past often involved highly corrosive methods such as incineration, cyanide, and aqua regia, which emitted toxic gases and waste solutions that significantly harmed the environment and human health. Only recently, eco-friendly, nontoxic methods were developed and promoted, such as eco-friendly chemical tin stripping technology, eco-friendly electrolytic gold stripping technology, and eco-friendly plating solutions. The plating solution, composed of organic substances, can be reused by simply filtering with filter paper after use and achieves environmental protection goals. Although these types of eco-friendly recovery methods come with high costs, their level of toxicity and corrosiveness is far lower than old methods and, therefore, are pretty friendly to the environment [3].

Since the 1960s, the third industrial revolution has spurred economic and societal growth. The advancement in production technologies has led to the democratization of luxury goods, with many brands venturing into the luxury market. However, environmental degradation has become a consequential issue during extraction in the quest to offer high-quality, rare, material-based luxury items. Luxury products, known for showcasing excellence and supreme quality, should not entail environmental and societal burdens. Among luxury categories, the jewelry industry, long perceived as a nexus of art and craftsmanship imbued with notions of value and perpetuity, has begun to contemplate the sustainability of its products and reflect on

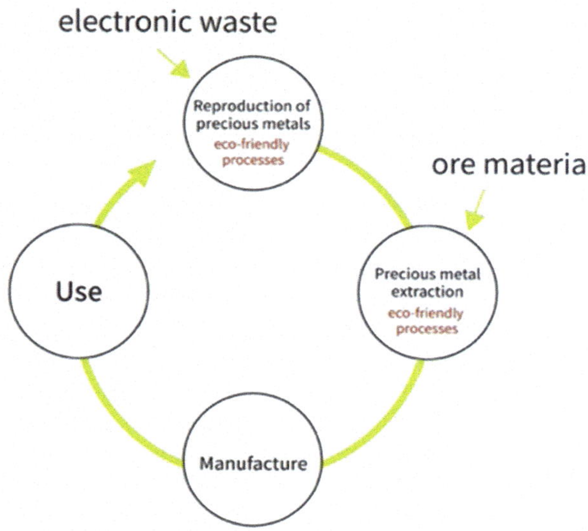


**Fig. 1** Jewelry circular economy

resource utilization, according to studies conducted by PwC Italia titled "La sostenibilità cattura Millennials e GenZ" and Ipsos' 2018 CSR Exhibition and Social Innovation survey, understanding consumer purchasing behavior revealed that sustainability is crucial for younger consumers. The data indicates that consumers are willing to pay an average of 10% more for products ensuring sustainability and high quality [5]. Considering this, unique jewelry items embody the essence of this research on urban mining and sustainability. This paper proposes a preliminary design strategy, encompassing precious metal recycling and green packaging design, to align the luxury jewelry sector with sustainable and circular economic principles.

Let us first explain the recovery and recurrence of precious metals: To implement the concept of recovery, electronic waste will be recycled and converted into secondary materials for use; low-carbon, eco-friendly, and organic, non-toxic processes or solutions should be used in the precious metal recovery; alternatively, corresponding production processes and special recovery solutions should be developed as much as possible for different metal materials, to minimize environmental harm generated during the recovery process. The following figure shows the jewelry circular economy in precious metal recovery (Fig. 1).

The feasibility of a circular economy in the jewelry sector is epitomized by Pandora, the world's largest jewelry manufacturer, which has embarked on a journey towards becoming a low-carbon enterprise. The company has publicly committed to achieving net-zero carbon emissions by 2025 and halving its greenhouse gas emissions across its operations by 2030. Pandora plans to manufacture jewelry using 100% recycled gold and silver by 2025 to fulfill these ambitions. Official reports in 2019 disclosed that 71% of the gold and silver used in Pandora's products were already sourced from recycled materials, significantly reducing the environmental footprint. The environmental benefits extend beyond the mere reduction of waste, also encompassing a lower level of water and air pollution compared to the conventional practices of metal extraction. Procurement of recycled materials is projected to cut carbon emissions from silver by two-thirds and gold by over 99% [6]. This case sheds light on how strategic sustainability initiatives can drive a circular economy in the jewelry industry, setting a precedent for other players in the luxury goods market to follow suit.

Next, we discuss green packaging design. Green packaging design primarily focuses on creating reusable packaging and its integration with the product. The core value of green packaging design is "cradle to cradle" [7]. The selection of packaging materials, structure, and recycling and regeneration methods should all be considered from the initial design stages. Scholars have differing opinions on the definition of "green design." For instance, Zeng Hanshou (2001) believes that design should have an endpoint as the goal, focusing on the product's life cycle and the direction of waste to avoid causing harm to the environment. To this end, he proposed the 5R principles: Reduce, Reuse, Recycle, Recovery, and Research as the design philosophy [8]. Another scholar emphasized 'Recyclable, low pollution, energy saving.' Green design should fully consider reducing waste generation and using the best materials to alleviate the exhaustion of the Earth's resources [9].

Among them, the methods of reducing packaging in green design can be divided into the following seven categories: simple packaging, reducing material usage, compression packaging, choosing biodegradable materials, unifying packaging materials, collective packaging, and reducing printing [10].

In recent years, biodegradable materials have become the primary packaging materials for designers pursuing circular sustainability. After being used to create packaging materials, the biodegradable materials can return to their original form once the packaging is recycled (see Fig. 2). The materials designers select should be used to create reusable packaging as much as possible, such as bottles and boxes, and refilled through bulk distribution. Achieving circular benefits from green packaging requires manufacturers and designers to consider packaging materials and

**Fig. 2** Example of material recycle process diagram [11]

appearance more and, even more importantly, the consumer's willingness to purchase. Consumers must easily distinguish between original and refillable packaging to increase their willingness to buy a product.

After understanding the characteristics of packaging materials, the diversity of packaging structures is also an aspect that designers can consider further. The concept of "the packaging is the product" can be used to increase the value and significance of the packaging. There are three methods: (1) following the original method: reusing in a way that is close to the original packaging, (2) the creative way: creative design transforms the function of the packaging [12], and (3) the instructional way: manufacturers clearly define the function of the packaging. Take the MINIWIZ research and development company as an example:

We are obsessed with making the circular economy a reality in everyday consumption. We live to support the mass adoption of a circular system whereby all the materials we use are reused, again and again, and again with zero waste.

This represents the business philosophy of MINIWIZ. They designed an eco-friendly wine bottle packaging called "RE-Wine Classic," made from recycled PET plastic and discarded rice straw fibers. The packaging sleeve is designed to become a stackable wine bottle rack, allowing consumers to assemble it freely. Clever structural arrangements can also transform the packaging into a table lamp by adding a light fixture. Integrating the concept of green design throughout the product design process avoids waste generation and extends the product's life cycle, achieving the goal of raising consumers' sustainability awareness (see Figs. 3 and 4) [13].

In the initial stage of designing jewelry packaging, the product's endpoint should be set as the design goal. Adhering to the 5R principles of green design, biodegradable materials should be selected, and the packaging should be integrated with the product design to avoid resource waste and consumption. This also endows the final packaging product with added value and the capability of recycling and remanufacturing. The following image represents the circular economy diagram of the jewelry packaging design (Fig. 5):

**Fig. 3** Storage rack for wine bottle packaging [14]

**Fig. 4** Wine bottle packaging lamp [14]

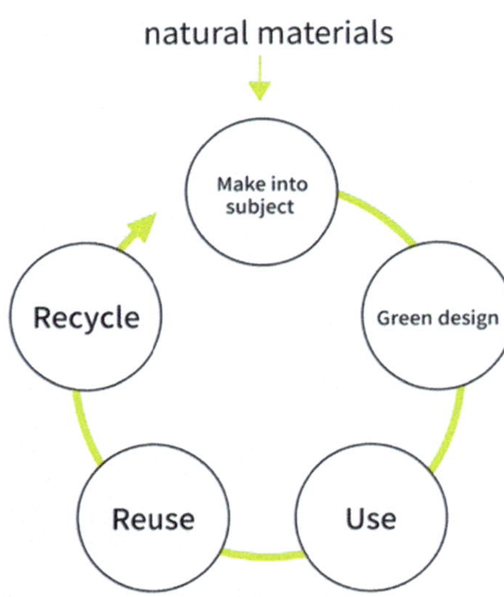


**Fig. 5** Circular economy of jewelry packaging

Precious metal recycling products and green packaging should provide clear instructions on disassembly, reuse, and recycling methods. Appropriate incentives should be given to consumers who participate in recycling to stimulate their willingness to repurchase and support sustainable products. This way, both the product and its packaging can ultimately achieve a bio-cycle and industrial cycle following the circular economy system (see Fig. 6).

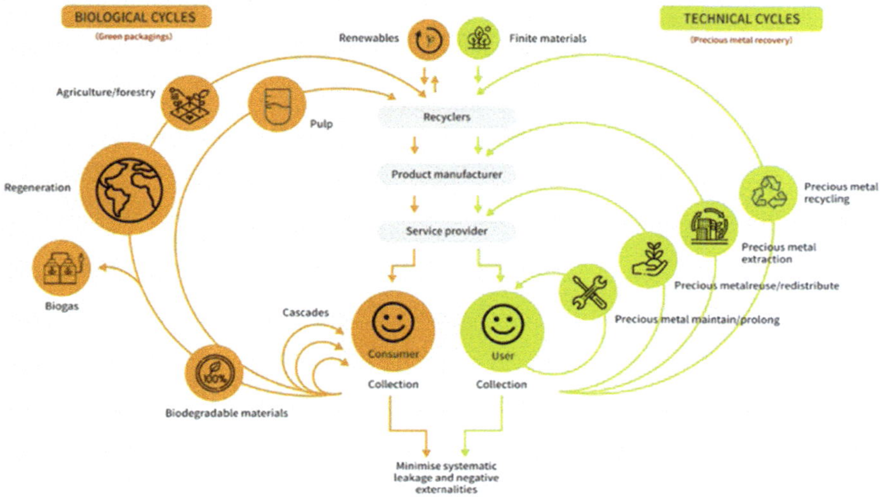


**Fig. 6** Circular economy diagram of green packaging reuse design with respect to sustainable jewelry

## 2 Method

This study was conducted through a literature review, expert interviews, and surveys. By collecting domestic and foreign academic literature and journals, the research explores the potential of combining urban mining with green packaging reuse design to achieve a circular economy. Expert interviews were also conducted on precious metal recycling to understand the current market conditions and future development trends. Finally, through an online survey targeted at Taiwanese females aged 19–30, who regularly purchase nonexpensive jewelry, the study delves into the future development possibilities of precious metal recycling, jewelry, and green packaging. This is accomplished through examining environmental awareness issues, purchase intention surveys, and attitudes toward green design topics (Fig. 7).

The expert interviewed in this study is Mr. Peng Kun-Zeng, an Associate Professor in the Department of Materials Engineering at the Ming Chi University of Technology. His main research areas include applying and analyzing electrochemical methods, environmentally friendly solutions for metal surface anodizing, metal smelting, and casting processes. He has published articles on bio-assisted, environmentally friendly electroplating solutions and related topics. In addition to his academic work, he has also gained industry experience, providing academic and industry insights on precious metal recycling for this study.

By collecting data on urban mining, green packaging, and other topics relevant to this study in advance, followed by a survey, we aim to understand the real-world scenario and the public's willingness to purchase green products (see Fig. 8).

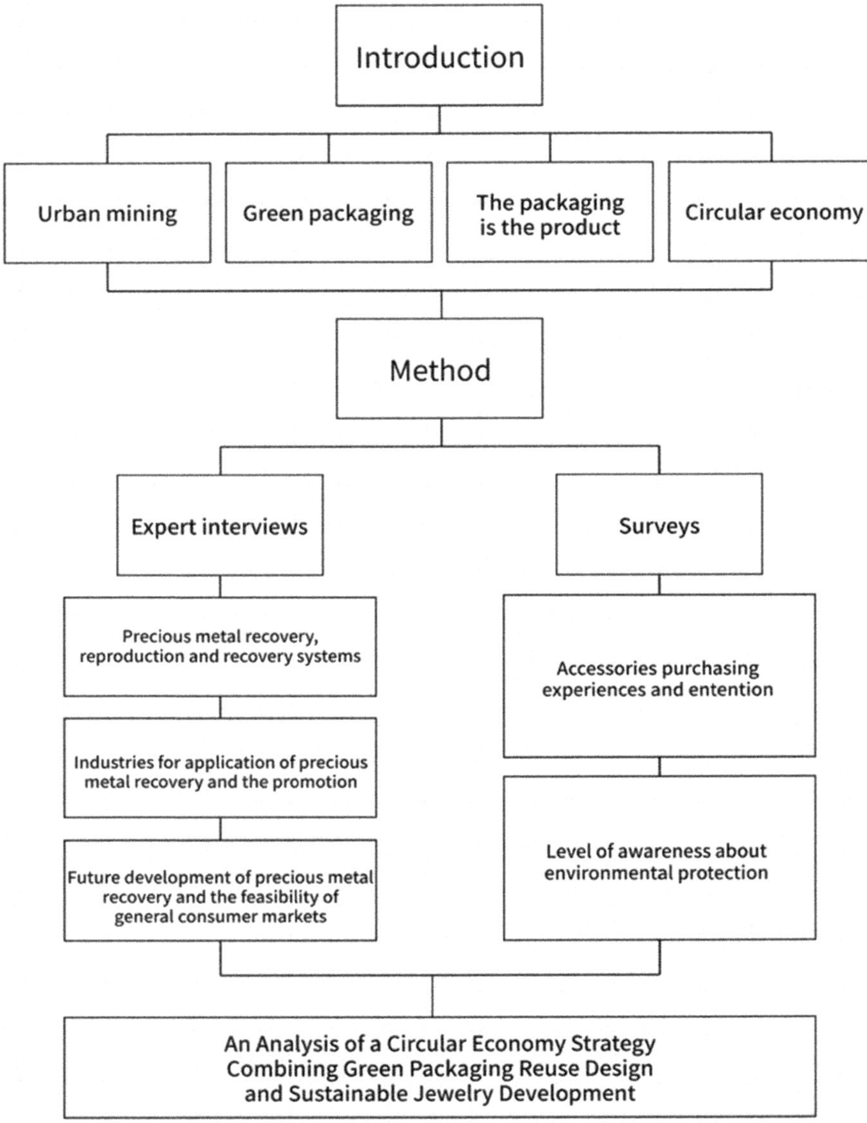


**Fig. 7** Research structure diagram

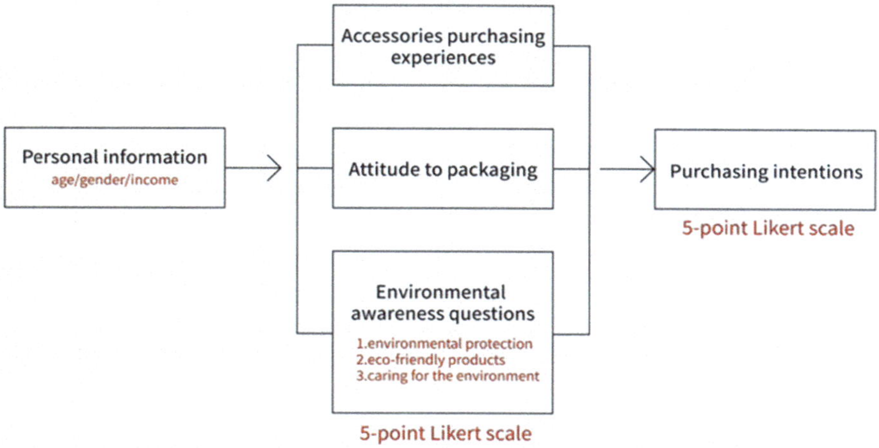


**Fig. 8** Structure diagram of survey

# 3 Results

This study uses data collection, expert interviews, and survey results to explore the feasibility of a circular economy strategy combining green packaging reuse design with sustainable jewelry. The following will be discussed in three sections.

## *3.1 Results of Data Collection*

To achieve resource sustainability, Taiwan began promoting "Green Living for All" in 2020, covering five aspects: green eating, green travel, green consumption, green home, and green office. Among them, green consumption aims to encourage the public to consume moderately, reduce the source, and prioritize green products with a lower environmental impact. Specific measures include reducing packaging on e-commerce platforms, increasing the number of green products, promoting green consumption, and promoting environmental reward programs. As of March 2021, 580,000 people have participated [15].

In line with government policies, the combination of precious metal recycling and green packaging design not only advocates the effective circular use of resources but also minimizes the waste from the manufacturing process and maximizes the packaging design features to attract consumers' support and purchase.

## 3.2 Expert Interview Results

After interviews with professors, this study has gained a deeper understanding of the subject of precious metal recovery, which is explored through the following three main aspects:

A. Issues related to precious metal recovery, reproduction, and recovery systems:

What are the types and necessary conditions for recovering and reproducing precious metals? How to dispose of waste liquids during the recovery process?

*For precious metals, it is mainly about the substance to be recovered. Solid metal recovery is generally recovered through the so-called "electroplating method,"* i.e., *the act of dissolving and analyzing the precious metal in an aqueous solution and then performing the so-called "extraction." As for the shape you would want it to take, it depends on the mold you use.*

*The critical difference lies in the purity of the extracted metal. The purer the metal, the better its processing properties. The rest depends on what you want and how pure you want it to be before you extract that type of metal.*

*There are designated waste liquid recovery units for precious metal recovery and special drug recovery. However, some manufacturers still choose to illegally discharge waste liquid, because the price of eco-friendly solutions is slightly higher than that of toxic substances, which is not cost-effective.*

B. Issues related to the industries for the application of precious metal recovery and promotion:

How is precious metal recovery promoted in the current market? What are the main recovery items?

*Currently, most of the precious metal recovery is still limited to gold. However, other precious metals are still recovered on a lesser scale. Silver, gold, platinum, copper, and nickel vary in recovery methods and costs. Therefore, recovering precious metals other than gold can be cost-effective is still under development.*

*As for the method of promotion, the one adopted by Uwin Nanotech Co., Ltd. is exposure, then authorization and patenting, conducting so-called "green environmental education," and actively participating in government competitions to increase exposure.*

C. Issues related to the future development of precious metal recovery and the feasibility of general consumer markets:

Which area should be the focus to achieve sustainable development and a circular economy?

*I once asked my students how long their mobile phones' tenure of use is. It appeared to be around three years. However, few people would recycle their phones—they'd be left at home or in the trash can. This is mainly due to the willingness of the public to recycle and the inadequate government recovery mechanism. There should be more promotions, and at the technical level, many solutions suitable for recovering different metals should be developed. Currently, in Taiwan, the market is not large enough. Despite the many 3C Products, the main customers that need to be targeted are not in Taiwan. Foreign manufacturers are the largest customers. No company can enact a recovery process if it does not have a certain scale and quantity.*

Based on the above interviews, it can be preliminarily inferred that recovering precious metals to reproduce jewelry or other items is feasible. However, it is not widely applied because the general market in Taiwan is not yet familiar with it, resulting in a generally low level of acceptance. The future of the precious metal recovery market will rely on improvement in technology at the enterprise level and promotion and cooperation with the government; enhancing public awareness is also a project worth in-depth exploration.

## 3.3 Survey Data Analysis

A total of 70 questionnaires were collected for this study, of which 42 valid questionnaires were from women between 19 and 30 years old. The results of the survey will be analyzed using descriptive statistics.

Most respondents' monthly income is under NTD 25,000, accounting for about 80%. However, over 90% had purchased jewelry within the past year, indicating that monthly income does not necessarily impact one's decision to purchase jewelry.

Regarding the questions about jewelry purchasing experiences, the survey results showed that most women purchase jewelry to match their outfits, followed by gifting. Over half of the women surveyed purchase jewelry 3–6 times yearly. Over 70% of women wanted different styles when buying new jewelry, and the amount spent on jewelry varied from NTD 500 to NTD 3000, accounting for over 60% of respondents. From this, it can be inferred that this demographic is indeed the primary consumer group for jewelry. The primary types of jewelry purchased were rings, followed by necklaces. These results can serve as a reference for future product development.

Next, the survey explored questions regarding jewelry packaging: According to the statistics, over 70% of women would be influenced by the packaging design when purchasing jewelry. Around 60% preferred sophisticated packaging over simple packaging. This indicates that most consumers still value the appearance and design of the packaging. Only a minority choose products that omit unnecessary packaging materials. When asked further about their views on eco-friendly packaging, 70% of respondents preferred packaging that could be disassembled for other uses. Over 60% preferred packaging that could be cleaned, disinfected, and reused.

Through a series of questions regarding environmental awareness, the survey investigated public opinions on "environmental protection," "eco-friendly products," and "caring for the environment." Generally, the public has a relatively high level of awareness about environmental protection, followed by an awareness of eco-friendly products. It can be inferred that the public still has concerns about the price and quality of eco-friendly products, green packaging, and precious metal recycling. Regarding caring for the environment, most people's awareness level was moderate, with fewer people actively caring for the environment and participating in activities.

Lastly, the survey on purchasing intentions found that the market consumers have a relatively high acceptance of green and "naked" packaging. Regarding recycled precious metal jewelry, even though over 60% of respondents had rarely purchased it before, over 60% were willing to try it in the future. Also, they were highly aware of sustainable enterprises and willing to share them with people around them. This shows that most people are willing to contribute to environmental protection as much as possible.

## 4 Discussion

After comparing and analyzing the literature review, expert interviews, survey results, and real-world cases, we have made several findings as follows:

Precious metal recycling in Taiwan mainly focuses on gold, silver, palladium, platinum, tin, nickel, etc.; the technology is well-developed (see Fig. 9) [16].

According to the results of expert interviews, many more environmentally friendly refining methods, such as eco-friendly and organic solutions, exist than those that use toxic solutions during the metal refining process. However, literature reviews show that many domestic operators still predominantly use toxic liquids for refining, indicating that the industry still needs further research and improvement regarding cost consumption and environmental impact during refining.

The burgeoning issue of electronic waste (e-waste) presents a global challenge with significant implications for sustainable and circular economic practices.

| Types of Recycling | Type | Number of vendors | Product |
|---|---|---|---|
| Permitted Recycling | Waste containing gold-palladium, waste solution, printed circuit board waste, spent filters, waste ion exchange resin. | 7 | Precious metal ingots (gold, palladium, platinum, rhodium, ruthenium, silver), potassium gold cyanide. |
| | Waste indium-tin solution, sludge, sandblasting waste. | 2 | Indium ingots. |
| | Waste developing solution, waste fixing solution, waste photographic films, waste silver photo paper, waste containing silver, waste solution. | 3 | Silver ingots. |
| | Copper sludge, waste tin-lead slag, waste nickel solution. | 4 | Copper oxide, tin ingots, nickel ingots. |
| Processing Facility | Waste containing gold-palladium, waste solution, printed circuit board waste, spent filters, waste ion exchange resin. | 12 | Precious metal ingots (gold, palladium, platinum, rhodium, ruthenium, silver). |
| | Waste developing solution, waste fixing solution, waste photographic films, waste silver photo paper, waste containing silver, waste solution. | 4 | Silver bullion, silver ingots. |
| | Copper sludge, waste tin-lead slag, waste nickel solution. | 8 | Copper oxide, nickel dimethylglyoxime, nickel ingots. |

**Fig. 9** Reuse of rare and precious metals in Taiwan [16]

According to global e-waste studies in 2019, the world generated 53.6 million tons of electronic waste, with projections surging to 74 million tons by 2030. Asia led the pack in 2019 by producing 24.9 million tons of e-waste, followed by the Americas, Europe, Africa, and Oceania. However, 17.4% of this e-waste was documented and recycled [17]. The unaccounted majority likely undergoes disposal or recycling through informal channels, such as incineration. The improperly managed e-waste represents a lost economic opportunity estimated at USD 62.5 billion [18]. Even though numerous valuable resources are present in waste, the lack of recycling willingness among consumers hinders the effective reutilization of these resources.

Finally, according to the survey results, 60% of the respondents need clarification about the 5R principles of green design, and they need to understand the purpose of precious metal recycling. Even as businesses and designers gravitate toward green design, consumers are merely aware of it and hold reservations about green products, thus naturally inhibiting the desire to purchase.

Thus, based on the literature and practical considerations, it was found that the public needs to gain awareness of green design and precious metal recycling. Therefore, this study proposes a people-centered green packaging design and a sustainable jewelry circular economy strategy. By using green packaging design and repurposing precious metals from commonly purchased jewelry types, consumers are provided with clear instructions and information about the production process. This approach assists consumers in implementing recycling or reuse methods after purchase and provides incentives. Apart from encouraging consumers to repurchase eco-friendly products, this strategy enhances the likelihood of their active interest in other sustainable products, thus expanding the field of precious metal recycling into the general consumer market.

The current discourse on precious metal recycling for jewelry production gravitates towards optimizing cost-effectiveness, with a concerted effort underway to unearth viable recycling methodologies for various metals. This initiative stems from the recognition that electronic waste harbors a higher concentration of precious metals than ores, making urban mining a more economical avenue for metal extraction than traditional mining. Moreover, recycled metals are more energy efficient than producing from virgin ore [19]. The Pandora's sustainability report indicated, approximately 90% of their emissions are attributed to the raw materials used in jewelry and the associated transportation emissions. Hence, collaborations with experts and suppliers have been fostered to explore avenues for reducing carbon emissions. Lifecycle assessments reveal that procuring recycled metals substantially reduces carbon emissions by about a third for silver and over 99% for gold, marking a notable reduction in Pandora's value chain carbon emissions [18]. This study posits "An Analysis of a Circular Economy Strategy Combining Green Packaging Reuse Design and Sustainable Jewelry Development "model for the relevant industries, aspiring to catalyze resource reuse and minimize the demand for new resources, aligning with the overarching objective of fostering a circular economy.

# 5 Conclusion

This study explores the current state and possibilities of urban mining in Taiwan, understanding consumer environmental consciousness and purchasing intentions, analyzing the approaches of globally renowned jewelers towards sustainable jewelry, and understanding the industry conditions through expert interviews, a set of strategies termed as "An Analysis of a Circular Economy Strategy Combining Green Packaging Reuse Design and Sustainable Jewelry Development" in response to sustainable development trends. It also proposes several resolutions based on the current market conditions:

1. Maximize recycling efficiency using eco-friendly or organic solvents:

   Due to insufficient funds, precious metal recycling companies often use acidic solids or essential solutions for metal refining. The waste solution produced not only requires a large amount of water to dilute it before being dumped into nature but may also be converted from liquid to solid form and arbitrarily discarded. To effectively reuse resources, some organic solvents are cheaper than eco-friendly solvents, making them a more viable investment for small and medium-sized enterprises.
2. Develop core technologies to enhance competitiveness:

   Precious metal recycling aspect: At this stage, in Taiwan, precious metal recycling only includes gold, silver, copper, nickel, tin, etc. Additionally, many rare metal resources are still sourced from overseas markets. If businesses devote their efforts to developing innovative solutions, and the government provides suitable subsidy programs, other rare precious metals in the electronic waste can be refined and remanufactured, enhancing Taiwan's competitiveness in the international market.

   Green packaging design aspect: Green packaging is one of the main projects of "National Green Living" currently promoted by the Taiwanese government [15]. Designers should choose eco-friendly, recycled materials from the initial stages, coupled with innovative design models, to integrate the packaging design with the product using minimal materials. The design should be interactive and capable of returning to the recycling system after use.
3. Raise public environmental awareness and promote the design strategy of "People-centered Green Packaging Design Reuse Combined with a Sustainable Jewelry Circular Economy".

   Most people generally have essential environmental awareness. The government and businesses should focus on developing people-centered green packaging designs combined with precious metal products. This study proposes several policy recommendations for this strategy:

   A. Clear instructions on green packaging, such as QR codes. The green packaging should comply with the 5R principles, enabling users to disassemble it for recycling or other purposes after use. Additionally, clear instructions should be provided, on the packaging, introducing the importance of precious metal recycling and green packaging for the environment in a simple and easy-to-understand way. This allows consumers to have greater recognition and confidence in green products.

B. Choose products with higher market acceptance for sustainable design. In the stage where the general consumer market is not very familiar with precious metal recycling, it is recommended that businesses use the product with the highest consumer acceptance for sustainable design. For example, survey results show that the most frequently purchased jewelry item is a "ring." Manufacturers can use this as a prototype, make recycled precious metal jewelry, and use the merchant's promotion to stimulate the leading consumer group to try new styles and types within the same jewelry category.

C. Sustainable product reward mechanism. It is recommended that businesses and governments jointly provide suitable discounts and offers to consumers who purchase sustainable products. They should also offer consumers the right to track the subsequent direction of the actively recycled packaging and precious metals. When consumers realize the impact of their behavior on the environment, reasonable offers can stimulate their willingness to purchase sustainable products again. It can also enhance public evaluations of green products, further proactively sharing with more people and ultimately achieving the goal of a sustainable circular economy.

## References

1. Kazançoglu Y, Ada E, Ozturkoglu Y, Ozbiltekin M (2020) Analysis of the barriers to urban mining for resource melioration in emerging economies. Resour Policy 68:101768
2. COVESTRO CEO Tang Peide: Solving E-Waste Starts with Material Disclosure (2017). https://www.cw.com.tw/article/5126119. Accessed 4 June 2023 [in Chinese]
3. Xiao Kang M, Jing Xiang X (2015) Green circular economy industry: recycling and reusing precious metals from electronic waste. https://www.uwin-nano.com/pdf/a10.pdf. Accessed 4 June 2023 [in Chinese]
4. Taiwan Biomedical Innovation Association (2023) The latest global electronic waste statistics for the year 2023. https://tibia.org.tw/archives/9980# [in Chinese]
5. Miguel Ángel Gardetti, Ivan Coste-Manière (2020) Sustainable Luxury and Craftsmanship, pp 2–4, 7
6. Pandora (2019) Sustainability report 2019, pp 14–15
7. Shi Lun C (2013) The research on green packaging design application- the 2012 "Packaging green miracle" In: Taipei product design improvement project (Master's thesis), China University of Technology [in Chinese]
8. Zeng H (2009) Green products and EuP and EPD. Sustain Ind Dev 47:29–33
9. Jui Che T, Jyh Her S (1999) The study of notebook personal Computer Design of Green Assembly and Disassembly. J Des 4(1):101–120. [in Chinese]
10. Zhong Hui C, Guan Ying L, Jia Yi C (2017) Application of green packaging design and printing materials. Print Technol 33(1):66–82. [in Chinese]
11. Zhu Z, Liu W, Ye S, Batista L (2022) Packaging design for the circular economy: a systematic review. Sustain Product Consum 32:817–832
12. Liu Y (2013) Product design and creation of integrated packaging concept, pp 26, 38–43
13. About MINIWIZ. https://www.miniwiz.com/about.php. Accessed 4 June 2023

14. Pro S (2011) Taiwanese team's smart-designed table lamp for wine bottle packaging, https://www.damanwoo.com/article/51468. Accessed 4 June 2023 [in Chinese]
15. Republic of China Executive Yuan (2021) Promoting green living for all. https://www.ey.gov.tw/Page/5A8A0CB5B41DA11E/8eba52d8-958e-4393-a5d7-11e6e757f63f. Accessed 4 June 2023 [in Chinese]
16. Fen Gqi Y, Liang Dong C (2014) Current status and development strategies of rare precious metal recycling in Taiwan. Ind Pollut Prev Control 130:88–108. [in Chinese]
17. The United Nations Global E-Waste Monitor (2020) A collaborative product of global E-Waste Statistics Partnership (GESP), the International Telecommunication Union (ITU), and the International Solid Waste Association (ISWA). Available online: http://E-wastemonitor.info/. Accessed 20 Sept 2023
18. Garam B et al (2019) A new circular vision for electronics, time for a global reboot. In: PACE and world economic forum, pp 5–6
19. Kamachi Mudali U et al (2021) Review on E-waste recycling, part I—A prospective Urban Mining opportunity and challenges. Trans Indian Natl Acad Eng 6:9–10

# Exploring Suitable Growing Parameters of Mycelium Prototyping for Art Installations

**Tarteel Abdelsadek, Kazutoshi Tsuda, and Tomohiro Inoue**

**Abstract** This paper aims to extend the limitations of the biomaterials field to be applied in the art/design field. Nowadays, art installations mean a pile of plastics and toxins in the atmosphere. However, have you ever thought of creating installations from bio-based natural microorganisms? The previous research in the past decade have shown clear hope in using mycelium in the design field. This study documents an experimental study to create art installations from the mycelium of *reishi* mushroom species belonging to the genus *Ganoderma*. We will go through creating the mycelium mixture and molding it then finally baking it. Furthermore, we will discuss the adequate growing parameters and the best culture media for its growth. First, we will use malt, yeast, and glucose culture media and then compare it to the flour cornstarch mixture. We concluded that the flour and cornstarch mixture had better and faster results. In conclusion, our findings indicate the accessibility of using mycelium in art installations and that it has several advantages over the plastic ones in terms of sustainability and creating a zero-waste society.

**Keywords** Mycelium · Art installations · Art/design · Biomaterials · Bio-design · Mushroom · Culture media · *Reishi* · Zero-waste design

## 1 Introduction

Waste produced by mankind globally has been on the increase since 2009, this is not only limited to plastic waste, but also including consumption waste and agricultural waste. According to the researchers, by 2050 based on the current trends the plastic waste alone is assumed to be around 26 billion metric tons [1, 2]. Needless to say,

T. Abdelsadek (✉)
Design department, Kyoto Institute of Technology, Kyoto, Japan

K. Tsuda
Center for the Possible Futures, Kyoto Institute of Technology, Kyoto, Japan

T. Inoue
KYOTO Design Lab, Kyoto Institute of Technology, Kyoto, Japan

 
S. Fukushige et al. (eds.), *EcoDesign for Circular Value Creation: Volume I*,
https://doi.org/10.1007/978-981-97-9068-5_10

with inefficient recycling, this is still considered a fetal soaring pace for the nondegradable material that was first introduced to humans less than a century ago. Consequently, such waste is forecast to end up in the landfills and various ecosystems such as the oceans and the fertile soils [3, 4].

Shedding the spotlight on Japan, it is safe to say that the country's efficient recycling system and overall waste management—which is ranked the second highest according to the Plastic Management Index (PMI)—did not save it from having the second highest per capita plastic packaging waste after the United States. Having a recycling rate as high as 85.8% and a collection rate of 93% ironically resulted in an enormous amount of 4.1 million tons of plastic waste in 2019 that mostly ended up in the marine ecosystems. This is perhaps due to incineration, which is used intensively for around 88% of the plastic waste, leaving a mere 22% segment for the actual recycling of the plastics [5].

Furthermore, public awareness is quite low among the Japanese, especially young students. According to a study, the average Japanese user consumes around 450 plastic shopping bags annually despite having the bags priced and not given for free [5].

Considerably, we argue that this is no longer the user's problem. Basically, everything around the Japanese people is made out of plastic, ranging from convenience store products and vending machine PET bottles—a total of 23.2 billion bottles annually—to the obentos (packed lunch) and even fresh vegetables and fruits. Ultimately, even if the user wishes to take positive steps towards the plastic consumption, he is not left with many options to select from [5]. It is also worth mentioning that currently we are experiencing the post-COVID-19 plastic packaging damages due to the increase of the disposable containers and cups usage during that time to limit the infection [6].

One of the contributors of plastic waste is art installations; a great part of the art installations and 3D sculptures use plastic heavily. For instance, Yayoi Kusama's fiber-reinforced plastic structure was installed in Naoshima, Seto Inland Sea in 1994 and has been on display ever since. Generally speaking, taking a look at the artworks' life cycle, after the artwork is created and exhibited, it can end up in roughly three cases: the artist's collection, acquired by another art collector, or discarded. When discarded, they are classified into recyclable and nonrecyclable materials. However, even the recyclable materials are often hard to be correctly organized and recycled. Consequently, artistic waste is piling up in landfills.

Beside the plastic waste, another huge factor of the general waste in Japan is the agricultural waste. A lot of enormous amounts of biomass resources such as saw dust, wood, paper, and others are usually discarded without utilizing them enough [7]. With two-thirds of its landmass occupied with forestland, Japan surely has plentiful forest resources, including mushrooms and other materials [8]. According to Harada, the available wood-based resources are approximately 42.7 Mt per year in Japan [9].

Consequently, innovative recycling systems will not efficiently cope with the people's usage, and thus biodegradable materials must be introduced for the public to use. One of the trending alternative biodegradable candidates for the plastic is mycelium. Initiating the usage of mycelium as an alternative material will not only reduce plastic waste but also utilize agricultural wastes. This is because agricultural wastes such as straw or sawdust or wood are mixed in the mycelium substrate [10]. Accordingly, this research paper explored making mycelium art installations using two media and compared their results to better assess mycelium growth in order to be used more in the artistic field.

## 2 Background

### *2.1 Mycelium*

Mycelium is the root-like structure of the fungi. Naturally, the mushrooms growing in the forests have the hyphae—which is the branching structure of mycelium—to digest and absorb surrounding biopolymers [11, 12]. After digesting it, a white fuzzy skin starts to form and later when completed the mushroom caps start to fruit above the ground [13]. On the other hand, laboratories and mycelium farms are ceasing that process mid-way just before the fruiting colonization phase by heating and then drying, thus yielding mycelium in the shape of brick, leather, and many more [14–16].

There are two types of mycelium materials: pure mycelium (PM) and mycelium-based composites (MBCs) [17]. PM is made out of liquid fermentation of the fungal-grown culture through the upper hyphae surface grown on top of the liquid. After drying, depending on various conditions, this can result in mushroom leather, plastic or paper-like texture [18, 19]. MBCs, on the other hand, are created through inoculating a substrate—usually an agricultural waste such as sawdust, wood pellets, or paper—with an individual strain of fungus and adding culture media to it. The hyphae then feed on the culture media, grow and create the filamentous white web, colonizing the organic substrate. Mycelium can have various properties. For instance, such properties of the PM depend on the fungal strain, growth conditions, and post-synthesis treatment. MBCs, on the other hand, depend on the substrate type, any additives, fungal strain, colonization conditions, culture media, and post-synthesis treatment [20–22]. This paper experimented creating art installations with the latter type of mycelium.

So what is special about mycelium? Mycelium bio-composites are not only biodegradable but are natural vacuum cleaners, meaning they help to recycle the neighboring agricultural accumulating by-products. These by-products are part of the agricultural waste due to the large consumption of human beings, for instance, saw dust, wood pellets, rice straw, and other materials [23]. Furthermore, it acts as a two-way purifier, as using mycelium will decrease the usage of other nonbiodegradable materials and their toxic pollutants by-products [24, 25].

## 2.2 Mycelium in Design

Recently, mycelium has been gaining a lot of attention from both the academia field and the industry field. Many researchers have been extending its limitations and exploring its other usages due to its zero by-product property and its relative low energy consumption during its growth process [26]. Specifically, in the artistic field, it has become very popular and has served outstanding results in creating artistic installations in several domains such as bio art, science art, and hybrid arts [27]. Another explored new usage is packaging, as it serves as one of the most eco-friendly alternative materials to the dominant expanded polystyrene [28].

The species that is tackled in this paper is *Ganoderma lucidum*; it has great benefits in the medicine field and is very popular in Southeast Asia. It is also known as *reishi* in Japan [29]. Figure 1 clearly indicates the study area.

## 2.3 Problem Statement

Although it has been receiving a lot of interest, the applications of the mycelium are still few in number and only a few limited startups have emerged in the industry such as the Indonesia-based Mycotech Lab (MYCL) and US-based MycoWorks. This is due to many reasons such as inadequate awareness among public, which plays a huge role in distancing them from the movement, and the disconnect between

**Fig. 1** Intended research area

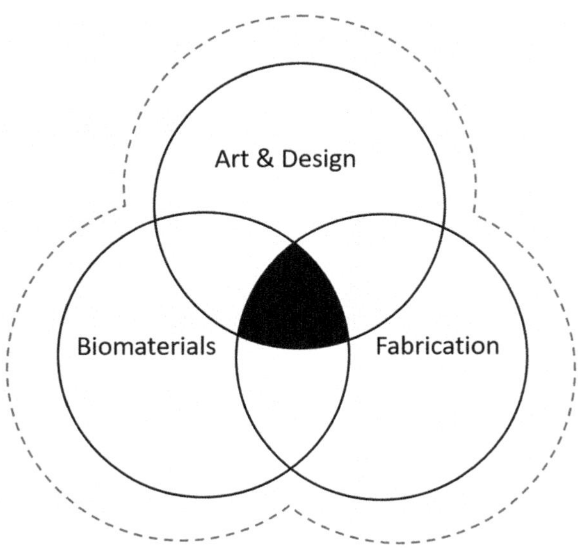

academia and everyday applications; a large number of research papers are being published, yet in reality, there is very limited impact of the research in terms of its application in daily life.

So the question is: how can we use fabricated mycelium to create zero-waste art installations?

### 2.4 *Objectives*

Our research aims at applying scientific knowledge of mycelia's growth to achieve novel artistic designs. For this, we needed the best media that is accessible and easy to work with for the artists and designers with little background in the bio-based microorganisms field. We believe that the art field is one of the most flexible fields that can easily reduce plastic waste by using alternative environmentally friendly materials.

## 3 Materials and Methods

### 3.1 *Experimental Workflow*

A ready inoculated *reishi* substrate kit was used as the main substrate for the mycelium production. The technique used in this paper is (molding and casting). First, the molds were conceptually designed and 3D printed using PLA filament—bioplastic producing 80% less greenhouse gas [30]. After the culture media was prepared, the substrate was mixed with it and casted in the molds. The experiment was divided into two phases. The first phase had two culture media prepared and compared for better growth results of mycelium. The second phase had the best culture used with the substrate and the molds were incubated and then dried in the oven. Figure 2 shows the overall workflow.

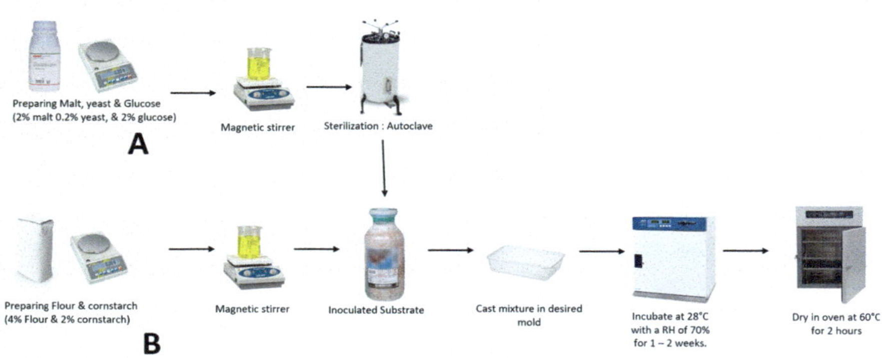


**Fig. 2** Experimental workflow

## 3.2 Phase 1: Culture Media Experiment

### MYG Experiment

The first basic liquid culture medium containing carbon, energy and nitrogen sources to be assessed was prepared. The MYG medium was mixed at the ratio (2% malt extract, 0.2% yeast extract, and 2% glucose). A total of 500 ml was prepared and magnetically stirred for around a few minutes. It was then autoclaved at 121 °C (15 psi) for 20 min. A 2415 µl of the sterile medium was added to 0.03 kg of the substrate using micropipette in a sterile clean bench, as seen in Fig. 3. This mixture was casted in a clean sterile mold and sealed with plastic wrap to properly lock the humidity within the mycelium. A few holes were poked to leave room for its breathing. The mold was then incubated at 28 °C with a relative humidity of 70% for 1–2 weeks in a dark static incubator. After 2 weeks, the mycelium was then harvested and dried in an oven at 60 °C for around three sets of an hour's time with an interval of 2–3 h in between. It was carefully observed to reach the desired firmness while controlling the growth of the fungi without being burnt.

### FCM Experiment

A flask containing 700 ml liquid FCM culture medium was prepared with the ratio of (4% flour, 2% cornstarch) and magnetically stirred for around a few minutes. For every 1 kg of the substrate 700 ml of the prepared media was thoroughly mixed with it. In a sterile clean bench, the mixture was casted in the sterile mold and like the previous MYG medium experiment, the mold was roughly sealed. The rest of the incubation and drying parameters were typically the same as the MYG's conditions.

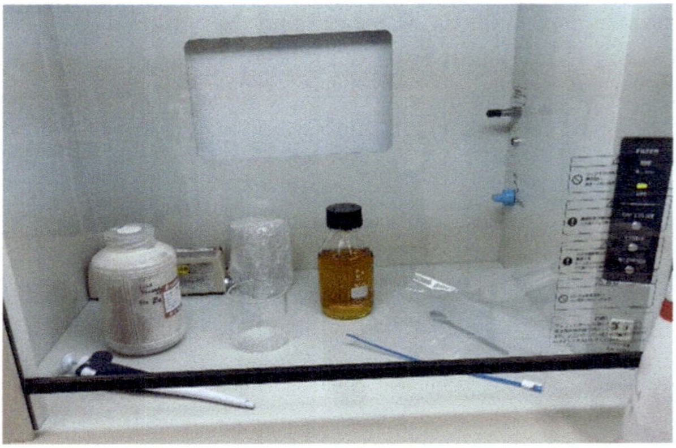

**Fig. 3** MYG culture media experiment

## 3.3 *Phase 2: FCM Medium Installations Experiment Execution*

### Conceptual Design of Molds

The concept of the artworks is inspired by the Japanese nature surrounding the Japanese people in their daily work or school commuting routine. The explored theme is nature taken for granted. It serves as a wakeup call for the endangered blessings that we enjoy or even worse, overlook every day freely while knowing that our actions are posing a huge threat to it. Sometimes, only imagining the worst-case scenario of having our environment ruined will make us nurture and tend to it. Taking everything for granted and not noticing the beauty around us every day puts us in what is called the "rat race," i.e., solely running every day to get our work done without taking a moment to contemplate or enjoying a break with ourselves and our loved ones.

### 3D Designing of Molds

The designs are specifically inspired by the Japanese huge oval lotus green leaves seen by the rivers and ponds as seen in Fig. 4. 2 molds were designed using the design software: Rhinoceros 3D. The molds were then exported as meshes ready to be 3D printed.

### 3D Printing of Molds

Using the slicing software: Simplify 3D, the meshes' printing parameters were adjusted. Including the nozzle diameter, the layers' settings and infills. The meshes were then exported as a g-code file and sent to the 3D printer to be printed. The 3D printer's bed was heated and prepared for the molds to be printed. This can be seen in Fig. 5, as the molds are pictured printed with their supports.

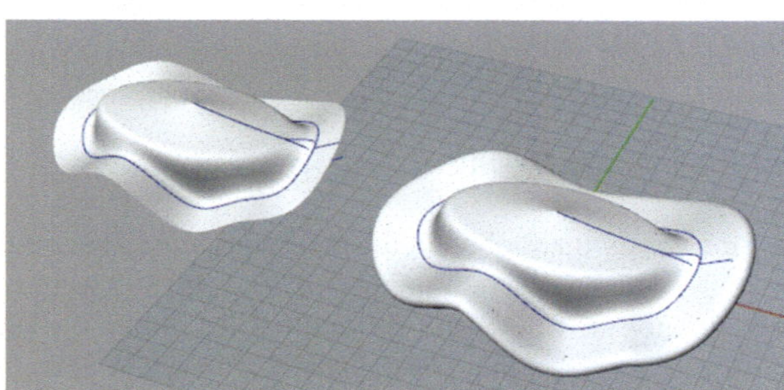

**Fig. 4** Molds' 3D designing

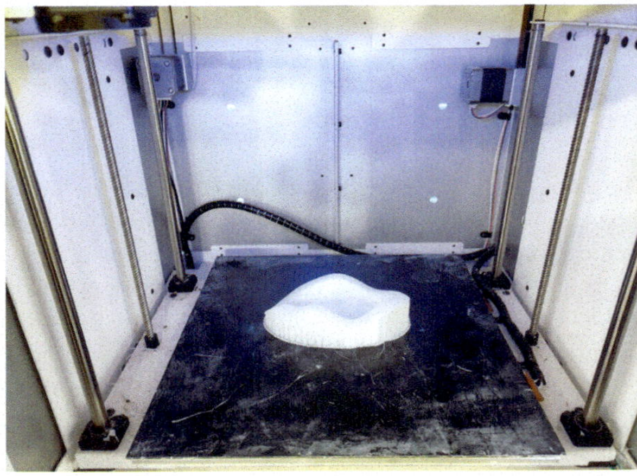

**Fig. 5** Three-dimensional printing of molds

**Fig. 6** Molds' casting with the substrate

## Substrate Preparation

In a sterile clean bench, as seen in Fig. 6, 0.25 g of the substrate was mixed with the correspondent ratio of the FCM medium and casted in the sterile molds. The substrate mixture was tapped and smoothed to fit the mold. It was then enclosed within a sterile Ziploc bag without zipping it all the way to leave some room for oxygen for the mycelium.

## Incubation and Drying

As previously mentioned in the first experiment, the casted mold was also incubated at 28 °C with a relative humidity of 70% for 1–2 weeks in a dark static incubator to prevent any growth of the fruiting bodies. After 2 weeks, the mycelium was then harvested and dried in an oven at 60 °C for around three sets of an hour-period with a rest interval of 2–3 h in between to stop the fungal growth. It was carefully observed to reach the desired firmness without being burnt.

# 4 Results and Discussion

## *4.1 Phase 1: Culture Media Experiment*

In order to find out the best liquid culture media to promote the highest growth rate and facilitate the execution of mycelium art installations for the artists with little background in bio-materials field, both MYG and FCM grown mycelium were compared. The *reishi* mycelium growth was estimated using qualitative characteristics such as the white color, mycelial density, and the presence of aerial or submerged spores [31].

After 2 weeks in the incubation, followed by the drying period in the oven, the white mushroom-like layer in the FCM medium was much denser than the MYG. Furthermore, the MYG mycelium tended to be sparse and infrequent, the substrate fibers texture could still be seen. On the other hand, the FCM samples' upper surface was fully white and fuzzy, as seen in Fig. 7.

After drying, there was a quite visible yellow layer on the MYG samples, as pictured in Fig. 7. As reported by Powrie, this could be due to the Maillard reaction, i.e., the reaction between the sugars and proteins in fungal cell walls due to the high temperature of the material while being dried in the oven. Another possible reason is the caramelization of the polysaccharides in the material and thermal organic matter degradation [32]. Besides the scientific explanations, speaking from an artistic point of view, this exterior feeling was not the best for art installations, as it felt too friable and too fragile. It also did not feel united. It was agreeable that most of the artists would prefer FCM over MYG for this reason.

The use of the novel FCM medium is very rare among researchers. Rather, it is common among the public interested to grow mushrooms in their homes, more than the academics themselves. Therefore, there are no sufficient number of previous research papers on it. Consequently, we wanted to experience it and shed light on it due to its accessibility among the public and artists, as it involves ingredients that can

**Fig. 7** MYG (left) and FCM (right) molds

be found anywhere and at the same time they have the needed energy sources for the mycelium to grow. Sophisticated chemicals can feel a little bit overwhelming for artists and locals to work with. Consequently, the adoption of this whole new material will seem a bit distant for them. Most significantly, environmentally speaking, the total carbon footprint emissions of FCM is 80% less than that of MYG [33–37].

## *4.2 Phase 2: FCM Medium Installation Experiment Execution*

Fabrication design has opened a whole world of exciting and endless possibilities for artists and designers to help them work on any material through fabricating it. Implementing fabrication with bio-materials gives us bio-fabrication, a new trend in which designers try to reduce the waste as much as possible.

On the other hand, as exciting as it might sound, mycelium will still be a little uncertain choice for some designers and artists. One significant factor is its flexibility in creating art installations. Nevertheless, its biodegradability is fascinating enough to tempt artists to take the first step and experience it. Consequently, such research should be done to help facilitate it for artists and designers [38].

The material along with FCM medium succeeded in creating artwork installations but it still is a little bit fragile in comparison to other materials. Typically, artists would not prefer it over other materials as it will pose a huge threat on the artworks while moving and installing it in the exhibitions. Furthermore, there are quite limitations for casting the mycelium in the mold, as sometimes it can get stuck. More research is needed into the density and the layer infills of the parameters of the 3D printing. The filament type is also suspected to have an effect (Fig. 8).

However, as far as the creativity of the artists goes, it is still possible to experiment with mycelium, and even though it can be a little bit fragile, this in itself has some artistic characteristics for artists and designers to experiment with. For

**Fig. 8** FCM installations

instance, some artists are experimenting with coloring mycelium with natural dye sprays. Three-dimensional printing molds with recyclable materials such as fiber and waste paper cups are also being negotiated [39].

## 5 Conclusion and Future Work

In the context of introducing new biodegradable materials, this paper provided an extensive overview of creating art installation designs using mycelium after establishing which liquid culture media out of MYG and FCM is better. The MBC artworks were successfully developed by incubating *reishi*-inoculated substrate with FCM medium. Nonetheless, mycelium products are still at the peak of the research trends and commercialization.

Future work is definitely needed to further assess the physical properties of the mycelium as an artistic medium to be used in artworks in exhibitions and product design. Moreover, we plan to research integrating media arts with mycelium installations such as projection mapping.

## References


1. Hoornweg D, Bhada-Tata P (2012) What a waste: a global review of solid waste management. Urban Dev Ser 15:87–88
2. Chattopadhyay S, Dutta A, Ray S (2009) Municipal solid waste management in Kolkata, India —a review. Waste Manag 29(4):1449–1458
3. Derraik JGB (2002) The pollution of the marine environment by plastic debris: a review. Mar Pollut Bull 44:842–852
4. Moore CJ (2008) Synthetic polymers in the marine environment: a rapidly increasing, long-term threat. Environ Res 108(2):131–139
5. Gamaralalage P, Abeynayaka A, Hayashi M (2022) Japan's plastic waste management – challenges and potential solutions, Plastic Atlas Asia Japan special edition, pp 1–2
6. Choudhury M, Sahoo S, Samanta P et al (2022) COVID-19: an accelerator for global plastic consumption and its implications. J Environ Public Health 2022:1–17
7. Minami E, Saka S (2005) Biomass resources present in Japan—annual quantities grown, unused and wasted. Biomass Bioenergy 29(5):310–320
8. Kousaikai R (2000) Directory of the forestry statistics. Forest Agency annual report, pp 6–45
9. Harada T (2000) Possibility of methanol production from wood-based waste materials. In: Evans PD (ed) Proceedings of the 5th Pacific rim bio-based composites symposium. Australian National University, Canberra, pp 635–637
10. Haneef M, Ceseracciu L, Canale C et al (2017) Advanced materials from fungal mycelium: fabrication and tuning of physical properties. Sci Rep 7:41292
11. Pelletier MG, Holt GA, Wanjura JD, Bayer E, McIntyre G (2013) An evaluation study of mycelium based acoustic absorbers grown on agricultural by-product substrates. Ind Crop Prod 51:480–485
12. Islam MR, Tudryn G, Bucinell R, Schadler L, Picu RC (2018) Mechanical behavior of mycelium-based particulate composites. J Mater Sci 53:16371–16382

13. Ghazvinian A, Benay G (2022) Mycelium-based composite graded materials: assessing the effects of time and substrate mixture on mechanical properties. In: Biomimetics, fungal architectures special issue
14. Jones M, Mautner A, Luenco S, Bismarck A, John S (2019) Engineered mycelium composite construction materials from fungal biorefineries: a critical review. Mater Des 187:108397
15. Elsacker E, Vandelook S, Van Wylick A, Ruytinx J, De Laet L, Peeters E (2020) A comprehensive framework for the production of mycelium-based lignocellulosic composites. Sci Total Environ 725:138431
16. Manan S, Ullah MW, Ul-Islam M, Atta OM, Yang G (2021) Synthesis and applications of fungal mycelium-based advanced functional materials. J Bioresour Bioprod 6(1):1–10
17. Karana E, Davine B, Erik-Jan H, Serena C (2018) When the material grows: a case study on designing (with) mycelium-based materials. Int J Des 12:119–136
18. Appels FV (2014) Mycelium design (Unpublished master's thesis). Utrecht University, Utrecht
19. Blauwhoff DRLM (2016) Mycelium based materials: A case on material driven design and forecasting acceptance (Unpublished master's thesis). Delft University of Technology, Delft, The Netherlands
20. Appels FVW, Dijksterhuis J, Lukasiewicz CE, Jansen KMB, Wösten HAB, Krijgsheld P (2018) Hydrophobin gene deletion and environmental growth conditions impact mechanical properties of mycelium by affecting the density of the material. Sci Rep 8:4703
21. Appels FVW, Camere S, Montalti M, Karana E, Jansen KMB, Dijksterhuis J, Krijgsheld P, Wösten HAB (2019) Fabrication factors influencing mechanical, moisture- and water-related properties of mycelium-based composites. Mater Des 161:64–71
22. Jones M, Huynh T, Dekiwadia C, Daver F, John S (2017) Mycelium composites: a review of engineering characteristics and growth kinetics. J Bionanosci 11:241–257
23. Peng L, Yi J, Yang X, Xie J, Chen C (2023) Development and characterization of mycelium bio-composites by utilization of different agricultural residual byproducts. J Bioresour Bioprod 8(1):78–89
24. Yang L, Park D, Qin Z (2021) Material function of mycelium-based bio-composite: a review. Front Mater 8:737377
25. Madurwar MV, Ralegaonkar RV, Mandavgane SA (2013) Application of agro-waste for sustainable construction materials: a review. Constr Build Mater 38:872–878
26. Jones M, Mautner A, Luenco S, Bismarck A, John S (2020) Engineered mycelium composite construction materials from fungal biorefineries: a critical review. Mater Des 187:1–16
27. Rapp R (2019) On mycohuman performances: fungi in current artistic research. Fungal Biol Biotechnol 6:1–22
28. Vandelook S, Elsacker E, Van Wylick A, De Laet L, Peeters E (2021) Current state and future prospects of pure mycelium materials. Fungal Biol Biotechnol 8(1):20
29. Jayasinghe C, Imtiaj A, Hur H, Lee GW, Lee TS, Lee UY (2008) Favorable culture conditions for mycelial growth of Korean wild strains in Ganoderma lucidum. Mycobiology 36(1):28–33
30. Islam S, Bhat G, Sikdar P (2022) Thermal and acoustic performance evaluation of 3D-printable PLA materials. J Build Eng 67:105979
31. Evangelista FR, Chairez I, Sierra S, Leal Lara H, Martínez-González CR, Garín Aguilar ME, Valencia Del Toro G (2021) A novel coconut-malt extract medium increases growth rate of morels in pure culture. AMB Express 11(1):167
32. Powrie WD, Wu CH, Molund VP (1986) Browning reaction systems as sources of mutagens and antimutagens. Environ Health Perspect 67:47–54
33. Arias A, Feijoo G, Moreira MT (2021) Evaluation of starch as an environmental-friendly bioresource for the development of wood bioadhesives. Molecules 26(15):4526
34. Zhao R et al (2021) A comparative study on carbon footprints between wheat flour and potato in China considering the nutrition function of foods. IOP Conf Ser: Earth Environ Sci 726:012004
35. Blanco J, Iglesias J, Morales G, Melero JA, Moreno J (2020) Comparative life cycle assessment of glucose production from maize starch and Woody biomass residues as a feedstock. Appl Sci 10(8):2946

36. Yawson DO, Adu MO, Armah FA (2020) Impacts of climate change and mitigation policies on malt barley supplies and associated virtual water flows in the UK. Sci Rep 10:376
37. Dunn JB, Mueller S, Wang M, Han J (2012) Energy consumption and greenhouse gas emissions from enzyme and yeast manufacture for corn and cellulosic ethanol production. Biotechnol Lett 34(12):2259
38. Barati B, Karana E, Foole M (2017) Experience prototyping smart material composites. In: Proceedings of international conference of the DRS special interest group on experiential knowledge (EKSIG). Delft University of Technology, Delft, pp 50–65
39. Dessi-Olive J (2022) Strategies for growing large-scale mycelium structures. Biomimetics 7: 129

# The Role of Shame and Guilt in Designing Anti-littering Interventions

June Kyong Trondsen, Casper Boks, Wanjun Chu, Babak Nemat, and Renee Wever

**Abstract** This paper reports on the practical application of a recently developed design tool, Shame Cues, which is a card deck created to help designers understand, reflect upon, and discuss social concepts related to shame and other self-conscious emotions. It consists of 64 cards divided into 16 categories, explaining how shame manifests itself in society and through concepts such as awkwardness, guilty pleasures, euphemisms and taboos. The tool has been tested in various workshops and for different contexts, mainly concerning social challenges focusing on stigma and taboos. Changing the focus toward sustainability, this paper reports on applying Shame Cues in the context of waste disposal and littering. While Design for Sustainable Behaviour provides a rich repertoire of design strategies to stimulate sustainable practices, scrutinizing the user research that has informed these littering cases reveals that aspects of shame, guilt, and hedonistic motivations have received limited attention. Addressing sustainable behaviour through a "shame lens" may lead to different insights about the suitability of Design for Sustainable Behaviour strategies and open for more empathic approaches rather than forceful ones. For this case study, the tool was made available to professional designers and design researchers without prior focused attention on shame as part of user research. The workshop was recorded and analysed, and the results were jointly discussed. This allowed for conclusions on two levels, considering both theme and method. Firstly, how can using the shame perspective inform the design process and expand the solution space in the context of littering? Secondly, what is the learning potential for designers when encouraged to look through a "shame lens", in comparison to the default situation of applying design for sustainable behaviour strategies in which shame does not play an explicit role?

J. K. Trondsen (✉) · C. Boks
Department of Design, Norwegian University of Science and Technology, Trondheim, Norway
e-mail: june.k.trondsen@ntnu.no; casper.boks@ntnu.no

W. Chu · B. Nemat · R. Wever
Department of Management and Engineering, Linköping University, Linköping, Sweden
e-mail: wanjun.chu@liu.se; babak.nemat@liu.se; renee.wever@liu.se

 
S. Fukushige et al. (eds.), *EcoDesign for Circular Value Creation: Volume I*,
https://doi.org/10.1007/978-981-97-9068-5_11

**Keywords** Design for sustainable behaviour · Littering · Design strategies · Shame · Case study

# 1 Introduction

Littering is a packaging-related problem where environmental, social and economic aspects meet [1]. It has been researched by several scholars from a behaviour-changing perspective, primarily focusing on packaging redesign, infrastructure perspectives, and awareness campaigns. Such interventions often rely on user research methods to understand habitual and intentional processes and situational influences [2, 3] and insights into users' efficiency, safety, and satisfaction [4]. While Design for Sustainable Behaviour provides a rich repertoire of design strategies to stimulate sustainable practices, scrutinizing the user research that has informed these littering cases reveals that aspects of shame, embarrassment, guilt, and hedonistic motivations have received limited attention. Addressing sustainable behaviour through a "shame lens", and recognising shame aversion as a contributing factor, may lead to different insights about the suitability of Design for Sustainable Behaviour strategies and open for more empathic approaches instead of forceful ones. For example, when considering our reluctance towards experiencing shame, one may include that instead of stubbornness, lack of knowledge, or inconvenience being the root cause of unwillingness to behave sustainably, insecurity of learning something new, the risk of being called out by others and shame of admitting to being wrong, might be equally important barriers for desired sustainable behaviour.

A few designers have engaged with taboos and shame through domains of norm-critical and social design [5, 6]. However, with a few exceptions [7], the implications of considering shame and taboos in the context of sustainability seem unexplored, leaving designers unequipped to recognize how taboos and feelings of shame can sustain social practices with a negative environmental impact.

Although shame might be a well-known emotion made familiar to most of us through speech and social interactions, it is also a complex emotion, with many and sometimes varying disciplinary definitions. While shame can be experienced as a social threat, uncomfortable and painful momentarily, it can also be a motivational force for prosocial behaviour [8]. Many classify shame as a self-conscious emotion, together with embarrassment, guilt, and pride, and it reflects how we see ourselves as well as how we think we are seen by others [9]. Some emphasize shame as a social emotion in that it depends on social conditioning [10], others define it as a moral emotion, as it tells us something about right and wrong [11]. Shame can also be regarded as a social control mechanism and a behavioural agent with both positive and negative effects [12]. When designing interventions, emphasizing shame in subtle or explicit ways can affect how we act and feel about certain topics. Such interventions can play a role in many contexts: for example, supporting sustainable practices, counteracting unhealthy behaviours, or criticizing social injustice. In some cases, fear of shame can nudge us into more socially healthy behaviours (e.g. refrain from littering in public), or it can act as a threshold to doing things differently (e.g. wearing the same clothes for multiple days).

## 2 Shame in Design for Sustainable Behaviour

Much of previous Design for Sustainable Behaviour research aims to categorise different aspects of behaviour in a way that it can facilitate application for designers. Examples are Lockton's Design with Intent method, based on 101 patterns for influencing behaviour through design [13] and the Dimensions of Behaviour Change by Daae and Boks [14]. To assess the potentially novel contribution of adding a "shame lens" to the design process, these tools are briefly reviewed here for occurrences and relevance to shame.

A number of patterns from the Ludic, Interaction and Cognitive lenses from the Design with Intent toolkit play with social elements of rewards, senses of achievement, peer behaviour, peer feedback and social proof. These evoke possible connections to pride, peer pressure and bragging, which in turn can be connected to some understandings of shame or to the opposite side of the coin. A few other Design with Intent cards (such as the "Emotional engagement" card from the "Cognitive Lens") address strategies which potentially could include using certain features of shame, but do not do so explicitly. The most obvious shame-related card in Lockton's Design with Intent toolkit is "Assuaging guilt" from the "Cognitive Lens" patterns. This card suggests reducing users' feelings of guilt about their behaviour, for example by pointing out unethical or unsustainable consequences or their behaviour and then offering a way out.

So, the notion of shame is represented in the Design with Intent toolkit, but only once explicitly, and it is left to designers themselves to interpret possible uses and implications of certain social mechanisms connected to shame in the context of a specific design challenge.

The Dimensions of Behaviour Change encompass eight, partly ambiguously named elements of design strategies: "Control", "Obtrusiveness", "Direction", "Obtrusiveness", "Direction", "Encouragement", "Meaning", "Importance", "Timing" and "Exposure". These provide designers with aspects to address when designing behaviour change interventions. In line with Lockton's patterns, the dimensions "Encouragement", "Meaning" and "Importance" carry connotations to the emotional aspects of behaviour. "Meaning", for example, represents the variation between, on the one end, motivating rationally by making the user feel it is the right thing to do and, on the other end, motivating emotionally by making the desired behaviour pleasurable. "Empathy" represents a scale of solutions to influence behaviour ranging from making the user aware of, or consider, others, to allow for a purely egocentric perspective. However, none of the dimensions explicitly addresses or connects to shame or other closely related concepts such as guilt, embarrassment, taboos, and stigma.

Looking beyond specific Design for Sustainable Behaviour methods, design literature has a strong tradition for understanding and evoking positive emotions [15], but negative emotions, particularly shame, are hardly considered with the same granularity and nuance [12].

# 3 Shame Cues

## *3.1 Development of Shame Cues*

Thus, in early 2022, a design tool named Shame Cues was conceived to fill the gap of addressing shame in design for behaviour change, and more in general to help designers understand, reflect upon, and discuss social concepts where shame and other self-conscious emotions may play an important role. Shame Cues was developed as one out of several tools contributing to understanding shame, offering a "shame lens" and an additional perspective to factor into the design process. Shame Cues specifically was developed with the intention to provide designers with a broad vocabulary to identify and articulate understandings of shame, and hence bridge socio-psychological theories of shame and guilt with applications relevant to design practice.

Shame Cues consists of 64 cards divided into 16 categories (see Table 1), with each card explaining how shame manifests itself in society and through concepts such as awkwardness, guilty pleasures, euphemisms, and taboos (see Fig. 1). Much of how the shame concept is captured, verbalized, illustrated, and described on the

**Table 1** Overview of categories and Shame Cues in the card deck

| | | | |
|---|---|---|---|
| **Exaggeration** | **Counter action** | **Humour** | **Vulgarity** |
| Sarcasm | Counterculture | Dad jokes | Taboos |
| Camp | Reappropriation | Dark humour | Profanity |
| Shamelessness | Subculture | Schadenfreude | Karen stereotype |
| Satire | Civil disobedience | Irony | Vulgarism |
| **Dark attraction** | **Pleasure** | **Secrecy** | **Softening** |
| Morbid curiosity | Benign masochism | Closeting | Euphemism |
| Forbidden fruit | Naughtiness | Snooping | Metaphors |
| Innuendo | Guilty pleasure | Eavesdropping | Stylizing |
| Neophilia | Eccentricity | Sweeping under the rug | Awkward turtle |
| **Covering** | **Awkwardness** | **Embarrassment** | **Slang** |
| Pardon my French | Oversharing | Pussyfooting | Cheesiness |
| Gedoogbeleid | Personal space invasion | Cringe | Tackiness |
| Sanitizing | Verbosity | Vicarious embarrassment | Corniness |
| Sugarcoating | Awkwardness | Overpraising | Geekiness |
| **Guilt** | **Moralising** | **Humiliation** | **Separation** |
| Confession | Sanctimommy | Public humiliation | Scapegoating |
| Guilt tripping | Political correctness | Walk of shame | Stigmatizing |
| Guilt hibernation | Holier-than-thou | Badge of shame | Cancel culture |
| Sympathy card | Outrage culture | Mockery | Shunning |

**Fig. 1** Two Shame Cues from the "Humiliation" category: "Badge of shame" (left) and "Mockery" (right)

cards stems from theory and observations from social interactions, cultural production, and public discourse, and in many cases, extracted from various non-academic and informal platforms such as Urban Dictionary, Wikipedia, Pinterest, Know Your Meme, Imgur, Bored Panda, Reddit and Quora.

The Shame Cues tool can be used in different ways, such as providing a vocabulary to describe how shame and related aspects manifest themselves in our daily lives, to broaden insight into understanding shame, act as a probe to spur discussion with others or be used as a source for inspiration for brainstorming ideas on how shame can be added or removed in search of a suitable design intervention. One recommended way of using the tool is tasking users of the tool to explore how a certain phenomenon (for example dumpster diving) connects to and can be studied in light of a specific card; for example, "Pussyfooting" may be connected as having a negative effect on dumpster diving, because people who may want to join a dumpster diving community may hesitate, thinking that they won't be accepted or considered sufficiently "poor". Similarly, "Vicarious embarrassment" may be relevant, considering that others may be embarrassed on your behalf and may therefore hinder people from dumpster diving. On the other hand, *"Snooping"*, *"Morbid curiosity"*, or *"Guilty pleasure"* may have a positive effect as it can be tempting or trigger curiosity to investigate others' (food) trash, bring associations to treasure hunting and the pleasure of finding something for free. A second part of the exercise is to take random cards which previously have not been selected and see if examples can be found (or created) to illustrate the potential relevance of the card. This part of the exercise focuses the designer's attention towards uncovering blind spots, increasing their understanding of how aspects of shame may play a role, and how adding or removal of shame can contribute to the desired behaviour.

## *3.2 Previous Testing of Shame Cues*

The tool has been tested in various workshops (see Fig. 2), in student master projects and for different contexts, mainly concerning social challenges focusing on stigma and taboos [7, 16–18]. Analysis of outcomes of these workshops and projects have brought important insights into the usefulness of applying this design tool, as well as input for future iterations. Participants generally experience the Shame Cues as very inspiring, as it sparks discussions concerning the meanings and definitions of shame, changing social norms surrounding shame, and the influence of these norms on individual feelings and behaviours. The tool helped participants spin off each other's associations and ideas, contributing to collective thinking. It was also clear that more experienced designers found it easier to tap into the domains of norm-awareness and critical reflections; for example, bachelor students seem to have more difficulties making quick associations between cards and the studied phenomenon in a shame context, where more experienced designers have less trouble handling such a meta-level of thinking. Our overall insights suggest that a better vocabulary and awareness of shame can support designers in identifying how this emotion can hinder and promote sustainable practices, and how they can more effectively design with shame in mind, resulting in ideas that attempt to challenge and counteract societal norms and taboos [7, 18].

**Fig. 2** Participants discussing and organizing their Shame Cue cards during a NORDES 2023 workshop. (Picture taken by the authors)

## 3.3 *Goal of the Workshop*

These conclusions were mostly derived from projects and workshops with a clear focus on shame and taboo. This provided the motivation for testing application of the tool in a context where shame, guilt, embarrassment, and other self-conscious emotions are less prominent, but still might play a role. We were also interested in learning more about the learning potential for designers when being encouraged to look through a "shame lens", in comparison to the default situation of applying design for sustainable behaviour strategies in which shame does not play an explicit role. Lastly, it was interesting to find out how a shame perspective may inform the design process and expand the solution space in a specific sustainability context, or even a specific very limited solution space. With these questions in mind and changing the focus toward sustainability, this paper reports on applying Shame

Cues in the context of littering and a packaging-related problem where environmental, social, and economic aspects meet. The next chapter describes the set-up of the workshop which was designed to help answer these questions.

# 4 Workshop

## *4.1 Workshop Topic: Snus Boxes and Cigarette Butts*

For this case study, the Shame Cues tool was made available to professional designers and design researchers with experience working with design and sustainability issues, but without prior focused attention on shame as part of user research or with previous experience working on/with Shame Cues (or any other shame tools). The participants each had a track record in research focusing on design for sustainable behaviour as a senior researcher or as PhD candidates. In their work, they all previously focused on packaging (date labelling, in-home waste sorting, food waste and littering). While all participants lived in Sweden, they hailed from different parts of the globe (the Netherlands, Iran, and China), and thus brought in different cultural backgrounds.

The case study took place as a 3-h workshop in June 2023, with three participants collaborating over Zoom. The participants were provided with the Shame Cues card deck together with a set of workshop exercises where they were requested to evaluate the usefulness of the Shame Cue cards and their relevance for past and potentially future design projects. The workshop was recorded and analysed, and the results were jointly discussed in a post-workshop interview.

The chosen context for this specific workshop was public littering caused by empty snus boxes and cigarette butts. Snus is a tobacco product originating from a variant of dry snuff which have become increasingly popular over the last few years, predominantly in Sweden and Norway.

Like cigarette butts, it is not uncommon to see snus pouches and empty snus containers, improperly disposed of in public space. Having already an interest in, and past experience with working on this topic, the participants were asked to revisit this waste problem using the Shame Cues card deck, to seek new perspectives.

## *4.2 Workshop Process*

After a short introduction to the workshop, the participants were left alone to independently explore the Shame Cues tool together as a group. During this session, the participants would go through and familiarize themselves with the different cards and pick out and discuss the ones they found relevant to the topic at hand. This could be both cards that reflected aspects that were made visible in previous projects or cards that reflected aspects they would want to include in future projects. Throughout the workshop, the three participants worked with a wide design space in light of design for behaviour change [19], while considering how different Shame Cues might be applied to littering. The discussion explored a broad design space, anything from the snus box itself to an entire public space. At the end of the workshop, the design space was deliberately narrowed. A typical place for snus boxes to be littered is on top of electricity boxes (the grey cabinets placed in the street which control the electrical supply to a number of households). The designers briefly explored ideas whilst purposely limiting their design space to working with the top of such electricity boxes. Here, the Shame Cues still worked, although other Design with Intent [13] strategies also quickly came into play.

The card sorting process started off slowly with the intention of reading and discussing each individual card, but it was eventually sped up by selecting cards based on category and based on inspiration. In the end, the participants were challenged to pick their favourite cards. These would be cards that they agreed on as being the most intriguing and worth focusing on for a future littering project. The card sorting session was followed by a group discussion summarizing the process and reflecting upon the usefulness of having Shame Cues available to designers engaging with sustainability issues. In addition to participants' written notes and video recordings of the workshop, a post-workshop interview was conducted, allowing for follow-up questions and for the participants to elaborate on their reasoning for selecting cards and their experience with using this new tool.

## 5 Discussion of Workshop Results

This workshop provided a platform for exploring the intersection of shame and sustainable design, focusing on the usefulness of using Shame Cues as a tool to guide designers working with anti-litter interventions. The output of the workshop was analysed and discussed in-depth with all participants, improving internal validity. The general feedback was positive from all three participants, as the cards were valuable as inspiration to seek new perspectives, make reflections and raise important questions. Since the workshop was not designed to create a concrete result (e.g. a concept sketch for an anti-littering intervention) but rather to discuss the usefulness of having such a design tool available, the remaining part of this article focuses on the main takeout from this discussion, together with some suggestions on how the tool can be modified to suit future projects better.

### *5.1 Using a "Shame Lens" as Point of Perspective*

By sorting through the different Shame Cues, the participants would use different understandings of shame to investigate their littering context. Not unexpectedly, the participants reported that approaching their context through a "shame lens" forced them to consider the social aspects to a much larger extent compared to Design for Sustainable Behaviour approaches or other sustainable design tools they had used previously. Unlike how they usually would go about when initiating such sustainability projects, often focusing on individual behaviours, user needs, health concerns and/or relations to key stakeholders, the Shame Cue cards prompted them to consider users' actions as part of a social (often unspoken or extraverbal) conversation. This inherent property of shame and its immediate entanglement to social practices, norms and perceptions drew the participants' focus from looking at behaviours isolated to an individual, to investigating how such behaviours are acted out on a societal level. Although such social aspects might reveal themselves in any design project, also without the Shame Cues to guide, the participants highlighted the value of considering this social complexity at the beginning rather than at the very end of a project. One recurring example that was brought up during discussions was a student project looking into introducing cigarette extenders as a solution to provide better waste disposal. Whereas the common behaviour would be to simply flick the cigarette away (and cause littering the streets), the students' idea was to break current smoking routines by introducing an accessory offering an Audrey Hepburn-inspired way of smoking (see Fig. 3). This example came up

**Fig. 3** Audrey Hepburn with a cigarette holder (image taken from the participants' workshop notes) (left). Shame Cues cards: "Eccentricity" (Pleasure category) (right)

many times during the workshop as the concept would align to many different cards (e.g. "Camp", "Eccentricity", "Cheesiness", and "Re-appropriation"). However, in the case of this student project, an unforeseen awkwardness took place the moment they asked people on the street to test the concept. People's reluctance to use this cigarette holder in public spaces ended up being a much bigger issue than the students initially thought of. Thus, the existence of a grey zone between something subtly camp to overly eccentric, and the need to address that, became clear, but at a very late stage of this project. With this experience in mind, the workshop participants recognized that using the Shame Cues cards and having a vocabulary to uncover and discuss these social thresholds early on could have benefited the overall outcome of this particular project.

As implied through the above example, the Shame Cues card deck supported the workshop participants with a vocabulary to identify and articulate different types of shame, enabling them to distinguish between different nuances of shame, which in turn could offer new and unexpected design directions. "Secrecy" was one of the card categories that the participants highlighted as particularly interesting as it depicted a special nuance of shame and a fruitful direction they previously never had thought of exploring. Despite the tendency of snus consumption to become part of the dominant culture in Sweden, when considering the "Secrecy" category, the participants would raise awareness of a very subtle nuance of shame, a nuance that could tell that despite its commonly wide use, people still feel a bit ashamed about using snus. Examples could be how people would use snus in inconspicuous ways, try to hide their snus box fear of being outed by their surroundings, make excuses in certain social settings, or apologize when being called out. Thus, on top of the descriptions given in each card, the Shame Cues revealed a new granularity for considering specific nuances of tobacco related shame, something the participants expressed as valuable in the context of littering, both to disclose hidden aspects and to open new design directions (Fig. 4).

In addition to providing a vocabulary into different types of shame, the card deck also relies on using imagery and examples taken from different parts of society and to some extent with inspiration from different cultures or social groups. This socio-cultural diversity was much appreciated by the participants (who were, as stated, of different cultural backgrounds themselves), with the "Gedoogbeleid" card (a Dutch expression for authorities tolerating something that is forbidden by law, often experienced by citizens as something that can be mocked and appreciated at the same time) and the "Cringe" card (depicting the Chinese basketball player Yao Ming as a meme) brought forward as two cards they fondly could relate to. On the other hand, the two "Trump cards" (both the "Vicarious embarrassment" and "Mockery" cards were exemplified by images of Donald Trump), and the "Karen Stereotype" card were partly criticized as references giving the overall tool a somewhat USA inspired and Western look. Nevertheless, as all 64 cards contain shame descriptions but without specific pointers, they are all open to personal interpretation. This was made visible during the workshop as the participants often would have their own take on how they would resonate with a card and contextualize its content. This could vary between connecting the cards to common experiences from everyday life,

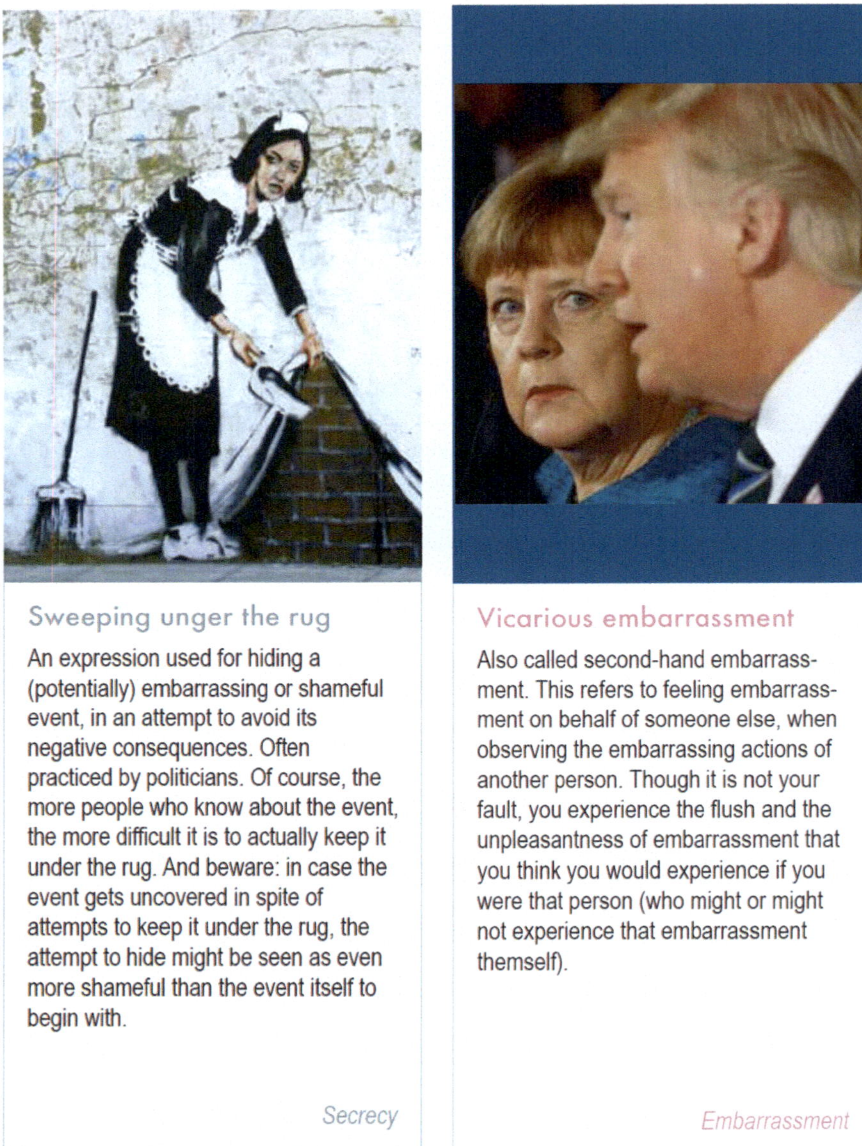


**Fig. 4** Shame Cues Cards: Left: "Sweeping under the rug" (Secrecy). Right: "Vicarious Embarrassment" (Embarrassment)

social phenomena they had observed, connotations to media trends and public discourse or juxtapositions in their own cultural experiences. In the post-workshop interview, the participants expressed that this level of granularity also added value to their discussions, using the "Personal Space Invasion" card to exemplify how the meaning of certain cards can unpack differences between individuals and between different cultures, allowing for multiple standpoints and experiences to become visible and take part in the discussion. During the post-workshop interview, these insights also fuelled a discussion on the appropriateness of using the Shame Cues by designers or researchers from or in high shame cultures, and that care should be taken in such cases.

## 5.2 Exploring Shame Strategies

Although not directly prompted to do so, the participants were eager to explore and evaluate the Shame Cues tool according to its ability to provide useful intervention strategies which previously had not been considered. As the card deck never was intended to be a set of strategies but rather focused on describing shame types and social phenomena and being a source of inspiration, it was not surprising that participants reported back that the cards lacked pointers or indications of a clear design direction. While many cards were less applicable for this specific purpose, some cards, either through pictures or due to the nature of their content (e.g. "Separation"), still were perceived to communicate in a clear direction. The overall feedback was that most cards would trigger multiple ideas and discussions that, in turn, could lead towards a specific design direction, but without having any clear guidelines for doing so. With this in mind, it was suggested that the card deck (and the introduction of the card deck) could be revised to accommodate certain settings better. One approach could be to complement the card deck with specific examples or questions to prompt the discussion (something which already had been tested in workshops with bachelor students to provide them more guidance [18]). Another could be to make the card sorting into gameplay to give more guidance and make it easier to work independently. A third suggestion was to combine the Shame Cues with other design tools that are more directional towards a design intervention. It was thought that the divergent nature of the Shame Cues and the multiple possible interpretations of the cards could provide valuable granularity to specific design strategies within other design tools such as Design with Intent [13] or Dimensions of Behaviour Change [14]. The openness of the cards was perceived to have both benefits and drawbacks. An initial broad divergence at the outset of a project can be both inspiring and mind-opening but also overwhelming or frustrating for its lack of clear directions. This strengthened conclusions from earlier workshops, which saw less experienced designers (bachelor students) struggle more with free form use of the cards in comparison with more experienced ones [7, 17], which suggests that the more experienced and design oriented the user, the less guidance the user needs to make the tool useful. This is also in line with the preferences of this workshop's experienced participants: they all acknowledged the importance of not having too many clear pointers, the valuableness of broad exploration and having the ability to engage in multiple and sometimes diverging discussions. This gave them room to reflect more deeply upon the meaning of certain cards and use the tools to reveal hidden aspects and raise questions about what to investigate.

The participants emphasised that engaging with the card deck as a whole and reflecting on the role of shame inspired them to think beyond their existing toolbox and come up with new design approaches. Much influenced by the "Vicarious embarrassment" card, the workshop revealed an idea for a new design strategy which considered involving third-party disapproval as an intentional part of a design intervention. This was a direction neither of the participants previously had ever thought as they normally would focus on understanding the user, the object and the litter and pay little attention to potential bystanders who are not actively involved. As a result of using the cards, the participants came to acknowledge our sensitivity towards our social surroundings, our concerns towards what others might think of us, and the potential of purposely involving bystanders by using second-hand embarrassment as a mechanism to influence first-person behaviour. This strategy became more strongly visible after using the cards, and as one participant enthusiastically expressed "*…could yield all kinds of potential strategies*". Furthermore, the post-workshop interview also revealed an approach that was focused on involving bystanders, but this time, not as a mechanism to influence others, but as an approach to place agency (or responsibility) with a bystander agent. Thus, opening to explore new and potentially even more disruptive interventions.

## 5.3 Evaluating Shamefulness

Much of the discussions during this workshop largely captured but also questioned the fluctuating character of shame. The participants discovered not only how something can move from being attractive to causing reluctance (as with the cigarette holder example), but also how certain shame types have the ability to move in the opposite direction. One discussion being raised was the risk of diminishing the attractiveness and extraordinary character of certain behaviours by making them more socially acceptable or integrated into the dominant culture. "Taboos" was one of the cards that was brought up to highlight this complexity of shame, its ability to inhibit counteractive forces and its potential to have both positive and negative effects. One participant expressed the potential pitfall of trying to make something "more taboo". Instead of making people reluctant, creating such prohibitions could potentially act as forbidden fruits increasing the temptation to commit to certain actions (e.g. young teenagers who think it is mature or cool to smoke). It is therefore important to avoid framing shame as just a negative force, and the cards may contribute to seeing nuances of how shame can move between positive and negative, attraction and aversion, and with effects that can differ between contexts. The cards' ability to articulate the production of shame, its nuances and potential effects were all appreciated as important functions in evaluating the design of new interventions in their risk of causing shame and undesired aversion (or undesired attraction). The cards were also considered as a possible "shame checker" to be applied to already existing solutions, with possibilities of analysing their shameful (or non-shameful) effects, look for potential blind spots, and provide input to ideas that could act as efficient "vaccines" in preventing future littering.

## 6 Summary

The workshop provided valuable insights into the usefulness and practicality of the Shame Cues in sustainability design challenges and in the context of anti-littering interventions. It also provided insight into the usability options for a user group that was not previously tested in prior workshops and projects. Regarding the theme of the study, it did prompt designers to explicitly focus on understanding the social aspects of the particular (un)sustainable behaviour in the case of littering. Many discussions around the experience of shame related to littering often extended beyond shame itself and touched upon other relevant social aspects such as demography, culture, community, and social relations. As a result, the "shame lens" provided by the Shame Cues cards, becomes an entry for designers to incorporate relevant social elements into their existing design interventions to address littering, which they had not considered before.

Moving beyond the theme, when compared to the default situation of applying design for sustainable behaviour strategies, we found that the shame cards can play two roles in expanding the solution space. Firstly, it serves as an instrumental tool that straightforwardly summarizes different shame types for designers to consider. Secondly, it also serves as a prompt for designers to connect their personal shameful moments, memories, and experiences with the phenomenon they are studying. For instance, instead of simply describing a solution as "*some people would not litter/smoke in public/waste food because they must feel shame about it*", the card enables designers to develop a more comprehensive and empathic understanding of the problem: "*Some people would not litter/smoke in public/waste food because they may fear being shamed of...* (e.g. a type of "Secrecy", "Embarrassment", "Awkwardness", as described in the card). *For example, once I/he/she/they did...*(e.g. a personal experience relating to the card description)". This approach supports designers in not only enriching the understanding of shame but also contextualizing it into something more tangible, describable and sharable. In this regard, the cards are effective in facilitating designers to better understand and contextualize shame as a sustainable design intervention. Further research will focus on exploring how to transform the rich understandings of shame as generated by the cards into concrete, implementable sustainable design strategies for products and services.

## References


1. Wever R, Onselen L, Silvester S, Boks C (2010) Influence of packaging design on littering and waste behavior. Packag Technol Sci 23:239–252
2. Daae J, Boks C (2015) A classification of user research methods for design for sustainable behaviour. J Clean Prod 106:680–689
3. Klöckner CA, Blöbaum A (2010) A comprehensive action determination model: toward a broader understanding of ecological behaviour using the example of travel mode choice. J Environ Psychol 30(4):574–586
4. Crilly N (2011) Do users know what designers are up to? Product experience and the inference of persuasive intentions. Int J Des 5(3):1–15
5. Wilde D (2022) Shitty food-based world-making: recasting human microbiome relationships beyond shame and taboo. Futures 136:102853
6. Torkildsby AB, Vaes K (2019) Addressing the issue of stigma-free design through critical design workshops. In: DS 95: proceedings of the 21st international conference on engineering and product design education (E&PDE 2019), University of Strathclyde, Glasgow. 12th-13th September 2019
7. Trondsen JK, Herweyers, L, Boks C, Dubois E (2023). Pee Poo Period: exploring the intersection between shame, bodily fluids, and sustainable design. Paper presented at Cumulus Antwerp 2023: Connectivity and Creativity in times of Conflict, Antwerp, Belgium, 12–15 April 2023
8. Scheff TJ (2003) Shame in self and society. Symb Interact 26(2):239–262
9. Lewis M (2008) Self-conscious emotions: embarrassment, pride, shame, and guilt. In: Lewis M, Haviland-Jones JM, Barrett LF (eds) Handbook of emotions. The Guilford Press, pp 742–756
10. Hareli S, Parkinson B (2008) What's social about social emotions? J Theory Soc Behav 38(2): 131–156
11. Tangney JP, Stuewig J, Mashek DJ (2007) Moral emotions and moral behavior. Annu Rev Psychol 58:345–372
12. Trondsen JK, Boks C (2022) Exploring the role of shame in design strategies. Proc Des Soc 2: 2233–2242
13. Lockton D, Harrison D, Stanton NA (2010) The design with intent method: a design tool for influencing user behaviour. Appl Ergon 41(3):382–392
14. Daae JZ, Boks C (2014) Dimensions of behaviour change. J Des Res 12(3):145–172
15. Desmet PM (2008) Product emotion. In: Schifferstein H, Hekkert P (eds) Product experience. Elsevier, San Diego, pp 379–397
16. Trondsen JK, Boks C (2023). Shame cues: detecting shame in disguise and playing with new perspectives to inform the design process. Workshop held at NORDES 2023: this space is intentionally left blank, Norrköping, Sweden 12–14 June 2023
17. Boks C, Trondsen JK (2022) Reflections on addressing taboo, shame and social stigma in design projects. In: DS 117: proceedings of the 24th international conference on engineering and product design education (E&PDE 2022), London South Bank University in London, UK. 8th-9th September 2022. The Design Society
18. Boks C, Trondsen JK (2023) A Toolbox for addressing shame in design projects. In: Proceedings of the 25th international conference on engineering and product design education (E&PDE 2023), 7th–8th September 2023. The Design Society
19. Selvefors A (2017). Design beyond interventions–supporting less energy-reliant activities in the everyday. Doctoral thesis, Chalmers University of Technology

# Enhancing Sustainable Behavior in Waste Management: A Social Perspective

**Siti Khatijah Zamhari, Ahmad Fariz Mohamed, and Azahan Awang**

**Abstract** Household waste management is considered as part of the municipal waste where it contributes to a large amount to municipal solid waste (MSW). It is crucial to identify the existing waste management behavior of the household to make adjustment and process improvement to achieve sustainable waste management. Therefore, the objective of this study is to examine the behavior and factors that influence community behavior in managing household waste in rural areas particularly in Kampung Sungai Midin, Kuching, Sarawak. The study focuses on physical environment as well as social and cultural aspects. This study employed both quantitative and qualitative approach. Quantitative data were collected by using questionnaire and analyzed using IBM SPSS Version 24.0 while qualitative data were collected through face-to-face structured interviews, semi-structured interview, and observation. These interviews were transcribed and analyzed using thematic analysis. The survey method involved 75 respondents, selected based on convenience sampling method. The findings show that respondents' behavior was highly influenced by physical environment (57.6%) and moderately influenced by social and cultural aspects (42.4%). The physical features of the river as well as the condition and characteristics of some houses built on stilts along the riverbank could be seen as a means of easing the household waste. Furthermore, lack of facilities is one of the barriers that hinders them from practicing waste segregation and recycling. Low social interaction affects the intention of community to engage in

---

S. K. Zamhari
Institute for Environment and Development (LESTARI), Universiti Kebangsaan Malaysia, Bangi, Selangor, Malaysia

Faculty of Social Sciences and Humanities, Universiti Malaysia Sarawak, Kota Samarahan, Sarawak, Malaysia

A. F. Mohamed (✉)
Institute for Environment and Development (LESTARI), Universiti Kebangsaan Malaysia, Bangi, Selangor, Malaysia
e-mail: fariz@ukm.edu.m

A. Awang
Faculty of Social Sciences and Humanities, Universiti Kebangsaan Malaysia, Bangi, Selangor, Malaysia

 
S. Fukushige et al. (eds.), *EcoDesign for Circular Value Creation: Volume I*,
https://doi.org/10.1007/978-981-97-9068-5_12

sustainable household waste management. There are no individuals or groups who can motivate community to manage their waste with the sustainability concept. In this context, technology advancement should not be viewed as the only means for waste management but an effective solid waste management requires active public involvement, awareness and acceptance at all level of societies. To inspire the community's sense of "mottainai," motivation for waste reduction and recycling should be done continuously with coordination from various stakeholders (both public and private sectors) with common vision and shared goals in support of sustainable growth. Provision of waste management facilities is also vital as well as incentive mechanism to promote 3R initiative among the community. Hence, a bottom-up approach together with top-down approach is essential in enhancing community's sustainable behavior in waste management.

**Keywords** Sustainable behavior · Waste management · Community empowerment · Sustainability

## 1 Introduction

All communities in every country generate waste in their daily activities. Annually, the world generates approximately two billion metric ton of municipal solid waste (MSW). High-income countries contribute around 34% of the world's waste, but the total quantity of waste generated in low-income countries is predicted to be more than triple by 2050 [1]. The high-income countries have succeeded in diverting a greater proportion of municipal waste away from landfill where only 2% of waste is openly dumped compared to low-income countries with 93% of its waste is dumped openly [1, 2].

Most of the current problems associated with waste management are the result of rapid urbanization in both high-income and low-income countries. The challenges in providing waste management services to all levels of society particularly the low-income group is exacerbated by the lack of financial resources as well as underutilization of individuals involved in the enhancement of waste management services as a whole [3]. In addition, the sense of responsibility toward sustainable waste management and the mentality toward cleanliness among the public are still lacking.

Household waste is one of the primary sources of municipal waste consisting of food waste, plastic, paper, and glass from residential areas. To reduce the increasing amount of household waste, several recycling associations in Malaysia have implemented their practices to improve waste management and ensure sustainability. Nevertheless, the effectiveness is still unclear in achieving sustainable household waste management [4]. Waste segregation at source has also been recognized as an effective strategy for recycling, and its potential is yet to be utilized by the public. The rate of public participation remains low in the state of Sarawak particularly in Kuching North City due to public perception and attitude toward it.

It is crucial to identify the existing waste management practices and behavior of households to make adjustments and improvements in achieving sustainable waste management. Although Industry 4.0 or smart technology contribute to the improvement of the solid waste management system, social aspects should not be ignored especially relating to issues with public acceptance, public involvement in planning and implementation, public behavior and shifting value system [5].

This paper attempts to answer questions on how people's behavior from different spatial locations manage their household waste and how they can enhance sustainable behavior in waste management. Therefore, this paper aims to examine the behavior and factors that influence community behavior in managing household waste in rural areas, particularly in Kuching North city. The assessment of the behavior among the community toward waste management is crucial to address their awareness and practices for planning and effective solid waste management. With the aims mentioned above, this explains the need for this study.

## 2 Background of the Study

### *2.1 Waste Management Issues in Sarawak*

Kuching is the capital city of Sarawak and was officially declared as a city on August 1, 1988. Kuching city is divided into two areas, namely, Kuching North and Kuching South. Although Kuching North city has undergone rapid development over the years, it still maintains some of the traditional villages in the area particularly in Petra Jaya, the suburb of Kuching North city. As Kuching North city is moving toward rapid development, solid waste management has become a great concern where landfill is the most widely used method for disposal of MSW. Sarawak has 46 landfills and dumpsites where 36 are open dumpsites. Most of the waste is disposed of mainly to open dumpsites with no pollution controls [6].

Other issues related to waste management in Kuching North city areas are illegal dumping and littering. According to Kuching North City Mayor, illegal dumping and littering are the most common waste management issues in the northern part of Kuching city. As the population in a city grows, the amount of waste generated increases as well. This situation puts pressure on the city's environment and worsens solid waste problems [7, 8].

The existence of new settlements in the city occurs because of migration to the city and within the town itself. The push and pull factors that cause the migration to urban areas put significant pressure on the urban environment. For instance, the occupants of low-cost apartments and slum dwellers are frequently linked to the issue of solid waste management, particularly in urban areas [9]. According to [10] the community of these informal settlement sites produces almost 200 metric tons of solid waste per day. Slums and other informal settlements are believed to contribute to environmental contamination due to poor solid waste management [11–13]. This condition is because of the informal settlement areas lack of facilities and poor attitudes toward managing household waste [13, 14]. Therefore, the study focuses on Kampung Sungai Midin, one of the villages in Kuching North city, which was gazetted as a Government (Village) Reserve in 2006. Although the community has yet to receive their land grant title, they continue to receive basic facilities from the state government, such as paved roads, electricity and clean water supply. There are no door-to-door waste collection services but the local authority provided communal bins to make it easier for the community to dispose of their waste.

The generation and disposal of household waste in Kampung Sungai Midin is one of the issues in solid waste management due to several factors, such as dense residential conditions with narrow alleys that are not accessible by collection trucks, as well as moral and attitude issues that tend to put the responsibility on the local authority. Inadequate consideration of behavior, attitude and perceptions toward waste management at source as well as ranking waste disposal issues in public's mind and official development plans has led to the recent increase in waste management problem in developing countries [15]. Communities always perceive that local authority is responsible for managing waste as long as the waste is collected from their surroundings.

## 2.2 *Solid Waste Management in Sarawak*

In Sarawak, the responsibility of municipal waste management is placed under local authority (LA) administration per the Local Authorities (Cleanliness) By-Laws 1999 which oblige the LAs to decide on, determine, establish or maintain a system for the collection, removal and disposal of waste from all premises within its jurisdiction [16]. The by-law requires LAs to provide solid waste disposal facilities to premises owners by charging a predetermined rate. Solid waste management in Sarawak is also guided by several laws as stated below:

(a) Natural Resources and Environmental Ordinance (Cap 84—Laws of Sarawak Edition 1985)
(b) Local Authorities Ordinance (LAO), 1996
(c) Local Authorities Cleanliness By-Law (LAC), 1998
(d) Environmental Quality Control (Control of Pollution from Solid Waste Transfer Station and Landfill) Regulation, 2009

The state has decided to privatize waste collection, transportation, treatment and disposal particularly in Kuching city to a German and Sarawak Government joint venture operating company named Trienekens (Sarawak) Sdn Bhd. Since then, the local authority has supervised solid waste management and has not been involved in the logistical activity [17]. The privatization of collection services falls under the jurisdiction of three councils, namely, Kuching North City Council (DBKU), Kuching South City Council (MBKS), and Padawan Municipal Council (MPP).

Under the privatization scheme, households are provided with 120 litre volumes of plastic bins. Waste collection is done twice a week on a fixed schedule basis. In some areas, the council supplied communal bins along the main road due to the site being not accessible by municipal trucks. The door-to-door collection is done by uniformed personnel using modern municipal trucks in all accessible areas. The collection zone is divided into 45 zones which cover approximately 135 areas [18].

Sarawak still uses landfills and dumpsites as disposal methods. MSW from Kuching city is mainly disposed of at the sanitary landfills. Most of the waste in rural areas are disposed of at open dumpsites where 78% of the available dumps are without any pollution control.

### 2.3 *Factors Influence Behavior*

Sustainable solid waste management is required for an effective and efficient SWM system, particularly in developing countries [19]. This study is an additional step in Malaysia's transition to sustainable waste management behavior, which is also consistent with the SDG3 (Good health and well-being), SDG6 (Sustainable cities and inclusive communities), and SDG 12 (Responsible consumption and production) of the United Nations Sustainable Development Goals.

New technologies have been developed to solve waste management issues, particularly involving waste generators to separate their waste. There is a concern about environmental sustainability, community involvement, and public awareness in waste management that continues to put pressure on Sarawak municipal waste management.

Many factors influence behavior toward sustainable solid waste management, including social, physical, infrastructure, institutional, and political factors [20]. According to [21], multiple factors might affect the development of the waste management system differently including legislative, social, environmental, health, and economic aspects. Some of these factors must be identified in understanding the behavior toward sustainable waste management whether the current system works according to local requirements and is feasible in terms of social, environmental, and economic aspects. This identification is important because the critical factors in influencing behavior toward the system could vary between individual city's characteristics [22].

## 3 Method and Materials

In this study, various research techniques were employed to obtain quantitative and qualitative data which include questionnaires, interviews, document reviews, and observation. Table 1 shows a summary of the methodology used in the planning and implementation of the data collection process.

A sample size of 75 respondents participated in the survey. The respondents were chosen based on convenience sampling. The researcher met the village headman and

**Table 1** Outline for the planning of data collection

| Techniques for data collection | Description | Corresponding aims |
|---|---|---|
| Observation (Qualitative approach) | Interacting and learning about the community waste management | Current practices used in waste management<br>Existing facilities<br>Physical characteristics |
| Household survey (quantitative approach) | The survey was conducted in Malay language<br>Questionnaire were distributed to household<br>The determination of sample size is based on determination table by [23]<br>75 households participated in the survey | Current practices used in waste management<br>Factors influence behavior toward waste management |
| Interview with local residents and local authority officers (qualitative approach) | Semi-structured interview<br>Face-to-face interviews were conducted on the issues related to waste management such as waste segregation and recycling | Difficulties in waste management<br>Identify waste management problem and proposed solutions |
| Reviewing documents and reports on waste management (qualitative approach) | Online research was done to gather information from official government websites, news portal, media resources and other online data | Existing legislation on SWM<br>Existing waste treatment in Sarawak |

committee to discuss the information needed for this study such as the village's history, demographics, economics and waste management. Interviews were also conducted with the officer from DBKU to gather information on the issues related to solid waste management. The community waste management practices, the village's natural environment, and physical characteristics are observed.

The reliability value of the survey instrument in this study is based on Cronbach's alpha coefficient which is at the value of 0.91. According to Berawi [24] a value over 0.6 indicates high reliability and consistency. For qualitative data, interviews were conducted through face-to-face structured interview, semi-structured interviews, and observation. The interviews were transcribed and analyzed accordingly.

## 4 Findings and Discussion

### *4.1 Knowledge on Sustainable Waste Management*

Waste segregation at source and recycling are two essential components in the hierarchy of integrated solid waste management [25]. Attitude and public awareness of waste management can influence individual willingness to participate in waste segregation and recycling practices [26, 27]. This is because knowledge and attitude

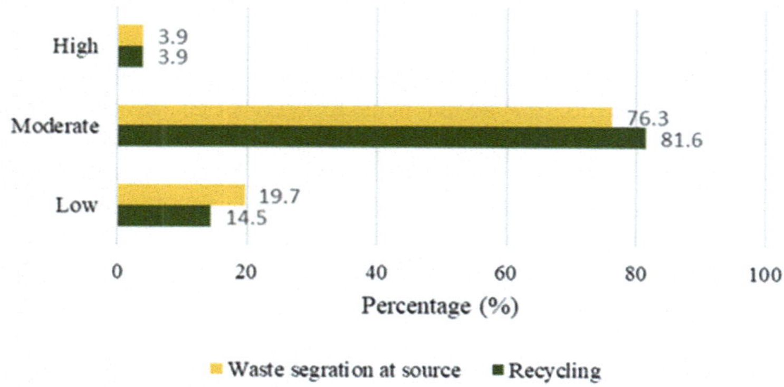


**Fig. 1** Knowledge on waste segregation and recycling

**Table 2** Source of information

| Media | Percentages (%) |
|---|---|
| Social media (Facebook, Instagram, WhatsApp) | 67.1 |
| Radio/television | 46.1 |
| Newspaper | 26.3 |
| Electronic billboard | 13.2 |
| Leaflet/brochure | 23.7 |

can predict an individual's behavior in practicing sustainable solid waste management [28, 29]. According to Otitoju and Seng [26], knowledge is an important predictor of waste segregation behavior, where knowledge can bring changes in behavior. For example, household waste segregation is about what, where, when, and how to practice in real life as mentioned by [30].

The responses received were examined and the results are shown in Fig. 1.

Around 76.3% of the respondents indicated moderate knowledge of waste segregation at source while 19.7% have low knowledge on waste segregation. Only 3.9% of respondents have high knowledge about waste segregation as well as knowledge of recycling. For recycling, the majority of respondents (81.6%) indicated that they have moderate knowledge about recycling followed by 14.5% of the respondents having low knowledge on recycling. The result shows that the respondents had moderate knowledge on waste segregation and recycling.

## 4.2 *Source of Information*

Exposure and awareness are vital to motivate and reinforce positive behavior toward sustainable waste management. Therefore, respondents were asked how they found information on waste segregation and recycling. The responses are shown in Table 2.

Results show that the primary source of information about waste segregation and recycling is through social media (67.1%), followed by radio and television with 46.1%. Respondents also indicated that information was also obtained through newspapers and articles (26.3%) and municipal leaflet (23.7%). Electronic billboards only constitute 13.2% of the responses. Respondents also pointed out that the information in social media and municipal leaflets needs to be improved especially in educating the younger generation on waste segregation and recycling.

Dewan Bandaraya Kuching Utara (DBKU) has done the engagement process to expose the community to the current situation of waste management and provide a platform for the community to convey their problems. This allows the community to learn more about sustainability in managing waste and to be aware of the current situation of waste management. Without public engagement and participation, it may be difficult to maintain the efficiency of MSW services, and if waste is not properly segregated, it may affect the resource recovery system [26].

## 4.3 Physical Environment and Waste Management Behavior

Kampung Sungai Midin, Kuching, is a small Malay village located on the outskirts of Kuching city with 208 houses and population of 1008 people. The community in Kampung Sungai Midin were not provided with individual bins. Still, they prepared their own bins to store their waste before disposing of the waste to the nearest communal bin provided by the local authority. It is very challenging to carry out a proper waste collection in this area due to its unplanned housing layout with narrow alleys that are not accessible by municipal trucks.

The physical aspects were examined to understand the behavior of the community in Kampung Sungai Midin in managing their waste, particularly the physical environment in the village area. Human behavior is affected by the environment that provides a variety of possibilities for its inhabitants. Humans also impact the environment by their presence and actions to adapt to the conditions that affect their lives [31–33]. Before the communal bins were placed in the village, poor social habits led to the household waste being disposed of into the waterways of the river in the village. This may cause environmental damage to the village and affect public health. According to Kendie [15], environmental damage is often caused by human behavior that is probably influenced by several factors such as availability of facilities. The result of this study revealed that the physical features of the river in Kampung Sungai Midin as well as the condition and characteristics of some houses built on the river's waterway were viewed as a way to get rid of household waste. It is easier for some communities to dispose of their waste into the river rather than taking steps to the communal bins. The respondents' believed that the river current would remove the waste away from their houses. The behavior of the respondents in disposing of household waste is presented in Table 3.

The result in Table 3 indicated that the village's physical feature influences most of the respondents' behavior toward waste management. Before they were provided with the communal bins, the community threw their waste into the river and the

**Table 3** Behavior toward waste disposal

| Behavior | Percentage (%) |
|---|---|
| I dispose household waste in designated place (communal bin) | 45.1 |
| I separate the waste before disposing of it to communal bin | 18.1 |
| Sometimes I throw waste into the river | 25.5 |
| I select the waste before disposing of it to the river | 19.1 |
| Sometimes I burn the waste as a way of disposing of household waste | 6.5 |

current will remove them. It was a concern for some communities as this way of disposal might affect their health and environment. Therefore, the local authority provides communal bins in two designated areas to make it easier for them to dispose of household waste.

The result also revealed an improvement in the community behavior in managing their waste after the local authority placed the communal bins in the area. Most respondents (45.1%) dispose of their waste in the communal bins, with 18.1% of respondents segregating the waste before disposing of it. However, 25.5% of the respondents indicated throwing their waste into the river because the distance between the communal bins and their house is quite far. For some respondents, carrying the waste to the communal bin is troublesome, especially perishable food waste.

The design and condition of some houses built on stilts along the river greatly influence respondents' behavior in managing waste they produce. The existence of space under their home seems to function as a waste dump. It would be easier for some of the respondents to dispose of their waste into the river as the river's flow is relatively stable. Table 3 also shows that 19.1% of the respondents select the waste before disposing it to the river. The respondents informed that selected waste, such as food waste is thrown into the river in a small amount. Only 6.5% of the respondents mentioned that they burnt the waste and this depends on the weather.

From the above result, it can be concluded that the physical aspect of the environment influences the behavior and motivation to manage their household waste.

## 4.4 Social and Cultural Aspects Toward Waste Management Behavior

Factors in the social environment are essential in determining the behavior that involves safety, social participation as well as the collective efficacy of the neighborhood environment [34, 35] speculated that social norms and perceived behavior control were not crucial in recycling behavior but [27] discovered that social norms were significantly important in curbside recycling programs where visual evidence of not participating is much higher. This study examined subjective norm criteria, focusing on social pressure and cultural norms [36].

**Table 4** Subjective norms

| Statement | Mean score | Standard deviation | Score interpretation |
|---|---|---|---|
| 1. My family support me for separating household waste | 3.44 | 0.826 | Moderate |
| 2. My neighbor think I should separate waste | 3.65 | 0.837 | Moderate |
| 3. Local authority provide me with the opportunity to participate in waste management program | 3.60 | 0.870 | Moderate |
| 4. I get support from the local NGO to separate waste and recycle | 2.20 | 0.800 | Low |
| 5. My neighbor think I should participate in recycling activities | 3.65 | 0.837 | Moderate |
| 6. I will only participate in waste segregation program if my neighborhood participate | 3.83 | 0.619 | High |
| **Overall mean score** | **3.39** | | **Moderate** |

The study revealed that subjective norms moderately influence respondents' behavior in waste management. Individual's background, culture, age, experience, and even position in life affect how they see and deal with social environments [31]. Based on Table 4, the interpretation of the mean score was adapted from [37]. The primary source of social influence is family, neighbor and local authority despite the mean score being moderate. As for nongovernmental organizations (NGOs), the mean score is low with a value of 2.20 because there are no local NGOs that come to the village to provide awareness programs related to waste management. The community depends solely on the local authority to provide them with knowledge and awareness in managing waste sustainably.

In this study, the subjective norm is related to social pressure that affects the behavior which may cause people to act in such behavior with the acceptance of their social peers [38]. Social pressure might encourage or discourage an individual's behavior [39]. Due to their hectic lifestyle, low social interaction has affected the community's intention to engage in sustainable household waste management. One of the respondents pointed out that if incentives are given, it could increase the participation of the community in waste management programs.

No individuals or groups in the village can initiate a community to actively participate in waste segregation and recycling. The analyses show that the respondents do not have sufficient knowledge of the benefits of waste segregation and recycling. From the above findings, the social aspect has a moderate tendency to influence behavior toward sustainable waste management. The community needs more social support from their local authority and other stakeholders to guide, educate, and monitor their involvement in sustainable waste management.

Meanwhile, cultural aspects in this study refer to the community's culture or habit that influence the behavior in waste management. The findings revealed that most respondents still believe that wastes are objects or products that need to be destroyed or disposed of without acknowledging they might have different ways to manage them sustainably. Based on observation at the communal bins area, a lot of waste that

**Table 5** Barriers to practice sustainable waste management

| Barriers | Percentage (%) |
|---|---|
| Lack of facilities | 38.2 |
| Lack of awareness/attitude | 27.6 |
| Non-strategic location of facilities | 22.4 |
| Misused facilities by others | 11.8 |

has an economic value such as paper, plastic bags, cardboard, bottle plastics, and tin cans is still being discarded. Due to poor social habits and attitude, they threw wastes into the river.

### *4.5 Barriers to the Practice of Sustainable Waste Management*

Identifying barriers to community participation in practicing sustainable solid waste management can help to alter programs that meet the community's needs. Table 5 shows the results on the barriers to practicing sustainable waste management. It was found that the majority of the respondents (32.8%) stated that lack of facilities was one of the barriers to practicing waste segregation and recycling. In addition, 27.6% of the respondents claimed that lack of awareness and attitude also affect respondents to practicing sustainable solid waste management. Other factors such as the non-strategic location of facilities (22.4%) and facilities misused by others (11.8%) were also pointed out as barriers to practicing waste segregation and recycling in the study area.

### *4.6 Suggestions to Improve Sustainable Behavior Toward Waste Management*

Respondents were asked what needs to be done to improve the residents' behavior toward sustainable waste management. Based on Fig. 2, 39.5% of respondents suggested the local authority should do more awareness programs. Respondents also mentioned that they need more educational programs focusing on sustainable waste management especially for the youth to create awareness among them. Meanwhile, 32.9% of the respondents want more waste management facilities. Although they were being provided with communal bins, some were reluctant to use them because of the distance, and the bins were always full as also being used by others outside of the village. Therefore, they suggested providing a communal bin for each alley of the village and a temporary recycling center.

A total of 10.5% of respondents suggested the local authority should also do monitoring and enforcement since the communal bins were misused by others not from the village while 9.2% mentioned that cooperation from all communities is

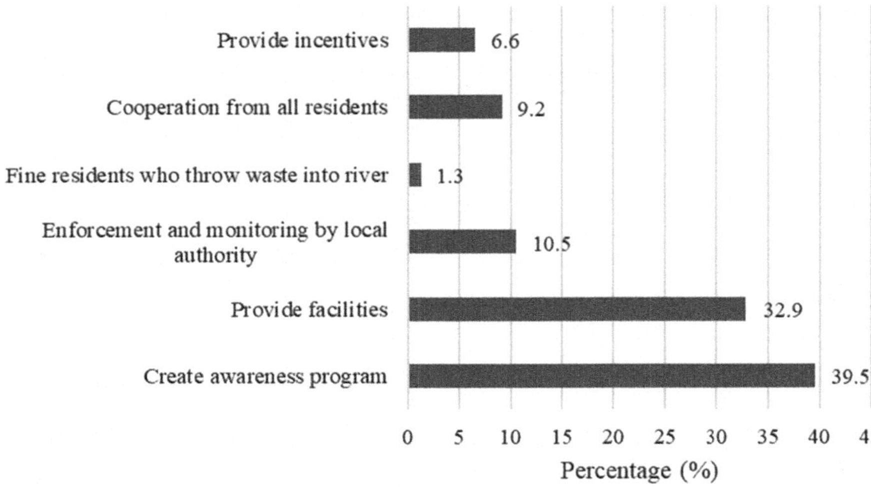


**Fig. 2** Suggestions to improve sustainable behavior

needed to ensure everyone practices sustainable waste management. Other suggestions made by the respondents include imposing fines on those who throw waste into the river and providing incentives to encourage the community to practice sustainable waste management.

## 5 Conclusion and Recommendations

Sustainable solid waste management is a concept that has been introduced here in Malaysia. The idea requires changes in practices and behavior where there are often overlapping responsibilities for task relating to waste management which hinder the effectiveness of sustainable waste management system. Even established technologies may only be suitable to be adopted with a proper modification with detail study and evaluation [19]. Although the effectiveness of solid waste management is a much affected by technological and financial issues, the social perspective can provide insight into understanding human behavior by identifying several factors, including social integration, social participation, and social relationship [34]. A social support system is crucial, and it significantly impacts how people behave.

The findings in this study also offer evidence that local authority does play vital roles in improving program or activities that can enhance community behavior toward sustainable waste management. The program or activities must comply with local conditions and technical, environment, sociocultural, financial considerations and the appropriateness with the local context [40]. The community needs to be supported to actively participate in sustainable waste management because the success in implementing a program or activity depends on the involvement of local community and other through a bottom-up approach. This approach focuses on the involvement of local community to determine a plan of action by considering the needs and problems within the community and its environment [7]. Top-down and bottom-up approaches should be coordinated to ensure waste management's sustainability [41].

After examining the behavior and factors that influence community behavior in managing household waste, several recommendations have been identified to enhance the community's sustainable behavior toward waste management:

- Increase knowledge and awareness

  Knowledge and awareness should be done continuously to educate and empower community behavior toward sustainable waste management. The sense of "mottainai" allows the community the importance of not wasting what nature has given by making the best of what we have. According to Wang et al. [42], a community's ability to appropriately sort waste can be hampered by the lack of knowledge and awareness about the process. The community needs to be involved in planning the program or activities according to their needs and be able to reduce the burden on the environment. Bortoleto et al. [43] believed that knowledge possessed by the community is an important factor in influencing behavior. Research done by Yukalang et al. [44] suggested that a continuous education program on sustainability behavior significantly contributes to sustainable waste management. Ramayah et al. [45] and Li et al. [46] also emphasized the importance of knowledge and awareness in practicing sustainable waste management. Considering that the respondent's waste segregation and recycling knowledge is between low and moderate, local authority, pro-environmental agencies, and policy makers need to take action to improve the community's waste segregation and recycling knowledge. This can be done by developing waste segregation manuals and guidelines, and giving them freely to the community [38]. The community can benefit from these manuals and guidelines by accurately identifying and segregating waste. Alternatively, the media also plays a vital role in providing knowledge and awareness to the community on the importance of sustainable behavior toward waste management. For instance, social media is an effective communication tool in the digital era for connecting people with common interests. The most popular social networking applications being used are Facebook, Instagram, YouTube as well as Twitter [47]. However, other electronic media such as radio, television, and electronic billboard should not be overlooked either because of their potential to promote sustainable waste management [48].
- Provide appropriate facilities and equipment

  Sustainable behavior in waste management can be influenced by the sufficiency and suitability of the infrastructure [49, 50]. According to Mukhtar et al. [19] the availability and suitability of facilities for waste management activities can lead to positive waste management behavior in the community. Insufficient waste management facilities and extra expenses hinder the community's intention to take actions [38]. Physical distance such as communal bins from individual households is one indicator to evaluate accessibility where shorter distance means easier access and saves time. Therefore, the local authority must expand the facilities for community participation. Ando and Gosselin [51] mentioned that adequate space for temporary storage of sorted material can increase waste segregation and recycling behavior. For instance, provide a bigger bin in each village alley for the community to empty waste. This will hinder the community from disposing into the waterways of the river in the village. Community waste segregation and recycling programs should also incorporate not only the procedure on segregating and recycling but also introduce kits or technique applicable to the community.

- Improve stakeholders' collaboration

  The critical factor in sustainable waste management is the involvement and participation of all stakeholders, such as waste producers, formal and informal agencies, nongovernmental organizations, financing institutions, and educational institutions [52]. They play an important role in supporting and guiding the community to practice sustainable waste management. Capacity-building programs and roundtable discussions can be organized to improve stakeholders' involvement and participation. By connecting, stakeholders can benefit from each other to enhance community sustainable behavior toward waste management [53]. They should also evaluate the successes and failures of the organized waste management program where findings can be used to strengthen the structure and produce efficient waste management programs within the community.
- Reward system

  Incentives were viewed as motivational tools to reward good practice and strengthen the effect of intention on behavior [19, 38]. According to Fadhullah et al. [41], providing economic incentive is a key factor in improving household waste management in Malaysia. The community can initiate monetary incentives to encourage others to be actively involved in segregating their waste as mentioned by White et al. [54] monetary incentives are the main motivation to engage in sustainable waste management. Welfens [55] found that monetary rewards in cash, coupons and lotteries can influence recycling behavior. However, Reid et al. [56] argue that the provision of monetary incentives is not sustainable and is suitable only for short-term goals [57]. But still, empirical evidence shows that monetary incentives for recycling play a crucial role in fulfilling the waste management goal [58]. Nonmonetary incentives such as honorary titles and environmental certificates can also be given to the community who made outstanding contributions to waste segregation and recycling [38].

**Acknowledgments** This paper is part of PhD research, funded under Academic Training Scholarship Scheme (SLAI) provided by the Ministry of Higher Education Malaysia (MoHE) and Universiti Malaysia Sarawak (UNIMAS). The study was also supported by funding from Universiti Kebangsaan Malaysia TAP K006974.

## References

1. Kaza S, Yao L, Bhada-Tata P, Van Woerden F (2018) What a waste 2.0: a global snapshot of solid waste management to 2050. The World Bank Group, Washington, DC
2. Ferronato N, Torretta V (2019) Waste mismanagement in developing countries: a review of global issues. Int J Environ Res Public Health 16:1–28
3. Jha AK, Singh SK, Singh GP, Gupta PK (2011) Sustainable municipal solid waste management in low income group of cities: a review. Trop Ecol 52(1):123–131
4. Cheng KM, Tan JY, Wong SY, Koo AC, Amir Sharji E (2022) A review of future household waste management for sustainable environment in Malaysian cities. Sustain For 14:1–127
5. Joos W, Carabias V, Winistoerfer H, Stuecheli A (1999) Social aspects of public waste management in Switzerland. Waste Manag 19(6):417–425
6. NREB Official Website (2021) Available at: https://www.nreb.gov.my/page-0-0-257-EIA-Landfill.html
7. Teuku Anwar TA (2016) Pengurusan Sisa Pepejal di Malaysia: Pendekatan Komuniti. Dewan Bahasa dan Pustaka, Kuala Lumpur
8. Ayeleru OO, Dlova S, Akinribide OJ, Ntuli F, Kupolati WK, Marina PF, Olubambi PA (2020) Challenges of plastic waste generation and management in sub-Saharan Africa: a review. Waste Manag 110:24–42
9. Murad MW, Siwar C (2007) Waste management and recycling practices of the urban poor: a case study in Kuala Lumpur city, Malaysia. Waste Manag Res 25:3–13
10. Komoo I (1996) Urban planning within the contexr of the environment: the Kuala Lumpur experience. Lestari Roundtable Dialogue 1. Institute for Environment and Development (LESTARI), Universiti Kebangsaan Malaysia, Bangi, Selangor Darul Ehsan
11. Abdullah H, Muhammad S, Jahi JM (1987) Squatter settlements and pollution along the Klang River. In: Sani S (ed) Tropical urban ecosystems studies, 2nd edn. Malaysian National Mab Committee, Bangi, pp 61–73
12. Jahi JM (1987) Pollution sources along the Sungai Klang and its tributaries. In: Sani S (ed) Tropical urban ecosystems studies, 2nd edn. Malaysian National Mab Committee, Bangi, pp 88–104
13. Ismail H, Syed Hussain TPR, Mariyappan R (2011) Sikap komuniti setinggan tebingan sungai: Pencemaran alam sekitar dan pengurusannya. Jurnal Perspektif 3(1):1–12
14. Gutberlet J, Kain JH, Nyakinya B, Oloko M, Zapata P, Zapata Campos MJ (2017) Bridging weak links of solid waste management in informal settlements. J Environ Dev 26(1):106–131
15. Kendie SB (1998) Do attitude matter? Waste disposal and wetland pollution in the Cape Coast municipality of Ghana. Malays J Trop Georg 29(2):61–81
16. Tang KHD (2020) Municipal solid waste management in the Sarawak state of Malaysia and the way forward. Asian J Environ Ecol 12(2):38–55
17. Lee S (2012) Challenges and strategies carrying out solid waste management services in Kampung Tabuan Melayu, Kuching, Sarawak, Thesis. Faculty of Social Sciences, Universiti Malaysia Sarawak
18. Trienekens Official Website (2023) Available at http://trienekens.com.my/
19. Mukhtar EM, Williams ID, Shaw PJ (2017) Visible and invisible factors of solid waste management in developing countries. In: Proceeding Sardinia. 16th international waste management and landfill symposium, 2–6 October 2017
20. Ali NE, Siong HC (2016) Social factors influencing household solid waste minimization. MATEC Web Conf 66:1–6
21. Mukhtar EM, Williams ID, Shaw PJ, Ongondo FO (2015) Evolution of waste management systems in developed and developing cities. In: Proceedings Sardinia. 15th international waste management and landfill symposium. CISA Publisher, Cagliari, pp 1–8
22. Bartlett JE, Kotrlik JW, Higgins CC (2001) Organizational research: Determining appropriate sample size in survey research. Information Technology, Learning and Performance Journal 9 (1):43–50

23. Contreras F, Ishii S, Aramaki T, Keisuke H, Connors S (2010) Drivers in current and future municipal solid waste management systems: cases in Yokohama and Boston. Waste Manag Res 28:76–93
24. Berawi MF (2017) Metodologi Penyelidikan: Panduan menulis tesis. Penerbit Universiti Utara Malaysia, Sintok
25. Sakawi Z (2011) Municipal solid waste management in Malaysia. Solution for sustainable waste management. J Appl Sci Environ Sanitation 6(1):29–38
26. Otitoju TA, Seng L (2014) Municipal solid waste management: household waste segregation in Kuching South City, Sarawak, Malaysia. Am J Eng Res 3(6):82–91
27. Barr S (2007) Factors influencing environmental attitudes and behavior: a U.K case study on household waste management. Environ Behav 39(4):435–473
28. Ahmad J, Noor SM, Ismail N (2015) Investigating students' environmental knowledge, attitude, practice and communication. Asian Soc Sci 11(16):284–293
29. Babaei AA, Alavi N, Goudarzi G, Teymouri P, Ahmadi K, Rafiee M (2015) Household recycling knowledge, attitudes and practices towards solid waste management. Resour Conserv Recycl 102:94–100
30. Sidique SF (2008) Analyses of recycling behavior, recycling demand and effectiveness of policies promoting recycling. PhD thesis, Michigan State University, Department of Agricultural Economics
31. Lindheim R, Syme SL (1983) Environment, people and health. Annu Rev Public Health 4:335–359
32. Saarloos D, Kim JE, Timmermans H (2009) The built environment and health: introducing individual space-time behavior. Int J Environ Res Public Health 6(6):1724–1743
33. Steg L, van den Berg AE, de Groot JIM (2012) Environmental psychology: history, scope and methods. In: Steg EM, van der Berg AE, de Groot J (eds) Environmental psychology: an introduction, 1st edn. Wiley-Blackwell, pp 1–12
34. Ahern J, Galea S (2011) Collective efficacy and major depression in urban neighbourhoods. Am J Epidemiol 173(12):1453–1462
35. Chan K (1998) Mass communication and pro-environmental behavior: waste recycling in Hong Kong. J Environ Manag 52(4):317–325
36. Fornara F, Carrus G, Passafaro P, Bonnes M (2011) Distinguishing the sources of normative influence on pro environmental behaviors: the role of local norms in household waste recycling. Group Process Intergroup Relat 14(5):623–635
37. Best JW (1977) Research in education. Prentice Hall, New Jersey
38. Wang S, Wang J, Yang S, Li J, Zhou K (2020) From intention to behavior: comprehending residents' waste sorting intention and behavior formation process. Waste Manag 113:41–50
39. Yuan Y, Nomura H, Takahashi Y, Yabe M (2016) Model of Chinese household kitchen waste separation behavior: a case study in Beijing. Sustain For 8:1–15
40. Abdul Rahman H (2017) Effort and challenges in the application of environmental education in Malaysian school system. Asian J Environ Hist Herit 1(2):61–70
41. Fadhullah W, Imran NIN, Ismail SNS, Jaafar MH, Abdullah H (2022) Household solid waste management practices and perception among residents in the East Coast of Malaysia. BMC Public Health 22(1):1–20
42. Wang S, Wang J, Zhao S, Yang S (2019) Information publicity and residents' waste separation behavior: an empirical study based on the norm activation model. Waste Manag 87:33–42
43. Bortoleto AP, Kurisu KH, Hanaki K (2012) Model development for household waste prevention behavior. Waste Manag 32(12):2195–2207
44. Yukalang N, Clarke B, Ross K (2018) Solid waste management solutions for a rapidly urbanizing area in Thailand: recommendations based on stakeholder input. Int J Environ Res Public Health 15:1–23
45. Ramayah T, Lee JWC, Lim S (2012) Sustaining the environment through recycling: an empirical study. J Environ Manag 102:141–147

46. Li J, Zuo J, Cai H, Zillante G (2018) Construction waste reduction behavior of contractor employees: an extended theory of planned behavior model approach. J Clean Prod 172:1399–1408
47. Muniandy S, Kuan-Siew K, Thurasamy R, Ai Ping T (2019) The role of social media on recycling behavior. Sustain Prod Consum 20:365–374
48. Moh YC, Manaf LA (2014) Overview of household solid waste recycling policy status and challenges in Malaysia. Resour Conserv Recycl 82:50–61
49. Taherzadeh MJ, Rajendran K (2014) Factors affecting development of waste management: experiences from different cultures. In: Ekström KM (ed) Waste management and sustainable consumption: reflection on consumer waste. Routledge, Croyton
50. Guerrero LA, Maas G, Hogland W (2013) Solid waste management challenges for cities in developing countries. Waste Manag 33(1):220–232
51. Ando AW, Gosselin AY (2005) Recycling in multifamily dwellings: does convenience matter? Econ Inq 43:426–438
52. Joseph K (2006) Stakeholder participation for sustainable waste management. Habitat Int 30(4): 863–871
53. Koiwanit J, Filimonau V (2023) Stakeholder collaboration for solid waste management in a small tourism Island. PLoS One 18(7):1–19
54. White K, Macdonnell R, Dahl DW (2011) It's the mind-set that matters: the role of construal level and message framing in influencing consumer efficacy and conservation behaviors. J Mark Res 48(3):472–485
55. Welfens MJ (2016) Drivers and barriers to return and recycling of mobile phones. Case studies of communication and collection campaigns. J Clean Prod 132(2):108–121
56. Reid D, Luyben P, Rawers R, Bailey J (1976) Newspaper recycling behavior: the effects of prompting and proximity of containers. Environ Behav 8:471–481
57. Iyer ES, Kashyap RK (2007) Consumer recycling: role of incentives, information, and social class. J Consum Behav 6:32–47
58. Abila B, Kantola J (2019) The perceived role of financial incentives in promoting waste recycling: empirical evidence from Finland. Recycling 4(1):1–11

# Factors Influencing Smart Technologies Social Acceptance: A Mini Review Study

**Andante Hadi Pandyaswargo, Meilinda Fitriani Nur Maghfiroh, Tanika Dewi Sofianti, and Hiroshi Onoda**

**Abstract** Smart technologies have become increasingly integrated into people's daily lives. Therefore, understanding the factors that influence their social acceptance is becoming crucial. This study compares findings from existing studies on the social acceptance of smart technologies with the recent developments in the field gathered through interviews with stakeholders from the industry. Although respondents agree with the literature's findings, especially on perceived value, the order of importance of trust in the technology, ease of use, and customers' personalities are context based. The identified differences between the two sources are as follows: (1) hedonic motivation and technology readiness were only emphasized by the literature; (2) customer partnership/relationship and customer readiness were only emphasized by some respondents; and (3) the importance of innovativeness was highly valued by the literature but poorly rated by the respondents both as an influencing factor and indicator of acceptance of the technology.

**Keywords** Smart technologies · Technology acceptance · Social acceptance · Stakeholder perception · Adoption

A. H. Pandyaswargo (✉)
Environmental Research Institute, Waseda University, Tokyo, Japan
e-mail: andante.hadi@aoni.waseda.jp

M. F. N. Maghfiroh
Industrial Engineering Department, Universitas Islam Indonesia, Jogjakarta, Indonesia

Faculty of Transport and Logistics, Muscat University, Muscat, Oman

T. D. Sofianti
Industrial Engineering Department, Swiss German University, Tangerang, Indonesia

H. Onoda
Graduate School of Environment and Energy Engineering, Waseda University, Tokyo, Japan

 
S. Fukushige et al. (eds.), *EcoDesign for Circular Value Creation: Volume I*,
https://doi.org/10.1007/978-981-97-9068-5_13

# 1 Introduction

The emergence and rapid expansion of smart technologies have changed the way people interact with the world around them. Smart technologies are used in homes [1], as wearable devices [2], autonomous devices [3, 4], and other artificial intelligence-powered platforms have allowed people to work more conveniently, efficiently, and connectedly [5]. However, investments made to develop smart devices are risky without proper knowledge of whether they will be successfully accepted and adopted among individuals and communities. Michael Spence, a Nobel laureate in economics, commented about investment in automation technologies during his recent visit to Japan [6]. He said, "Like many technology investments, automation technologies will scale up very quickly. However, the returns depend on the market." It is then important for technology developers and policymakers to understand the factors that influence the social acceptance of smart technologies for smooth integration into the market and society.

Like any new technology, social resistance toward accepting something radically innovative is expected [7]. When conventional automobiles were first invented between 1910 and 1930, people reacted violently toward automobiles' operation on the streets of American cities [8]. Although smart technologies were developed gradually and more positively received, acceptance is not guaranteed. In this study, technology acceptance is considered as the user's attitude toward a technology, which is the first step of technology adoption [9]. We will also discuss factors that indicate the success of technology adoption at the end of the study.

The organization of this study is as follows: First, we review the literature on the trends of smart technology development and summarize the identified factors of smart technology acceptance and existing methods of smart technology acceptance. Second, we interviewed five smart technology developers from three countries (Malaysia, Japan, and Indonesia) to gain their insights into the current practice and acceptance of smart technology in their countries. Lastly, we compare our findings from the literature with the insights gained from the technology developers to see if there are any new developments in the field that previous studies have yet to capture.

# 2 Materials and Methods

There are two kinds of materials used in this study. The first is from reviewing previous literature, and the second is interviewing smart technology developers.

The literature reviewed in this study was searched through the Google Scholar search engine. The keywords used for the search were "smart technology social acceptance." We only review papers published within the last 10 years to ensure we review the latest publications. Twenty-one papers were systematically reviewed under this condition, most published after 2020. Through the literature review, we distilled factors influencing the social acceptance of smart technologies.

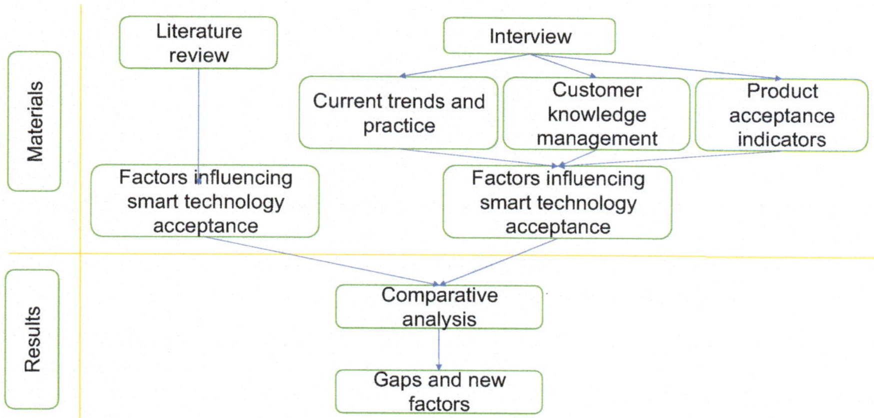


**Fig. 1** Research framework

The companies where this study's interviewees work are located in Japan, Indonesia, and Malaysia. Although it would be interesting to analyze the cultural aspects of the findings based on the country of origin, that would be outside this study research boundary due to confidentiality protection purposes. The interview lasted for 1 h with each interviewee. The first half of the interview contained open-ended questions about the recent developments in the smart technology industry and the company's approaches to smart technology development. The second half of the interview contained multiple-choice questions about customer knowledge.

We compared the results from each source of materials and identified gaps or new factors influencing smart technology acceptance that have yet to be captured in the previous literature. The framework of this study is summarized in Fig. 1.

# 3 Literature Review

## *3.1 Definition, History, and Trends of Smart Technologies*

While there is no official definition of smart technology, there is a consensus on the major characteristics of the technology and a common understanding about what type of elemental technologies are being used. For example, smart technologies are characterized by their real-time ability to collect and analyze data, enabling intelligent decision-making and automation [10]. Some popular elemental technologies of smart technologies include the Internet of Things (IoT), artificial intelligence (AI), machine learning (ML), and data analytics [11]. These elements are often interconnected to produce intelligent responses from the acquired data through intelligent algorithms. Smart technology is also sometimes equipped with physical elements such as sensors, robotic arms, or mobility devices to achieve the goal of intelligent tasks [12].

While the sectors of use of smart technology vary greatly, the goals of its use are commonly to improve efficiency, convenience, and user experience, monitor, control, or optimize various processes [13]. Sectors that benefit from the use of smart technologies include agriculture [14], homes [15], cities [16], transportation [17], waste management [18], health care [19, 20], and manufacturing [10].

The history of smart technology dates to the invention of its basic elements: the digital computers in the 1940s [21] and the internet in the 1980s [22]. While the concept of IoT emerged around the same time as the internet, the term was coined by Kevin Ashton, a Massachusetts Institute of Technology (MIT) lab executive director, in 1999. He described the role of IoT as follows.

> ..., people have limited time, attention, and accuracy. All of which means they are not very good at capturing data about things in the real world. If we had computers that knew everything there was to know about things, using data they gathered without any help from us, we would be able to track and count everything and greatly reduce waste, loss, and cost.

The progress of IoT was further boosted by the popularity of smartphones in the 2000s [23], wearable devices such as fitness trackers [2] and smartwatches [24] in the 2010s. These developments made people's engagement with smart technologies increasingly personal. Between the 2010s and the 2020s, the increased advancement of AI and machine learning algorithms significantly influenced the shaping of smart technologies [25]. These technologies enable devices and systems to analyze large amounts of data, learn from patterns, and make personalized intelligent decisions [26]. Around the same time, smart city initiatives are taking place globally. Examples of smart technologies used in smart cities are smart grids [16, 27], intelligent transportation systems [28], and smart waste management [18].

Several sources have defined "smartness" in a smart city context. For example, a deliverable from the European Union (EU) member countries [29] considers a smart city as one of the stages of cities that have been enabled through the adoption of IoT. The stages are (1) The instrumented cities, (2) The connected city, (3) The smart city, and (4) The responsive city. In a smart city stage, a city typically has a data deployment center where all information about demand and supply is available. However, a smart city has yet to reach the ultimate responsive city stage, where everything in the city is in sync. In a responsive city, the IoT acts as a sensor and an actuator for dynamic repurposing of the city's infrastructure and services. However, such an aspect is already included in the PricewaterhouseCoopers (PWC) definition of smart cities in one of their recent reports about the potential role of smart cities in Japan's future. PWC defines a smart city as, "A city that has the mechanisms for solving social issues, that utilizes new technologies, and that continuously improves resident satisfaction" [30]. This comprehensive definition reflects the expected "smartness" of any ideal smart technologies. It is also in line with the paradigm change from a technology-centered approach to a more inclusive social-issues-centered that was also the key factor of smart city scalability in Japan [16]. Therefore, in this study, we define a "smart technology" as:

A technology that has the mechanisms for solving social issues, utilizes new technology elements such as IoT, AI, ML, and data analytics, and continuously improves user satisfaction.

## 3.2 Factors Influencing Social Acceptance of Smart Technologies

Smart technologies in the market, both in digital platforms and physical devices, have experienced various degrees of success, failures, and sometimes "death and rebirth." Some notable ones are ChatGPT [31], which has been gaining recent mass attention; Facebook [32], which has replaced Myspace and Friendster for its cleaner interface but has to work hard to regain users' trust due to a data breach event; and pagers device that that the popularity of smartphones has replaced [33] but then repurposed in medical facilities.

Considering the various rates of success and failures of smart technologies, the question arises then, "What factors are influencing the social acceptance of smart technologies?"

There have been studies exploring factors influencing social acceptance of smart technologies. Some of these studies were conducted in a smart city or home setting. For example, the study about smart homes in Italy [1, 34] a smart city in Malaysia [35], and a smart city in the United States [36]. The Italian study identified that attitude and perceived behavioral control positively impact the intention to adopt smart grid technologies. Among the behavioral beliefs considered in the study, knowledge about smart grid play a positive role in the intention of technology adoption. On the other hand, values about environmental sustainability were found to be less important than personal financial benefits from energy savings [1].

The Malaysian case study indicated that the intention to use smart technologies is positively associated with trust, self-efficacy, time-saving potential, and facilitating conditions [35]. Moreover, trust in the technology is led by perceived usefulness [37], perceived ease of use, relative advantage, and reliability [35].

The American case study also mentioned trust in the technology, self-efficacy, effort expectancy, and price value to positively impact the behavioral intention of smart technology adoption [36]. Trust was also a moderating factor between perceived security and perceived privacy on behavioral intention. Like the Italian study, price value was a moderating factor between respondents' trust in the government and their intention to use smart-city services [36].

Other studies explored smart technology acceptance factors for specific technology contexts. For example, in autonomous vehicles [17], fitness trackers [2], medical treatment monitoring devices [19], and smart meters [5, 38].

The study by [4] employed survey results from 624 respondents in Germany in the context of three smart technologies usage: autonomous vehicles, fitness trackers, and a medical treatment monitoring device. The study argued that the central keystones for technology acceptance are the trade-off between perceived benefits

**Table 1** Factors influencing smart technology acceptance based on the literature reviews

| Influencing factors | Reference(s) |
|---|---|
| Trust to the technology/Perceived trust/Trust | [3, 27, 35, 36, 39–42] |
| Ease of use/Self-efficacy/Effort expectancy | [20, 35, 36, 38, 41, 43–45] |
| Price/Financial benefits/Financial values | [1, 36] |
| Reliability | [46] |
| Perceived usefulness | [3, 20, 35, 37] |
| Hedonic motivation | [37, 38] |
| Technology readiness | [43] |
| Innovativeness | [20, 42, 44] |
| Quality of service/Customer service | [41, 42, 47] |
| User's personality/attitude | [1, 41, 48] |

and privacy concerns. This finding was particularly important for the context technologies in this study as they require users' willingness to disclose their personal information to deliver output.

The study on autonomous vehicles (AV) highlighted why people might have a hard time trusting a smart technology; the autonomous function of AI is heavily based on machine learning algorithms that are characterized as black boxes because the deeper layer of how the algorithms work is not easily explainable. Furthermore, how artificial agents learn and produce answers may not be comprehensible to humans. These factors create user uncertainties and may cause perceived risks [3].

Finally, the smart meter study mentioned three influencing factors for acceptance: hedonic motivation, performance expectancy, and effort expectancy [38]. These findings align with other studies that proposed perceived usefulness and ease of use as influencing factors. We summarized findings from the literature reviewed in Table 1.

Not all studies agreed with all the previously discussed findings. A survey conducted in a tourism-oriented town in Italy [34] rejected reliability as something that would correlate positively with smart technology adoption. The study highlighted the specific characteristics of the 216 respondents from the responding community that suffer from an aging population, desertification, lack of infrastructure, and digital divide as the factors influencing the extraordinary results. Moreover, sustainability value, which was considered a weak factor by another Italian study [1], was considered a positive factor in this study, along with ease of use and perceived usefulness.

Another opposing study came from a survey of over 3000 residents from various cities across Denmark. The study showed that the intention to use smart technologies is determined by trust, policy, awareness, social influence, and knowledge about the technology. However, the study revealed that perceived usefulness has no influence [39]. The rejection of perceived usefulness as something that could influence technology adoption is also supported by [34, 40], especially for some technologies requiring human contact.

Divergent findings in previous studies may indicate that influencing factors of smart technology may depend on the characteristics of the potential technology users. This argument is supported by Zitkiene et al. [41], who noticed two dominant consumer categories: the early adopters and the late majority. According to the study, financial expenditure is not a problem for some people, and expensive investments can be planned responsibly. The consumers' personality was also mentioned by Sindermann et al. [42] as something behind the factors that influence smart technology acceptance. For example, the agreeability traits of a person are correlated with their perceived usefulness of a technology. Furthermore, Sindermann et al. [42] argued that the characteristics of AI could not determine people's attitudes toward AI technology, but users' characteristics may. Therefore, knowledge of the user's characteristics would be more useful in predicting an individual's attitude toward an AI product.

Other studies mentioned innovativeness that may indirectly influence customers' attitudes toward smart technologies [43, 44]. However, this factor may entail further investigation, such as its technology readiness level. New technologies with low readiness levels are usually innovative but expensive. Prices drop only when the technology achieves high mass production and commercialization [45]. Other than price, Norouzi et al. [46] argued that communication is key to the social acceptance of innovative technologies.

A multicountry review study on AI acceptance was conducted by Kelly et al. [40]. The study covered the United States, India, Canada, Lithuania, Finland, Spain, France, Germany, Singapore, Sweden, Britain, Italy, Russia, Australia, the Netherlands, United Arab Emirates, and Spain. It also covered several industrial contexts, such as telecommunication, healthcare, customer service, the wine industry, disaster relief operations, and mental health treatment. Supporting other studies, the study found perceived usefulness, performance expectancy, attitudes, trust, and effort expectancy as positive predictors of AI technology acceptance. However, the study also highlighted that cultural factors play a role in the acceptance or rejection of AI.

### *3.3 Existing Methodologies to Measure Social Acceptance of Smart Technologies*

Previous studies have used several tools to measure the acceptance of smart technologies. Ji et al. [47] used factor analysis to find key performance indicators of smart home energy technology (SHET) to find the technology's user expectations. Prassida and Asfari [48] used the Unified Theory of Acceptance and Use of Technology (UTAUT) to suggest the key determinants of intention and usage behavior of new technology systems. UTAUT was also used by Gumz and Fettermann [38] to measure the acceptance of smart meters. Ponce et al. [27] used a modified signal detection theory (SDT) with fuzzy logic-2 to capture end users' perceptions about smart grid technology.

Developments of new methods have also emerged. Based on dissatisfaction with the existing methods' comprehensiveness, Occhipinti et al. [49] proposed a new method called the Social Acceptability of Technology (SAT). The method integrates multiple qualitative and quantitative methodologies that evaluate stakeholder perceptions of user experience, values, and trust. This development justifies the use of stakeholders' perceptions and qualitative approaches in the study.

# 4 Results and Discussion

## *4.1 Interview Results*

Interview respondents in this study work as an engineer, a project manager, a chief of business, a chief of AI, and a cofounder of the company. The type of products and services their companies offer varies. Examples of products made by the interviewee's companies in this study are manufacture and logistic automation (autonomous guided vehicle (AGV), autonomous mobile robots (AMRs), robotic arms, artificial intelligence-based visual inspection, special purpose machine (SPM)), digital payment applications (electronic money, electronic wallet, white labeling), smart vision and control systems, face detection for surveillance or forensic purposes, traffic monitoring, data logging applications for healthcare and childcare facilities, and a chatbot. The number of employees of each company ranges from about 50 to 400 people. The business style ranges from business to customer (B to C), business to business (B to B), business to business to customer (B to B to C), and business to government (B to G). Interviewees' answers to five open-ended questions about recent trends in their practices are summarized in Table 2.

Most respondents agreed that the recent surge of interest in smart technologies had impacted their businesses positively as market interests and new opportunities emerged. However, they have also foreseen negative impacts on society in general, such as some work skills that technologies may replace, causing a loss of employment.

Smart technology design inspirations came from different outlets for each respondent. Some examples are needs in society, insights from experts, inspiration from abroad, and suggestions from various divisions in the company. The way inspiration is gained varies through consultation with customers, interviews with experts, participatory observation on the field, social interaction with co-workers, and visits to technology exhibitions abroad.

On the topic of the necessity of demonstration and training of the technology versus a highly intuitive technology [42], it seems to depend on the complexity and the nature of the technology context. The higher complexity and specialized technologies usually require assistance for their usage. On the other hand, common and simple ones are expected to be intuitive.

**Table 2** The open-ended question answers summary

| | Respondent 1 | Respondent 2 | Respondent 3 | Respondent 4 | Respondent 5 |
|---|---|---|---|---|---|
| Impact of the surge of public interest in smart technologies | 1. Newly opened opportunities<br>2. Increased affordability and efficiency of smart technologies | The false expectation of what AI can do now | 1. Optimization of the current product and services<br>2. Some jobs have been replaced by smart technology | Increased market interest in smart products offered | 1. New and more complex demands from customers<br>2. New competitors |
| Source of product inspiration | Needs identification through consultation with customers | 1. Gain insights from subject matter experts<br>2. Participatory observation | 1. Problem identification from the customer's perspective that needs to be solved<br>2. Ideas from various teams in the company (e.g., business team, operation team, and others.) | 1. Needs identification from the domestic market<br>2. Observing newly developed products abroad | Combining the available human capacity with computer capacity to solve problems experienced by customers |
| Why customers choose your company | 1. Experienced people<br>2. Technology mastery<br>3. Affordable solutions<br>4. Product durability | 1. Compliance with governmental regulations<br>2. Excellent customer service and accessibility | 1. A match between demand with the available product/services offered.<br>2. Competitive pricing<br>3. Relationship<br>4. Use case | 1. Word-of-mouth<br>2. Risk taker to solve problems with new technologies | 1. Customer service accessibility<br>2. Effective problem identifier |
| Main factors of your product acceptance | Customers' needs are being met by the product offered | 1. It answers the customer's specific needs<br>2. Reasonable pricing that can be responsibly justified | Personalized features to the users' needs | 1. Market readiness<br>2. How forward-thinking the customers are toward new technologies | Products appropriately meet the identified needs and requested order |

(continued)

**Table 2** (continued)

| | Respondent 1 | Respondent 2 | Respondent 3 | Respondent 4 | Respondent 5 |
|---|---|---|---|---|---|
| Demonstration and training necessary for customers | 1. For B-to-C products: Should be intuitive without instructions<br>2. For B-to-B products: Demonstration and training are necessary | Demonstration and training necessity depends on the complexity of the product offered | Not necessary for our products. We highly value ease to use | Smart products in our field are not commonly used in society. Therefore, demonstration and training are necessary | Demonstrations are provided, but users are encouraged to explore by themselves too |

## 4.2 New Regulations on Smart Technologies

Most of the respondents in this study referred to the context-specific laws and regulations of their developed methodologies. These laws may apply internationally or nationally. For example, for facial recognition technology (FRT), an international standard made by the National Institute of Standards and Technology (NIST) [50] must be complied with. Then, there is ISO27001 [51], which regulates data security and privacy that must be followed before the developed FRT can be offered to potential customers. The government of Indonesia recently released a national regulation on personal data handling in 2022 [52], and it must be complied with by all technologies dealing with personal data in the country. A respondent suggested the need for specific laws that regulate smart technologies in general, for example, the handling of large language models, limits of the size of data that may be processed, and appropriateness of the developed models to the local culture and social values.

Japan has initiated an international standard to ensure the safety and security of IoT systems in 2021. These standards are the "IoT Security Guidelines" and "IoT Safety/Security Development Guidelines." The government of Japan expected this standard to be widely used in developing and maintaining IoT products and services and to improve the safety and security of the more connected world [53].

## 4.3 Customer Knowledge Management

According to the book developed by a Silicon Valley product group, *INSPIRED: How to Create Tech Products Customers Love* [54], deep customer knowledge is one key success factor of a technology product. In this study, we developed and presented the dimensions of customer knowledge management based on our previous study [55] to the respondents and asked them to choose the relevant practices in their company (Table 3).

We found that all the respondents already practice safe data storage, commonly on safe cloud data storage, and use the information gained from customers in all stages of product development. This aligns with the *INSPIRED* book that emphasized the need for frequent customer consultation through prototype demonstrations to get frequent feedback throughout the product development stages [54]. We also found that all respondents provide customer feedback channels, either through email addresses, phone numbers, social media channels, or face-to-face communications. Most respondents also encourage informal communications with customers, using social media, shared activities on common interests and hobbies. Although considered effective in maintaining partnerships, these practices are not commonly officially organized by the company. The customers' feedback gained through these formal and informal channels are usually used to develop new products or add features to existing products.

**Table 3** Dimensions of customer knowledge management

| |
|---|
| **Input dimension** |
| Knowledge of customer |
| Data storage |
| Data usage in product development stages |
| Knowledge from customer |
| Feedback channels |
| Feedback prompting strategies |
| Feedback use |
| Knowledge for customer |
| Product promotions and marketing |
| Product usage demonstration and training |
| **Output dimension** |
| Knowledge co-creation |
| Quality of innovation |
| Quantity of innovation |

Through exhibitions, media, digital promotion, and networking with potential customers and stakeholders, marketing is done. One of the respondents provides educational training based on an invitation or a proposal to raise awareness about the products offered.

Regarding knowledge co-creation with customers, a five-level Likert scale was given to the respondents to rate their company's innovation level. Interestingly, most respondents were reluctant to choose "very good." In their argument, a dramatic innovation is not necessarily something desirable. They argue that changes in society take time and ensuring that the technology is appropriate to the current needs and readiness of the people to use and maintain the technology is more important than innovation.

## *4.4 Comparison Between the Literature and the Interview Results on Indicators of Smart Technology Acceptance*

We asked the respondents to rate how strong the factors [55] summarized in Fig. 2 are in indicating a smart technology product acceptance. Respondents were asked to choose from five levels: very strong, strong, neutral, weak, and very weak. We weighted their answers for each option with 3, 2, 1, −1, and −2 points, respectively. Results of the weighted answers are summarized in Fig. 2. The results show that the strongest indicators of smart product acceptance are the "number of repeated purchases," "the number of new projects resulting from word-of-mouth," and "good relationship with the customer." On the other hand, "high level of innovation" and "decision speed from offer to deal to purchase" are considered weak.

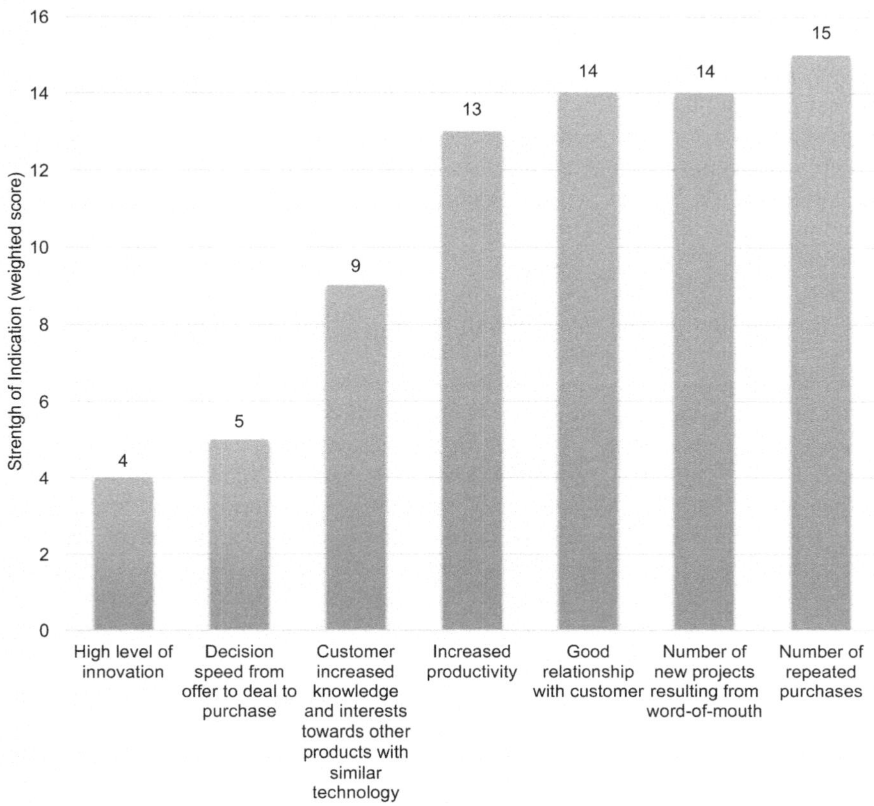


**Fig. 2** Indicators of smart technology acceptance, weighted by respondents

We also asked if other factors may be considered good indicators of smart product acceptance. The following are some of the responses; "new feature requests from the customer," "increased media interests," "interview or presentation invitations," and "number of active users." A respondent mentioned that for app-based products, it is easy to see how many users have downloaded the app for app-based products. However, it does not necessarily reflect how many people use the app regularly. Therefore, it is necessary to look specifically at the number of "active users" to rate how accepted the product is.

## 4.5 *Comparison Between the Literature and the Interview Results on Influencing Factors of Smart Technologies Acceptance*

We presented some findings from the literature about factors influencing the acceptance of smart products to the interviewees. Specifically, we asked the respondents to rank from the most important to the least important factors influencing their product acceptance among "trust in the technology," "perceived usability," "ease of use," and user's personality." Although respondents provided varying ways of ranking these factors, "perceived usability" came up as the most mentioned factor to be on the first position, followed by "trust to the technology," "ease of use," and "users' personality". Trust to the technology was found to be highly critical in privacy-sensitive products such as financial service systems and government level surveillance products. Ease of use, on the other hand, is often placed in the middle as it is commonly assumed to be something that comes together with trust to the technology or perceived usability. In the literature itself, ease of use was found to be not directly correlated with "willingness to pay" while perceived usability is [20].

Comparing the overall responses from the respondents with the summary of factors in Table 1. The gaps identified are as follows: none of the respondents mentioned "hedonic motivation" or "technology readiness." On the other hand, the literature did not emphasize the importance of "relationship/partnership with customers" and "market readiness" the way the respondents did. For example, the literature noted that some customers do not mind paying more for an innovative product due to hedonic motivation [38, 56]. However, almost all the respondents agree that price competitiveness or affordability is a major determinant of smart product acceptance. Regarding technology readiness, all respondents never mentioned it as a prohibiting factor. Instead, the maturity process tends to be worked out for, together with the customers, within the technology development process.

For customer relationships, although each respondent shows a varying degree of eagerness toward the topic, most agree that it is an important aspect in determining acceptance. A respondent argued that being accessible for customer support is highly important; some customers want to reach out for immediate troubleshooting support through social media on direct communication with the key person. Another respondent also supported this factor by saying that as customers already spend a significant amount of resources on the product, it is acceptable if they would like to gain the full benefits, even if it requires personal communication. In the absence of customer support accessibility, for example, due to buying imported goods without international warranty or customer support that does not speak the local language or understand the local culture, product acceptance can be hindered. Lastly, market readiness, rather than technology readiness, was considered a factor of whether the customer is ready to invest in new technology and commit to maintaining its operation. Introducing a novel smart technology that is considered "too soon" for the market capacity and interest may backfire with the non-acceptance of the technology.

There is also an opposing view between the literature and the respondents. While innovativeness was argued to be a positive factor in the literature [20, 42, 57], it was rated poorly by most of the respondents. Respondents argue that the development of needs in society does not evolve dramatically or too fast. If there is a need for dramatic innovation, it might even imply a failure of the previous product. While small changes such as added features are desirable, innovation does not necessarily indicate acceptance of technology.

## 5 Summary

This study reviewed the existing literature on factors of smart technology acceptance and interviewed key people working in the industry. Results from both sources were compared to see gaps and new knowledge from the currently accelerated growth of the industry.

The influencing factors identified from the literature are summarized as follows: trust in the technology/perceived trust/trust, ease of use/self-efficacy/effort expectancy, price/financial benefits/financial values, reliability, perceived usefulness, hedonic motivation, technology readiness, innovativeness, quality of service/customer service, user's personality/attitude.

Respondents in this study were asked about the current trends and practices, their customer knowledge management practices, and their opinions about some of the factors identified in the literature. We found that the recent surge of interest in the smart industry can positively and negatively impact the industry. Positively, there is an increase in demand and open opportunities in the industry. Negatively, there may be false perceptions of AI development and the reduced workforce needed for some jobs in society. Furthermore, while existing regulations in each sector where smart technology can help remain in place, international regulations specific to IoT have emerged. Regarding how the industry gains inspiration for product development, the sources varied from internal teams, customers, and international trends.

Respondents agree with the literature's findings, especially on perceived value. However, the order of importance of trust in the technology, ease of use, and customers' personalities are context-based. The identified gaps and differences between the two sources were identified as follows: (1) hedonic motivation and technology readiness were emphasized only by the literature; (2) customer partnership/relationship and customer readiness were emphasized only by some of the respondents; (3) the importance of innovativeness, which was highly valued by the literature, was poorly rated by the respondents both as an factor influencing and indicator of acceptance of the technology.

Finally, although a study reviewed in this study indicated users' higher interest in financial savings over environmental sustainability, orienting smart technology developments to the identified acceptance factors in this study will improve the effectiveness and efficiency of resource expenditure and marketing. Such practice will eventually contribute to both environmental sustainability and financial benefits.

This study has limitations due to confidentiality protection. Therefore, we did not conduct in-depth discussions on cultural and specific technology contexts. However, we acknowledge the importance of these aspects as results may vary widely depending on the customer's location and the specific technology discussed. Therefore, our future study will explore those areas from the users' perspective, ideally by employing larger data that is quantitatively analyzed.

**Acknowledgments** This study was supported by the Japan Society for the Promotion of Science (JSPS) KAKENHI grant number JP23K11548 titled "Improving the social acceptance of smart technologies: A people-centric approach for sustainable smart waste management and smart mobility technologies." The authors express tremendous gratitude to the respondents who contributed to this study; Mr. Thalhah Hanif from Asperio, Mr. Sandi Fajariadi from MotionPay and Association of Payment Gateway Indonesia; Dr. Adhiguna Mahendra; and Mr. BARROUX Guillaume from ExaWizards Inc.

## References

1. Perri C (2020) Smart users for smart technologies: investigating the intention to adopt smart energy consumption behaviors. Tech Forecast Soc Change 155:119991
2. Santo B (2019) The consumer electronics hall of fame: Fitbit – IEEE Spectrum [cited 2023 Jun 6]. Available from: https://spectrum.ieee.org/the-consumer-electronics-hall-of-fame-fitbit
3. Choung H, David P, Ross A (2022) Trust in AI and its role in the acceptance of AI technologies. Int J Hum Comput Interact 39:1–13
4. Schomakers EM, Lidynia C, Ziefle M (2022) The role of privacy in the acceptance of smart technologies: applying the privacy calculus to technology acceptance. Int J Hum Comput Interact 38(13):1276–1289
5. Humeres M (2022) "Meters can be wonderful": the relevance of democracy and technical aspects for user acceptance of smart meter policy in Chile. Energy Res Soc Sci 90(September 2021):102613
6. Waseda University International Office (2023) UBS & WASEDA: Nobel laureates symposium 「Nobel perspectives live!」 Special Event in Japan – Waseda University [Internet]. [cited 2023 Jun 9]. Available from: https://www.waseda.jp/top/en/other/77702
7. König M, Neumayr L (2017) Users' resistance towards radical innovations: the case of the self-driving car. Transp Res Part F Traffic Psychol Behav 44:42–52
8. Norton PD (2008) Fighting traffic: the dawn of the motor age in the American City – MIT Press Scholarship. MIT Press [cited 2022 Jul 21]. Available from: https://mitpress.universitypressscholarship.com/view/10.7551/mitpress/9780262141000.001.0001/upso-9780262141000
9. Granić A (2022) Educational technology adoption: a systematic review. Educ Inf Technol 27(7):9725–9744
10. Thoben KD, Wiesner S, Wuest T (2017) "Industrie 4.0" and smart manufacturing-a review of research issues and application examples. Int J Autom Technol 11(1):4–16
11. Shah SFA, Iqbal M, Aziz Z, Rana TA, Khalid A, Cheah YN et al (2022) The role of machine learning and the internet of things in smart buildings for energy efficiency. Appl Sci 12(15): 7882
12. ts2space (2023) The advantages of robotics and automation for recycling and waste management [Internet]. [cited 2023 May 18]. Available from: https://ts2.space/en/the-advantages-of-robotics-and-automation-for-recycling-and-waste-management/
13. Harvard Business Review (2014) How smart, connected products are transforming competition [cited 2023 Jun 6]. Available from: https://hbr.org/2014/11/how-smart-connected-products-are-transforming-competition
14. Mahmud MS, He L, Heinemann P, Choi D, Zhu H (2023) Unmanned aerial vehicle based tree canopy characteristics measurement for precision spray applications. Smart Agric Technol 1(4): 100153
15. Wilson C, Hargreaves T, Hauxwell-Baldwin R (2017) Benefits and risks of smart home technologies. Energy Policy 103:72–83
16. Trencher G (2019) Towards the smart city 2.0: empirical evidence of using smartness as a tool for tackling social challenges. Technol Forecast Soc Change 142:117–128
17. Pandyaswargo AH, Maghfiroh MFN, Onoda H (2023) Global distribution and readiness status of artificial intelligence application on mobility projects. Energy Rep 9:720–727

18. Abdullah N, Alwesabi OA, Abdullah R (2019) IoT-based smart waste management system in a smart city. Adv Intell Syst Comput 843:364–371
19. Alshamrani M (2021) IoT and artificial intelligence implementations for remote healthcare monitoring systems: a survey. J King Saud Univ – Comput Inf Sci 34(8):4687–4470
20. Bettiga D, Lamberti L, Lettieri E (2020) Individuals' adoption of smart technologies for preventive health care: a structural equation modeling approach. Health Care Manag Sci 23(2):203–214
21. Williamson T (2021) History of computers: a brief timeline (2021) Live Science [cited 2023 Jun 6]. Available from: https://www.livescience.com/20718-computer-history.html
22. Brahambhatt R (2021) Origin of the Internet: who invented the World Wide Web? [cited 2023 Jun 6]. Available from: https://interestingengineering.com/culture/origin-of-the-internet-who-invented-the-world-wide-web
23. Montgomery A, Mingis K (2022) The evolution of Apple's iPhone | Computerworld [cited 2023 Jun 6]. Available from: https://www.computerworld.com/article/3692531/evolution-of-apple-iphone.html
24. Lamkin P. (2023) Smartwatches 1927–2023: The devices that paved the way for the Apple Watch – Wareable [cited 2023 Jun 6]. Available from: https://www.wareable.com/smartwatches/smartwatch-timeline-history-watches
25. Tang X, Li X, Ding Y, Song M, Bu Y (2020) The pace of artificial intelligence innovations: speed, talent, and trial-and-error. J Informetr 14(4):101094
26. Burns E (2023) What is artificial intelligence (AI)? – AI definition and how it works [cited 2023 Jun 6]. Available from: https://www.techtarget.com/searchenterpriseai/definition/AI-Artificial-Intelligence
27. Ponce P, Polasko K, Molina A (2016) End user perceptions toward smart grid technology: acceptance, adoption, risks, and trust. Renew Sust Energ Rev 60:587–598
28. Visan M, Negrea SL, Mone F (2021) Towards intelligent public transport systems in Smart Cities; Collaborative decisions to be made. In: Procedia computer science. Elsevier B.V, pp 1221–1228
29. Miaoudakis A (2018) Ideal cities: intelligence-driven urban internet-of-things ecosystems for circular, SAfe and IncLusive smart CITIES, Report No.: H2020-MSCA-RISE-2017. European Comissions. Available from: https://www.ideal-cities.eu/wp-content/uploads/2022/10/IDEAL-CITIES-D1.1_FINAL.pdf
30. PWC (2021) Smart Cities in 2050: rebuilding the future of Japanese cities [cited 2023 Sept 4]. Available from: https://www.pwc.com/jp/en/knowledge/thoughtleadership/assets/pdf/smart-city2050-en.pdf
31. Chow AR (2023) Why ChatGPT is the fastest growing web platform ever, time. [cited 2023 Jun 6]. Available from: https://time.com/6253615/chatgpt-fastest-growing/
32. Nix N (2021) Facebook promised to gain the Public's Trust – Bloomberg [cited 2023 Jun 7]. Available from: https://www.bloomberg.com/news/newsletters/2021-09-16/facebook-s-promised-to-gain-the-public-s-trust
33. Beier (2011) Life and death of pagers: Haoran's marketing blog [Internet]. [cited 2023 Jun 7]. Available from: https://blogs.ubc.ca/beier0816/2011/02/22/life-and-death-of-pagers/
34. Baldi G, Megaro A, Carrubbo L (2023) Small-town citizens' technology acceptance of smart and sustainable city development. Sustainability 15(1):325
35. Gunabalan JJH, Ooi SK, Yeap JAL (2022) Assessing residents' receptiveness towards Smart City technologies. Proc Natl Int Conf 15:18
36. Habib A, Alsmadi D, Prybutok VR (2019) Factors that determine residents' acceptance of smart city technologies. Behav Inf Technol 39(6):610–623
37. Lenz J, Bozakov Z, Wendzel S, Vrhovec S (2023) Why people replace their aging smart devices: a push–pull–mooring perspective. Comput Secur 130:103258
38. Gumz J, Fettermann DC (2022) What improves smart meters' implementation? A statistical meta-analysis on smart meters' acceptance. Smart Sustain Built Environ 11(4):1116–1136

39. Billanes J, Enevoldsen P (2022) Influential factors to residential building occupants' acceptance and adoption of smart energy technologies in Denmark. Energ Buildings 276:112524
40. Kelly S, Kaye SA, Oviedo-Trespalacios O (2022) What factors contribute to the acceptance of artificial intelligence? A systematic review. Telemat Inform 77:101925
41. Zitkiene R, Markeviciute G, Mickeviciene M (2017) Factors influencing consumer behaviour and decision to use smart technologies perspectives of business and entrepreneurship development in digital age. Perspectives of business and entrepreneurship development in digital age [Cited 2023 June 6] Available from: https://conference.fbm.vutbr.cz/ic/index.php/ic/article/view/60/15
42. Sindermann C, Yang H, Elhai JD, Yang S, Quan L, Li M et al (2022) Acceptance and fear of artificial intelligence: associations with personality in a German and a Chinese sample. Discov Psychol 2(1)
43. Park I, Kim D, Moon J, Kim S, Kang Y, Bae S (2022) Searching for new technology acceptance model under social context: analyzing the determinants of acceptance of intelligent information technology in digital transformation and implications for the requisites of digital sustainability. Sustainability 14(1):579
44. Roy SK, Balaji MS, Quazi A, Quaddus M (2018) Predictors of customer acceptance of and resistance to smart technologies in the retail sector. J Retail Consum Serv 42:147–160
45. Ihara I, Pandyaswargo AH, Onoda H (2018) Development and the effectiveness of the J-TRA: a methodology to assess energy technology R&D programs in Japan. In: EcoDePS 2018 proceedings Tokyo, pp 109–117
46. Norouzi F, Hoppe T, Elizondo LR, Bauer P (2022) A review of socio-technical barriers to smart microgrid development. Renew Sust Energ Rev 167:112674
47. Ji W, Chan EHW (2020) Between users, functions, and evaluations: exploring the social acceptance of smart energy homes in China. Energy Res Soc Sci 69:101637
48. Prassida GF, Asfari U (2021) A conceptual model for the acceptance of collaborative robots in industry 5.0. Procedia Comput Sci 197:61–67
49. Occhipinti C, Carnevale A, Briguglio L, Iannone A, Bisconti P (2022) SAT: a methodology to assess the social acceptance of innovative AI-based technologies. J Inf Commun Ethics Soc 21(1):94–111
50. NIST (2020) Facial Recognition Technology (FRT) | NIST [cited 2023 Jun 12]. Available from: https://www.nist.gov/speech-testimony/facial-recognition-technology-frt-0
51. ISO (2022) ISO/IEC 27001 Standard – Information Security Management Systems [cited 2023 Jun 12]. Available from: https://www.iso.org/standard/27001
52. BPKRI (2022) UU No. 27 Tahun 2022 tentang Pelindungan Data Pribadi [JDIH BPK RI] [cited 2023 Jun 12]. Available from: https://peraturan.bpk.go.id/Home/Details/229798/uu-no-27-tahun-2022
53. METI (2021) New International Standard for Safe Use of IoT Products and Systems Issued [cited 2023 Jun 12]. Available from: https://www.meti.go.jp/english/press/2021/0621_003.html
54. Cagan M (2018) INSPIRED: how to create tech products customers love. Wiley, New Jersey
55. Sofianti T (2014) Framework for measuring knowledge management for service development in industrial distributor. In: Proceedings of the international MultiConference of engineers and computer scientists 2014. IMECS, Hong Kong
56. Özbek V, Alnıaçık Ü, Koc F, Akkılıç ME, Kaş E (2014) The impact of personality on technology acceptance: a study on smart phone users. Procedia – Soc Behav Sci 150:541–551
57. Beştepe F, Yildirim SÖ (2021) Acceptance of IoT-based and sustainability-oriented smart city services: a mixed methods study. Sustain Cities Soc 80(103794):103794

# Part III
# Business Models & Policies

# Plastic Packaging Policy in Malaysia: An Analysis Towards the Implementation of Extended Producer Responsibility (EPR)

**Khalilulnisha Abu Bakar, Ahmad Fariz Mohamed, and Yusuke Kishita**

**Abstract** Malaysia has one of the highest plastic packaging consumptions among Southeast Asian countries. However, the amount of plastic pollution and the lack of capacity to manage plastic waste effectively has led Malaysia to adopt a mandatory EPR for plastic packaging in 2026. Under the federalised and privatised system of solid waste management in Malaysia, policy challenges and opportunities in implementing the said EPR were discussed and analysed. Documents and reports between the years 2007 and 2021 endorsed by government agencies, brand owners and associations were used and analysed using thematic analysis. Data from survey feedbacks from consumers were also analysed to supplement the discussion. Even though there are many gaps in this area, only legislation and institutional components were discussed in this study. Findings from the consumer survey showed that 95.4% of the respondents were supportive of the implementation of EPR programme and 80.3% agreed for it to be carried out on a mandatory basis. However, the federalisation system and related Act only covers certain states in Malaysia and therefore insufficient to support the establishment of a nationwide EPR policy. Referring to federal, state, and local governments' structure of governance, there is separation and overlapping of powers between these agencies portraying a disjointed approach in managing solid waste, thus creating grey areas in certain aspects of solid waste management. This sparked concerns among producers, brand owners and relevant NGOs especially in the view of implementing a mandatory EPR. In conclusion, the existing legislations and institutional set-up form a solid foundation for an adoption of EPR policy in Malaysia. However, harmonisation and streamlining of these elements, having a clearly defined roles of the stakeholders

K. Abu Bakar (✉) · A. F. Mohamed
Institute for Environment and Development, Universiti Kebangsaan Malaysia Malaysia, Bangi, Selangor, Malaysia

Y. Kishita
Department of Precision Engineering, School of Engineering, University of Tokyo, Tokyo, Japan

 
S. Fukushige et al. (eds.), *EcoDesign for Circular Value Creation: Volume I*,
https://doi.org/10.1007/978-981-97-9068-5_14

and taking consumer behaviours into policy considerations are vital in ensuring the effectiveness and positive impacts of EPR for sustainable plastic packaging management in Malaysia.

**Keywords** Extended Producer Responsibility · Plastic packaging · Circular economy · Policy analysis · Sustainable consumption and production

# 1 Introduction

The current production and consumption of plastic is not sustainable in the long run and this have already contributed greatly to plastic pollution. It is estimated that 11 million tonnes of plastic is being discharged into the ocean and without significant interventions, plastic leakage into the ocean will triple by the year 2040 [1]. Malaysia has the highest plastic use per capita at 16.78 kg annually, compared to other ASEAN countries such as Indonesia, Philippines, Thailand and Vietnam [2] and one of the highest plastic per capita consumption in the world [3]. Apart from that, according to World Bank [4], Malaysia consumed 1.41 million tonnes of four key plastic resins (PET, HDPE, LDPE and PP) but only 24% of key plastic resins are being recycled in Malaysia. If it were fully recovered, the total material value could amount to USD 1.3 billion per year. However, only about 19% of the total material value of plastics is unlocked, resulting in a potential loss of about USD 1 billion to the Malaysian economy. To avoid from losing out from a potential resource and economic standpoint, there is a strong need to look into plastic management in Malaysia, especially plastic packaging.

Plastic packaging makes up 40% of the global plastic consumption [5] and Malaysia's consumption of plastic packaging has increased dramatically in recent years due to population growth, economic growth and change in consumption and production behaviour. Regrettably, there is a lack of capacity and capability to manage plastic waste effectively, contributing towards plastic pollution and inefficient resource recovery. Thus, reducing the amount of plastic waste in the system is imperative. In order to reduce plastic waste generation and disposal, a more holistic approach and systemic shifts in the ecosystem and supply chain are required.

This was attempted to be addressed as part of the United Nation's Sustainable Development Goals (SDG) 2030, under SDG 12–Ensure sustainable consumption and production patterns. Two of the initiatives related to plastic waste under this SDG are to a) substantially reduce waste generation through prevention, reduction, recycling and reuse and to b) implement 10 Year Framework of Programmes (10YFP) on Sustainable Consumption and Production (SCP) Patterns [6].

Malaysia has committed to the 2030 Agenda for Sustainable Development through the adoption of SDG implementations by formulating actionable initiatives as outlined in the National SDG Roadmap Phase 1 2016–2020 [7] and focused on SCP with the production of The National SCP Blueprint 2016–2030 [8] to guide government agencies with the establishment of SCP strategies and assist in mainstreaming SCP in Malaysia. These are necessary efforts in changing current practices, consumption behaviours and improving governance as a whole especially for plastic waste.

## *1.1 Plastic Waste Management in Malaysia*

Plastic is the second highest in composition where it comprised of 13% of solid waste generated in Malaysia [9]. Even though Malaysia's waste coverage for waste collection is at 80% [10], the collection, recycling and disposal methods vary between states. This is because, Malaysia practices two types of management methods where some states are under the federal government's jurisdictions and some still rely on state/local municipalities to manage their waste. Seven out of fourteen states have embraced federalisation and privatisation of solid waste management under the Solid Waste and Public Cleansing Management Act (SWPCM) 2007, namely Johor, Kedah, Federal States of Kuala Lumpur and Putrajaya, Melaka, Negeri Sembilan, Pahang and Perlis. The stakeholders involved and their roles are summarised in Table 1.

Under this scheme, households receive a 2 + 1 collection system where household solid wastes are collected twice weekly while garden, bulky waste and recyclables are collected once a week [11]. For industrial, commercial and institutional areas, (ICI) waste collection frequency is dependent on the agreed concession agreement. To further strengthen the implementation and enforcement of this Act and to promote resource recovery and circular economy, Separation at Source (SAS) Regulation was also enforced in 2015 [12].

For states that have not agreed to the SWPCM 2007 Act, the cost, frequency, quality and methods of solid waste handling will depend on the state/municipality and many local authorities do not make it mandatory to separate recyclables from normal household waste, nor have a separate system of collection for recyclables. Thus, recyclable waste from these states such as plastics are co-mingled with other wet waste, contaminating them and rendering these resources unrecyclable and

**Table 1** Related stakeholders and summary of their roles under federalisation and privatisation of solid waste management in Malaysia

| Agency | Roles |
|---|---|
| Ministry of Housing and Local Government (MHLG) | Policy formulation and agencies coordination related to solid waste management |
| Department of National Solid Waste Management (JPSPN) | Implementation of policies and legislations |
| SWCorp | Enforcement and monitoring of policies, legislations and operations of concessionaires |
| Concessionaires (Alam Flora, SWM Environment and Environment Idaman) | In charge of waste collection, transportation and disposal |

forced to be discarded to landfills. Data for waste composition and recyclables are also not readily available for most of the states.

Realising the dire implications of plastic waste mismanagement, the Government of Malaysia has taken a few significant actions to manage plastic waste more sustainably by establishing a Roadmap Towards Zero Single-Use Plastics 2018–2030 [13] and this is further translated into a more detailed, action-oriented approach under the Malaysia Plastics Sustainability Roadmap 2021–2030 [14]. Under this, the absence of Extended Producer Responsibility (EPR) scheme has been identified as one of the key challenges in plastic circularity. Hence, the Malaysian Government has proposed the implementation of EPR by first focusing on the packaging sector.

## 1.2 Extended Producer Responsibility (EPR)

According to OECD [15], EPR can be defined as an environmental policy approach in which the producer's responsibility for a product is extended to the post-consumer stage of a product's life cycle, where EPR policies aim to promote waste prevention and reduction, increase resource efficiency and the usage of recycled materials. The implementation of EPR schemes are often a joint collaboration between government and the business sector that includes the application of instruments such as fees or subsidies, standards, labels, restrictions, permits, communication tools etc. In a way, EPR provides a new outlook at plastic waste management issues, by treating it as a materialisation/externalisation of environmental consequences predetermined by how products and systems are designed [16]. In recent years, many countries especially developing countries have opted to implement EPR in light of increasing marine pollution and the high volume generated from plastic packaging [17]. The most common approach for EPR is take back requirements, followed by advance disposal fees (ADF) and deposit refund system (DRS). Even though there is no one-size-fits-all to implement EPR, therein lies its attraction as due to its flexibility to be molded and adapted according to local settings and values. Hence, a combination of instruments is often implemented for a more effective execution of an EPR scheme [18]. It can be integrated with economic instruments that complement the plastic packaging waste ecosystem such as pay-as-you-throw (PAYT) schemes and deposit refund systems (DRS) where by sorting, reducing and returning their packaging waste can indirectly contribute towards the increase of collection of recyclables under the EPR scheme. Aside from economic instruments, regulatory frameworks also play a vital role in influencing EPR implementation such as setting targets or imposing bans on certain types pf plastic packaging [19]. To date, there are currently more than 400 EPR systems in place [18] with more than 30 countries implementing plastic packaging-related EPR [19]; [2]. Many scholars have discussed the direct and indirect impacts of EPR, that often has positive outcomes [20] such as increasing recycling rates [19], therefore it will not be discussed in detail here.

### 1.3 *Current Update Regarding Extended Producer Responsibility (EPR) in Malaysia*

EPR for e-waste has already been implemented in Malaysia under the Environmental Quality Act 1974 [21] but there has yet to be an EPR scheme for solid waste such as plastic packaging. Matters related to plastic packaging waste under the federalisation of solid waste management is governed by the SWPCM Act 2007. In order to implement EPR for plastic packaging, the provisions in the Act will be referred to. However, there is no mention of EPR within the document, but there was mention of take back system and deposit refund system [11].

In terms of related policies for EPR, it was mentioned in several policy documents. The more recent documents were Roadmap Towards Single Use Plastics 2019–2030 [13] and the actions were further detailed out in the Plastics Sustainability Roadmap 2021–2030 [14]. The aim of these documents were to provide a clearer policy direction so that a unified and collective actions regarding plastic waste management can be taken. The Malaysian EPR system aims to address at least four major issues for plastic waste management including waste avoidance and reduction, collection, recovery, recycling and reuse, while continue to carry out proper treatment and disposal of waste. Parties involved will contribute to an eco-modulated fee fund to support the operational cost of running an EPR scheme, which will be handled by a Producer Responsibility Organisation (PRO). The governance framework will also be developed in line with the mandatory implementation of EPR to collate and streamline existing policies and frameworks between ministries and sectors and provide a variety of capacity building programmes to prepare industry players for the adoption of voluntary EPR in the year 2023 before the nationwide implementation of mandatory EPR in the year 2026.

In anticipation of voluntary EPR starting from 2023, a group of private companies from the fast-moving consumer goods (FMCG) sector have taken proactive actions by setting up a PRO called Malaysia Recycling Alliance (MAREA), with its members consisting of multinational companies such as Coca-Cola, Nestle, Unilever and Mondelez [22]. However, the members of MAREA forms a fraction of the FMCG sector players where the bulk of them are often from small-medium enterprises (SMEs) and the volume of their plastic waste are unaccounted for. As of the time of writing, there has yet a finalised version of EPR governance framework to guide the implementation of EPR for plastic packaging in Malaysia despite voluntary EPR starting in 2023. But many organisations such as WWF Malaysia [2] and MAREA [22] have proposed how EPR and DRS can be carried out in Malaysia and the suggestions can be adopted and adapted to suit Malaysia's needs.

Taking these factors into consideration, this paper will discuss on the policy and institutional challenges of implementing EPR in Malaysia for plastic packaging while also taking into consideration the preferences and willingness of consumers to participate in the EPR scheme to ensure an effective execution of the programme in the near future.

## 2 Materials and Methods

Firstly, a literature review via desktop study was carried out on published official documents (policies, legislation and reports) and journal papers related to EPR especially on municipal and plastic waste and were analysed using thematic analysis. Secondly, we also carried out a survey on consumer readiness towards the proposed implementation of EPR for plastic bottles in the beverage sector between September 2021 and March 2022. The survey was distributed via Google Form in both English and Malay language using random sampling and snowball sampling to reach a wider audience. The survey was answered by 395 respondents living in Greater Klang Valley area consisting of the states Kuala Lumpur, Putrajaya, Selangor and Negeri Sembilan, achieving the confidence level of >95%. The questionnaire was mostly conveyed using dichotomous questions and Likert scale of 1–5, with 1 being least important and 5 being most important to simplify the survey experience. In addition, open-ended questions were also used to gather a more in-depth information. The questionnaire assessed respondents' readiness towards plastic beverage packaging and willingness to pay in light of future EPR implementation, but only some of the results will be discussed here. The result was analysed using descriptive and simple statistical analysis and thematic analysis was used to analyse the open-ended answers. This survey was used to support the findings and discussions gathered from this literature review and is also part of a larger research on the development of governance framework for EPR in the beverage sector for plastic packaging.

## 3 Results and Discussion

### *3.1 Review of EPR-Related Legislations and Policies*

We decided to look at documents between the years of 2007 and 2021 because the federalisation and privatisation of solid waste management started in the year 2007 and after 2021, no new policies or regulations regarding EPR were published. We have identified these phrases referring to EPR in these documents using the keywords 'plastics', 'Extended Producer Responsibility', 'take back', 'buy back' and 'deposit refund system'. A summary of the phrases are as shown in Table 2.

As previously mentioned, there was no specific mention of EPR within the SWPCM Act 2007, except for take back system and deposit refund system [11]. However, under the provision for Reduction and Recovery of Controlled Solid Waste, the Act does provide powers to the Minister to carry out a scheme such as EPR if it contributes towards reduce, reuse and recycle of solid waste, where the Minister may gazette an order for:

(a) Any waste generator to reduce their waste generation.
(b) Any person to use environmental-friendly material.
(c) Any person to use specified amount of recycled materials in products.

**Table 2** Policy documents related to EPR for solid/plastic waste management (between years 2007 and 2021)

| Year | Document | Keywords/phrases |
|---|---|---|
| 2007 | Solid Waste and Public Cleansing Act 2007 | 102. Take back system and deposit refund system "(1) The Minister may, by order published in the Gazette establish take back system which..." "(2) The Minister may, by order published in the Gazette establish deposit refund system and determine ..." |
| 2011 | Tenth Malaysia Plan 2011–2015 | "As part of this effort, a deposit refund scheme and a take back system will be implemented" |
| 2016 | National Solid Waste Management Policy 2016 | "Minimise solid waste generation and optimise resource by implementing deposit refund system" |
| 2016 | Eleventh Malaysia Plan 2016–2020 | "Implementation of existing mechanisms such as Polluter Pays Principle and Payment for Ecosystem Services will be expanded to support funding needed..." |
| 2018 | Roadmap Towards Zero Single-Use Plastics 2018–2030 | "Introduction of legal framework on single-use plastics." |
| 2020 | National Cleanliness Policy 2020–2030 | Strategy 3.4: Implement EPR to promote recycling by: 3.4.1 Encourage the involvement of industry to implement Extended Producer Responsibility (EPR) through the EPR Implementation Plan. 3.4.2 Create an EPR Roadmap as a guide and reference for stakeholders. 3.4.3 Introduce the Reverse Vending Machine (RVM) to encourage recycling. 3.4.4 Create a platform for engagement sessions between the government and industry, NGOs and educational institutions. |
| 2021 | Malaysia Plastics Sustainability Roadmap 2021–2030 | 2.4 Transitioning to a Sustainable Plastics Economy 2.4.1 Extended Producer Responsibility– "The implementation of EPR will begin with voluntary scheme before transitioning to a mandatory scheme..." "Obliged parties will contribute to an eco-modulated fee to fund the operational costs..." "To ensure smooth transition... an EPR governance framework will be developed starting from 2022 and take into consideration existing policies, acts, and frameworks across ministries and sectors..." |
| 2021 | Twelve Malaysia Plan 2021–2025 | "The EPR approach will also be extended to cover other types and streams of waste, particularly packaging materials and single-use plastics. Accordingly, relevant regulations, economic instruments and monitoring mechanisms will be put in place to enable producers and retailers to implement the EPR for other type of waste. Implementation of EPR will be supported through the take-back system by producers or brand owners as well as the adoption of the user-pay and polluter-pay principles." |

(d) Any person to limit generation, import, use, discharge or disposal of products/ materials.
(e) Implementation of labelling system to promote recycling.
(f) Use of any method to reduce solid waste impact on the environment.
(g) Use of any method to reduce, reuse and recycle solid waste.

This is a strong enough indication that the mechanisms of an EPR scheme for plastic packaging can still be implemented despite not being mentioned specifically in the legislation. As EPR is gaining traction globally and here to stay with the mandatory implementation, it would be better to enhance the existing legislation framework to include EPR in the legal provisions. In addition, for the purpose of uniformed and concerted approach of EPR, a separate regulation on EPR and its guidelines should be developed so that states not under the federal jurisdiction can adopt and adapt the regulation and guidelines to suit their respective legislations in order for EPR to be implemented throughout Malaysia uniformly.

Apart from that, discussions have highlighted the implementation of take back and DRS in the Tenth Malaysia Plan 2011–2015 and National Solid Waste Management Policy 2016, and the conversation started to steer towards EPR beginning with the Eleventh Malaysia Plan 2016–2020 and continued to be discussed in the Twelfth Malaysia Plan 2021–2025. In total, EPR methods and strategies have been discussed for the past 15 years but have yet to be materialised for solid waste such as plastic packaging. This clearly shows that the pace of policy making has not caught up with efforts to implement them effectively [23] and this ambiguity is often considered the reason that policy implementations fail [24]. Also, due to the complexity of its implementation and the lack of attention and deliberate consideration on it, the scheme has not taken off. Thus, key agencies should work together to develop and finally materialise the EPR scheme.

## 3.2 *Review of Institutional and Stakeholders Involvement*

Implementation of EPR for plastic packaging involves two key ministries, namely Ministry of Housing and Local Government (MHLG) and Ministry of Natural Resources, Environment and Climate Change (NRECC). This is evident where power to establish an EPR scheme lies with MHLG but NRECC has a more international overview of the issues on plastic waste therefore driving the discussion as its mitigation strategies goes beyond EPR and the local waste scenario. Hence, the development of the two roadmaps on plastic initiated by NRECC. However, implementation involving different agencies and tiers of government along with different sets of authorities and responsibilities for management, operational and planning functions will often create complexity [23]. Since developing an EPR scheme for plastic packaging is often crosscutting and a massive undertaking, key ministries might be lacking in terms of manpower and expertise. Therefore, it is suggested that a special taskforce or joint committee to be established between the two ministries

**Table 3** Other agencies directly and indirectly related to the implementation of EPR for plastic packaging in Malaysia

| Agency | Roles |
|---|---|
| Ministry of Natural Resources, Environment and Climate Change (NRECC) (formerly known as MESTECC and KASA) | Focus on policy formulation for plastic waste pollution, reduction and circular economy in general. |
| Ministry of Finance | Plan and implement direct and indirect taxation policies and incentives. |
| Ministry of Investment, Trade and Industries | Work with industries to promote ESG and coordinate, develop and provide license for manufacturing activities |
| Ministry of Domestic Trade and Cost of Living | Encourage good governance among businesses and enhance programmes for consumer awareness regarding consumer goods. |
| Economic Planning Unit | Promote, develop policies and carry out research related to environmental economic instruments |
| MAREA | Improve packaging waste and build awareness on circular economy approach for packaging waste management (Voluntary PRO) |

and to include other related ministries, agencies and non-governmental bodies to assist in the establishment of an overarching EPR governance framework and a suitable EPR policy and scheme for plastic packaging in Malaysia. To ensure that the committee can play its roles effectively, it is required that a timeline is set to ensure that mandatory EPR will take off in 2026.

Apart from the previously mentioned agencies directly related to solid waste management, below are the other agencies that are indirectly related in the implementation of EPR for plastic packaging. A summary of their roles is briefly explained in Table 3.

For the agencies not directly related with plastic packaging, the involvement of these multiple stakeholders in the creation and implementation of this policy is critical to ensure successful execution of the EPR scheme.

## 3.3 *Review of Consumer Readiness for EPR*

Based on the survey that were carried out on consumer readiness on the implementation of EPR, the general demographic data were compiled from the respondents such as age, location, academic background and average household income. A summary of the respondents' demographic background is provided in Table 4.

65.3% of the respondents are female and 34.7% are male. 78% of the respondents were between the age of 31 and 50 years. Among the respondents, 89.1% have gone through tertiary education and about 40.5% of the respondents have household income of RM10,000. In summary, our sample consist of a majority of working

**Table 4** Summary of respondents' demographic background

| Criteria | Percentage |
|---|---|
| Gender | |
| Female | 65.3% |
| Male | 34.7% |
| Age | |
| 21–30 | 7.1% |
| 31–40 | 35.2% |
| 41–50 | 42.8% |
| 51–60 | 10.9% |
| 61 and *above* | 4.0% |
| Academic background | |
| Primary | 0.0% |
| Secondary | 10.9% |
| Diploma/bachelor's degree | 46.1% |
| Master's degree | 35.4% |
| PhD and above | 7.6% |
| Average household income | |
| RM2,000 and below | 4.6% |
| RM2,001–RM5,000 | 15.2% |
| RM5,001–RM8,000 | 22.8% |
| RM8,001–RM10,000 | 17.0% |
| RM10,001 above | 40.5% |

**Table 5** Results on willingness to participate and pay

| Willingness to participate and pay | Yes | No |
|---|---|---|
| 41. Are you willing to pay a minimal deposit fee (that you will get back) for drinks in plastic bottles? | 87.6% | 12.4% |
| 44. If Malaysia implement a DRS system for drinks in plastic bottles, will you make the effort to clean and segregate the plastic bottles for recycling before claiming your deposit? | 86.6% | 13.4% |
| 45. If Malaysia implement a DRS system for drinks in plastic bottles, will you make the effort to return the plastic bottles for your deposit? | 95.9% | 4.1% |
| 49. Will packaging label that states the deposit amount encourage you to return the plastic bottle to a collection point? | 92.9% | 7.1% |

women in their middle-age who are well educated, within the middle- to high-income bracket and has strong purchasing power. Based on this background, it is also assumed that the overall sample is made up of respondents who has high environmental awareness.

In this survey, the respondents were given an explanation on DRS and EPR and upon understanding the concept, 81% feels that DRS is suitable to be implemented in Malaysia and 80.3% feels that EPR should be made mandatory in Malaysia. The survey also seek to find out consumers' willingness to participate and pay in a DRS/EPR scheme. The results are as shown in Table 5.

If a deposit refund system (DRS) is implemented, 87.6% of the respondents are willing to pay a minimal fee for their plastic packaging and 86.6% of the respondents will make the effort to clean and segregate their plastic bottles before returning and claiming their deposit. 95.9% mentioned they will make the effort to return them for their deposit. In addition, 92.9% of the respondents mentioned that they will be encouraged to return the plastic bottles if the deposit amount is stated on the packaging label. When asked whether they hope EPR is implemented for plastic packaging in the next 5 year, 94.5% said yes. The responses indicated that consumer buy-ins towards EPR implementation for plastic packaging is already there. Although the sample might not be representative of the whole of Malaysia, it can give a good indicator to policy makers that the people are ready to be actively involved in the EPR scheme. However, the scheme should not be designed in such a way that it inconvenience the consumers. Other than provide continuous knowledge and awareness to the consumers, labelling helps to notify them of EPR/DRS and serves as a reminder for them to clean and return their plastic packaging later for their deposit. It will also be helpful if suitable infrastructures within close proximity of households be provided to create a condusive environment for consumers to return their plastic packaging. Since the survey only covers Greater Klang Valley area, it is suggested that a nationwide survey is carried out to cover a more diverse background of consumers for a more accurate representation of the population.

## 4 Conclusion

In conclusion, despite the legislative and institutional challenges, the existing legislations and policies already provide a solid foundation for the development of EPR scheme in Malaysia for plastic packaging, especially in the FMCG sector. With the consumers being onboard with EPR/DRS implementation, the uptake will not be an uphill task; however, the implementation must be carried out effectively and transparently. There is also no detailed documentation of how best to move forward with the existing legislations and policies nor a clearly defined roles of related stakeholders in ensuring the effectiveness and impacts of EPR for sustainable plastic packaging management in Malaysia. On top of that, there is a need to streamline and harmonise between the legislations; policies; institutions and stakeholders; federal, state, local governments; private sectors; and non-governmental organisations. Therefore, the need to develop a governance framework for EPR is pressing in order to set the direction and implementation of EPR. When this is materialised, only then can everybody play their role effectively in making EPR for plastic packaging a success.

**Acknowledgements** This paper is part of PhD research, funded under Hadiah Latihan Persekutuan (HLP) provided by the Department of Public Service Malaysia. The study also supported by fund from Universiti Kebangsaan Malaysia TAP K006974.

## References

1. The Pew Charitable Trust and SYSTEMIC (2020) Breaking the plastic wave: a comprehensive assessment of pathways towards stopping ocean plastic pollution. The Pew Charitable Trust, Philadelphia, USA
2. WWF Malaysia (2020) Study on EPR scheme assessment for packaging waste in Malaysia. World Wildlife Fund Malaysia, Petaling Jaya
3. Jambeck JR, Geyer R, Wilcox C, Siegler TR, Perryman M, Andrady A, Narayan R, Law KL (2015) Plastic waste inputs from land into the ocean. Science (New York) 347(6223):768–771
4. World Bank Group (2021) Market study for Malaysia: plastics circularity opportunities and barriers. Marine Plastics Series, East Asia and Pacific Region, Washington DC, USA
5. OECD (2022) Global plastics outlook. Policy Scenarios to 2060. OECD Publishing, Paris, France
6. United Nations (2023) Goal 12: Ensure sustainable consumption and production patterns. Goal 12 targets. Retrieved from https://www.un.org/sustainabledevelopment/sustainable-consumption-production/. Accessed on 14 September 202)
7. Economic Planning Unit (EPU) Malaysia (2016) SDG Roadmap for Malaysia: Phase 1: 2016–2020. EPU, Putrajaya, Malaysia
8. Economic Planning Unit (EPU) Malaysia (2016) The National SCP Blueprint 2016–2030–The pathways for sustainable consumption and production (SCP) in Malaysia. EPU, Putrajaya, Malaysia
9. Ministry of Housing and Local Government (MHLG) Malaysia (2013) Survey on solid waste composition, characteristics & existing practice of solid waste recycling in Malaysia: final report. Department of National Solid Waste Management Malaysia, Putrajaya
10. World Bank (2018) What a Waste 2.0: A global snapshot of solid waste management to 2050. International Bank for Reconstruction and Development/The World Bank. World Bank Publications, Washington DC, USA
11. Sdn AF, Bhd. (2021) Alam Flora: beyond waste management in creating a sustainable future. Alam Flora Sdn. Bhd, Selangor, Malaysia
12. Ministry of Housing and Local Government (MHLG) Malaysia (2015) Separation at source regulation. Putrajaya, Malaysia
13. Ministry of Energy, Science, Technology, Environment and Climate Change, (MESTECC) (2020) Roadmap towards zero single use plastics. Putrajaya, Malaysia
14. Ministry of Environment and Water (KASA) (2021) Malaysia plastics sustainability roadmap 2021–2030. Putrajaya, Malaysia
15. Organisation for Economic Co-operation and Development (OECD) (2001) Extended producer responsibility–a guidance manual for governments. OECD Publishing, Paris, France
16. Manomaivibool P (2010) Extended producer responsibility in East Asia: approaches and lessons learnt from the management of waste electrical and electronic equipment. In: East Asian review, vol 13. Asian Research Institute, Osaka University of Economics and Law, Osaka, pp 267–286
17. Geyer R, Jambeck J, Law K (2017) Production, use, and fate of all plastics ever made. Sci Adv 3/7:e1700782
18. Kaffine D, O'Reilly P (2015) What have we learned about extended producer responsibility in the past decade? A survey of the recent EPR economic literature. Working Party on Resource Productivity and Waste OECD
19. Watkins E, Gionfra S, Schweitzer J, Pantzar M, Janssens C, Brink P (2017) EPR in the EU plastics strategy and the circular economy: a focus on plastic packaging. Institute for European Environmental Policy, Belgium
20. Organisation for Economic Co-operation and Development (OECD) (2016) Extended producer responsibility–updated guidance for efficient waste management. OECD Publishing, Paris, France

21. Government of Malaysia (GOM) (1974) Environmental Quality Act 1974. Percetakan Nasional Berhad, Kuala Lumpur, Malaysia
22. Malaysia Recycling Alliance (MAREA) (2022) Recommendations for an effective extended producer responsibility implementation and policy in Malaysia. Retrieved from https://marea.com.my/wpcontent/uploads/2022/08/MAREA-Position-Paper.pdf. Accessed on 14 July 2023
23. Read AD (1999) Making waste work: Making UK national solid waste strategy work at local scale. Resources, Conservation, and Recycling 26:259–285
24. Paudel NR (2009) A critical account of policy implementation theories: status and consideration. NJPG 15(2):36–54

# Second Life of Electric Vehicle Lithium-Ion Batteries from a Sustainable Business Model Perspective

**Saad Ahmed, Elli Verhulst, and Casper Boks**

**Abstract** The exponential growth in electric vehicles (EV) sales has created a challenge of managing the amount of end of life (EOL) EV lithium-ion batteries (EV LiBs) expected in the future. Using these batteries for second-life battery applications (SLB) can potentially prolong the use phase of the batteries and delay recycling. The SLB has paved way for new value chain structures and emergence of new business models integrating sustainability at their core. This study investigates the current SLB value chain structures and the associated sustainable business models (SBM) present in the Norwegian market. The results show two distinct SLB value chain structures, and a typology of associated SBMs. However, there are various SBM barriers that need to be addressed to make SLB viable and sustainable. The study thus discusses various strategies that can be used to overcome these barriers.

**Keywords** Sustainable business models · Electric vehicles lithium-ion batteries · Battery second use · Value chain

## 1 Introduction

There has been an exponential rise in the sales of electric vehicles (EVs) globally, driven by sustainability trends, technology advancements and policies for decarbonization [1–3]. The number of EVs on the roads globally was 16.5 million by the end of 2021, which is triple the amount compared to 2018 [4]. This rapid growth has also increased the demand and production of lithium-ion batteries (LiBs), which are widely used in mobility solutions because of their high density, high capacity, and longer life span [5, 6]. However, according to industry standards these

S. Ahmed (✉) · E. Verhulst
Department of Industrial Economics & Technology Management, NTNU, Trondheim, Norway
e-mail: saad.ahmed@ntnu.no

C. Boks
Department of Design, NTNU, Trondheim, Norway



S. Fukushige et al. (eds.), *EcoDesign for Circular Value Creation: Volume I*,
https://doi.org/10.1007/978-981-97-9068-5_15

batteries are no longer used in vehicles once they reach 70%–80% of their capacity, and are taken out of the EVs at their first end of life (EOL) [7]. LiBs contain hazardous materials and if treated irresponsibly at their EOL, can result in more wastage and pollution [8]. According to estimates, approximately 11 million metric tonnes of LiBs will have to be recycled globally by 2030, triggered by the rise in EVs in the market [9, 10]. However, current recycling processes are considered to be inefficient and unsustainable where the LiBs are recycled at a meagre rate of 5% in the European Union [11–13]. Other challenges related to recycling include supply chain bottlenecks, energy inefficiency, inefficient materials recovery, and waste issues [12]. Therefore, there is a need to prolong the life of spent EV LiBs to potentially postpone the recycling phase until more sophisticated recycling technology and methods are in place.

Using spent batteries in second life (SLB)—including reusing, remanufacturing, and repurposing—can potentially prolong the life of the EOL EV LiBs, depending on their state of health (SOH). Reuse and remanufacturing are sometimes used interchangeably in the literature, but for this study "reuse" is defined as an EOL battery used in another vehicle as is, while remanufacturing is when the EOL battery is repaired and tweaked before it is used for the same application [12]. Repurposing, on the other hand, is when the EOL battery is used in a less demanding application, usually an energy storage system (ESS) [14]. For this study, our focus will remain on "repurposing."

SLBs can partially close the value chain loop increasing the economic and social values, while decreasing the environmental impact [15, 16]. However, it comes with numerous barriers that need to be addressed. These barriers include but are not limited to collection of the EOL batteries, testing the SOH, disassembly of the battery pack from module to cell level, pricing, and other social and regulatory challenges related to safety etc. [17]. Innovating business models, and integrating sustainability into the core activities of the businesses across the value chain is considered a potential way to address these challenges [18]. However, the EV LiB value chain consists of multiple stakeholders including miners, cell manufacturers, original equipment manufacturers (OEMs), system integrators, users, repurposers, and recyclers. Moreover, these value chains are structured in different ways depending on the regional context. This increases complexity as multiple business models with different value chain structures and separate value propositions interact either positively, negatively, or symbiotically with each other [19].

Business models that include social, environmental, economic, intellectual, and human values are called sustainable business models (SBMs), while circular business models are a subtype of SBMs that are focused on circularity. Nevertheless, both these BMs can close, slow, and narrow resource loops [20], and hence both are considered in this study.

To date, one of the biggest costs for SLBs is battery disassembly, readjustment of the battery management system (BMS), and repackaging catered to the type of SLB application [12, 21]. This is due to nonstandardized battery designs that make it difficult and costly for SLB use. SBMs and circular design considerations can supplement each other [22, 23]. Studies suggest that these SBMs can be complemented by design considerations that make the EOL EV LiBs better fitted for SLB application from the start of the battery life cycle. These design considerations can be related to design for disassembly, reassembly, life extension, remanufacturing, etc. [20]. Furthermore, SBMs can help overcome technological, economic, social, and environmental barriers, and make SLB viable [18].

SBM is an emerging field in the literature [24, 25], however initial investigations reveal that there is scarcity in the SBM literature considering SLB [7]. As such, there is a need to consider SLB from an SBM perspective to help businesses make strategic decisions to become more circular. Addressing this gap, this study investigates the role of SBMs in making SLBs more viable and sustainable. The Norwegian market, because of the rapid uptake of EVs and an emerging battery value chain, provides a good ground for such a study. The following research questions are addressed:

1. How is the SLB value chain in Norway currently structured?
2. What SBMs are present for the SLB in Norway?
3. How can SBM barriers for SLB be overcome?

## 2 Methodology

This study used a multi-case study research method [26]. The method helps in understanding complex problems in depth by providing an opportunity to conduct a detailed analysis [26]. As SBM is an emerging field, and the SLB market is rapidly developing, consisting of a complex value chain structure, this method is viewed as the most suitable. The data for the study was collected from semi-structured interviews conducted with four different companies in the EV LiB value chain. Three companies are located in Norway, while one is in the UK. An overview of the case companies and the respective interviewees are provided in Table 1. Apart from interviews, additional data comes from the literature, including peer-reviewed journal articles, conference papers, and to some extent grey literature such as news articles and press releases.

The emphasis in the study is on the Norwegian context mainly because Norway is aiming to establish a complete battery value chain and is thus ideal for such a study,

**Table 1** Company profiles included in the study

| Company | Interviewee position | Position in value chain | Context |
|---|---|---|---|
| Alpha | Compliance Manager | Producer responsibility | Norway |
| Beta | Product Manager | Anode producer | Norway |
| Gamma | Cofounder | Repurposer | UK |
| Delta | Sales Rep | Repurposer | Norway |

in addition to the authors being situated here, which makes it easier to get access to the stakeholders.

# 3 Results: Theoretical Insights

## *3.1 Second Life for EV LiBs (SLB)*

An EV LiB is typically composed of hundreds or thousands of individual cells depending on the LiB types, to create multiple modules that are combined to make a battery pack. These batteries consist of specific lithium-ion chemistries, of which two are the most common: nickel–manganese–cobalt (NMC) and lithium iron phosphate (LFP). Additionally, an EV LiB contains a battery management system (BMS) and a thermal management system (TMS) catered to the kind of application they are used for [27]. LiBs are no longer used in EVs once they reach 80% or lower of their capacity, but they can still be repurposed in less demanding applications like ESS for a second life [6, 12, 18, 28–30]. This gives an additional 8–10 years of life to the EOL battery and delays the recycling phase which is considered to be unsustainable and inefficient as of yet [31]. However, these batteries have to go through SOH tests, disassembly, module modifications, BMS adjustments, and repackaging for SLB application [7, 32]. These steps are usually performed by OEMs or repurposers depending on the business models of the stakeholders and the value chain structure.

## *3.2 Sustainable Business Models and SLB*

A business model illustrates how an organization creates, delivers, and captures value [33]. Value creation determines the value proposition and what problem is being solved, Value delivery shows how this problem is solved (service channels, etc.) and Value capture determines what revenue and profits are being generated by the organization [33, 34]. However, as the world has become more aware of the climate crisis, in academia and in practice, sustainability has grown to become an integral part of a business model. In this case, sustainable values such as environmental values, social values, and human values are also included in addition to the economic values [35, 36]. Moreover, they also promote proactive stakeholder engagement where both systems and firm-level perspectives are considered [20, 37, 38]. It has become essential for businesses to create sustainable values to attain competitive advantage for a longer period of time [39].

With an increase in the EV market, the SLB market is also growing, giving rise to new business models [39]. In literature, multiple SBM archetypes have been proposed for the SLB market where three fundamental actors are considered in the value chain: OEMs, Users, and Repurposers [7, 40, 41]. These archetypes include but are not limited to maximizing material and energy efficiency, closing and slowing resource loops, inclusive value creation through stakeholder collaboration etc. [18]. Jiao and Evans [8] and Rufino Júnior et al. [42] have proposed three types of SBMs related to EV LiBS: (1) Standard business model, (2) Collaborative Business model, and (3) Integrative business model. In (1) the OEMS sell EOL batteries to repurposers, in (2) the OEMS and repurposers collaborate to create a solution, while in (3) a joint venture is typically formed by the OEM and repurposer, and the responsibility is shared.

## *3.3 SBM Barriers for SLB*

There are various SBM barriers for SLB discussed in the literature. These include technological, market, social and legislative barriers. Technological barriers can include variable battery design, chemistry, management system, data traceability, unclear SOH etc. Market barriers may include supply chain risks, price reduction on new batteries, and transportation costs, while social and legislative barriers may be related to safety concerns, lack of awareness, and lack of regulations and standards [12, 16, 43–45]. Barriers are usually context-dependent based on the region the battery value chain is in.

## *3.4 SBM Strategies to Overcome Existing Barriers*

Two of the main circular design resource cycles can be employed to overcome the SBM barriers mentioned above. These are slowing and closing resource loops that provide a combination of product and business model design strategies that can promote circularity [22].

### Slowing Resource Loop and Associated SBMs

Slowing resource loops entails having strategies that typically extend the life of the product. These design strategies include but are not limited to batteries designed for repairability, maintenance, and in some instances modifications for future use [46, 47]. This means that they are designed in a way that makes them easy to repurpose at the end of their first life in another application [48]. The associated SBMs that can complement these strategies include business models such as batteries as a service, in which the batteries are leased instead of being sold to the customer or battery swapping [49, 50]. This usually means that the responsibility of the batteries remains with the owner of the batteries: it provides flexibility and control over the entire life cycle of the battery.

**Closing Resource Loop and Associated SBMs**

Closing resource loops typically entails creating a product that is made for circularity and enables recycling. This includes using design strategies that make it easier for batteries to get disassembled and reassembled, making it easier for repurposing and then recycling [51]. It also entails extending the resource value for the materials present in the battery so that they can produce more value for the stakeholders and for the environment [40]. This requires collaboration across the value chain.

# 4 Results: Insights from Practice

There is a potential to capture economic, social, and environmental value considering the number of EOL batteries expected to be in the market by 2030. However, the interview data shows some skepticism from stakeholders because of the multiple barriers that can hinder SLB adoption.

## *4.1 SLB Value Chain in the Norwegian Context*

In the Norwegian context, the SLBs are handled by third-party companies in the majority of the cases. Generally, the EOL battery pack is retired from an EV, and some basic SOH tests are performed. Depending on their SOH, these batteries are then sent to either a repurposing or a recycling facility. For example, Alpha, a producer responsibility company, has contractual agreements with LiB producers, and thus is responsible for the collection, basic testing, and transportation of the EOL battery packs. However, Alpha does not usually work with the EV LiB packs. This is due to economies of scale: these EOL batteries are not available in large numbers, and it is costly to check their SOH and transport them because of stringent safety regulations and the labor needed for the task. The market for EOL battery collection, transport, and to some extent recycling is mainly handled by Batteriretur, a large Norwegian collection and recycling company that has managed to make contracts with several stakeholders in the battery value chain and municipalities [52]. This leaves little room for small players such as Alpha.

However, in some cases, battery repurposers have built contractual agreements with OEMs. Two kinds of such agreements come forward from the interviews and the grey literature.

1. The OEM collects, tests, disassembles, and repurposes the batteries. The repurposer then takes these batteries and makes their own products. The responsibility of the product then lies with the repurposer. In Delta's case, they have a contractual agreement with Nissan, where Nissan takes control of the take-back system and repurposing of the batteries. These repurposed batteries are then delivered to Delta which makes SLB products.

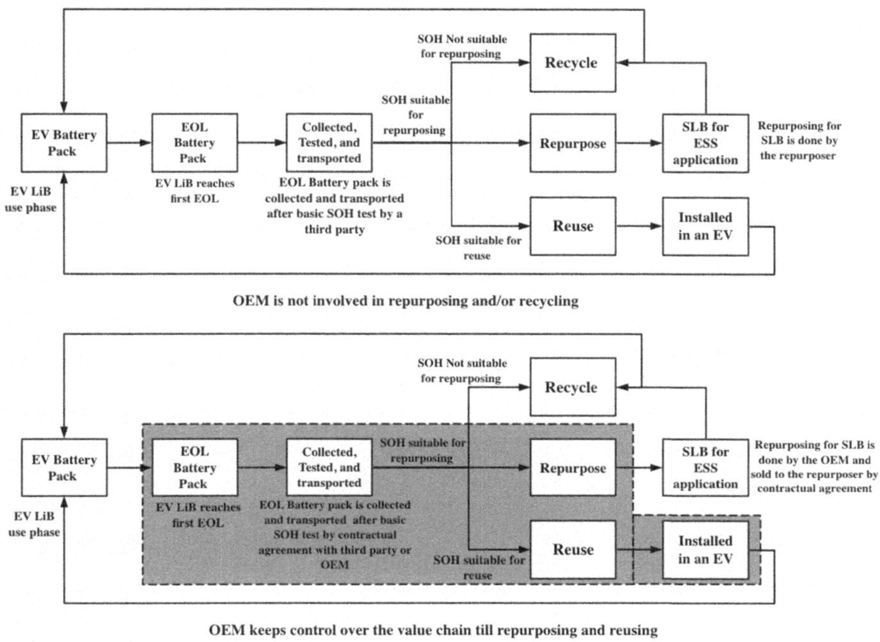


**Fig. 1** Value chain structures for SLB in Norway

2. The repurposer controls the SLB value chain, where they or the third party collects EOL batteries. The repurposer then tests, disassembles, and repurposes the battery for SLB application. The OEM in this case is not involved in the process, except for providing expertise in some cases. Gamma, for example, is a battery repurposer that receives EOL EV LiBs from dismantlers in the UK.

In the Norwegian context, Delta is an example of (1) while Ecostor, an ESS company, is an example of (2) [53]. Both value chain structures are illustrated in Fig. 1.

## *4.2 SLB from a SBM Perspective*

There is an increasing need to understand the SLB value chain from a SBM perspective. SBMs determine how stakeholders create, deliver, and capture value, and that determines the type of SLB value chain the stakeholders adopt. For example, the value chain where the OEM controls the batteries until repurposing, gives an opportunity to the OEM to increase traceability of the SOH of the battery through its life cycle until the SLB application. A collaborative business model is thus formed in partnership with the repurposer where both the companies generate economic, social, and environmental values from the EOL battery pack, as is the case with Delta. However, additional investments are needed by the OEM in collection, repurposing infrastructure, and SLB development. Moreover, there is increased liability in this kind of business model for the OEM.

On the other hand, the stakeholders' business models change significantly when the OEM is not involved. The collection, transportation, and testing in this case is generally handled by a third-party company responsible for EOL battery collection while the repurposers themselves disassemble and repurpose batteries for the SLB application. In this case, the OEM loses the opportunity to capture value from the EOL battery pack for a longer span of time but does not have to invest in other infrastructure development either. However, it also causes problems for the repurposers where they must invest significantly in testing, disassembly, and repurposing facilities. It also creates challenges related to data traceability of the batteries, where the OEMs are reluctant to share data with repurposers. Another challenge that comes forward with this model that was mentioned by all interviewees, was the variable battery designs. The lack of standardization in the battery design makes it difficult to automate the disassembly and reassembly process of the batteries. Although most of these batteries can be disassembled easily, except from some battery packs, they require manual labor. This raises the cost of repurposing and increases safety hazards, as an EOL EV LiB is considered a dangerous good.

These challenges can be mitigated with pro-active stakeholder engagement, partnerships, and collaborations throughout the SLB value chain. Depending on the collaboration, companies usually go for a standard, collaborative, or integrative business model.

**Typology of SBM Designs for the SLB in the Norwegian Context**

The data gathered from the interviews and the companies' website analysis shows that there are generally three types of business model designs present in the Norwegian context. Figure 2 illustrates these BM types. The repurposer driven SLB application and OEM driven repurposing business model design are currently present in the Norwegian market, but the OEM controlled SLB application is yet to take form. This BM is considered to be ideal to ensure circularity of the EV LiBs. However, from a stakeholder perspective, it would drive out the repurposers and car dismantlers from the market. This concern was shown by representatives from Gamma and Delta in their respective interviews.

## *4.3 Barriers to SBM Design in SLB*

The interviewees highlighted several barriers that can hinder the adoption of the SLB applications. This confirmed some of the barriers also mentioned in Sect. 3.3 in this paper.

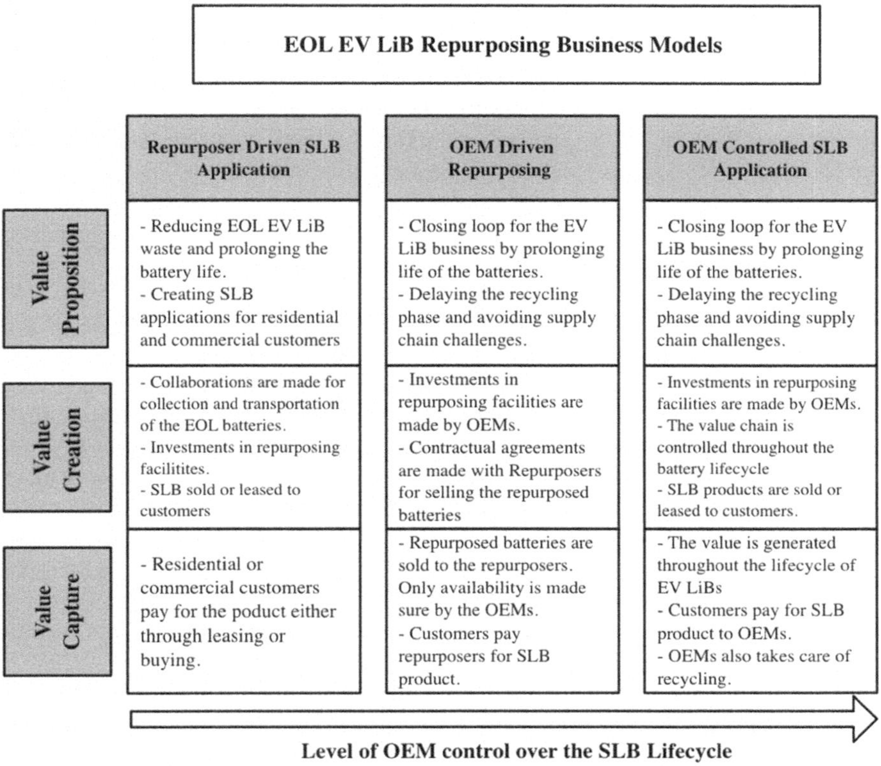


**Fig. 2** Typology of SBM designs for SLB

## Collection and Transport of EOL Batteries

The materials present in the EV LiBs can be hazardous and the EOL batteries are considered as dangerous goods. Moreover, they weigh between 450 and 900 kg, making it very difficult to transport them. The safety regulations only allow the EOL battery to be transported in specific protective packaging that is very costly. This raises the overall cost of the repurposed battery and drives out the small players in the market. This concern was particularly highlighted by Alpha.

## Testing the State of Health (SOH)

The basic SOH test is used to determine whether the battery is fit for repurposing or if it should be sent to recycling right away. This is a complex task where a lot of effort is spent in disassembly, BMS test, and reparations if needed. This is mainly done by manual labor and can cost a lot of time and money.

### Battery Availability

At present, the supply of EOL EV LiBs is limited. This creates an economies of scale challenge which entails that the repurposers and/or OEMs do not have enough batteries to use for SLB applications. This restricts value capture for the stakeholders.

### Variable Battery Design

Although the EV LiBs are easy to dismantle, they are not standardized. They have variable battery chemistries, module assemblies, battery management systems, and thermal management systems. As mentioned by both Alpha and Gamma, this makes it harder for repurposers, if they do not have a contractual agreement with an OEM, to standardize the process. Variable battery designs means that the disassembly and reassembly must be customized according to the type of battery through manual labor. This raises the costs and slows down the process, again affecting the value creation and capture.

### Pricing

Various factors including increasing costs, regulations, and battery availability is driving the SLB price upwards, while technology advancements are lowering the price of the new EV LiBs. This disparity is increasingly seen as a barrier for SLBs in the future by the stakeholders, as the end users would rather buy a new LiB for an ESS instead of a SLB.

### Safety

A perception exists in the market that there are safety hazards related to SLBs, and there may be a chance of fire hazard if certain standards are not employed by the repurposer. Moreover, user awareness and somewhat basic knowledge of the SLB application is also highlighted to make sure these hazards can be avoided. Maintaining safety standards and providing warranties are some ways through which this can be avoided. Delta highlighted that these safety regulations are somewhat vague in Norway, where no guidelines are provided to ensure safe installations of SLB, especially in the household context.

## 5 Discussion and Analysis

This study aims to contribute to the body of knowledge concerning SBM design for SLBs. It addresses the paucity in literature concerning the SBMs in this industry in the Norwegian context. The theoretical insights and the semi-structured interviews conducted with four companies in the battery value chain provides insights into how the value chain is structured in Norway and what kind of business models exist in the market.

The results show that currently two dominating value chain structures exist for the SLBs in the Norwegian context. In one structure, the repurposers source the EOL batteries through contracts with the third-party companies like car dismantlers, insurance companies, producer responsibility companies, and then repurpose the batteries for new SLB applications. While in the second structure, the OEM is responsible for the repurposing of the EOL batteries where they test, repurpose, and sell the batteries to repurposers for SLB applications. The value chain structures then also determine the kind of SBMs for the stakeholders. A typology of three SBM designs for the SLB market are presented namely (1) Repurposer driven SLB application, (2) OEM driven repurposing, and (3) OEM controlled SLB application. The type 1 and 2 already exist in the Norwegian context, but type 3 is a proposed SBM that may exist in the future. In both 1 and 2, a collaborative business model exists for both the OEMs and the repurposers. On the other hand, from a future perspective, an integrative business model is proposed, where the OEM is responsible for the whole life cycle of the batteries including the SLB application and recycling. Although this model creates a closed loop for the EV LiBs, it also drives out the repurposers from the market, which might seem as a negative effect for the value chain.

Moreover, several barriers related to SLB from SBM perspective are presented. These barriers confirm and compliment the barriers and challenges presented in the literature and specified in Sect. 3.3 in this paper. These barriers relate to market, technological, legal, and social aspects. The main elements that come forward from the interviews are collection and transportation of the EOL batteries, SOH testing, battery availability, variable battery design, pricing, and safety. From the market point of view, unavailability of the EOL batteries, non-standardized battery designs and unclear regulations concerning SLB applications remain the biggest barriers. The limited supply of EOL batteries, and non-standardized battery designs also give rise to the pricing barrier, where the price of the SLBs increases because of the supply risk and the manual labor employed for battery repurposing. Additionally, the lack of collaboration and communication between the stakeholders especially OEMs and the repurposers creates challenges for data traceability regarding the SOH of the EOL batteries. This again raises the cost of SOH testing, consequently increasing the SLB price. Interestingly, while the price of the SLBs is increasing mainly because of the barriers mentioned, the price of new batteries made for ESS is decreasing because of technological advancements and market factors. This shows that these barriers are interdependent.

To make the business models of the OEMs and repurposers viable and SLB market sustainable over a longer period, the mentioned barriers need to be overcome and an optimal value chain structure must be determined. For example, a standardized battery design made for repurposing and/or lasting longer, can effectively make repurposing cheaper through automation, and ultimately drive down the SLB prices. Circular resource cycle view and associated SBM design strategies can be employed to support this, of which two are presented briefly in Sect. 3.4 in this paper. These cycles are slowing and closing resource loops, which can help stakeholders employ design strategies like design for disassembly, repurposing etc. while complimenting SBM strategies for product life extension, take-back systems, and repurposing. These strategies can then support the standardization of the EV LiB design, and also enhance the collaboration between the stakeholders. However, these discussions concerning SLBs are lacking in literature even though the design elements have come forward in the interviews with stakeholders.

### *5.1 Limitations and Future Work*

The study is conducted with an emphasis on the Norwegian context, and the results obtained might not be relevant to other regional contexts. The interviews conducted are also limited to only a few stakeholders and only one representative from each, in the EV battery value chain. This may provide a limited overview of the EV battery value chain from an empirical perspective. However, this study lays the foundation for further research looking at SLBs from a SBM perspective in other regional contexts and provides insights into how SBM barriers could be overcome. Further research is proposed related to circular design strategies and how they can support SBM innovation in the SLB industry. This could lay down the foundation for making the SLB sustainable.

## 6 Conclusion

This study investigates the SLB value chain from a SBM perspective, focusing on the Norwegian context. Based on the theoretical insights and the semi-structured interviews conducted through a multi-case study analysis, it was observed that two main SLB value chain structures exist in Norway: (1) The repurposer collaborates and makes contractual agreement with third party companies to collect and transport batteries, while handling the disassembly and other repurposing processes themselves to create SLB application, and (2) The OEM takes control of the EOL battery until the repurposing stage where it disassembles, tests, and repurposes the battery while the repurposer creates the SLB application. Based on the value chain structures, a typology of SBMs for the SLB is created which shows two existing SBMs for the stakeholders and a proposed SBM that might come into existence in the future. The study also identifies several SBM barriers that need to be overcome to make the SLB market viable and sustainable, and suggests some strategies that can be used to support it.

## References


1. Fridstrøm L (2021) The Norwegian electrification policy and its implicit price of carbon. Sustain For 13:1–14
2. Muratori M, Mai T (2020) The shape of electrified transportation. Environ Res Lett 16(1): 011003. https://doi.org/10.1088/1748-9326/abcb38
3. Yang Y, Okonkwo EG, Huang G, Xu S, Sun W, He Y (2021) On the sustainability of lithium ion battery industry – a review and perspective. Energy Storage Mater 36:186–212. https://doi.org/10.1016/j.ensm.2020.12.019
4. IEA (2022) Global electric car sales have continued their strong growth in 2022 after breaking records last year – News. IEA. https://www.iea.org/news/global-electric-car-sales-have-continued-their-strong-growth-in-2022-after-breaking-records-last-year. Accessed 8 Sept 2022
5. Winslow KM, Laux SJ, Townsend TG (2018) A review on the growing concern and potential management strategies of waste lithium-ion batteries. Resour Conserv Recycl 129:263–277. https://doi.org/10.1016/j.resconrec.2017.11.001
6. Wrålsen B, Prieto-Sandoval V, Mejia-Villa A, O'Born R, Hellström M, Faessler B (2021) Circular business models for lithium-ion batteries – stakeholders, barriers, and drivers. J Clean Prod 317:128393. https://doi.org/10.1016/j.jclepro.2021.128393
7. Reinhardt R, Christodoulou I, Gassó-Domingo S, Amante García B (2019) Towards sustainable business models for electric vehicle battery second use: A critical review. J Environ Manag 245: 432–446. https://doi.org/10.1016/j.jenvman.2019.05.095
8. Jiao N, Evans S (2018) Business models for repurposing a second-life for retired electric vehicle batteries. In: Pistoia G, Liaw B (eds) Behaviour of lithium-ion batteries in electric vehicles: battery health, performance, safety, and cost, Green energy and technology. Springer International Publishing, Cham, pp 323–344. https://doi.org/10.1007/978-3-319-69950-9_13
9. Makuza B, Tian Q, Guo X, Chattopadhyay K, Yu D (2021) Pyrometallurgical options for recycling spent lithium-ion batteries: A comprehensive review. J Power Sources 491:229622. https://doi.org/10.1016/j.jpowsour.2021.229622
10. Yu X et al (2022) Current challenges in efficient lithium-ion batteries' recycling: a perspective. Glob Chall 6(12):2200099. https://doi.org/10.1002/gch2.202200099
11. Heelan J et al (2016) Current and prospective li-ion battery recycling and recovery processes. JOM 68(10):2632–2638. https://doi.org/10.1007/s11837-016-1994-y
12. Börner MF et al (2022) Challenges of second-life concepts for retired electric vehicle batteries. Cell Rep Phys Sci 3(10):101095. https://doi.org/10.1016/j.xcrp.2022.101095
13. (2019) Recycle spent batteries. Nat Energy 4(4):Art. no. 4. https://doi.org/10.1038/s41560-019-0376-4
14. Hossain E, Murtaugh D, Mody J, Faruque HMR, Sunny MS, Mohammad N (2019) A comprehensive review on second-life batteries: current state, manufacturing considerations, applications, impacts, barriers & potential solutions, business strategies, and policies. IEEE Access 07:73215–73252. https://doi.org/10.1109/ACCESS.2019.2917859
15. Chen M et al (2019) Recycling end-of-life electric vehicle lithium-ion batteries. Joule 3(11): 2622–2646. https://doi.org/10.1016/j.joule.2019.09.014
16. Montes T, Etxandi-Santolaya M, Eichman J, Ferreira VJ, Trilla L, Corchero C (2022) Procedure for assessing the suitability of battery second life applications after EV first life. Batteries 8(9): Art. no. 9. https://doi.org/10.3390/batteries8090122

17. Kumar P, Singh RK, Paul J, Sinha O (2021) Analyzing challenges for sustainable supply chain of electric vehicle batteries using a hybrid approach of Delphi and Best-Worst Method. Resour Conserv Recycl 175:105879. https://doi.org/10.1016/j.resconrec.2021.105879
18. Reinhardt R, Christodoulou I, García BA, Gassó-Domingo S (2020) Sustainable business model archetypes for the electric vehicle battery second use industry: Towards a conceptual framework. J Clean Prod 254:119994. https://doi.org/10.1016/j.jclepro.2020.119994
19. Bocken N, Boons F, Baldassarre B (2018) Sustainable business model experimentation by understanding ecologies of business models. J Clean Prod 208C:1498–1512. https://doi.org/10.1016/j.jclepro.2018.10.159
20. Geissdoerfer M (2018) Sustainable business model innovation: A review. J Clean Prod. Accessed 30 Sept 30 2022. [Online]. Available: https://www.academia.edu/39963494/Sustainable_business_model_innovation_A_review
21. Beaudet A, Larouche F, Amouzegar K, Bouchard P, Zaghib K (2020) Key challenges and opportunities for recycling electric vehicle battery materials. Sustain For 12(14):Art. no. 14. https://doi.org/10.3390/su12145837
22. Bocken NMP, de Pauw I, Bakker C, van der Grinten B (2016) Product design and business model strategies for a circular economy. J Ind Prod Eng 33(5):308–320. https://doi.org/10.1080/21681015.2016.1172124
23. Chesbrough H (2010) Business Model Innovation: Opportunities and Barriers. Long Range Plan 43(2):354–363. https://doi.org/10.1016/j.lrp.2009.07.010
24. Lüdeke-Freund F (2020) Sustainable entrepreneurship, innovation, and business models: Integrative framework and propositions for future research. Bus Strategy Environ 29(2):665–681. https://doi.org/10.1002/bse.2396
25. Geissdoerfer M (2019) Sustainable business model innovation. Sustain Bus Modelorg. https://sustainablebusinessmodel.org/2019/01/12/sustainable-business-model-innovation/. Accessed 12 Aug 2022
26. Yin RK (2018) Applications of case study research. Sage
27. Saw LH, Ye Y, Tay AAO (2016) Integration issues of lithium-ion battery into electric vehicles battery pack. J Clean Prod 113:1032–1045. https://doi.org/10.1016/j.jclepro.2015.11.011
28. Bobba S et al (2018) Life cycle assessment of repurposed electric vehicle batteries: an adapted method based on modelling energy flows. J. Energy Storage 19:213–225. https://doi.org/10.1016/j.est.2018.07.008
29. Chirumalla K, Reyes LG, Toorajipour R (2022) Mapping a circular business opportunity in electric vehicle battery value chain: A multi-stakeholder framework to create a win–win–win situation. J Bus Res 145:569–582. https://doi.org/10.1016/j.jbusres.2022.02.070
30. Islam MT, Iyer-Raniga U (2022) Lithium-ion battery recycling in the circular economy: a review. Recycling 7(3):Art. no. 3. https://doi.org/10.3390/recycling7030033
31. Golmohammadzadeh R, Faraji F, Jong B, Pozo-Gonzalo C, Banerjee PC (2022) Current challenges and future opportunities toward recycling of spent lithium-ion batteries. Renew Sust Energ Rev 159:112202. https://doi.org/10.1016/j.rser.2022.112202
32. Wrålsen B, Faessler B (2022) Multiple scenario analysis of battery energy storage system investment: measuring economic and circular viability. Batteries 8(2):Art. no. 2. https://doi.org/10.3390/batteries8020007
33. Osterwalder A, Pigneur Y (2010) Business model generation: a handbook for visionaries, game changers, and challengers. Wiley. https://www.wiley.com/en-am/Business+Model+Generation%3A+A+Handbook+for+Visionaries%2C+Game+Changers%2C+and+Challengers-p-9780470876411. Accessed 29 Nov 2021
34. Osterwalder A, Pigneur Y, Tucci C (2010) Clarifying business models: origins, present, and future of the concept. Commun AIS 16. https://doi.org/10.17705/1CAIS.01601
35. Lüdeke-Freund F (2010) Towards a conceptual framework of 'Business models for sustainability'. In: Proceedings of the knowledge collaboration & learning for sustainable innovation ERSCP-EMSU conference. https://doi.org/10.13140/RG.2.1.2565.0324
36. Bocken N (2023) Business models for Sustainability

37. Bocken N (2021) Sustainable business models, pp 963–975. https://doi.org/10.1007/978-3-319-95867-5_48
38. Boons F, Lüdeke-Freund F (2013) Business models for sustainable innovation: state-of-the-art and steps towards a research agenda. J Clean Prod 45:9–19. https://doi.org/10.1016/j.jclepro.2012.07.007
39. Jiao N, Evans S (2016) Secondary use of electric vehicle batteries and potential impacts on business models. J Ind Prod Eng 33(5):348–354. https://doi.org/10.1080/21681015.2016.1172125
40. Bocken N, Short SW, Rana P, Evans S (2014) A literature and practice review to develop sustainable business model archetypes. J Clean Prod 65:42–56. https://doi.org/10.1016/j.jclepro.2013.11.039
41. Klör B, Beverungen D, Bräuer S, Plenter F, Monhof M (2015) A market for trading used electric vehicle batteries – theoretical foundations and information systems. In: Proceedings of the twenty-third European Conference on Information Systems (ECIS 2015)
42. Rufino Júnior CA et al (2023) Towards a business model for second-life batteries – barriers, opportunities, uncertainties, and technologies. J Energy Chem 78:507–525. https://doi.org/10.1016/j.jechem.2022.12.019
43. Ahmed S, Verhulst E, Boks C (2023) Sustainable business models for the second use of electric vehicles lithium ion batteries in an ecosystem context: a review (Upcoming), Presented at the PLATE 23
44. Haram MHSM, Lee JW, Ramasamy G, Ngu EE, Thiagarajah SP, Lee YH (2021) Feasibility of utilising second life EV batteries: applications, lifespan, economics, environmental impact, assessment, and challenges. Alex Eng J 60(5):4517–4536. https://doi.org/10.1016/j.aej.2021.03.021
45. Lee JW, Haram MHSM, Ramasamy G, Thiagarajah SP, Ngu EE, Lee YH (2021) Technical feasibility and economics of repurposed electric vehicles batteries for power peak shaving. J Energy Storage 40:102752. https://doi.org/10.1016/j.est.2021.102752
46. Moss MA (1985) Designing for minimal maintenance expense: the practical application of reliability and maintainability. CRC Press
47. Chapman J (2015) Emotionally durable design: objects, experiences and empathy. Routledge
48. British Standard BS 8887-2 (2009) Design for manufacture, assembly, disassembly and end-of-life processing (MADE). Part 2: terms and definitions. BSI
49. Nedelea A (2022) Here's another look at a Nio battery swapping station in Norway. InsideEVs. https://insideevs.com/news/591837/nio-battery-swapping-station-norway-5-minutes/. Accessed 29 June 2023
50. Epiroc, "Batteries as a Service," *Epiroc*. https://www.epiroc.com/en-ke/products/parts-and-services/batteries-as-a-service. Accessed 29 June 2023
51. Bakker C, den Hollander M, van Hinte E, Zijlstra Y (2014) Products that last: Product design for circular business models. TU Delft Library, Delft. Accessed 29 June 2023. [Online]. Available: https://www.bispublishers.com/products-that-last-93487267.html
52. "Hva vi gjør," Batteriretur. https://batteriretur.no/en/hva-vi-gjor/. Accessed 29 June 2023
53. Ecostor (2021) https://www.eco-stor.com/news/nissan-eco-stor-norsk-gjenvinning-and-agder-energi-signs-mou-around-re-use-of-ev-batteries. Accessed 25 Mar 2023

# European Union Policy Developments Related to Ecodesign, Green Claims, and Eco-labeling: Implications for Voluntary Circular Labeling Initiatives

**Karolina Vikingsson, Carl Dalhammar, and Leonidas Milios**

**Abstract** The European Union (EU) has recently presented several new legal proposals for promoting ecodesign and product labeling. The aim is to harmonize national practices and rules in the EU, to regulate issues like product lifetime and repairability in a more comprehensive way, to introduce new tools like digital product passports, and to introduce labeling schemes for repairability and durability. In parallel, the EU has proposed rules to regulate green claims and ecolabeling schemes to effectively address greenwashing. In this paper, we outline these rules, and discuss the challenges emerging for new eco-labeling schemes. We further discuss the challenges for a proposed new scheme for circular products: Certified to LAST.

**Keywords** Ecodesign regulation · Product regulation · Product labeling · Product lifetime

## 1 Introduction

Currently, the European Union (EU) is regarded as the global green leader [1], having introduced and adopted the most progressive environmental policies in many areas of policy making, including laws that are directly influencing the environmental performance of products [2]. The EU has adopted binding Directives influencing the energy efficiency and chemical content of products and has also pioneered rules on extended producer responsibility. Often, progressive policies and laws adopted by the EU Member States (MS) have triggered the formulation and adoption of EU laws, as the European Commission strives for a harmonized regulatory framework within the EU [3]. Recently, the EU has started to set new requirements under the

K. Vikingsson
RISE, Research Institutes of Sweden, Gothenburg, Sweden

C. Dalhammar (✉) · L. Milios
IIIEE, Lund University, Lund, Sweden
e-mail: carl.dalhammar@iiiee.lu.se

 
S. Fukushige et al. (eds.), *EcoDesign for Circular Value Creation: Volume I*,
https://doi.org/10.1007/978-981-97-9068-5_16

Ecodesign Directive (2009/125/EC), including rules on minimum lifetimes for two product categories, and rules that support repairs for several product groups (most notably white goods) [2].

In the last couple of years, some EU MS, most notably France, have adopted several national policies to support Circular Economy (CE) developments. These include the mandatory French Repair Index and France's mandatory information scheme for products put on the French market [3]. These national rules are upsetting EU industries who want to have similar rules in all EU countries and to avoid having to deal with deviating national rules in each of the MS [3]. The EU has recently announced a large package of product laws under its Sustainable Products Initiative, which will provide an all-encompassing legal framework for products and harmonize some national laws. The cornerstone of the initiative is the "Ecodesign for Sustainable Products Regulation" (ESPR); a Regulation that will replace the current Ecodesign Directive and will apply to all products except medical products and foodstuff. It aims to set holistic ecodesign requirements for products, and addresses issues such as products' lifetimes, repairability, and recyclability; also aiming to introduce requirements on recycled content in new products.

Meanwhile, there are significant initiatives in the EU aimed at addressing "greenwashing" and false green claims and ensuring that product labeling schemes fulfil certain quality criteria. There is ample evidence showing that greenwashing is an increasing problem, and that there must be initiatives to guarantee the quality of ecolabeling schemes [4]. Central to these efforts is the proposed Directive on regulating so-called Green Claims, with rules on the communication and verification of voluntary environmental claims and environmental labels used by traders that market products toward EU consumers.

The traditional way to view the relationship between mandatory ecodesign regulations and labels, respectively, is that laws mainly address "laggard" products/producers, as it sets a base of criteria that all must pass, whereas labeling allows consumers—or professional procurers—to differentiate between products on the market [5]. As we now see efforts in the EU to regulate the environmental aspects of products in a more integrated way, there is a need to analyze the implications for labeling, and the need for, and role of, voluntary labeling schemes in the future.

## 2 Objective, Methods, and Outline

### *2.1 Objective*

The aim of the paper is to discuss the recent developments in EU regulatory initiatives for ecodesign, green claims and ecolabels, and to identify and discuss implications for future voluntary labeling schemes, exemplified by the proposed labeling scheme "Certified to LAST." This research is conducted as a part of the Certified to LAST project, which aims to develop a new voluntary labeling scheme for circular products.

### 2.2 Methods

The research is based on (1) a review of relevant policies and relevant literature; and (2) a questionnaire with five companies, which was part of the Certified to LAST project. The questionnaire included questions to companies that relate to inter alia their views on how proposed EU policies would drive ecodesign practices; if they believe that consumers will increasingly look for durable and repairable products in the future; and the need for voluntary ecolabels as a means of differentiation between products on the market.

### 2.3 Outline

The next section will account for recent EU legal developments related to rules that promote ecodesign. Section 4 will discuss the proposed new EU legal framework for Green Claims and voluntary labels. Section 5 discusses the need for voluntary labeling, how current developments in the EU influence the context of labeling, and industry's need for labeling developments. Section 6 looks at some of the implications for the Certified to LAST project, and accounts for the survey made with companies. The study ends with the conclusions.

## 3 The Emerging EU Legal Framework for Ecodesign

### 3.1 EU Policy Developments

The EU already regulates the chemical content of many product groups and sets rules mandating collection and recycling of a large number of product groups, through extended producer responsibility schemes [6]. Through the Ecodesign Directive (2009/125/EC), minimum mandatory energy efficiency standards have been set for a wide range of product categories. The Ecodesign Directive works in synergy with the mandatory EU Energy Label, which obliges manufacturers to label their products according to energy efficiency performance. More recently, ecodesign requirements for products' lifetimes have been set for vacuum cleaners and lighting products, and "right to repair" obligations—mandating manufacturers/importers to provide spare parts, manuals, and tools to independent repairers and consumers—have been adopted for some product groups [7]. In 2022, the European Commission proposed a new Ecodesign Regulation that will replace the current Ecodesign Directive, which has now been adopted, the Ecodesign for Sustainable Products Regulation (ESPR) [8]. This regulation is applicable to all product groups except medical products and foodstuff, and it is expected that ecodesign requirements will be set for, e.g., textiles and furniture; these are product groups where it is less relevant to regulate energy efficiency in the use phase, but very relevant to incentivize longer product lifetimes through legal standards. The Regulation also outlines a more comprehensive approach to ecodesign, where requirements that may increase products' lifetimes and repairability will always be considered for all product groups. It also introduces digital product passports (DPPs). The aim of the DPP is to provide relevant information about products to producers and other supply chain actors such as recyclers, but also consumers and market surveillance authorities, to further support the sustainable management of products. The proposed Regulation also has provisions that can be used to introduce mandatory labeling of products, e.g., related to their lifetime and repairability. The main aim is to set common EU rules for this and harmonize practices in EU MS; for instance, France has introduced a mandatory Repair Index for some product groups and increasingly EU companies are asking for them to be replaced by an EU-wide scheme (for a general discussion on the need for harmonization see [3]). Further, the proposed Regulation includes provisions that could force manufacturers to report data on the destruction of unsold and/or returned consumer goods. The ESPR contain as direct ban on the destruction of new unsold and/or returned clothes and footwear. Also, on this issue, there is a need to harmonize practices among EU Member States [3], where countries like France and Germany have already adopted national rules aimed at addressing product destruction [9].

The European Commission has also proposed a new Battery Regulation [10]. The regulation will introduce DPPs for large batteries and rules to support battery repurposing for prolonging their lifetimes. The Regulation also contains requirements about portable batteries in smaller appliances, which must be designed so that consumers can easily remove and replace them.

It can be expected that the EU will introduce more rules in the coming years that relate to ecodesign, which means it will become increasingly challenging for products to qualify for the Internal Market of the EU. We also see how the EU tries to adopt rules at the higher supranational EU level to harmonize various national rules (these are discussed in the next section).

## *3.2 Policy Developments Among EU Member States*

EU MS have adopted a number of national policies and laws that relate to ecodesign [3, 9, 11]. These include the French Repair index, a mandatory labeling program, which is planned to expand into a national Durability index by 2024. Several other EU MS have similar plans for national repair indexes. France has also introduced new rules that require manufacturers of certain product groups to label their products in accordance with several criteria, including repairability, sustainability, recycled materials, and microplastics. This is considered to be a challenging trade barrier by many EU companies [3]. Regarding destruction of unsold and/or returned products, some countries have adopted measures that restrict such practices, or require companies to report on them [9]. Some EU MS have also introduced national policies to support repairs, e.g., through tax breaks [12]. Some leadership has also been showcased at the local level, as some cities have introduced repair vouchers [13].

The above are just some examples of national policies. There are many more national policies of relevance, not least related to labeling of packaging. These national policies have led to increasing calls from companies to harmonize national practices at the EU level.

## 4 The Emerging EU Legal Framework for Green Claims and Ecolabels

### *4.1 Emerging Concerns with Green Claims and Eco-labeling Schemes*

Claims like "green," "sustainable," and "eco-friendly" have become increasingly valuable for companies on the market, as green products often experience greater sales growth than conventional ones. Companies communicate claims like "25% reduced $CO_2$ in our packages," "Made of 50% recycled materials," and "Responsible fashion"; however, what they claim is not always what consumers get. Many of the environmental claims made in the market today have no solid or measurable foundation and, therefore, cannot be trusted.

The European Commission has conducted two inventories of environmental claims, one in 2014 [14] and another in 2020 [4]. These studies examined 150 environmental claims relating to various products, against the principles laid out by the Directive on Unfair Commercial Practices (2005/29/EC), namely, clarity, unambiguity, accuracy, and verifiability. The 2020 inventory revealed that over 50% of environmental claims provided vague, misleading, or unsubstantiated information about the environmental attributes of products. Deficiencies were found in green claims in a wide range of product categories across the EU. The same study also analyzed the substantiation and verifiable evidence behind the claims and found that 40% of the claims were unsubstantiated.

The ecolabels currently existing on the EU internal market are subject to different levels of robustness, supervision and transparency. Increasingly, ecolabels cover different aspects, use different methods, and are being subject to varying levels of scrutiny. This diversity causes additional confusion among consumers and other market actors.

In a preparatory study with the objective of identifying relevant actions to "empower consumers in the green transition" [15], an assessment of 232 active ecolabels in the EU was conducted in relation to the "substantiation," "verification," and "certification" aspects of ecolabeling. The study concluded that almost half of the labels' verification was either weak or not carried out at all. Moreover, consumers are often unaware of the distinction between labels governed by third party certification schemes and those based on "self-certifications," meaning they lack verification by a competent third party.

Despite the existence of numerous policies to prevent false environmental claims, green advertising, and other types of greenwashing, both in the EU and at national level, misleading information seems to be everywhere. If green claims or ecolabels are not reliable, consumers are tricked into thinking that they are making an environmentally better choice. In practice, inaccurate labeling and baseless claims can lead consumers to choose a less sustainable product than they would have done otherwise, had it not been for the label [15].

Further, greenwashing is described by the EU as a factor that can hamper the green transition and prevent the reduction of negative environmental impacts. Therefore, the need to address greenwashing was set as a priority both under the New Circular Economy Action Plan [16] and the New Consumer Agenda [17].

## 4.2 New EU Proposals to Deal with the Identified Problems

The proposal by the European Commission for a new Directive on substantiating and presenting explicit environmental claims, also known as the "Green Claims Directive" [18], is part of a set of initiatives aiming at establishing a policy framework that will "make sustainable products the norm." This is anticipated to exert a substantial impact on what claims companies can legitimately make about the environmental performance of the company, the brand, or the product in the future.

To prevent consumers from being exposed to misleading commercial practices related to product sustainability, the proposal suggests that environmental claims that do not fulfil a minimum set of criteria are prohibited. This would protect consumers from greenwashing, as green claims must meet specific standards.

The Directive proposal also aims to ensure that consumers are protected from being misled by labels and similar tools by proposing a prohibition on ecolabels that do not meet minimum transparency and credibility requirements.

According to another legislative EU proposal, the proposal for the Directive on empowering consumers for the green transition [19], sustainability labels that are not based on certification schemes, or approved by public authorities, are considered unfair commercial practices. This means that self-certified sustainability labels, without third-party verification and regular monitoring to ensure compliance with requirements, are prohibited [15].

Generic environmental claims, such as "green," "nature's friend," and "eco-responsible" must be based on recognized environmental performance criteria relevant to the related claim, according to the Directive proposal.

# 5 The New Situation for Ecolabeling

## *5.1 The Interactions Between Legal Requirements and Ecolabeling*

As presented above, a new complex landscape of product regulation is emerging in Europe, which includes both EU policies and national policies. In this landscape, it is crucial to distinguish between four types of key policies: (1) mandatory laws on minimum ecodesign standards, and related standards on, e.g., provision of spare parts (setting legal requirements that all regulated products must comply with); (2) Mandatory information provision schemes, e.g., the EU's Energy Labeling Regulation and the French Repair Index (where products must be correctly labeled, otherwise various sanctions may apply); (3) voluntary labeling schemes with third party verification and; (4) Green Claims without third party verification.

At the same time, related to CE developments, new types of issues are emerging. This includes a focus on "right-to-repair" obligations in legal standards [20] and an emerging trend focusing on novel aspects/criteria in labeling, including product repairability and information on product lifetime [21].

Regarding interactions between legal requirements on products, and the mandatory/voluntary labeling schemes and green claims, a few issues are important to consider [5]. First, mandatory requirements apply to all products, whereas the labeling schemes—and possibly, green claims—allow for consumers (and professional procurers) to differentiate among products on the market. Thus, voluntary labeling can provide incentives for change among progressive manufacturers, while mandatory product laws provide incentives for change also among "laggards." Therefore, it is crucial that companies do not use unsubstantiated claims of legal compliance to show that their products are green, e.g., "We comply with all requirements in the Ecodesign Directive," a practice that is actually not uncommon. Second, it is important that mandatory legal requirements and voluntary labeling are synchronized. In some cases, labeling requirements have not been updated to ensure that a labeling scheme sets more stringent requirements than the law, which undermines the potential for mandatory and voluntary schemes to work in tandem. Third, labeling schemes could address issues that are of great importance for private and/or professional procurers, in cases that it would take several years before it is possible to address them in laws. This way, labeling can incentivize change on the market, and in some cases also "pave the way" for legal standards by showing that some companies can comply with stringent criteria. An example of this is when some labeling schemes for electronics addressed conflict minerals before the issue was addressed in law.

## 5.2 *The Emerging Policy Mix: Implications for New Ecolabeling Schemes*

An environmental labeling scheme is a certification scheme which certifies that a product, a process, or a trader complies with the requirements laid out by an environmental labeling authority. The proposed Green Claims directive requires environmental labeling schemes to provide transparent and accessible information about ownership and decision-making bodies as well as the objectives and compliance monitoring procedures. The schemes must also present proportionate conditions for companies to join, considering their size and turnover. The requirements must be developed by experts and consulted with stakeholders for scientific robustness and societal relevance and provide a complaint and dispute resolution mechanism. The scheme also needs to develop procedures for addressing noncompliance, including the case of label withdrawal or suspension. In the proposal for a Green Claims Directive no new national/regional environmental labeling schemes shall be established by the public authorities of EU MS. However, both EU and non-EU schemes established prior to the date of transposition can continue to award the environmental labels on the EU internal market, if they meet the requirements of the Green Claims Directive [18].

New environmental labeling schemes, established by public authorities in non-EU countries for awarding ecolabels on the EU internal market, shall be subject to approval by the Commission prior to entering the market. The aim is to ensure that these labels provide added value in terms of their environmental impacts, environmental aspects, or environmental performance at a certain product group or sector, as compared to existing schemes.

New schemes established by private organizations, in the EU or in third countries, are only approved if they provide added value in similar terms as presented above, and according to their ability to support the green transition of SMEs, as compared to the existing schemes, and meet the requirements of the Green Claims Directive.

Considering the upcoming regulations, including the requirements of third-party certification for ecolabels in the EU, the number of different ecolabels is likely to decrease. The third-party certification, where compliance with the requirements for labeling has been verified by independent experts, is supposed to lead to more trusted and relevant labels, strengthening the labels potential to make a bigger impact on consumer choices in the future.

Since new ecolabels must add new value, new labeling schemes are expected to set new and higher requirements compared to the labels that already exist. New labels with challenging requirements may, in the best case, enable companies that make an effort to deliver "real improvements" to products to gain a competitive advantage in the market. It can also guide consumers to find the best available options. This is intended to push and motivate producers to invest even more in developing long-lived products and new circular business offerings.

The labels developed today can be designed right from the start to meet and support future circular criteria, stricter rules, and a high level of transparency. At the same time, new labels can ensure that the consumer understands what kind of environmental aspects the label applies to, and that the information is useful or "in demand" among private and professional procurers, supporting resource efficiency and circular economy objectives. Another possible advantage of new labels is that they may have better conditions to build trust compared to existing labels because some of these, to some extent, might already be associated with greenwash and self-certification processes.

# 6 Case Study: Certified to LAST

## *6.1 About Certified to LAST*

To address the urgent need for increased resource efficiency and to ensure product readiness for a long life span in a circular economy, RISE, Research Institutes of Sweden, have developed the "Certified to LAST" concept [22]. The acronym LAST represents the four main criteria of the certification system:

1. *Lifetime:* Emphasizes the product's durability and longevity.
2. *Accessibility:* Highlights the availability of information, services, and spare parts that extend the product's life span.
3. *Sustainability:* Evaluates the product's environmental performance, including social aspects.
4. *Transparency:* Focuses on providing users and stakeholders with access to essential information for responsible product use, maintenance, servicing, refurbishment, dismantling, and recycling. This criterion also encompasses details about the scheme, compliance monitoring, and complaint procedures (Fig. 1).

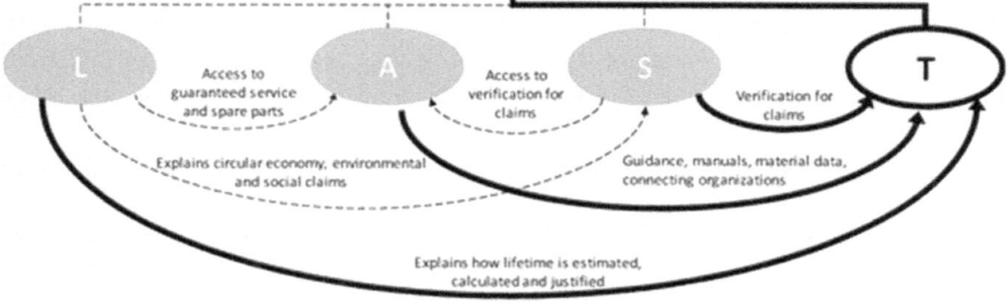


**Fig. 1** The role of (T)ransparency in a verified circularity product declaration based on (L)ong lifetime, (A)ccessible service and spare parts, and (S)ustainable materials and life cycle [22]

The initiative can be seen as an example on how to handle the complexity in developing ecolabels that deliver added value according to the upcoming EU regulations.

The LAST concept and certification scheme is being developed to push and motivate companies to put long-lived sustainable products on the market, and to guide and enable consumers to make conscious choices. It also aims to provide information on how to take care of the products. The whole array of interrelated EU initiatives to make sustainable products the norm, as well as relevant global product standards must be considered in the development process to reflect legal and market developments.

Certified to LAST aims to deliver a holistic certification and a global-reaching label to recognize sustainable long-lived products that are designed for a circular economy. That means that the whole value chain must be assessed, and all significant trade-offs must be accounted for. Spare parts, services, and information that can contribute to a longer service life must be available to the relevant stakeholders such as consumers, repairers, and recyclers. To put a "Certified to LAST" product on the market has to be seen as a long-term commitment by the producer. For the consumer, buying one means that their willingness to take care of the product is enabling an optimized lifetime.

The Certified to LAST requirements are developed by experts and consulted with producing companies and other stakeholders to ensure scientific robustness and societal relevance. This method aligns with the ISO standard Ethical claims and supporting information (ISO/TS 17033), as well as the upcoming Green Claims directive.

In the development stage of the "Certified to LAST" label, a number of companies with global brands contribute with product specific information as well as insights on impact, relevance and practice. In the end of 2021, the companies took active part in a workshop with the aim to identify obstacles and opportunities regarding a comprehensive approach for Certified to LAST. One of the key takeaways from the workshop was the request for an increased focus on product lifetimes, both from customers and internally in the corporations. The lack of standards on measuring product lifetime was highlighted as a challenge which could complicate further work.

As a part of the following step in the development work, a canvas tool for mapping the performance of products based on the four dimensions: *Lifetime*, *Accessibility*, *Sustainability*, and *Transparency* was created. The LAST-Canvas [23] can be used to assess any product and to identify lifetime and sustainability risks over the whole product life cycle. Participating companies have already made assessments of their products using the canvas. The results identified weaknesses and risks connected to the specific product, but also added valuable information to the process of setting criteria and verifiable requirements for different product categories.

## 6.2 Certified to LAST: Added Value for Companies

In the context of the "Certified to LAST" development, a qualitative survey was carried out among some of the participating companies. The sample is small ($n = 5$) and includes companies from quite different sectors, so the results should be seen as merely indicative.

Most companies saw policies as strong drivers of change on the market. Among the policies that were considered important are Digital Product Passports (DPPs), rules on producer responsibility, rules on Product Environmental Footprint (PEF), rules on green claims, building codes, the forthcoming EU Battery Regulation, and the forthcoming Ecodesign Regulation.

The participating companies also believed that consumers will increasingly start to look for products with longer lifetimes and/or products that are more repairable in the near future. In general, European consumers state that they would like to be able to purchase products with longer lifetimes, which are repairable, but it is questionable if consumers in general are prepared to pay the higher price often associated with such products [24, 25]. Likewise, consumers may state a willingness to repair products, but various factors—such as uncertainty of repair outcomes, and the residual value of used product—may undermine their willingness to do so [25]. In order to increase consumers' willingness to repair, there is a need to strengthen the three key drivers or repair behavior, including environmental protection, economic considerations, and social norms [26].

The companies also noted that the "Certified to LAST" concept has potential to fill a gap when it comes to labeling and product sustainability indicators, as noted by one interviewee:

> In the /.../ retail area different retail chains are requesting brands to present a sustainability certificate. Most of them have a specific certification in mind; for example /.../. Retails requests are however very different and in different level so it is very difficult for a brand to decide what certification to choose. If Certified to LAST would be seen as an authorized certificate to go for, this position would make the situation easier.

Certified to LAST is also expected to support the design of more circular business offerings, which is highlighted by the Commission as an important strategy for making sustainable products the norm. This was also described as an advantage with the concept, by companies in the survey:

> Even stronger positioning for our quality product—Perfect input for our circularity strategy—Useful for developing long term financing solutions (help customers finance high quality products).

The ongoing EU work to create a standard for DPPs can strengthen concepts like Certified to LAST, as the DPP is likely to become mandatory for several product groups. Certified to LAST can offer a possibility to producers to communicate a DPP for products that have a very high performance when it comes to lifetime and other relevant sustainability indicators. It can also guide consumers to make more sustainable choices and how to take care of the product. Producers of long-lived products are likely going to seek confirmation for their efforts to be able to strengthen their position on the market. These producers can be used as role-models, both to push their competitors to do the same but also to motivate and justify sharper sustainability requirements in the future.

However, some companies also indicate that the Certified to LAST concept is complex, and probably it will create a challenging effort to keep updated with the fast-moving development for more circular business. As Certified to LAST has the ambition to include all aspects of sustainability in the assessment of a product and develops criteria for several different types of product categories, it would be challenging. At the same time it might be necessary to go through some level of complexity in order to be able to have an impact on the circular economy and create a global label that will set a high standard for long-lived sustainable products. As the certification rolls out and its merits and mechanisms are scrutinized, further opportunities will be identified to streamline and simplify some procedures.

## 6.3 Challenges for a New Ecolabeling Scheme

The 'Certified to LAST' concept often implies the involvement of multiple competences within the company to gather the necessary information and specifications of the product and its supporting system. This is identified as a challenge but also presents a great opportunity to engage various teams such as design, purchasing, compliance, sustainability and marketing in the process, as noted by one interviewee in the survey:

> The main insights are that different departments get to know where the restrictions are with regard to products lasting a certain life.

This "side effect" may increase general awareness and knowledge about circular design, resource efficiency and circular business models, and thereby enhance the company's product development, competitiveness and ability to support the green transition.

One of the major challenges for any new ecolabel scheme on the EU market is expected to be the actual "proof" that it contributes a substantial added value to society. While the EU authorities will determine the validity of a new label, it is worth considering this issue from a more "philosophical" standpoint as well.

In the ISO/DIS 59004 draft "Circular Economy – Terminology, Principles and Guidance for Implementation" [27], value in the context of circular economy is defined as the "gains or benefits from satisfying needs and expectations in relation to the use and the conservation of resources." Value creation is a fundamental principle to define the concept of Circular Economy. This implies that only organizations that effectively contribute to socioeconomic and environmental value by recovering, retaining, or adding value, and by using resources efficiently, can be considered circular. This dimension needs to be dealt with when creating new eco-labels, at least in the long term. Because, even if a product meets the sustainability and durability requirements, it might be unsustainable if it does not create enough "value for resources."

## 7 Conclusion and Discussion

We may conclude that the EU has more and more policies that regulate the environmental impacts of products. The new rules aiming to ensure the quality of green claims and eco-labeling schemes pose a challenge for existing labeling schemes, but also for new ones. The Certified to LAST project needs to consider this when the main setup and criteria are developed.

More attention must be paid to the role of the consumer. In a sustainable and circular economy, the customer is not merely a passive consumer or "buyer." The consumer must actively participate in the transition toward a more resource-efficient circular economy—e.g., by selecting sustainable products, and by maintaining and repairing them—and the producers need to provide the necessary support to enable customers to take responsibility for their products and contribute to prolonging their life span.

Regulators and producers are important enabling actors but ultimately the consumers need to use their power.

Customer communication encompasses various ways to encourage consumers to buy new products. "Green claims" are sometimes used to strengthen a producer's brand and to increase willingness to buy. However, there is a risk of rebound effects associated with "green marketing," as it may stimulate consumers to buy and create continuous demand for new "sustainable" products, potentially leading to unsustainable consumption patterns [28, 29]. This risk needs to be considered in the development of Certified to LAST and in the use of all ecolabels and "green claims."

When the LAST-concept was created, the four letters was not only chosen to highlight product lifetime. It was also intended to visually prompt consumers to consider a product as something valuable that they need to take good care of. Providing guidelines on how consumers can contribute is a crucial aspect of the transparency criterion.

A recent inspiring example of how communication is highlighted as an enabler for green transition is The Sustainable Fashion Communication Playbook [30], co-published by UNEP and UN Climate Change.

The Playbook recognizes the immense influence of fashion's marketing engine and its profound impact on shaping desire and demand through compelling narratives and visual language. It emphasizes the importance of redirecting these efforts toward sustainable and circular solutions, presenting it as a creative opportunity. In a time when regulations governing environmental claims communication are gaining traction. The playbook aims to empower communicators to explore strategies that separates value creation from resource extraction and volume growth, all the while promoting improved well-being.

There is still much research needed to fully understand the impacts of marketing as a contributor to the Circular Economy, something that is also highlighted in literature [31]. The exploration of how customer communication and green marketing in the form of labeling could influence consumers to contribute to a more circular and resource-efficient economy will be part of the continued work to develop Certified to LAST.

**Acknowledgments** The research was funded by the Swedish strategic innovation programme Production2030, supported by the Swedish Innovation Agency VINNOVA, The Swedish Energy Agency and Formas, and by the Mistra REES (Resource Efficient and Effective Solutions) programme, funded by the Swedish Foundation for Strategic Environmental Research—Mistra (Stiftelsen för miljöstrategisk forskning).

## References

1. Bradford A (2020) The Brussels effect: how the European union rules the world. Oxford University Press, New York
2. Dalhammar C, Maitre-Ekern E, Richter JL, Svensson-Hoglund S, Milios L (2023) Regulation with a lasting impact? Policies for product durability. In: Jaeger-Erben M, Wieser H, Marwede M, Hofmann F (eds) Durable economies: managing the material foundations of wealth and prosperity. Transcript Verlag, Bielefeld
3. Dalhammar C (2023) Making sustainable products the norm on the internal market: an assessment of the proposal for a new ecodesign regulation. Europarrättslig Tidskrift 1:43–67
4. European Commission (2020) Environmental claims in the EU: inventory and reliability assessment. Final report. European Commission, Brussels
5. Dalhammar C (2007) An emerging product approach in environmental law: incorporating the life cycle perspective. Lund University, Lund
6. Dalhammar C, Milios L, Richter JL (2021) Ecodesign and the circular economy: conflicting policies in Europe. In: Kishita Y, Matsumoto M, Inoue M, Fukushige S (eds) EcoDesign and sustainability I. Sustainable production, life cycle engineering and management. Springer, Singapore, pp 187–198
7. Dalhammar C, Milios L, Richter JL (2021) Increasing the lifetime of products – policies and consumer perspectives. ER 2021:25. Swedish Energy Agency, Eskilstuna
8. Regulation (EU) 2024/1781 of the European Parliament and of the Council of 13 June 2024 establishing a framework for the setting of ecodesign requirements for sustainable products, amending Directive (EU) 2020/1828 and Regulation (EU) 2023/1542 and repealing Directive 2009/125/EC (Text with EEA relevance)
9. Roberts H, Milios L, Mont O, Dalhammar C (2023) Product destruction: exploring unsustainable production-consumption systems and appropriate policy responses. Sustain Prod Consum 35:300–312
10. European Commission (2020) Proposal for a regulation of the European Parliament and of the council concerning batteries and waste batteries, repealing directive 2006/66/EC and amending regulation (EU) No 2019/1020. COM(2020) 798 final. European Commission, Brussels
11. Dalhammar C, Hartman C, Larsson J, Milios L, Mont OK, Jarelin J (2022) Moving away from the throwaway society: five policy instruments for extending the life of consumer durables. Chalmers University of Technology, Gothenburg
12. Almén J, Dalhammar C, Milios L, Richter JL (2021) Repair in the circular economy: towards a National Swedish Strategy. In: Proceedings of the 20th European round table on sustainable consumption and production, Graz University of Technology, Graz, 8–10 September 2021

13. Lechner G, Wagner MJ, Tena AD, Fleck C, Reimann M (2021) Exploring a regional repair network with a public funding scheme for customer repairs: the 'GRAZ repariert'-case. J Clean Prod 288:125588
14. European Commission (2014) Consumer market study on environmental claims for non-food products. DG for Justice and Consumers. European Commission, Brussels
15. European Commission (2021) Preparatory study to gather evidence on ways to empower consumers to play an active role in the green transition. DG for Justice and Consumers. European Commission, Brussels
16. European Commission (2020) A new circular economy action plan for a cleaner and more competitive Europe. COM(2020) 98 final. European Commission, Brussels
17. European Commission (2020) New consumer agenda – strengthening consumer resilience for sustainable recovery. COM(2020) 696 final. European Commission, Brussels
18. European Commission (2023) Proposal for a directive of the European Parliament and of the council on substantiation and communication of explicit environmental claims (Green Claims Directive). COM(2023) 166 final. European Commission, Brussels
19. European Commission (2022) Proposal for a directive of the European Parliament and of the council amending directives 2005/29/EC and 2011/83/EU as regards empowering consumers for the green transition through better protection against unfair practices and better information. COM(2022) 143 final. European Commission, Brussels
20. Svensson-Hoglund S, Richter JL, Maitre-Ekern E, Russell JD, Pihlajarinne T, Dalhammar C (2021) Barriers, enablers and market governance: a review of the policy landscape for repair of consumer electronics in the EU and the U.S. J Clean Prod 288:125488
21. Milios L, Dalhammar C (2023) Consumer perceptions of product lifetimes and labelling: implications for introducing a durability label. Circ Econ 1
22. Carlsson R, Nevzorova T, Vikingsson K (2022) Long-lived sustainable products through digital innovation. Sustainability 14:14364. https://doi.org/10.3390/su142114364
23. Carlsson R, Nevzorova T, Vikingsson K (2022) LAST CANVAS – principles and guidelines. RISE – Research Institutes of Sweden, Gothenburg
24. Šajn N (2022) Right to repair briefing, PE 698.869. European Parliamentary Research Service, Brussels
25. van den Berge R, Thysen T (2020) State-of-the-art knowledge on user, market and legal issues related to premature obsolescence. In: PROMPT (premature obsolescence multi-stakeholder product testing) deliverable 2.6. Technical University of Delft, Delft
26. Fachbach I, Lechner G, Reimann M (2022) Drivers of the consumers' intention to use repair services, repair networks and to self-repair. J Clean Prod 346:130969
27. ISO/DIS 59004 draft, Circular Economy – Terminology, Principles and Guidance for Implementation
28. Berkhout PH, Muskens JC, Velthuijsen JW (2000) Defining the rebound effect. Energy Policy 28:425–432
29. Kopnina H (2016) Animal cards, supermarket stunts and World Wide Fund for Nature: exploring the educational value of a business-ENGO partnership for sustainable consumption. J Consum Cult 16(3):926–947
30. UNEP and UNFCCC (2023) The sustainable fashion communication playbook, Nairobi
31. Rejeb A, Rejeb K, Keogh J (2022) The circular economy and marketing: a literature review. Etikonomi 21:153–176

# Factors Affecting Local Digital Transformation Toward the Sustainable Local Development

**Takaharu Takikawa and Yuna Seo**

**Abstract** The purpose of this study was to examine strategies for the smooth implementation of Digital Transformation (DX) in municipalities using two approaches. First, a model of local DX adoption was developed by exploring factors that influence Acceptance of local DX by platform users. Second, the relationship between local DX and community satisfaction was clarified for sustainable local development. It suggests that basically, Interest in UI/UX was important in the model. By advancing this research, we will contribute to business transformation and the addition of new value in the public sector.

**Keywords** Local government · E-government · Digital civic engagement · Digital transformation · Online platforms · Structural equation model

## 1 Introduction

Over the past decade or so, digital transformation (DX) has been advancing in various fields, utilizing digital technology to transform business and create new value, and business efficiency has been improved using digital technology, Robotic Process Automation (RPA), Artificial Intelligence (AI), and other technologies. DX by the private sector has been largely driven by changes in consumer behavior due to the nest egg of consumers following the spread of the novel coronavirus infection. DX related to public administration has also advanced along with the promotion of DX by private companies. Looking at the world, in Estonia in Eastern Europe, the e-government project started in 2000 enabled civil servants to access documents of the entire administrative organization. In the 2010s, Estonia has evolved to an integrated e-government where citizens can handle various administrative

T. Takikawa
Department of Industrial and Systems Engineering, Tokyo University of Science, Tokyo, Japan

Y. Seo (✉)
Tokyo University of Science, Tokyo, Japan
e-mail: yuna.s@rs.tus.ac.jp



S. Fukushige et al. (eds.), *EcoDesign for Circular Value Creation: Volume I*,
https://doi.org/10.1007/978-981-97-9068-5_17

procedures directly via the Internet, without the need for civil servants, and has moved to full. This was the transition to full digital access. In Denmark, an e-government system has been established and functioning since the 1970s, and in 2010, a national e-government concept was promoted through a citizen portal for social security information, various benefit-related procedures, medical services, and tax return procedures.

In Japan, the Ministry of Internal Affairs and Communications (MIC) has formulated the Digital Transformation (DX) Promotion Plan [1] for Municipalities for the period from January 2021 to March 2026, and in line with this plan, the creation of a new mechanism for citizen participation using digital technology in decision making in local governments has begun to progress through collaborative activities between local governments and companies. Specifically, regional portal sites that digitally realize citizen contact points and Digital Participatory Platforms (DPPs), which are tools for citizens and governments to work together to promote city planning, are examples. For example, "Decidim" in Spain, "vTaiwan" and "Join" in Taiwan, and other online platforms for public administration have been introduced outside of Japan. For example, "Decidim" in Spain, "vTaiwan" in Taiwan, and "Join" in Japan have been introduced as online platforms for public administration. In Japan, an online platform for citizen participation called "Liqlid" has begun to be introduced, but compared to other countries, it is still limited to the introduction of ICT technology, and there is still a delay in the movement of change, such as the establishment of rules for city planning with the cooperation of residents and the government.

Thus, while new platforms such as those described above are beginning to be introduced for DX in public administration, there are many unclear issues such as what conditions will increase user satisfaction when using online platforms and what points to keep in mind when introducing such platforms.

According to Toda et al. (2017) [2], those who think that their own skills will be insufficient or no longer applicable in the future and those who have a flexible attitude toward new technologies are more likely to be anxious about the evolution of AI and other technologies. In addition, Goto et al. (2016) [3] stated that the viewpoint of choosing the form is important in the introduction of cloud computing in large-scale municipalities and presented criteria for deciding about the choice. In relation to the introduction of resident-autonomous online platforms, Nakayama et al. (2014) [4] showed the effectiveness of P2M platform management in promoting regional revitalization, albeit partially, and Shikita et al. (2012) [5] also revealed that regional platforms are needed. And in Germany, Bernd et al. (2017) [6] examined the antecedents of citizens' use of open government data, and the results showed that ease of use, usefulness, intrinsic motivation, and Internet competence significantly determine citizens' intention to use open government data as shown in the findings.

The existing studies mentioned above are limited to research on digital support, research on institutions to realize digital citizen participation, and research targeting a single region. Therefore, this study will conduct a survey of residents in Japan to determine the platform improvements required by different regional attributes and how the introduction of digital technology will affect local government residents. To do so, it is necessary to identify differences among regions and gain knowledge that will contribute to more effective introduction of digital technology.

Therefore, the purpose of this study was to examine strategies for the smooth implementation of DX in communities using two approaches. First, the relationship between local DX and Community satisfaction for sustainable local development was identified. Second, a model of local DX adoption was developed by exploring the factors that influence the Acceptance of local DX by platform users. The model considers Community participation, satisfaction, and affinity with the Internet.

The rest of this study includes a hypothesis for each model, along with previous research, a description of the methodology, a description of the analysis and research results, followed by a discussion, and finally a summary, the challenges of this study, and directions for future research.

## 2 Literature Review and Hypothesis

Based on a review of previous studies, this paper presented hypotheses and designed two types of conceptual models: One is the model (Fig. 1) was designed for the introduction of a platform for sustainable regional management, and the other (Fig. 2) was designed for the introduction of a platform among regions. In Fig. 1, we focused on the level of *Community satisfaction*, which was set as a latent variable, and considered that increasing the level of *Community satisfaction* would lead to sustainable management of the region by increasing residents' satisfaction

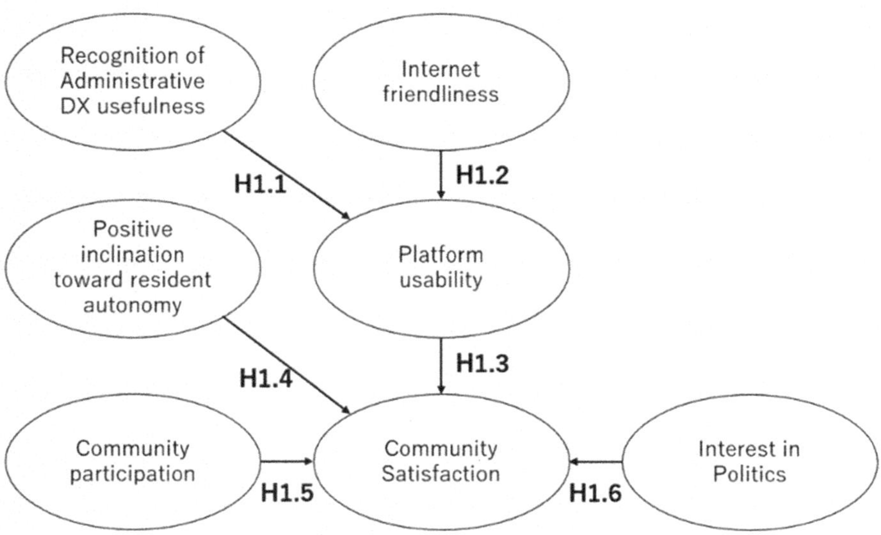


**Fig. 1** Introduction model of a platform for sustainable regional management

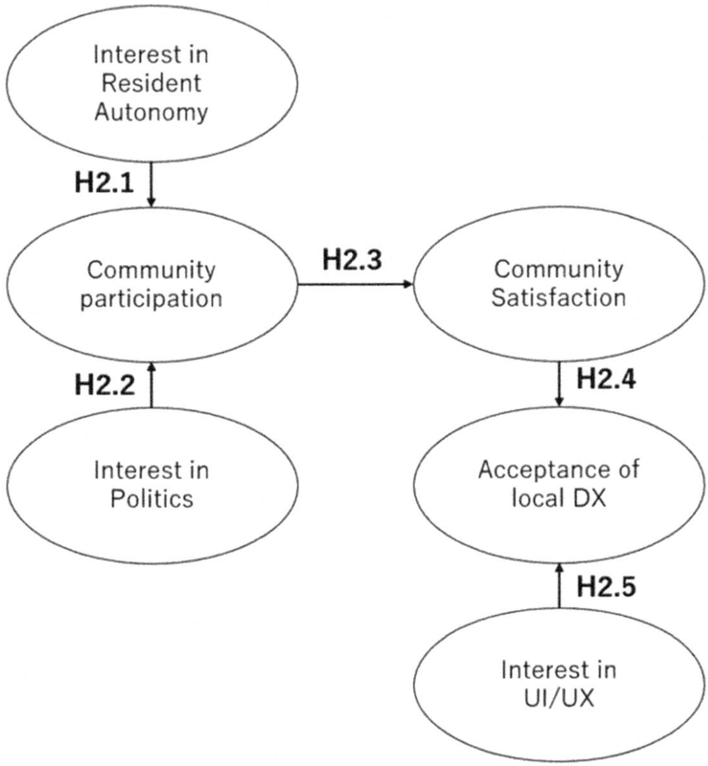


**Fig. 2** Introduction model of a platform among regions

with the region and local government. In Fig. 2, we considered that raising the level of *Community satisfaction*, which was the focus of the first model, would not only lead to the development of app design for administrative DX, but also to the satisfaction of residents, who would be the users of the app.

In this study, three major hypotheses were formulated based on a comprehensive review of previous studies and aligned with the study's research design (Figs. 1 and 2), H1.1–1.6 and H2.1–2.5 for a total of 11 hypotheses. The three hypotheses are discussed in detail below.

## *2.1 Usability of the Platform*

There have been many studies on DX and sustainable development in local governments, and the study by Narisako et al. (2020) [7] assumes that DX literacy is IT literacy, such as the ability to operate equipment, and information literacy, such as the ability to process information and solve problems, at the very least, it can be thought of as the ability to respond to organizational and business change. In other words, by understanding the usefulness of DX, we can also understand the importance of IT literacy, such as ease of use. In addition, based on Koltay's (2011) [8] study, media literacy can provide media affinity users with the ability to be transmitters equally, so that affinity for the Internet makes the platform easier to use.

Many studies have been conducted on the impact of app design on users, and in particular, the impact of user interface design (UI), user experience (UX), and features of app design on users are often mentioned. For example, Burn et al. (2022) [9] suggest that enhancing the UI of an app and conducting usability testing, which is the ease of use by the user, increased the acceptability of the app in schools, and in a study by Alsswey et al. (2020) [10], the UI design of mHealth (mobile health) is culturally significant, and therefore, enhanced UI design is more likely to be accepted by the elderly. In particular, the study by Yamada et al. (2011) [11] mentions usability as one of the factors behind the low usage rate of e-government initiatives, and the relationship between government and app design can be seen.

Based on the above background, this study considers that *Internet friendliness* and *Community satisfaction,* play an important role for user accessibility to local DX. This study assumed that app design is a significant in promoting DX in public administration. From this analysis, the following hypotheses are assumed in this model.

H1.1 *Recognition of Administrative DX usefulness* affects *positively Platform usability.*
H1.2 *Internet friendliness* affects *positively Platform usability.*
H1.3 *Platform usability* affects *positively Community satisfaction.*
H2.5 *Interest in UI/UX* affects *positively Acceptance of local DX.*

## 2.2 Connecting Residents to the Community

Studies have been conducted on the level of Community satisfaction and participation in community activities, focusing on the ability of local self-governing organizations in local cities and other areas to implement policies. For example, Kamiya (2013) [12] notes that to promote citizen participation in policy making in local cities, it is important to successfully combine the power of traditional community organizations such as neighborhood associations and community centers with the power of network-type voluntary organizations such as NPOs. The more deeply involved residents are in local self-governance, the more satisfied they are with their community, as they are involved in local events and policy making.

In addition, Shishido (2006) [13] found that the characteristics of the community in which individuals' lives influence their participation in activities, and that participation in community volunteer activities, one of the community activities, has the effect of increasing *Community satisfaction.* Yamamoto et al. (2016) [14] reported that the proportion of elderly people who have participated in community activities increases with the level of interest in community activities.

Furthermore, a study by Mahmood (2016) [15] shows that citizens' lack of knowledge about how government transformation can improve services leads to lower e-government usage. And the study by Sukma et al. (2022) [16] mentions that digital technology is important for increasing local community expectations and residents' satisfaction with e-services. It is also known that users participate in e-services with satisfaction and expectations, and the relationship between government DX and relationship between *Community satisfaction* can be seen.

A study by Gad et al. (2021) [17] reported a negative correlation between perceptions of organizational politics and job satisfaction, and their study makes recommendations for NGOs to improve organizational politics and increase employee satisfaction in the workplace. The results should inform future research on NGOs and the politics of managing social employees. In addition, Neal et al. (2012) [18] study finds that individual satisfaction with the community plays an important role in urban politics, and that the quality of public schools is an important predictor of satisfaction with the community.

Based on the above background, this study considers that interest in resident autonomy plays an important role in *Community participation* and satisfaction, *Interest in Politics* plays an important role in *Community participation* and satisfaction, and *Community Participaction* plays an important role in acceptance of municipal DX. From this, the following hypotheses are assumed in this model.

H1.4 *Positive inclination toward resident autonomy* positively affects *Community satisfaction*.
H1.5 *Community participation* positively affects *Community satisfaction*.
H1.6 *Interest in Politics* positively affects *Community satisfaction*.

H2.1 *Interest in resident autonomy* positively affects *Community participation*.
H2.2 *Interest in Politics* positively affects *Community participation*.
H2.3 *Community participation* positively affects *Community satisfaction*.

## *2.3 Relationship Between* **Community Satisfaction** *and Transition to Administrative DX*

A study by Zhen et al. (2019) [19] found a relationship between community ICT usage and community satisfaction. The study results showed that residents with high levels of community ICT use were more satisfied with their communities than those with low levels of community ICT use. In addition, a study by Chompis et al. (2014) [20] investigated virtual community satisfaction in financial services and found that user satisfaction was influenced by social connectedness, content, and technology.

Based on these previous studies, the following hypotheses can be made in the relationship between community satisfaction and migration to administrative DX

H2.4 *Community satisfaction* positively affects *Acceptance of local DX*.

# 3 Study Design and Data

## 3.1 *Data Collection*

The target population of this study is potential users of online platforms, which consists of about 30 men and 30 women aged 20 or older in each gender and age bracket. The data were collected through a 4-day web-based survey from July 8 to July 11, and a total of 366 valid responses were obtained.

## 3.2 *Questionnaire*

The questionnaire (Table 1) consisted of 31 items (four items of *Interest in administrative DX*, 4 items of *Interest in UI/UX*, four items of *Interest in resident autonomy*, three items of *Interest in resident autonomy application*, five items of *IT literacy*, four items of *Interest in Politics*, three items of *Participation in local activities*, and four items of *Local satisfaction*) in eight scales based on related studies. All the responses were obtained by a five-field method. The questions Q1 ~ Q6, which were related to the understanding and agreement of the questionnaire contents and the personal attributes of the respondents, were excluded from the factor analysis because they were considered too difficult to be included as factors in the factor analysis.

## 3.3 *Factor Analysis*

An exploratory factor analysis is conducted for the items Q7-Q37, except for the questions on understanding and agreement of the questionnaire content and Q1-Q6 regarding individuals, using the maximum likelihood method to extract factors and to obtain factor loading matrices. Then, the correlation coefficients among the factors are interpreted by using oblique rotation (Promax rotation), which allows correlation among the factors. Factor loadings of 0.7 or higher, which are high in practical terms, are used as a criterion for interpreting the factors, and factors are classified and aggregated into groups of factors that have the same meaning. Cronbach's alpha coefficient is also calculated to confirm the internal consistency of each item.

In this analysis, SPSS Statistics 28.0, a multivariate statistical software, was used to clarify the potential factor structure of the recommendation for the introduction of the application for local self-governance.

**Table 1** The questionnaire

| Item | Contents |
|---|---|
| Q7 | Do you think DX (Digital Transformation) should be promoted more in public administration? Please tell us why you answered this question (free answer). |
| Q8 | Do you think that DX will make people more interested in public administration? |
| Q9 | Do you think that DX will improve the efficiency of work and make life more convenient? |
| Q10 | Do you think that Municipality DX should be accelerated more than it is now? |
| Q11 | Have you ever found the use of colors in apps difficult to use? |
| Q12 | Have you ever found the screen transitions of the application difficult to use? |
| Q13 | Do you feel that registration and login are cumbersome? |
| Q14 | Have you ever found the button layout of the application difficult to use? |
| Q15 | Have you ever submitted your opinions to the city/town hall through the website of the city/town hall? |
| Q16 | Have you ever submitted your opinions directly to the city or town hall by writing them on a form? |
| Q17 | Do you have any doubts or feel that there is room for improvement in the way the government interacts with residents? |
| Q18 | Have you ever participated in a briefing session hosted by the city or town hall? |
| Q19 | Do you often see applications related to resident autonomy on SNS (Twitter/Instagram/Facebook, etc.), newspapers, or in the news? |
| Q20 | Do you think that the use of the application will increase the participation of residents in local government? |
| Q21 | Do you think you would like to use applications for local self-governance such as posting and voting? |
| Q22 | Do you collect information from SNS (Twitter/Instagram/Facebook, etc.)? |
| Q23 | How often do you use online shopping? |
| Q24 | How often do you post on SNS (Twitter/Instagram/Facebook, etc.)? |
| Q25 | How much do you use your smartphone per day? |
| Q26 | Do you use search engines (Yahoo/Google, etc.) to gather information? |
| Q27 | Have you ever gone to an election? |
| Q28 | Do you read political news? |
| Q29 | Do you register and follow the social networking accounts (Twitter/Instagram/Facebook/YouTube, etc.) of politicians? |
| Q30 | Are you interested in politics? |
| Q31 | Have you ever attended a neighborhood association meeting? |
| Q32 | Have you ever participated in community-sponsored life environment improvement (traffic safety, crime prevention activities, cleanups)? |
| Q33 | Have you ever participated in community-sponsored recreational activities (local traditional events, public relations activities)? |
| Q34 | Are you satisfied with the public facilities (railroads and buses) in your current area? |
| Q35 | Are you satisfied with the public facilities (gymnasium, park, library, community center) in your current community? |
| Q36 | Is the area where you live a safe place to live? |
| Q37 | Do you want to continue to live in the area? |

### 3.4 *Structural Analysis of Covariance (SEM)*

Based on the latent factor structure revealed by the factor analysis, we constructed a model using the structural analysis of covariance (SEM). This model visualizes the structure of the positive attitude toward administrative DX as a multiple indicator model. SPSS Amos 28.0 was used in the analysis. Structural equation modeling (SEM) techniques were performed to examine the model fit of each component (evaluation of the measurement model) and to verify the relationship between components (verification of hypotheses in the structural model). The GFI as a measure of goodness of fit, the CFI as a measure of comparative fit, and the root mean square error of approximation (RMSEA) were used as goodness of fit indices.

## 4 Results

### 4.1 *Influence of* Community Participation *and Satisfaction to the* Acceptance of Local DX

We investigated the relationship among factors affecting *Community satisfaction* and illustrated especially the contribution of *Platform usability* on *Community satisfaction* (Fig. 3). As can be seen in Fig. 3, the GFI, CFI, and RMSEA of this model are 0.908, 0.913, and 0.068, respectively, all of which are good values. This indicates that the model is a good fit. In addition, among the multiple hypotheses,

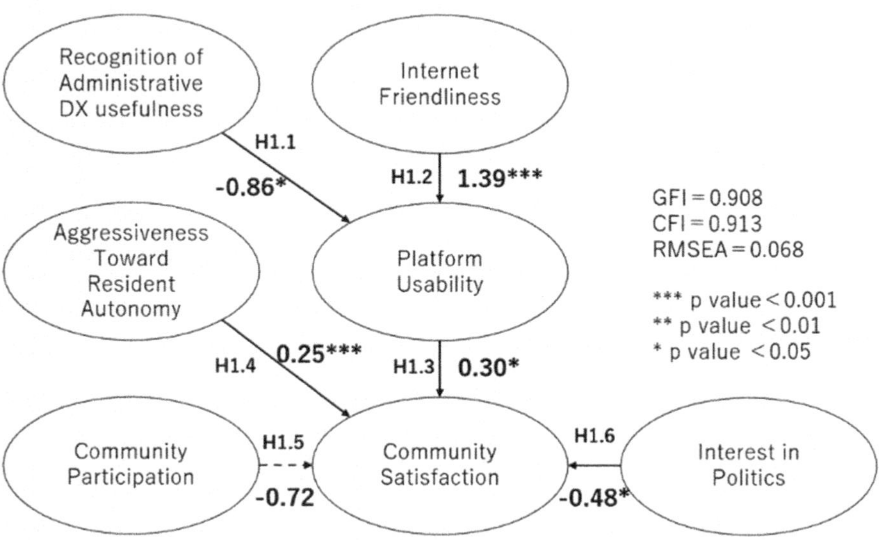


**Fig. 3** Model of platform implementation for sustainable regional management

H1.1, H1.2, H1.3, H1.4, and H1.6 are supported, H1.5 is rejected, and five out of six hypotheses are supported. The path coefficients and significance levels are 1.39 for Internet friendliness toward *Platform usability*, which is 0.1% significant, –0.86 for *Recognition of Administrative DX usefulness*, which is 5% significant, and –0.86 for *Positive inclination toward resident autonomy*, which is 0.1% significant. *Positive inclination toward resident autonomy* is 0.25, which is 0.1% significant for *Community satisfaction*, and *Platform usability* is 0.30, which is 5% significant for *Community satisfaction*. Finally, there is a significant effect of –0.48 on *Community satisfaction* from *Interest in Politics*.

### 4.2 *Influence of* Community Participation *and Satisfaction to the* Acceptance of Local DX

We investigated the relationship among factors affecting *Acceptance of local DX* and illustrated especially the contribution of *Interest in UI/UX* on *Acceptance of local DX* (Figs. 4 & 5). As can be seen from Figs. 4 and 5, both models categorized by the number of respondents below and above 200,000 showed good values for goodness of fit indices (GFI) of 0.903 and 0.915, comparative fit indices (CFI) of 0.931 and 0.940, and root mean square error of approximation (RMSEA) of 0.082 and 0.07, respectively. This indicates that the goodness of fit is also high for this model. Of the multiple hypotheses, H2.1, H2.2, and H2.5 were supported in Fig. 4, while H2.3 and H2.4 were rejected. Also, in Fig. 5, H2.1 and H2.2 were supported, while H2.3, H2.4, and H2.5 were rejected.

The path coefficient of 0.90 is significant at 1% for Interest in Resident Autonomy to *Community participation*, and 0.90 is significant at 1% for *Interest in Politics* to *Community participation* to *Community satisfaction* is not significant at 1.00, *Community satisfaction* to *Acceptance of local DX* is not significant at 0.00, and *Interest in Politics* to *Community participation* is not significant at 0.44. *Community satisfaction* to *Acceptance of local DX* is not significant at 0.01, and *Interest in UI/UX* to *Acceptance of local DX* is significant at 1.00. On the other hand, the model with more than 200,000 respondents has a significant effect of 0.1% at 0.44.

On the other hand, in the model with more than 200,000 people, Interest in Resident Autonomy has a significant effect on *Community participation* at 0.25, *Interest in Politics* has a significant effect on *Community participation* at 0.97, and *Interest in UI/UX* has a significant effect on *Acceptance of local DX* at 1.00. and not significant at 1.00 for *Community participation* to *Community satisfaction*, not significant at 0.20 for *Community satisfaction* to *Acceptance of local DX*, and not significant at 0.20 for *Interest in UI/UX* to *Community participation*. *Interest in UI/UX* to *Acceptance of local DX* was 0.98, not significant at 1%.

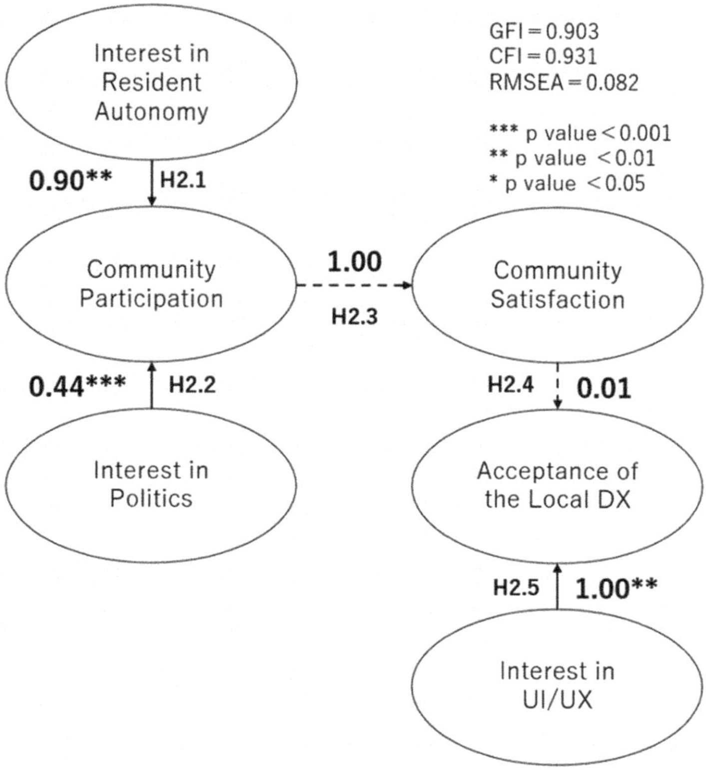


**Fig. 4** Model of the *Acceptance of local DX* platform (residents less than 200,000)

## 5 Discussion

In this study, we set out to find and model the factors that influence administrative DX from two perspectives: sustainable regional management and acceptance of local DX platform. As a result, we found the following three findings.

In the model of platform implementation for sustainable regional management (Fig. 3), we found that ease of use of the platform and proactiveness toward resident autonomy are important for improving local satisfaction that leads to local sustainability, and that Internet friendliness has a significant impact on the ease of use of the platform. From this, it can be said that it is important for local governments and public administrations to work on raising residents' proactiveness toward local government when conducting sustainable regional management in promoting administrative DX, and to provide training and other education for residents who are not very familiar with the Internet to help them become familiar with the platform.

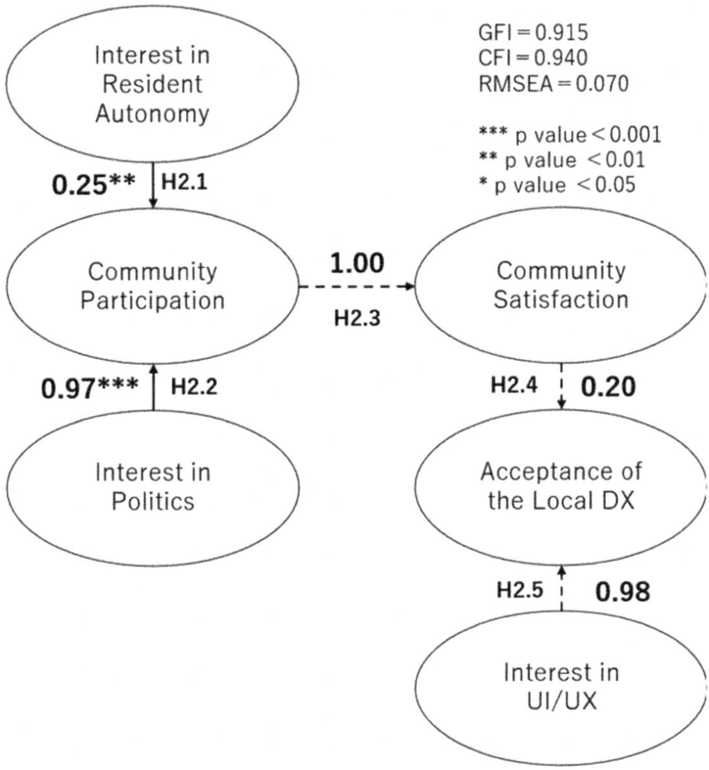


**Fig. 5** Model of the *Acceptance of local DX* platform (residents more than 200,000)

In the model of the *Acceptance of local DX* platform (Fig. 4), which concerns people living in municipalities with less than 200,000 inhabitants, we searched for factors that are important in improving the proactiveness of administrative DX and found that *Interest in UI/UX* had the greatest impact among these factors. On the other hand, the hypothesized level of residents' satisfaction did not have a significant impact, indicating that the level of *Community satisfaction*, which was considered important in the sustainable community management model, was not very relevant in promoting administrative DX.

One possible reason for the significant UI/UX interest in the model for areas with less than 200,000 residents is that many of the residents in areas with less than 200,000 are elderly. This is because the elderly is less likely to use a platform if the text size and operability are not well designed (2016) [22]. Based on these results, it can be suggested that the introduction of a user-friendly platform should be the priority for areas with many elderly residents.

In the model for less than 200,000 (Fig. 5), concerning people living in cities and towns with more than 200,000 inhabitants, *Interest in UI/UX*, which had influenced the proactiveness of administrative DX in the model with less than 200,000 inhabitants, had no impact, and, as in the previous model, regional satisfaction also had no impact. This suggests that there are other factors that are more important in promoting administrative DX in cities with more than 200,000 people.

One of the reasons why the model with more than 200,000 residents failed to identify any influencing factors is that the characteristics of medium-sized cities with more than 200,000 residents vary widely, making it difficult to narrow down the characteristics compared to those of cities with less than 200,000 residents. Therefore, we believe that significant influencing factors will emerge if we proceed with a survey targeting residents of large cities such as Tokyo and Osaka, rather than grouping them together as cities with more than 200,000 residents. In addition, we believe that "recognition of new values created by DX" is another possible factor. This is so because Watanabe et al. (1996) [23] suggest that conservation and management policies in rural areas will proceed smoothly if managers and residents share a common understanding of the value of wildlife in terms of social value.

## 6 Limitations and Future Perspectives

Criticisms and comments from local government officials are beginning to be seen that the introduction of a platform for administrative DX will only result in ongoing implementation costs and will be meaningless unless the local government improves the issues it faces through the introduction of the platform. Therefore, in the future, we would like to include a question asking whether both residents and local governments share a common understanding of the value of administrative DX that will be created through the introduction of the platform to promote administrative DX. We believe it is necessary to hypothesize new factors by including such items in the survey. However, since the ease of use of the platform had some influence in the model of sustainable regional management, and *Interest in UI/UX* had some influence in the model targeting residents living in municipalities with less than 200,000 people, this factor should continue to be taken into consideration going forward.

In addition, since the definition of "platform" was not clearly defined this time, it is necessary to specify "digital citizen participation platform (consensus building between citizens and local government)" in the future and build an evaluation model after narrowing down the scope of the platform.

Incidentally, while the data we collected for this study was evenly distributed in terms of age, the remaining demographic factors such as occupation, income, place of residence, and political views may be involved, but according to a study by Darko Stefanović (2016) [21], demographic factors do not affect platform use and purpose of use. However, we do not believe that this is necessary to consider, given the research findings by Darko Stefanović that demographic conditions do not affect platform use and purpose of use.

## 7 Conclusion

In this study, two types of models were constructed for residents in Japan against the backdrop of DX in public sector administration, which is in progress outside of Japan. The first type of model analysis was conducted on the theme of sustainable community management. The results showed that the ease of use of the platform had the greatest impact, as well as the positive attitude toward resident autonomy.

The second model analysis showed that *Interest in UI/UX* had a significant impact on the proactiveness of administrative DX in the model for residents living in municipalities with less than 200,000 inhabitants, while neither local satisfaction nor *Interest in UI/UX* had an impact on the proactiveness of administrative DX in the model for residents living in municipalities with more than 200,000 inhabitants.

For future research, it is important to consider including a question that asks whether both the citizens voicing the importance of the administrative DX that occurs during platform implementation and the administration, which takes a leading role during platform implementation by incorporating citizens' voices into policy. Such a question should be added to the survey to incorporate new factors into the hypothesis.

## References


1. Ministry of Internal Affairs and Communications (2020) Municipal Digital Transformation (DX) Promotion Plan. https://www.soumu.go.jp/main_content/000726905.pdf (accessed July 13, 2022)
2. Toda A, Chuma H, Hayashi S, Kume K (2017) Background analysis of anxiety caused by technology evolution. In: RIETI Discussion Paper Series
3. Goto M, Onishi K, Nakano H (2016) On the optimal form of cloud computing deployment in a large-scale municipality. J Inform 1(13):2016
4. Nakayama M, Kameyama H (2014) A case analysis of regional revitalization through P2M platform management. J Int Assoc P2M 8(2):71–82
5. Shikita M, Morishige M, Nakamura S (2012) Necessity of regional platforms with intermediate system roles and its structural analysis. Int J Public Relat Media Tour 14:23–42
6. Wirtz BW, Weyerer JC, Rösch M (2017) Citizen and open government: an empirical analysis of antecedents of open government data. Int J Public Adm 41:308–320
7. Narisako T (2020) Human resources, organization, and project structure for promoting digital transformation (DX): organizational culture and human resource development in traditional Japanese companies. Digit Pract 11(2):298–306
8. Tibor K (2011) The media and the literacies: Media literacy, information literacy, digital literacy. Media Cult Soc 33(2):211–221
9. Burn A-M, Ford TJ, Stochl J, Jones PB, Perez J, Anderson JK (2022) Developing a web-based app to assess mental health difficulties in secondary school pupils: qualitative user-centered design study. JMIR Formative Research 6(1):e30565
10. Alsswey A, Al-Samarraie H (2020) Elderly users' acceptance of mHealth user interface (UI) design-based culture: the moderator role of J Multimodal. User Interfaces 14:49–59

11. Yamada K, Hirasawa N (2011) Comparison of e-government policies in different countries based on usability guidelines. In: Research report information systems and social environment (IS), May 2011, pp 1–8
12. Kamiya H (2013) Current status of citizen participation in policy making and execution by local government. In: Proceedings of the General Meeting of the Association of Japanese Geographers Annual Meeting of the Association of Japanese Geographers, Autumn 2013. The Association of Japanese Geographers, 2013
13. Shishido K (2006) Factors promoting community volunteer activities and community life satisfaction among middle-aged and older adults, 2006
14. Yamamoto T, Kita T, Nagasawa T (2010) A study of factors promoting participation in maintenance work of agricultural water facilities. J Rural Plan Soc 29(2):101–106
15. Mohamed M (2016) Enhancing citizens' trust and confidence in government through digital transformation. Int J Electron Govt Res (IJEGR) 12(1):99–110
16. Sukma Narongsak, Adisorn Leelasantitham (2022) The influence and continuance intention of the e-government system: a case study of community water supply business. Front Environ Sci 2022 770
17. Gad S, Elsayed W (2021) Relationship between perceived organizational politics and job satisfaction among NGO social workers. Public Organ Rev 2021:1–12
18. Zachary NP, Neal JW (2012) The public school as a public good: Direct and indirect pathways to community satisfaction. J Urban Aff 34(5):469–486
19. Feng Z, Xiao Q, Yupei J, Hui C (2019) Effects of community communication technology (ICT) usage on community satisfaction: case study in Nanjing, China. Chin Geogr Sci 29:834–847
20. Evangelos C, Bons RWH, van den Hooff B, Frans F, Hans H (2014) Satisfaction with virtual communities in B2B financial services: social dynamics, content and technology. Electron Mark 24:165–177
21. Darko S, Marjanovic U, Delić M, Culibrk D, Lalic B (2016) Assessing the success of e-government systems: an employee perspective. Inf Manag 53:717
22. Ito S, Sakata M, Ishikawa M, Tamano K, Nakanishi M (2016) A study of UI design for portable touch terminals from the perspective of older users' willingness to use. Ergonomics 52:346–S347
23. Watanabe O, Ogura S (1996) Perceptions of wildlife values and intentions toward conservation and management policies in rural areas: a survey of value perceptions in agricultural 'damaged' areas in Eastern Aichi Prefecture. Wildlife Conserv 2(1):1–15

# Analysis of the Effects of Adopting "Imaginary Future Generations" on the Design of Technology Development and Business Proposal: Case Study of Workshop at a Plating Processing Company

**Ken Fujita, Tetsusei Kurashiki, Keishiro Hara, Junji Ikeda, and Takashi Nakamura**

**Abstract** In this study, we analyzed business proposals of three technologies possessed by the manufacturing industry related to the plating process by adopting Future Design. We conducted a business proposal workshop from the two standpoints of the current generations and the Imaginary Future Generations (IFGs) to discuss business strategies for future social issues. Then, the differences between the two generations were analyzed by using a 5-point scale questionnaire, focusing on the four categories of "individual," "technology," "company," and "society." We found that while the current generations considered solutions to immediate technological issues, the IFGs showed other viewpoints in terms of technological expansion methods and globalization, which were not proposed in the current generations. We also looked into the words in the workshop and analyzed the changes in the content between the two generations. While the current generations were concerned only with "individual" and "company," the IFGs thought not only about "individual" and "company" but also about "society" and "company." The results of the questionnaire showed an overall increase in points for the questions focusing on "society," "company," and "technology" when the IFGs were adopted. In addition, the questions that focused on "individual" indicated that the IFGs considered the impact of "individual" perspectives and actions on "society," "company," and "technology," which was not considered by the current generations. These results suggested that a business proposal that introduces the viewpoints of future generations adopting Future Design is effective for the technology possessed by the company.

K. Fujita (✉) · T. Kurashiki · K. Hara
Graduate School of Engineering, Osaka University, Osaka, Japan
e-mail: fujita-k21@mit.eng.osaka-u.ac.jp

J. Ikeda · T. Nakamura
Teikoku-ion Co., Ltd, Higashiosaka City, Osaka, Japan



S. Fukushige et al. (eds.), *EcoDesign for Circular Value Creation: Volume I*,
https://doi.org/10.1007/978-981-97-9068-5_18

**Keywords** Future design · Imaginary future generations · Workshop · Technology innovation · Business proposal

# 1 Introduction

## *1.1 Background*

Plating technology is being applied across a wide range of fields where further functionalization and electronization are demanded, including in consumer electronics, automobiles, solar cells, fuel cells, and medical equipment. It can be said that a superior coating technology is essential in the manufacturing of these devices. Therefore, it is crucial for companies to establish advanced plating technologies that can handle complex requirements, which are difficult to address with the current plating technology. However, the process of developing a new technology requires a great deal of cost. First, there is the search for ideas, screening, and development of new technologies. Even if a new technology is developed, it is necessary to study marketing strategies such as searching for needs and determining concepts, and to study strategic hypotheses such as business economic analysis. Then, product development, test marketing, and production are conducted toward product commercialization, finally leading to the introduction of the new product to the market [1]. For a company to continue to grow and survive in today's rapidly changing market, it is necessary to secure the above costs, generate profits, and consider the future cost and profits. However, it is difficult for firms to remain stable, as disruptive innovations [2] are taking away major customers in conventional industries. Therefore, companies always try to continue by considering the demands of future markets. Conventionally, scenario analysis such as forecasting [3] and backcasting [4] have been used to predict the future. There exists research on technology roadmaps using these scenario analyses. Phaal et al. have categorized numerous roadmaps [5], arguing that "the roadmapping process should accommodate the uncertainties associated with future forecasts and aspirations, and where appropriate to communicate these in the roadmap itself." They propose that scenario planning can be applied in this regard [6]. Moreover, Okada et al. have introduced a method based on backcasting to design roadmaps for envisioning a sustainable future [7]. However, these methods have their limitations. The difficulty arises when envisaging the future from the standpoint of the current generations, as it can be challenging to accurately grasp the needs of the future. This is due to the difficulties in overcoming shortsightedness [6] or intergenerational conflicts of interest. Therefore, it is considered necessary to incorporate the perspective of future generations.

In recent years, research on Future Design has been conducted as a means of designing and implementing mechanisms to pass on a sustainable society to future generations. According to Saijo [7], a person exhibits "Futurability" when he or she experiences an increase in happiness as a result of deciding and acting to forego current gains in order to enrich future generations, and the design and praxis of a society generating "futurability" is called Future Design. However, the practice of Future Design in corporate technological development is limited, and its effects need to be verified.

## 1.2 Purpose of This Study

In this study, we conducted a workshop to examine business strategies for future social issues in the field of plating processing, incorporating the Future Design concept. Specifically, we introduced Imaginary Future Generations (IFGs) which is an effective method to create a group of people who act as representatives for the benefits of future generations in actual, present-day decision-making situations. In addition, four other parameters were set in relevant discussion: "society" focusing on "Environment, Resources, and Energy," "company" centering on, "technology" that supports the "company," and "individual" who lives in the world. We investigated the effects of adopting IFGs on the direction of technology development and these parameters by looking into the discussions results both as current generations and the IFGs, and a questionnaire survey administered to workshop participants.

## 1.3 Future Design: Previous Studies

In the initial laboratory-scale experiment related to Future Design by Kamijo et al. [8], it was demonstrated that by introducing one IFGs member into a group of three, there was a shift towards sustainable decision-making, choosing to leave resources for future generations, even at the cost of reducing their own immediate rewards. Subsequently, social practice workshops focusing on the vision for Yahaba Town, Iwate Prefecture, in 2060 were conducted by Hara et al. [9] and Hara et al. [10]. Discussions were held separately between groups representing the current generations and groups representing IFGs as representatives of the inhabitants in 2060. The content of the discussions varied significantly between the two, with it being clear that the IFGs, compared to the current generations, proposed unique ideas, and had different priorities in their value judgments. Furthermore, the process of the current generations accepting the future generation's ideas became apparent as a result of consensus-building between the two generations.

From another Future Design practice conducted by Hara et al. [11] targeting the same town, it has been suggested that the coexistence of perspectives of the current generations and the IFGs by the same individual is possible (viewpoint sharing). This hints at the potential to distance oneself from shortsighted thinking and adopt a more overarching perspective. Hiromitsu et al. [12] assert that when national government employees, town hall employees, and ordinary residents form the same group and implement Future Design, they acquire a broader perspective that transcends individual positions by discussing from the viewpoint of the future.

Nakagawa et al. [13] have suggested that deliberation experiments in Future Design, providing participants with topics about past events and making them deliberate on decision-making about past events from the current perspective, can potentially influence the way of thinking and preferences for bettering the future. This sort of training or education to acquire a retrospective perspective may affect future-oriented thinking and preferences. It is reported that participants who have acquired such a retrospective perspective tended to prefer policies that fundamentally change the status quo, compared to untrained participants.

In the industrial business sector, workshops aimed at formulating research and development strategies and discussing the direction of sustainability-conscious technological innovation are being held under the concept of Future Design. In the field of water engineering, workshops have been conducted where research and development strategies are examined from a future perspective [14]. In the food business sector, the introduction of IFGs has yielded significant differences in terms of acquiring a long-term perspective and publicness in potential food needs, as well as the feasibility of ideas generated [15].

## 2 Methods

### *2.1 Setting of Workshop*

In this study, we conducted a workshop targeting Teikoku Ion Corporation, a surface treatment processing company engaged in the research, development, and business of plating. The workshop was held over 4 days, on November 13, 20, 27, and December 4, 2021, with twelve employees working in the manufacturing field (throughout all the workshop sessions, three groups of four people each, with the participants remaining the same over the four days and each group consistently focused on one technology) as the targets. Each discussion session was three hours long and took place at Teikoku Ion Co. Ltd. In this workshop, the design was divided into three stages: the current generations, past analysis, and IFGs. The details of the workshop will be discussed in the following. The results were organized and recorded on worksheets using sticky notes.

### *2.2 Workshop Design*

In order to generate futurability effectively, this workshop was conducted following the process used in the workshop by Hara et al. [11], which was conducted for local governments. First, participants engaged in discussions from the perspective of the current generations, followed by a shift in perspective to the IFGs for further discussions.

Moreover, based on the research by Nakagawa et al. [13], it has been shown to be effective to adopt a retrospective perspective to fully embody the role of the IFGs. Applying this, before the participants in this workshop fully become IFGs, we introduced a process where they look back on the past and discuss, within each group, their evaluation, and opinions from the current perspective on past events and decision-making, as a message to the past. The workshop design is described below.

#### Discussion on the Current Generations (Day 1)

Initially, to consider the societal image from the present to 2040, spanning from individuals to societal and global scales, we divided the discussion into three topics: "Environment, Resources, and Energy," "Science, Technology, Industry, and Labor," and "Lifestyle." Using a brainstorming format, participants freely generated ideas. Afterwards, they determined the values that were considered important in their ideas.

Based on this, they discussed the given theme (technology/device) for about 5 years from 2021, considering it from various perspectives such as products, services, ideas, and research and development. Finally, they selected one or two from the proposed ideas, and summarized the reasons for their selection, the target group, prospects, effects of the proposed ideas, and a 5-year plan.

#### Discussion and Redesign Through Past Analysis (Day 2)

The groups discussed evaluations and opinions from the current perspective on past events and decisions, reflecting on the past and taking it as a message to the past. Specifically, they selected several turning points in the history of each technology, evaluated the decisions made at the time, and recorded the messages.

#### Discussion on the IFGs (Day 3, 4)

After the past analysis and redesign, basic information about Future Design was provided to the participants, and it was explained that the perspective should be shifted to the IFGs. In assuming the role of the IFGs, the participants were asked to imagine a situation in which they were living in the world 20 years later with their current age and position. To constantly remind them that they were the IFGs during the discussion, a paper writing "It is now 2040" was placed on the desk where the participants were seated.

On the third day, from the perspective of the IFGs, they first proposed ideas freely about the social background in 2040. After the free brainstorming, they defined the society in 2040. Then, they considered the state of the plating industry in 2040 and proposed ideas for the ideal and better state of the plating industry after 2040.

On the fourth day, considering a 1-week gap since the third day, the same tasks of the third day were performed to have them become the IFGs again. Then, they defined the situation of the plating industry in 2040 and a further ideal based on the ideas proposed within each group. After that, they considered what they would have wanted to be done from 2021 to 2026 from the perspective of the IFGs, choosing one or two opinions, explaining the reasons for their choice, the target audience, future prospects, the effects of the proposed ideas, and outlined a 5-year plan. In this aspect, the same content was arranged for the current generations (Day 1) and theIFGs, allowing us to observe the effect of IFGs by comparing the results.

## 2.3 *Plating Technology on This Study*

This study was conducted in collaboration with Teikoku Ion Corporation (https://teikoku-ion.co.jp/). Teikoku Ion Corporation has fused its 54-year legacy of plating technology know-how with 23 years of advanced mechanical engineering, challenging themselves to develop entirely new functional plating technologies—termed as "Future Plating"—with unprecedented concepts. Through the development of this Future Plating, they have acquired patented plating technologies, including plating technologies that were selected as excellent in the 2020 Kansai Manufacturing New Selection and plating technologies that resulted from their Sapo-In efforts. In this section, we will describe the three Future Plating technologies used as themes in the workshop.

**(A) Plating Technology Aiming Wear Resistance (Ion Diamond Plating)** Ion Diamond Plating is a new coating that achieves low friction and high wear resistance by aligning and structurally controlling particles of diamond, the hardest material with Hv7000–15,000, using plating technology. The coefficient of friction does not increase even in the low rotation boundary lubrication region, and it is possible to maintain a friction coefficient of 0.05 or less, which can be used to prevent wear on sliding parts.

**(B) Plating Technology Aiming Corrosion Resistance (Ion Hard ® Chrome)** Ion Hard ® Chrome plating demonstrates a barrier effect with a non-cracking coating and can be used for corrosion resistance in antibacterial and sterile environments. Surface treatments used in antibacterial and sterile environments in the food and clothing industries require a high level of corrosion and heat resistance, and in recent years, a barrier effect with resistance to chemicals is necessary. In this regard, Ion Hard ® Chrome plating is well suited.

**(C) Small Lot Plating Equipment (Single-item Plating Equipment)** The single-item plating equipment deviates from the conventional mass plating equipment and, by changing to a single item approach, it makes it possible to customize each item with plating solutions or current, thus ensuring quality stability and energy savings.

### 2.4 Analysis of Worksheet Descriptions

To analyze the participants' awareness, we aimed for the changes in the participants' statements. Specifically, opinions and ideas put forth on the worksheet were analyzed using KH Coder. In this study, co-occurrence networks, which allow for the analysis of the relationship between words and between words themselves, were used.

### 2.5 Survey Using Questionnaires

To analyze the change in participants' perceptions, we conducted a questionnaire survey using a Likert scale (1: Strongly disagree, 2: Slightly disagree, 3: Neither agree nor disagree, 4: Slightly agree, 5: Strongly agree) at the end of each day's workshop. The questions were as listed in Table 1 (In the table, "✓" indicates the applicability of the question to the perspective.). The questions were designed with different focuses for each category. The questions (I) ~ (IV) were primarily about the company and technology. From there, we branched out to questions concerning the relationships between "the company, technology, and society," "the company, technology, and the individual," as well as those between "society, technology, and the individual, excluding the company." The Category(V) was about individual behavior.

## 3 Results and Discussion

### 3.1 Contents of Opinions Based on Worksheet

From the results of the worksheets, the main points are described from the opinions that participants considered.

**(A) Ion Diamond Plating** For the current generations, the novelty of ion diamond plating as a technology led to proposals emphasizing the importance of gaining credibility from other companies. On the other hand, the results for the IFGs advanced to a step further suggesting that we should develop our own products and engage in promotional activities to ensure credibility. When comparing the two, the current generations were considering the immediate issue of 'credibility', while the IFGs assumed credibility and focused on more concrete aspects.

**(B) Ion Hard ® Chrome** For the current generations, the values deemed important in the image of society in 2040 included "decarbonization," "consideration for environmentally harmful substances," and "automation by robots." Meanwhile, for the IFGs, technical proposals were made for the improvement of plating solution and

**Table 1** Questionnaires

| Category | Inquiry | Society | Company | Technology | Individual |
|---|---|---|---|---|---|
| (I) Solving Technological Challenges | (I-1) Our company's technological issues can be solved by the current generation | | ✓ | ✓ | |
| | (I-2) Our company's technological issues can be solved by the future generation | | ✓ | ✓ | |
| (II) Company's Sustainability | (II-1) Our company's technology and business are expected from the current society | ✓ | ✓ | ✓ | |
| | (II-2) Our company's technology and business are expected from the future society | ✓ | ✓ | ✓ | |
| (III) Relationship of Individual's Research and Development with the Company | (III-1) Individual's research and development contribute to the foundation of the current company | | ✓ | ✓ | ✓ |
| | (III-2) Individual's research and development contribute to the foundation of the future company | | ✓ | ✓ | ✓ |
| (IV) Relationship of Individual's Research and Development with the Society | (IV-1) Individual's research and development contribute to the contribution to the current society | ✓ | | ✓ | ✓ |
| | (IV-2) Individual's research and development contribute to the contribution to the future society | ✓ | | ✓ | ✓ |
| (V) Individual's Actions | (V-1) For the sake of the current company, I am thinking about what I should do and acting accordingly | | ✓ | | ✓ |
| | (V-2) For the sake of the future company, I am thinking about what I should do and acting accordingly | | ✓ | | ✓ |

the development of own products, targeting the establishment of the company's brand. Additionally, on the sales side, proposals were made to gain new customers by organizing the sales system. When comparing the two, the current generations were only discussing ideals without considering risks, while the IFGs were considering the product's reliability to realize the ideals.

**(C) Single-Item Plating Equipment** For the current generations, proposals were made to sell single pick-up plating equipment for the purpose of improving in-house productivity and expanding new businesses. On the other hand, the results for the IFGs shifted to the idea that the company's internal structure should be changed first. When comparing the two, the current generations had many abstract opinions and were thinking about efficiency and sales, while the IFGs started to consider quality, cost, and risk aspects.

## *3.2 Analysis of Worksheet Description Content*

Below, the co-occurrence network for the current generations are shown in Fig. 1, and the co-occurrence network for the IFGs is shown in Fig. 2. In the diagram, there are solid and dotted lines. Words included in the same group are connected by solid lines, while words belonging to different groups are connected by dotted lines.

From Fig. 1, the opinions on each technology in Sect. 3.1 are reflected in the co-occurrence network diagram. Additionally, when looking at the relationships among the four items: "individual," "company," "technology," and "society (environment)," it was found that while there were connections between individual and company at the time of the current generations, technology and society had a weak connection (dotted connection). From this, it can be inferred that at the end of the current generations' workshop, participants were considering matters related to individuals and companies, but they did not see their personal involvement in larger perspectives such as technology and society. Furthermore, for "technology," there were many proposals on what should be done to solve current technical issues, such as improving the productivity of products and automating factories.

From Fig. 2, it was observed that the opinions in Sect. 3.1 are reflected in the co-occurrence network diagram. Furthermore, when looking at the relationships among the four items: "individual," "company," "technology," and "society," it was found that while there were connections among "company," "technology," and "society," the "individual" remained independent. In the IFGs, the introduction of robots and AI was frequently considered, causing a significant change in the concept of "individuals," making it independent. It can also be inferred from the results that thinking as the IFGs allowed participants to adopt a long-term perspective, and they started to consider the connections between "society" and "technology" and "company," which were not connected in the current generations.

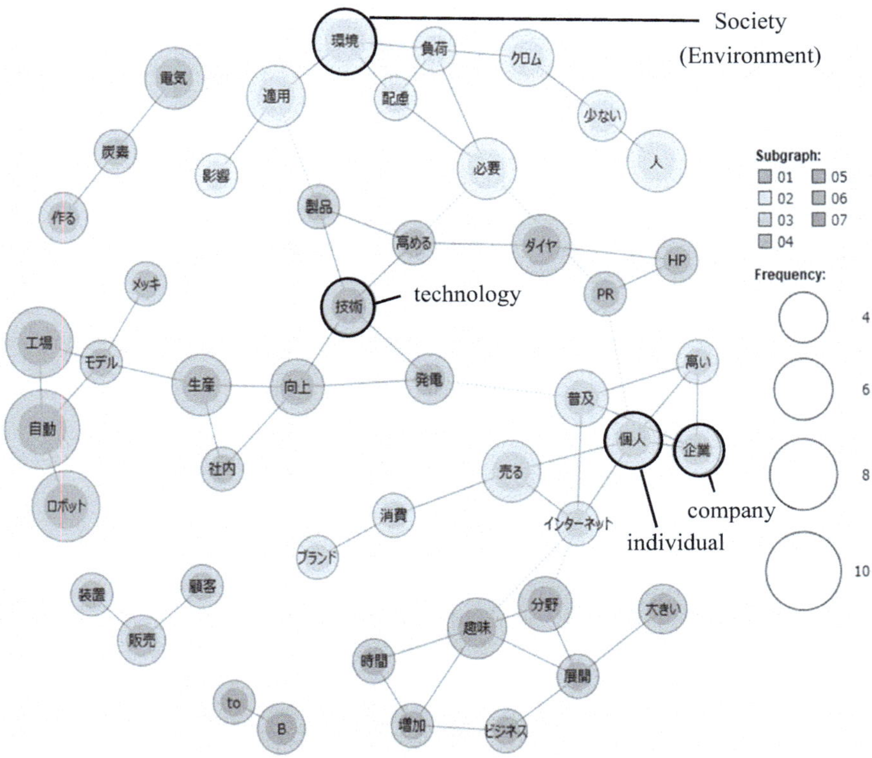


**Fig. 1** Co-occurrence network for the current generations (in Japanese)

Additionally, for "technology," it is thought that there were opinions with a long-term perspective and broader vision compared to the current generations, such as nurturing talent, creating unique devices, and considering overseas perspectives.

## *3.3 Results of Questionnaires*

With regard to the Questionnaires outlined in Sect. 2.5, we conducted a comparative analysis between the current generation and the IFGs based on the responses collected from the 12 participants. The average values and p-values of the participants' survey results discussed in Sect. 2.5 are recorded in Table 2.

The results from Table 2 show that the introduction of the IFGs led to improvements in most areas. As discussed in Sect. 2.5, participants improved in their considerations of "company and technology" (Category (I)), "company, technology and society" (Category (II)), and "company, technology and individual" (Category (III)) through the introduction of IFGs. All questions related to the company and

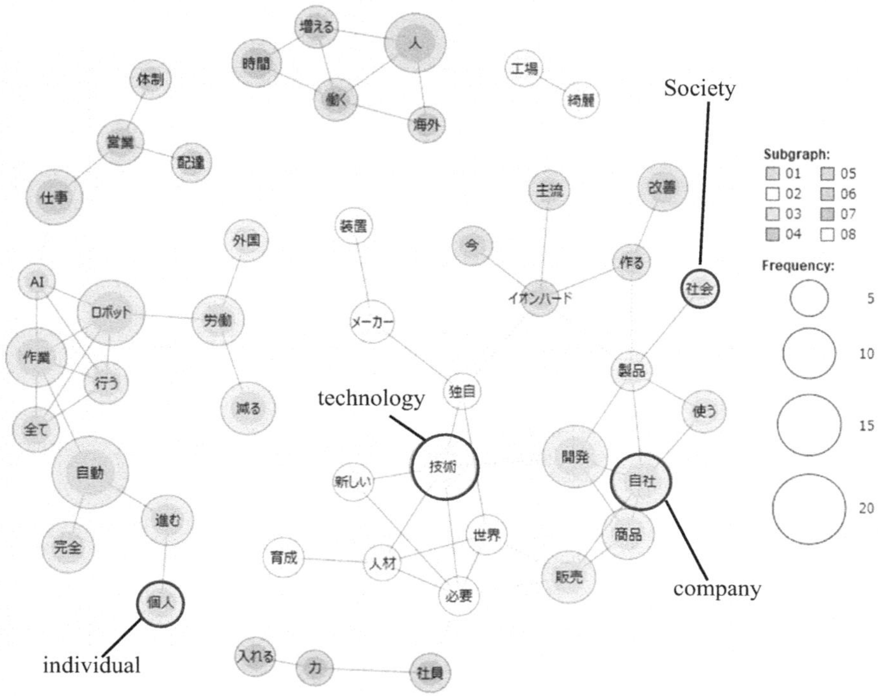


**Fig. 2** Co-occurrence network for the IFGs (in Japanese)

technology showed an increase. Thus, it is not only the company and technology that are rising, but also the individual and society which include them, are increasing. This suggests that by introducing IFGs, as noted in Sect. 3.2, one can become more holistic and think in detail about connections beyond just company and technology. However, Category (IV), "society, technology, and individual," did not show an improvement. This could suggest that despite significant efforts to introduce the IFGs in the context of the individual and society, there may still be a substantial gap.

Bar graphs for category (III) and category (IV) are shown in Figs. 3 and 4. Regarding category (III), both (III-1) and (III-2) increase the number of people answering "5: Strongly agree." As for question (IV-1), while there was an increase in people responding with "5: Strongly agree," there was also a rise in those choosing "2: Slightly disagree." Moreover, as for question (IV-2), people who answered "5: Strongly agree" increased, there was an increase in respondents selecting "3: Neither agree nor disagree" on the other hand.

(V-1) was designed as a dummy question. As the survey results show, there was not much change in this area. On the other hand, there was a significant improvement in the second question (V-2). The introduction of the IFGs seems to have stimulated individuals to think more about the future.

**Table 2** Survey results

| Category | Inquiry | Current generation | IFGs | *P*-value | |
|---|---|---|---|---|---|
| (I) Solving Technological Challenges | (I-1) Our company's technological issues can be solved by the current generation | 3.44 | 3.56 | 0.364 | – |
| | (I-2) Our company's technological issues can be solved by the future generation | 3.89 | 4.00 | 0.341 | – |
| (II) Company's Sustainability | (II-1) Our company's technology and business are expected from the current society | 3.44 | 3.78 | 0.040 | * |
| | (II-2) Our company's technology and business are expected from the future society | 3.56 | 3.67 | 0.297 | – |
| (III) Relationship of Individual's Research and Development with the Company | (III-1) Individual's research and development contribute to the foundation of the current company | 3.67 | 3.78 | 0.297 | – |
| | (III-2) Individual's research and development contribute to the foundation of the future company | 4.00 | 4.33 | 0.040 | * |
| (IV) Relationship of Individual's Research and Development with the Society | (IV-1) Individual's research and development contribute to the contribution to the current society | 3.67 | 3.67 | 0.5 | – |
| | (IV-2) Individual's research and development contribute to the contribution to the future society | 3.67 | 3.67 | 0.5 | – |
| (V) Individual's Actions | (V-1) For the sake of the current company, I am thinking about what I should do and acting accordingly | 3.33 | 3.22 | 0.341 | – |
| | (V-2) For the sake of the future company, I am thinking about what I should do and acting accordingly | 2.89 | 3.33 | 0.085 | ** |

–, non-significant; *, $p < 0.05$; **, $p < 0.10$

## 4 Conclusion

In this study, we designed a workshop introducing the IFGs, referencing methods from previous studies, and conducted the workshop in the field of plating processing. The insights gained are discussed below.

(1) Taking ion-diamond plating as an example, it was found that the present generation places a considerable emphasis on securing trust from external entities. However, the introduction of IFGs shifted this perspective, suggesting the necessity for internal product development and proactive public relations activities as mechanisms to ensure and maintain trust. Hence, where the present

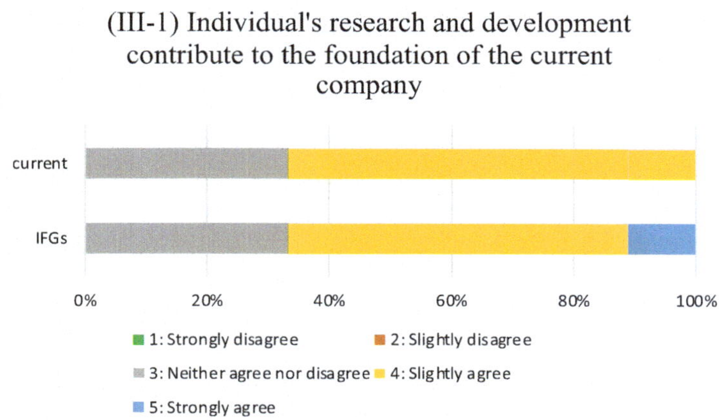


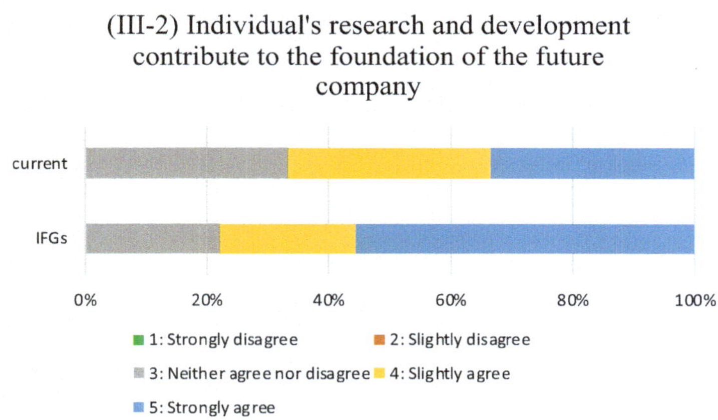


**Fig. 3** Bar graph of Category (III)

generation focused primarily on resolving technical issues, the incorporation of IFGs prompted a more comprehensive approach, considering more concrete actions to be taken.

(2) In terms of individuals, society, companies, and technology, the focus at the current generations stage was solely on the relationship between individuals and corporations. However, through the introduction of IFGs, not only did the relationship between individuals and corporations evolve, but connections to society and technology also became more prominent.
(3) On the survey questionnaire, the average scores for most questions increased. Specifically, when we measured changes in participants' thinking before and after the introduction of Future Design, there was a general trend of increase for questions primarily related to "companies, society, and technology."

From the above, it can be concluded that introducing the IFGs in technology development or technology business proposals is effective.

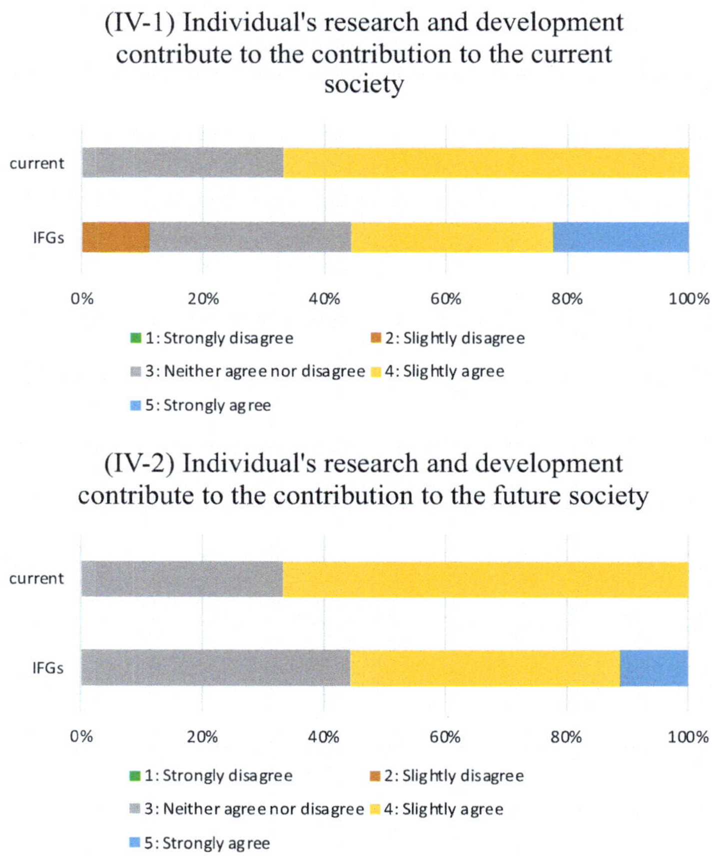


**Fig. 4** Bar graph of Category (IV)

In this study, we conducted surveys after each workshop, but we have not been able to investigate how to keep the participants' thinking in terms of the current generations and the IFGs. In addition, we conducted surveys using Likert scales to reduce the burden on the participants, but it is a future plan to obtain more detailed feedback such as through written responses.

## References

1. (2005) GLOBIS MBA Marketing, DIAMOND, Inc.
2. Christensen CM (2001) The Innovator's Dilemma. SHOEISHA
3. Swart (2004) The problem of the future: sustainability science and scenario analysis. Glob Environ Chang 14:137–146
4. Robinson (1982) Energy backcasting A proposed method of policy analysis. Energy Policy 10: 337–344

5. Phaal R, Farrukh CJP, Probert DR (2004) Technology roadmapping: a planning framework for evolution and revolution. Technol Forecast Soc Change 71:5–26
6. Phaal R, Muller G (2009) An architectural framework for roadmapping: towards visual strategy. Technol Forecast Soc Change 76(1):39–49
7. Okada Y, Kishita Y, Nomaguchi Y, Yano T, Ohtomi K (2020) Backcasting-based method for designing roadmaps to achieve a sustainable future. IEEE Trans Eng Manag 69(1):168–178
8. Saijo T (ed) (2015) Future Design. Keiso Shobo (in Japanese)
9. Saijo T (2018) Future Design. Soc Environ Econ Policy Stud 11(2):29–42. (in Japanese)
10. Kamijo Y, Komiya A, Mifune N, Saijo T (2017) Negotiating with the future: incorporating imaginary future generations into negotiations. Sustain Sci 12:409–420
11. Hara K, Saijo T (2017) Future design: evidence and insights from participatory deliberation. Japan Soc Water Environ 40(A):112–116. (in Japanse)
12. Hara K, Yoshioka R, Kuroda M, Kurimoto S, Saijo T (2019) Reconciling intergenerational conflicts with imaginary future generations: evidence from a participatory deliberation practice in a municipality in Japan. Sustain Sci 14:1605–1619
13. Hara K, Kitakaji Y, Sugino H, Yoshioka R, Takeda H, Hizen Y, Saijo T (2021) Effects of experiencing the role of imaginary future generations in decision-making—a case study of participatory deliberation in a Japanese town. Sustain Sci 16:1001–1016
14. Hiromitsu T, Kitakaji Y, Hara K, Saijo T (2021) What do people say when they become "future people"? —Positioning Imaginary Future Generations (IFGs) in general rules for good decision making. Sustainability 13:6631
15. Nakagawa Y, Kotani K, Matsumoto M, Saijo T (2019) Intergenerational retrospective viewpoints and individual policy preferences for future: a deliberative experiment for forest management. Futures 105:40–53
16. Hara K, Kuroda M, Nomaguchi Y: How Does Research and Development (R&D) Strategy Shift by Adopting Imaginary Future Generations? – Evidence from Future Design Practice in a Water Engineering Company, available at SRRN
17. Hosomi T, Kondo G, Wakamoto K, Hara K, Kurashiki T (2022) Proposal of a method for searching latent needs in the food field based on future design. J Sci Policy Res Manag 37:63–77

# Exploring Participatory Approaches for Facilitating ESG Initiatives in Organizations

**Kamila Kunrath and Devarajan Ramanujan**

**Abstract** This chapter presents environmental, social, and governance (ESG) categories distilled from five major compliance standards and proposes a framework for assessing them across an organization's strategic, tactical, and operational layers. The use of change management and participatory design approaches to facilitate the integration of ESG considerations into an organization's decision-making processes is also investigated. This chapter explores ESG implementation through the lens of participatory design to identify opportunities for employees to contribute to an organization's ESG strategy, across different organizational levels and functions. Results from a study conducted with a medium-sized global information technology company show the potential of participatory approaches for implementing ESG and highlight the importance of creating consensus and promoting open spaces for sharing information as the first step toward successful ESG implementation.

**Keywords** Collaborative design · Sustainability · ESG · Organizational planning · Change management

## 1 Introduction

Environmental, Social, and Governance (ESG) considerations have gained significant attention from investors and stakeholders in recent years. Companies are being evaluated not only on their financial performance but also on their ability to meet environmental and social obligations while ensuring good governance [1]. Regulatory frameworks and changes in the market are also requiring organizations of all sizes and sectors to determine metrics and processes that allow for effective ESG implementation and reporting. However, the implementation of ESG can be challenging, due to the complex and dynamic nature of organizational practices and decision-making processes [2].

K. Kunrath · D. Ramanujan (✉)
Department of Mechanical and Production Engineering, Aarhus University, Jutland, Denmark
e-mail: devr@mpe.au.dk



S. Fukushige et al. (eds.), *EcoDesign for Circular Value Creation: Volume I*,
https://doi.org/10.1007/978-981-97-9068-5_19

Accounting, documenting, and reporting on ESG can be extremely complex and time-consuming. Many standards, regulatory frameworks, and methodologies can be utilized to relate industry operations with global and regional targets and different purposes. The lack of a single consolidated approach, aspects of data transparency, and diversity in reporting and communication strategies pose several challenges, which often result in shallow and opportunistic reports leading to greenwashing [3]. ESG frameworks could serve as a useful tool to motivate and guide further improvements in organizations. Prior research has shown the efficacy of eco-feedback in promoting sustainable behavior [4]. Hence, feedback mechanisms such as establishing a baseline, setting targets for ESG performance, tracking progress, and encouraging continuous improvement, may have the potential to promote pro-sustainable behavior within organizations.

The implementation of ESG initiatives is of strategic relevance to most companies and is usually approached in a top-down manner in organizations [5]. However, aspects of individual employee contributions and the possibility of using workforce expertise as an asset for ESG intelligence, reporting, and drive are only starting to be acknowledged. There is a need for moving out of a reporting-centric approach when framing ESG initiatives to a collaborative effort, where individual employees are seen as valuable assets for facilitating and driving implementation (bottom-up) of new initiatives that align with a company's ESG ambitions and strategies [5].

The implementation of ESG initiatives across an organization—into their strategic, tactical, and operational layers—requires a clear vision of the existing constraints within the company's hierarchical structures, along with careful planning and execution [6]. Therefore, there is a need for approaches that can support change management efforts to implement ESG actions successfully. Among diverse approaches, participatory design (PD) is known to foster collective learning and enable reflexivity and iteration throughout the implementation of processes that involve diverse stakeholders [7]. The literature on PD focuses on various methods employed to facilitate participation, including workshops, charettes, and games [8]. These are now widely recognized as established approaches to participatory practice and the application of this comprehensive set of methods in specific contexts has been extensively documented and used as experiential knowledge for informing user engagement across diverse scenarios and is also known to support engagement with new initiatives in industry [9]. In this study, we explore the use of PD as a support mechanism for change management related to ESG implementation.

PD's history emphasizes the importance of shifting from designing objects to "infrastructuring" design, enabling ongoing participation in designing solutions for complex issues and envisioning sustainable futures [8]. The notion of ESG data commons focuses on creating common good [10] (e.g., climate action) by pooling and utilizing available data, which also refers to the infrastructure that supports the collection, storage, sharing, analysis, and management of data within a specific community [11]. Research suggests that applying a practice-based approach to ESG data commons reveals three important features of the ESG reporting process: (1) it requires a change in stakeholders' understanding; (2) the consolidation of meanings about ESG data and how it can facilitate change; (3) the use of data objects helps understand the collective becoming as part of change processes [12]. These constructs could differ based on the role of a stakeholder within an organization. Hence, investigating how different job functions relate to ESG can provide insights into how participatory approaches can be successfully deployed for implementing ESG initiatives.

This chapter addresses how organizations can implement ESG, via corresponding indicators and categories, into their decision-making and governance structure and specifically focuses on the role of PD in doing so. Herein, an ESG category is defined as an area of interest used to evaluate a company's ESG performance. This study aims to address the following research questions:

*RQ1) Can mapping ESG categories to specific job functions and organizational hierarchies uncover potential initiatives that can be supported by participatory approaches in organizations?*

*RQ2) Can participatory approaches be implemented based on these mappings of ESG categories across functions and hierarchies?*

To address these questions, the paper first defined a set of ESG categories through comparisons across compliance standards. Furthermore, we conducted individual workshops with the leadership team of a medium-sized information technology company to map these categories to company-specific factors, i.e., job roles and hierarchy (Strategic, Tactical, Operational). In this way, this article investigates PD-oriented opportunities for employees' contributions to ESG initiatives and provides an initial understanding of how PD can support ESG implementation in real-world organizational contexts. Our work focuses on change management and explores the potential benefits of PD methods in facilitating ESG implementation. By investigating the application of participatory approaches in the context of ESG, we contribute to the existing literature in PD and expand our understanding of how such methods can benefit organizations on sustainability challenges.

## 2 Methods

The methodology outlines the phases utilized to initially define common ESG categories from five major compliance standards, which serve as the basic understanding of how ESG indicators are approached. Further, selected ESG categories were defined to compose the research framework for this study, taking into consideration intra-organizational and contextual aspects. The selected indicators were investigated through a range of individual workshops and interviews [13], resulting in a framework for PD for ESG. Figure 1 illustrates the methodological phases undertaken:

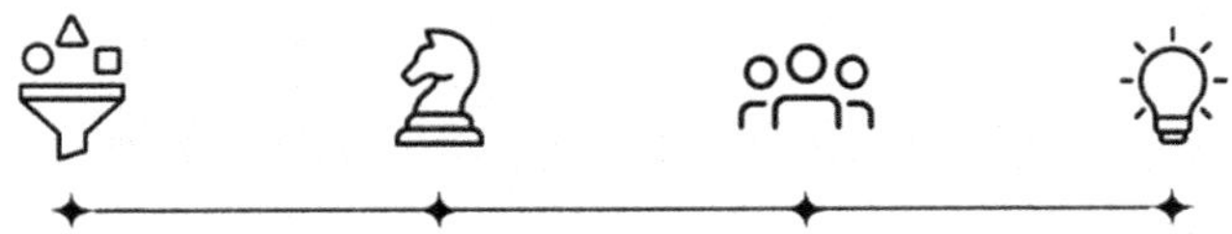


**Fig. 1** The four methodological phases utilized in this study: (1) Sorting and Selecting indicators from literature to construct the ESG categories, (2) Workshops with the ESG and Innovation leaders in the case company, (3) Interviews with Senior Leadership across diverse functions, and (4) Data analysis and interpretation of results in relation to designing Participatory Design approaches tailored for ESG

**Table 1** Mapping of core ESG categories across popular compliance standards

| | ESG categories | GRI | SASB | TCFD | UN-SDGs | PRI |
|---|---|---|---|---|---|---|
| *Environmental* | *Greenhouse gas emissions* | x | x | x | x | x |
| | *Energy consumption* | x | x | | x | x |
| | *Water usage* | x | x | | x | |
| | *Waste generation* | x | x | | x | |
| | *Air pollution* | x | x | x | x | |
| | *Biodiversity* | x | x | | x | x |
| *Social* | *Employee turnover rate* | x | x | | x | |
| | *Diversity and inclusion in the workforce* | x | x | | x | x |
| | *Health and safety performance* | x | x | | x | |
| | *Community engagement* | x | x | | | |
| | *Human rights* | x | x | | x | |
| | *Product safety and quality* | x | x | | x | x |
| *Governance* | *Board diversity and independence* | x | x | x | x | x |
| | *Executive compensation* | x | x | | | |
| | *Shareholder rights* | x | x | x | | x |
| | *Transparency and disclosure of information* | x | x | x | x | x |
| | *Ethical behavior* | x | x | | | x |
| | *Risk management* | x | x | x | x | x |

## *2.1 Sorting and Selecting ESG Categories*

In this work, an ESG category is defined as an area of interest used to evaluate a company's ESG performance. These areas of interest encompass carbon emissions, labor practices, diversity, inclusion, etc. (see Table 1), and are typically assessed using one or more ESG indicators.

The use of ESG indicators and categories is not standardized across companies, which makes it challenging to assess and compare sustainability performance as well as the integration of initiatives in an organization. To address this issue, we conducted a review of broadly used ESG indicators and categories based on five major standards commonly used for assessing ESG metrics: (1) Global Reporting Initiative (GRI) [14], (2) the Sustainability Accounting Standards Board (SASB) [15], (3) the Task Force on Climate-related Financial Disclosures (TCFD) [16], (4) The United Nations Sustainable Development Goals (SDGs) [17], and (5) The Principles for Responsible Investment (PRI) Reporting Framework [18]. The review does not intend to be exhaustive, but rather to identify a point of consensus for the investigation carried on in this work. Table 1 presents the list of 18 ESG categories and their presence across these five major standards.

The ESG categories identified in Table 1 were assessed in terms of relevance to the case company participating in this study and compared with the targets defined by the existing ESG strategy and initiatives adopted by this specific company. Six preselected ESG categories were seen as relevant to the existing efforts and therefore, utilized for an in-depth exploratory approach in the individual workshops and investigated further through dedicated interviews with senior leadership (SL).

We hypothesized that each ESG category may be impacted differently at each level of the company hierarchy (strategic, tactical, and operational) and may require the involvement of different job functions. Research on agent-based influence on ESG structures suggests that mission-based role assignments appear to be a favorable strategy for assigning leadership roles, as it simplifies personnel training and organizational planning [19]. Hence, by understanding these relationships, organizations could integrate ESG considerations into decision-making and engage relevant stakeholders effectively in a participatory process.

In this study, we use a research framework where the ESG categories, company hierarchy, and job functions are assumed to be interrelated, with different levels of the organization responsible for various ESG categories, and job functions requiring specific ESG knowledge and skills. The exploration of these relations was at the core of the investigation and the selection of both ESG categories and the specific job functions was heavily influenced by the available data and company structure in the study case. Figure 2 illustrates the research framework utilized, including the six ESG categories, the overview of the hierarchical structure and the investigated job functions.

## 2.2 Data Collection and Analysis

The data collection took place in 2023 and makes use of an industry case situated in a medium-sized global information technology company. Empirical data were collected in two stages: (1) two individual workshops with company experts, and (2) 10 interviews, with Senior Leadership (SL). The workshops and interviews took place remotely and lasted for about one hour, with moderation by the authors. Prior

**Fig. 2** Research framework utilized for in-depth discussions during workshops

research suggests that focused and in-depth exploratory studies with a limited number of domain experts are necessary to gain valuable insights into real-world challenges [8], which supports the single-case study design of this research. The study adopts an abductive design to explore and explain the phenomena while acknowledging that context specificity is created by the meaning attribution inherent to intersubjectivity and a case-oriented approach. We employ double hermeneutics to evaluate the relations between theoretical approaches and empirical data, recognizing its relevance for a deeper understanding of the subjectivities and contextual factors shaping organizational practices and decision-making processes.

The two *workshops* were conducted individually with the Head of ESG (HE) and Head of Innovation (HI) to evaluate the opportunities for participatory interventions across the company. These two key figures were responsible for initiating and implementing the ESG strategy in the organization and were identified as the most qualified employees to provide a general overview of how sustainability actions could impact or take place in diverse functions and activities inside the company. The participants were asked to reflect on "what kind of influence the selected ESG indicators have" across different functions related to (a) Executive & Legal and (b) Sales & Marketing. The participants elaborated their thoughts using the thinking-out-loud protocol [13].

Further data was acquired through *interviews* with SL across diverse company functions (including Executive, Finances, Product, Information, Sales, and Delivery) The interviews aimed to investigate the opportunities for employee contribution to the overall ESG strategy, highlighting perspectives from different job families and hierarchies inside a company. The 10 participants occupy positions at both strategic level ($n = 5$) and tactical level ($n = 5$) and were composed of seven males and three female employees. The selection of participants intentionally covers only the top two levels of the organization hierarchy structure presented in Fig. 2, in order to provide an overview from the strategic and tactical levels as well as the opportunities for buy-ins that could lead to larger deployment of future initiatives.

The data collected from the interviews and workshops were analyzed using a qualitative approach [13]. The findings are presented and discussed in Sect. 3, summarizing the insights gathered during the individual sessions and including recommendations for the use of participatory approaches across different organizational levels, building capability and intelligence for ESG reporting, as well as identifying opportunities for innovation. In Sect. 3.1, we highlight qualitative information and insightful quotes derived from the case study, shedding light on the research findings. Section 3.2 discusses some of the challenges encountered during the implementation of the participatory framework with the company, offering valuable insights into the practical aspects of the study. Section 3.3 describes the methodological implications and lessons learned from the study in relation to existing PD approaches. Finally, in Sect. 3.4, we dive into the prospects of this participatory methodology, exploring potential alterations or adaptations based on the case study's outcomes. Section 3.4 provides a forward-looking perspective on how the methodology can be refined and optimized for future research endeavors.

## 3 Results and Discussion

We explored the connections between selected ESG categories, organization hierarchy levels, and job functions (Executive & Legal, and Sales & Marketing) using the described research framework (see Fig. 1). In considering the use of participatory approaches cross-hierarchically for facilitating the implementation of ESG initiatives in an organization, the case analysis highlighted the complex nature of the interconnectivities between different job functions and the challenges related to simplifying activities involved in long-term strategic thinking to create accessibility for multiple stakeholders.

### *3.1 Mapping of ESG Categories to Job Functions and Hierarchies*

The mapping of ESG categories within the organizational context aimed to investigate RQ1, providing a deeper understanding of the company's existing ESG initiatives, how these actions are recognized across different job functions and hierarchies, and uncovering potential initiatives that can be supported by participatory approaches in organizations.

During the workshops, each of the six selected ESG categories was discussed separately and in relation to the predetermined job functions. Activities were mapped for each ESG category as providing a positive, negative, or neutral contribution to the overall capacity of the company to improve sustainability. Figure 3 illustrates the data collection process for the ESG category Greenhouse Gas Emissions (GHGe), taking into consideration the possibilities for different hierarchical levels (strategic, tactical, and operational).

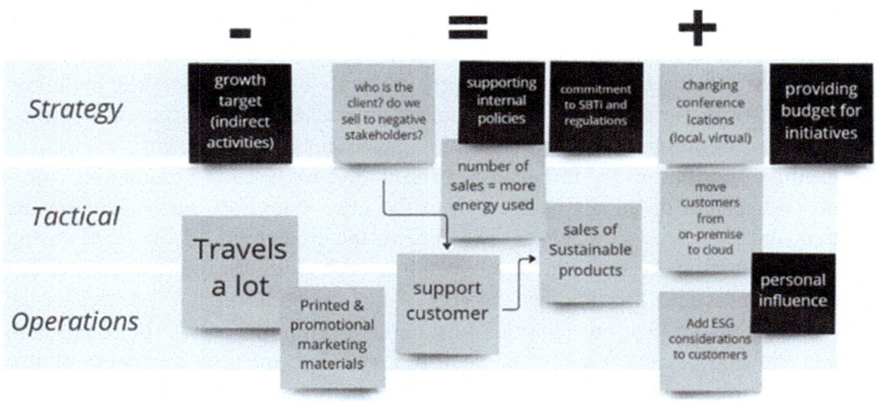


**Fig. 3** Representation of data collection for a single ESG category, including two job functions (**a**) Executive & Legal (dark notes) and (**b**) Sales & Marketing (bright notes)

This specific example illustrates how different functions hold intrinsically different potentials for contribution in the participatory processes and unveil latent aspects for the holistic implementation of actions. HE reflects on the major negative aspect of the activities related to Sales and Marketing, which is the large number of travels performed by employees to reach customers, close deals, and create exposure in events and conferences globally. In this sense, the reduction of travel and digitalization of events and meetings could be opportunities for reducing overall emissions. However, due to the nature of these activities such decisions could only be enforced top-down (from the strategic level). When asked about how personnel at the strategic level could contribute to the ESG initiatives around reducing GHGe, they said.

> I'm thinking again about the business travel, of course, but it could have, especially at a strategic level, actions to say we're going to do our sales conference regionally or we're gonna do it virtually. That's where they can definitely have a big impact. (HE)

When compared with the impressions from the Executive & Legal job function, we can see the change in orientation toward strategic and budgetary aspects, rather than materialistic aspects. Despite aspects related to the allocation of resources and commitment to major regulatory frameworks, strategic synergies and dilemmas demonstrate the interconnectivity of the two job functions. An interesting example of this kind of dilemma refers to the ethics of company growth and the impact of business relations with other companies and customers that are either not committed to ESG or share different values. In this line of thought, HE reflects on the impact of the value chain on the overall sustainability of the company as well as the business opportunities associated with supporting customers in their own ESG transitions.

> Who do we sell to? and who are the companies that we don't want to sell to? For example, if it is an oil and gas company, they could become more and more successful with our software. We could say that we don't want to indirectly increase the emission [by selling products or working with companies/clients that don't have a responsible take]…. (HE)

The category of *biodiversity* goes in the opposite direction and mainly reflects upon purchase and procurement. In general terms, this aspect is not heavily related to either of the job functions analyzed but provides insights into the relevance of other functions inside the organization. Most of the actions in this category relate to high-level decision-making, especially about investment in new infrastructure (e.g., solar panels) and budget allocation for secondary activities (e.g., maintaining a garden, planting trees, and voluntary actions). The strategic take into contributing to this ESG category also involves donating to organizations that support biodiversity e.g., World Wild Foundation (WWF), and having a garden around the company buildings and facilities.

Similar discussions were carried on for the other five ESG categories. For the social aspect, the dialogue around *diversity and inclusion in the workforce* was mainly centered on Human Resources and hiring processes. HE illustrates multiple examples of team members on multiple job functions across the organization that are especially conscious about gender balance. However, they also acknowledge a systemic aspect of this problem in the IT sector where companies face challenges around balanced recruitment for technical and sales functions. Challenges refer to IT being both a historically very male-dominated industry which creates a range of unconscious biases, as well as a possible shortage of qualified available candidates. To facilitate inclusion and diversity across functions, the company has included metrics to regulate gender balance as well as providing training on this topic. These efforts illustrate a space for open dialogues and participatory approaches concerning recruitment and career progression.

> We talk about inclusion. So, for everyone to be conscious about this, I would say following the training as well. That's one thing, making sure that they are trained to understand how their actions impact this, and that there's no unconscious bias and discrimination. (HE)

The category of *community engagement* was mainly approached by internal policies such as Corporate Social Responsibility (CSR), where multiple actions are led by driven employees and volunteers. These actions can be seen as grassroots initiatives with predetermined support from management, which may be pushed back due to the time employees spend on these extra-work activities or to have lower employee engagement when there is a high workload, deadlines, or deliverables to be considered.

The aspects of governance related to *Board Diversity and independence* were considered at SL Management—not even people in strategic positions had a say on it. It was estimated that with the eventual increase in representation, people could start recognizing or talking about it. However, it was uncertain which repercussions this could have on the overall structure of the company. The interviews, however, illustrate the need for the adoption of best practices at the high leadership levels to express the values and the viability of the proposed actions across the organization.

> You need to get the senior leaders to emulate whatever you're asking. So, if you put out an action or a strategy that you don't believe our senior leaders can actively digest and do, then there's no point in putting it out. It may be the absolutely right thing to do, but it will be a no-go for the other employees. (SL1)

Finally, the category of *transparency and disclosure of information* was highly discussed in relation to the Sales & Marketing functions once data transparency is also part of the value created by the company's products. Internal data, which is not necessarily related to the products or sales pitch, also makes use of marketing strategies for internal disclosure and is approached confidentially during the multiple calls and webinars hosted by the company. The idea of ESG data commons (i.e., consolidating a data pool on sustainability metrics related to both employees and organizational aspects) was highlighted during the interviews as an asset that could be used to create awareness and engagement for specific actions across the company, and ultimately facilitate change management. The role of strategic levels in this category reverberates with trustworthiness conveyed by leading figures, such as "if they walk the talk" and deliver on the vision disseminated, also touching upon sensitive topics and being open about areas of improvement.

For all the selected ESG indicators, a potential contribution was highlighted from different functions in the organization. The interviews suggested that there is, however, a need for greater standardization of ESG to properly communicate actions with diverse stakeholders and align outcome expectations with the overarching company strategy and vision.

RQ1 inquired about mapping ESG categories to specific job functions and hierarchies to uncover potential initiatives to be supported by participatory approaches. The results suggest that the inclusion of diverse stakeholders in mapping opportunities is key for (a) understanding functional capabilities within the organization, (b) determining employees' behavioral and cognitive barriers, (c) encompassing dissonances, and (d) promoting inclusion.

We observed that understanding functional capabilities and their relation to the overarching ESG strategy can support the general impressions and the representativity of ESG actions across the organization. One of the participants reflected on how functional data could be gathered and utilized as a mechanism to enhance communication and awareness of the new initiatives.

> Potentially, we could create a program that we look into these **small actions that each person is doing right** and then it becomes a bigger story. (SL2)

In this way, determining patterns of the employees' behavioral and cognitive barriers can help the management to navigate uncertainty and consolidate structured approaches that foster participation and engagement. The interviews highlighted the need for guidance in small and manageable steps to overcome preconceptions and false judgements about ESG initiatives and incurred organizational changes.

> It can't be complex; it's got to be concise and easily digestible. For what we're really looking for, I think trying to find the small steps to get people going will be important. Keeping it easy, boiling it down and finding those small things that we collectively can kind of do and move in a particular direction, those are the things. **(SL1)**

Finally, the aspects of encompassing dissonances and promoting inclusion were mainly discussed in relation to the consolidation of shared understanding, consensus, and values via internal educational programs related to ESG. These aspects were also highlighted as part of the company culture and the nurturing of these values to be embedded in the company DNA.

> ESG should be blended into what we do. It shouldn't be the driver because it won't be a driver with our customers, right? Find out how to integrate it more into that story of who we are and what we offer. So, let's find a place where it lives organizationally, right, and then really have these connections. Right, so that you could articulate to people, you know, our investment is 1% and here's what we're doing. So, you want to have that dimensionality of where ESG is embedded in the company's DNA. (SL3)

Mapping ESG categories to specific job functions and organizational hierarchies uncovered diverse initiatives that can be supported by participatory approaches inside the organization, especially for categories that are better understood in general terms. However, there is a high degree of hierarchical codependency in relation to decision-making associated with the establishment of new initiatives. These codependencies are often unidirectional, where the strategic levels regulate the pace of implementation based on a shared vision and access to resources.

## 3.2 *Participatory Approaches for Facilitating ESG Implementation*

The workshops and interviews also aimed to investigate RQ2, exploring the opportunities for developing participatory approaches based on the mapping of ESG categories across functions and hierarchies, which revealed that specific methodological considerations were necessary to take place to support larger inclusion in the organization for ESG initiatives.

One of the main takeaways was that even though the mapping exercise helped identify opportunities across functions and hierarchies, the multitude of ESG standards and frameworks, as well as different perceptions and objectives among the participants, create dissonance and influence the implementation of initiatives inside organizations. In fact, the workshops and interviews highlighted that assessing ESG, even at categorical levels, poses significant challenges.

ESG was viewed as a new topic in the company and in the vocabulary of the employees, considering that the establishment of ESG strategies in the case company started in 2021–2022. As the whole narrative around sustainability and ESG starts to consolidate, different understandings and perceptions about the meaning and nature of ESG indicators and categories could contribute to the dissonance in responses, frameworks, and behavioral approaches. We observed that different ESG categories also contributed to distinct types of dissonances in responses, where we identified the following:

1. Agreement/Confrontation to categories considered more tangible, e.g., *the implementation of restrictive travel policies that aim to reduce GHGe.*

2. Empowerment/Demotivation to categories considered intangible such as Board Diversity, e.g., *the inspirational role and endorsement of company visions based on the structure of the leadership.*
3. Optimism/Burden to categories with potential extra demands in terms of workload or working hours, e.g., *the requirement to participate or contribute to new activities or extra data collection initiatives.*

The outset of opening possibilities for participatory approaches is establishing consensus on the interpretation and value of each specific ESG category. Defining specific meanings and aligning the definitions and expectations about ESG strategies could create an open space that supports mature and relevant contributions. Without consensus, participatory approaches can only provide limited contributions and hold little influence on the decision-making process [20]. Fostering participation and negotiations on meanings can result in increased trust and learning as it enables participants to listen to a wider range of perspectives, fostering changes in attitudes that align more with common values. Moreover, decision-making outcomes perceived as fair and legitimate are more likely to be accepted by stakeholders in the long term [21]. In an interview with an SL participant, they reflected on the importance of clear communication of progress for creating engagement across the organization.

> Make the communication clear and try to get the employees involved in a way where they feel that they are engaged and can take part in it. And then make sure it's not about taking part in something that is silly, but with some real projects or some real things that can contribute to something bigger. (SL4)

However, the results also revealed that a high level of understanding and participation in ESG initiatives is needed to enable the participants to consistently reflect, envision the influence of different activities, and therefore correlate specific functions with possible contributions toward the company's sustainability-oriented goals. Participants in the strategic and tactical levels provide a comprehensive overview of the evaluated indicator, while technical and operational participants provide narrower contributions with a higher level of functional expertise. Furthermore, participants of higher hierarchical levels also have the opportunity to steer and influence the activities in their teams, consolidating actions and initiatives—e.g.,

> "I like to hope that I'm fairly influential within my team, and even within the company. So, I would say that I could use this influence and give ideas to my team. It should not be just something extra to add to their plates, it has to make sense. I think I could share it in such a way with my team and influence them by saying that it's important for us and that it is something we need to do moving forward." (SL5)

Functions and tasks are, therefore, shown to be intrinsically connected, while opportunities for contribution are codependent. In this sense, the decision-making processes for deploying initiatives that enforce general behavioral change across the company, transformation of company culture, or infrastructure changes still rely on authority (top-down approaches), especially for actions seen as unpopular or restrictive at a personal level, e.g., reducing meat consumption at the canteen as the means for reducing the company's overall GHGe.

Similarly, cross-hierarchical consensus is required for initiatives that drive, support, or accommodate the transitions proposed by the implementation of ESG strategies. However, these cross-hierarchical consensus are not obvious and require further investigation on what aspects could facilitate proper exchange of information and increase perceived value for different stakeholders. Future research is needed to explore other approaches to achieving consensus and facilitating participatory approaches in ESG implementation.

ESG-specific problems were observed to be fundamentally different from other PD problems presented in the literature. As a relatively new field of study, ESG has a wicked nature, which suggests the need for an initial task of standardization of indicators and definition of meaning, overall education of participants, and cross-hierarchical consensus. This aspect can be a limitation to the full implementation of PD methodologies within the scope of ESG, requiring dedicated actions that promote literacy-targeted applications and tailored action roadmaps within each function inside the company. Acknowledged PD methods have explored the variations in perception and understanding, recognizing the interconnectedness of tacit and explicit knowledge [8]. These methods assert that knowledge is not only individually held but also collectively created and expanded through social interaction [20]. Through social interaction and collaborative processes, PD bridges gaps in understanding, promotes dialogue, and ultimately works toward a shared consensus so that participants can appropriate the design process [7].

Additionally, there is a need to create a safe and open space for participation that is not limited by specific knowledge constraints. This aspect reflects upon the differences in the participants' competency levels around ESG and their ability to provide meaningful insights. In an initial assessment, open brainstorming provides the participant with the opportunity to convey information on areas of expert knowledge, e.g.,

> If I'm tied to think about greenhouse gas emission, I'm blocked because I don't know anything about it. And I don't know anything about the biodiversity, but I do know something about saving water. (HI)

Our study points to the fact that ESG definition is a fundamental aspect of including and engaging participants in new initiatives, engaging stakeholders in the design and decision-making processes, and allowing for diverse perspectives to be considered. Participatory approaches facilitate dialogue and collaboration, leading to the co-creation of shared meanings and definitions of actions related to ESG, as reflected below:

> We have environments that can be used to discuss what employees can do about ESG, and that's probably one of the best things here. We should have something to support it, some policies, and doing good business of course. Other than that, we should leave much of it to the local partners because they need to do something to feel ownership. (SL6)

This consensus-building process can be deployed as multiple interventions that tackle specific ESG challenges, promote transparency, create employee engagement, and lead to a common understanding of ESG goals and categories. Hence, it ensures that multiple perspectives and priorities are integrated, fostering ownership and alignment for the effective implementation of new initiatives.

Furthermore, the interviews highlighted the importance of embedding ESG as part of the company culture and as a collective mindset. Participatory approaches can support a shift toward a more sustainable mindset [22] and therefore change management in the organization by establishing a dialogue with the workforce, as reflected below:

> ESG is a mentality. More than a function, more than programs or strategies or anything, it's a mindset. It's the way that the people who work within the organization do work, the way they get relationships done and in being responsible. (SL2)

RQ2 inquired about the possibility of participatory approaches being implemented in organizations based on mappings of ESG indicators across functions and hierarchies. The results suggest that by involving diverse stakeholders, *PD could indeed assist the assessment of ESG indicators and aid the definition of actions tailored to specific functions*, once it facilitates collective learning, reflexivity, and iterative processes throughout the implementation journey [8]. However, different levels of literacy in relation to ESG were observed in the participants of both the workshops and interviews, which affected their ability to articulate ideas and therefore could also reduce the capacity for engagement and the quality of the outcomes from the participatory approaches. Therefore, a specific approach is needed to handle these perceived differences and gather the best outcome from the participatory approach, where participants feel comfortable and engaged within their realm of expertise.

## Methodological Framework for ESG

Based on the mappings of ESG indicators across multiple functions and hierarchies within the company, we observed a need for tailored approaches and participatory methodologies for promoting engagement and successful implementation of ESG initiatives in a global organization, featuring differences across functions.

We found, given the wicked nature of ESG as a subject, the use of a deterministic approach to investigate specific aspects at once could lead to diversification of perception and meaning attributions, which can create blockages and bias in responses. Therefore, a methodological approach for promoting reflection on the participants of different natures would benefit from following a funnel approach, where information is first expressed in an open environment and further classified from broader to narrower.

Based on findings from both RQ1 and RQ2, we propose a three-step PD approach for onboarding diverse knowledge regarding ESG actions, which corresponds to the three-step process for knowledge acquisition in participatory approaches and classifies information based on the level of understanding that participants have. Figure 4 outlines the PD approach for ideation sessions and categorizes information obtained as general, hierarchical, or functional knowledge based on participant perspective and competency level.

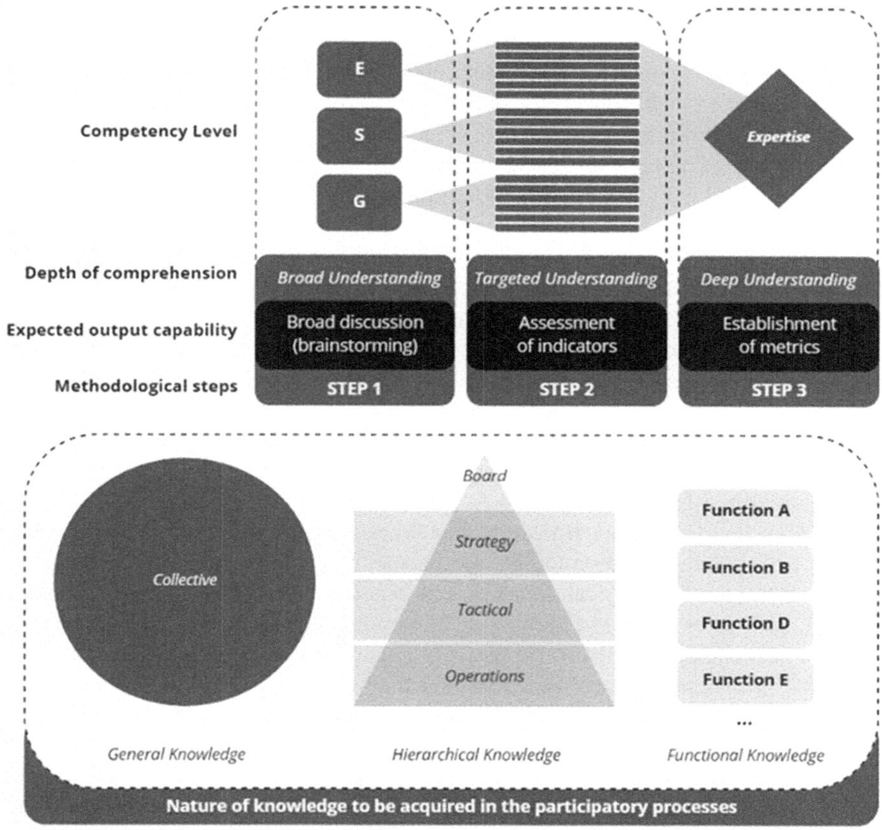


**Fig. 4** Methodological framework for PD focused on ESG data collection and categorization of knowledge from ideation sessions in organizations

In step 1, participants can provide generic insights based on a broad understanding of ESG targets and goals, which can contribute to mapping the different perspectives and competency levels across the organization. In step 2, participants are required to have a defined understanding of ESG indicators, which allows them to provide more meaningful contributions. In step 3, participants are expected to have specialized knowledge about the subject or indicators being evaluated, which support the establishment of metrics based on specific areas of expertise. These steps can be taken separately and across multiple rounds with the goal of refinement and classification of information.

The information acquired through participatory approaches can be further classified based on its nature such as general, hierarchical, or functional knowledge according to the lens and competency attribution of the participant. In this framework, general information provides a broad understanding of ESG indicators or concepts related to sustainability without specific details, such as definitions or introductory impressions. Hierarchical information follows a layered structure, arranging inputs based on a participant's role in the company, such as managerial positions or decision-making attributions. Functional information categorizes data by a participant's role or function within the company, focusing on practical aspects, such as job attributions or process workflows. These categorizations may overlap, and the classification depends on the purpose of the acquired ESG information.

The simplified methodology described above aims to acknowledge the complexity imposed by the diversification of inputs in the process of building capacity and intelligence for ESG implementations through PD. Furthermore, it highlights the need for consensus about the meaning of ESG indicators and categories before the integration of multiple perspectives. The overall goal is to increase participation from consultative toward co-design or shared decision-making, in a collective learning-by-doing process of building ESG competencies across the organizations [7]. For this to happen, the nature, quality, and type of input acquired during participatory approaches need to be addressed. The use of PD can be seen as a mechanism to access differences in perception across the company and unveil innovation opportunities.

The proposed framework is designed to support the inclusion of multiple stakeholders with diverse understandings and perspectives concerning ESG indicators. Therefore, the methodological approach exhibits the characteristics that align with the core principles of PD. Firstly, the framework places a strong emphasis on inclusive stakeholder involvement, actively engaging a wide range of individuals across multiple organizational functions and hierarchies, ensuring their insights are incorporated into the design process. Secondly, the framework promotes collaborative decision-making, creating an environment where stakeholders have opportunities to contribute their knowledge and expertise.

By fostering dialogue, deliberation, and consensus-building, the framework encourages a sense of shared ownership and responsibility among participants. Additionally, the framework supports co-creation and iteration, allowing stakeholders, designers, and decision-makers to collaborate in developing solutions and interventions iteratively, incorporating feedback loops based on stakeholder input. Overall, the framework empowers multiple stakeholders to take part in ESG initiatives, promotes reflexivity and learning, and is sensitive to the context in which it operates.

We propose the use of the framework presented in Fig. 4 as a prelude to guide design thinking approaches that integrate participatory guidelines and address specific challenges related to ESG. This framework could be used to explore the application of PD as a support mechanism for change management in relation to ESG implementation. By involving stakeholders with diverse perspectives and knowledge, organizations can gain a deeper understanding of ESG issues and develop more effective strategies to address them. Ultimately, this can lead to a shift toward a more sustainable mindset within organizations and a greater commitment to ESG [22].

## 3.3 Methodological Implications

The methodology deployed in this study targeted an associative exploration of PD practices with the emergent and timely area of study about ESG. Organizational transformation and deployment of sustainability actions is a topic especially difficult to study comprehensively given the highly integrative nature of the work when in concert with actual companies. Design projects aimed at promoting participatory actions on ESG indicators need to address the challenges posed by the unpredictable structuration of the appropriation processes within an organization's own social practices [20], which cannot be adequately anticipated. Furthermore, it must actively work toward bridging gaps of knowledge and creating consensus around the meaning of targets and goals. We view the tight integration with an industry case to be a strength of this research, however, due to the context-sensitivity of organizational studies, different scenarios could apply to other industry cases. Results from our study also highlight the importance of first exploring functional aspects and getting the buy-in from SL in an organization before proposing further actions. Buy-in is a personal and professional commitment to actively engage in a process, task, or initiative. Without buy-in, employees are less likely to commit to a level of change which results in active engagement [23]. Therefore, the contributions of this article are centered on acquiring an in-depth understanding of contextual aspects that facilitate the implementation of initiatives in the organizational context. Furthermore, we identified that the first action should be focused on creating consensus across the organization and open spaces for sharing information.

Based on the learnings from Sect. 3.2, we believe that designing successful participatory mechanisms for ESG in organizations should be conceived as expertise-based and utilize a tailored methodology to assist interventions that stimulate collective learning processes. These contextual conditions expand the various methods employed to facilitate PD approaches [7] and reflect upon unavoidable challenges that represent an inescapable aspect of social practice in design research. By targeting the wicked nature of sustainability data, design science can support change management and facilitate ESG-oriented transformations within organizations, offering methods to organize, design, and implement participatory processes inclusively and reflexively, also enabling iteration across different hierarchical levels. This work adds to the existing literature on PD by exploring the relationship between ESG categories, organizational hierarchy, and job functions. The study provides a research framework and a PD approach that can aid in obtaining knowledge and insights from stakeholders and in consolidating the ESG data commons.

## 3.4 Opportunities for Further Research

Mapping ESG categories to specific job functions and organizational hierarchies can help identify opportunities for participatory approaches. However, challenges such as the multitude of ESG standards and frameworks, as well as different perceptions and objectives, need to be addressed by defining meaning and creating consensus. Based on the identified challenges of consolidation of functional ESG strategies, this study suggests a need for greater standardization of ESG indicators. At the current stage of implementation of ESG practices within the organization, and due to the lack of infrastructuring of ESG-related data across the different company functions, it is not yet possible to provide any reliable quantitative assessment derived from ESG-oriented organizational change, e.g., cost/time reduction for ESG indicators evaluation after organization change. Further research can aim to develop standardized ESG reporting frameworks and to identify indicators relevant to specific sectors and regions.

Further work should aim to investigate the transition of ESG from a metrics-oriented system intended for reporting into a framework for a proactive means of internal guidance. In this sense, it is also important to quantify the influence of specific job functions for the implementation of initiatives and correlate with sustainability needs at sectorial, regional, and global levels.

This study is limited by its focus on a single company. Nevertheless, the chosen company was appropriate for the study, given its current phase of ESG implementation. Future studies should aim to include a diverse range of companies as the use of multiple cases could contribute to further developing the methodology.

## 4 Summary

This study highlights the use of Participatory Design approaches across different company levels for acquiring and classifying *Environmental, Social, and Governance* (ESG) information. Exploring specific job functions also provides insights into deploying participatory approaches effectively and reveals key aspects of participatory approaches in this context: (a) creating consensus in stakeholders' understanding and (b) utilizing expertise-based interventions for integrating cross-organizational insights into the ESG processes. The proposed framework offers an approach for engaging diverse stakeholders within the company. Results from the study provide insights into how participatory approaches can enhance intelligence and capacity for ESG implementations and reporting.

**Acknowledgments** Thanks to all participants and the case company. This research was partially sponsored by Innovation Fund Denmark as an industrial post-doc project with the case company. The opinions presented are of the authors alone and do not represent the sponsoring parties.

## References

1. Dye J, McKinnon M, Van der Byl C (2021) Green gaps: firm ESG disclosure and financial institutions' reporting requirements. J Sustain Res 3(1)
2. Drempetic S, Klein C, Zwergel B (2020) The influence of firm size on the ESG score: corporate sustainability ratings under review. J Bus Ethics 167(2):333

3. Gatti L, Pizzetti M, Seele P (2021) Green lies and their effect on intention to invest. J Bus Res 127:228
4. Bao Q et al (2019) Investigating user emotional responses to eco-feedback designs. J Mech Design 141(2):021103
5. Kezar A (2012) Bottom-up/top-down leadership: contradiction or hidden phenomenon. J High Educ 83(5):725
6. Schulte J et al (2020) Strategic sustainability risk management in product development companies: key aspects and conceptual approach. Sustainability 12(24):10531
7. Robertson T, Simonsen J (2012) Participatory design: an introduction. In: Routledge international handbook of participatory design. Taylor & Francis
8. Luck R (2007) Learning to talk to users in participatory design situations. Des Stud 28(3):217
9. Sanoff H (2000) Community participation methods in design and planning. John Wiley & Sons Inc, New York
10. Ostrom E (1990) Governing the commons: the evolution of institutions for collective action. Cambridge Press
11. Grossman RL et al (2016) A case for data commons. Comput Sci Eng 18(5):10
12. Wang C et al (2022) Becoming sustainable together: ESG data commons for Fintech startups. In: 43rd international conference on information systems (ICIS)
13. Miles MB, Huberman AM (1994) Qualitative data analysis: an expanded sourcebook. Sage
14. GRI, "Global Reporting Initiative" https://www.globalreporting.org/ (March 13, 2023)
15. SASB, "Sustainability Accounting Standards Board" https://www.sasb.org/ (March 13, 2023)
16. TCFD, "Task Force on Climate-related Financial Disclosures" https://www.fsb-tcfd.org/ (March 13, 2023)
17. United Nations, "Sustainable Development Goals" (SDGs) https://sdgs.un.org/goals (March 13, 2023)
18. United Nations, "The Principles for Responsible Investment Reporting" (PRI) https://unpri.org/signatories/reporting-and-assessment (March 13, 2023)
19. Meirina C, Yu F, Pattipati KR, Kleinman DL (2006) Model-based organization analysis and design for an ESG organization. Model Simul
20. Kyakulumbye S, Pather S, Jantjies M (2019) Knowledge creation in a participatory design context: the use of empathetic participatory design. Electron J Knowl Manag 17(1):49–65
21. Winschiers-Theophilus H, Bidwell NJ, Blake E (2012) Community consensus: design beyond participation. Des Issues 28(3):89
22. Kunrath K, Ramanujan D (2021) Fostering sustainable mindsets in engineering education. Des Soc 1:1597
23. French-Bravo M, Crow G (2015) Shared governance: the role of buy-in in bringing about change. Online J Issues Nurs 20(2):8

# Impact Assessment of Low-Carbon Electricity Transition by Applying Input–Output Analysis: A Case Study of the Impacts on Natural Gas Power Plants in Taiwan

**Ping-Hung Chen and Yuh-Ming Lee**

**Abstract** In recent years, the world has been affected by climate change, so countries have set net-zero carbon emission goals and reduced greenhouse gas emissions through carbon reduction policies. In 2022, Taiwan will amend the Greenhouse Gas Reduction and Management Act, adding the 2050 net zero carbon emission target into the law. Taiwan's greenhouse gas emissions caused by electricity production account for more than 60%. Therefore, Taiwan's energy department is actively planning a low-carbon energy path, including expecting to replace coal-fired power plants with natural gas power plants and to increase the proportion of natural gas power plants to 50% of power generation in 2025. It is expected that by 2050, natural gas power plants can be combined with hydrogen energy and CCUS technology to achieve the zero-emission goal of natural gas power plants, and with renewable energy power generation to achieve the goal of net zero carbon emissions in electricity.

For this reason, this study expects to evaluate the impact of natural gas power plants on the national economy under the net-zero path. Although there are previous literatures evaluating the cost-effectiveness of thermal power plants in Taiwan, the time is long, and Taiwan's low-carbon energy path and net-zero carbon have not been considered. Therefore, this study will use the input–output model to analyze the economic impact of changes in natural gas prices in the future under the transformation of electricity from coal to natural gas in Taiwan, so as to assist the Taiwan government to examine the suitability of the future low-carbon energy transition. This study can therefore contribute to foster the natural gas power plants in Taiwan with both characteristics of net-zero transition and sustainable innovation.

**Keywords** Input–output model · Low-carbon electricity transformation · Natural gas power plant

P.-H. Chen (✉) · Y.-M. Lee
National Taipei University, Taipei, Taiwan

 
S. Fukushige et al. (eds.), *EcoDesign for Circular Value Creation: Volume I*,
https://doi.org/10.1007/978-981-97-9068-5_20

# 1 Introduction

## *1.1 Low-Carbon Transition and Pathway to Net-Zero Emission in Taiwan*

In recent years, global climate change has received significant attention, and Taiwan, being an island nation, is particularly vulnerable to its impacts. Therefore, Taiwan is committed to promoting greenhouse gas reduction. In 2015, Taiwan established the "Greenhouse Gas Reduction and Management Act" (referred to as the GHG Act), which is the primary legislation for managing greenhouse gases in the country. Article 4 of the law sets the national greenhouse gas reduction target, specifying that by 2050, greenhouse gas emissions should be reduced to below 50% of the emissions in 2005. Additionally, Article 11 of the law stipulates the establishment of 5-year phased control targets. Based on these conditions, Taiwan has set the second-phase control target for 2025, planning to reduce greenhouse gas emissions by 10% compared to 2005 levels. By 2025, Taiwan's total greenhouse gas emissions will be reduced to 241.011 $MtCO_2e$.

However, there was a significant shift in Taiwan's 2050 reduction target during the years 2020–2021. As the world began to recognize the slow pace of carbon reduction, which could lead to an irreparable global crisis, many Asian countries started proposing pathways to achieve net-zero carbon emissions. For instance, mainland China pledged at the United Nations General Assembly in September 2020 to achieve net-zero carbon emissions by 2060 [7]. In October 2020, South Korea committed to reaching net-zero carbon emissions by 2050 [6]. In May 2021, Japan even incorporated the net-zero carbon emissions goal directly into the amendment of its "Act on Promotion of Global Warming Countermeasures" [4]. With competing countries declaring or enacting relevant legislation, Taiwan was compelled to expedite the planning of its net-zero pathway.

Therefore, in 2022, Taiwan, in response to the net-zero carbon emissions goal, amended the GHG Act to the "Climate Change Response Act" and revised Article 4 to establish a vision of achieving net-zero greenhouse gas emissions by 2050, affirming Taiwan's commitment to move towards the net-zero carbon emissions target.

In this situation, although the government considered it challenging to adjust the 2025 reduction target due to time constraints, adjustments were made to the 2030 target. Under the original NDC target declared by Taiwan in 2015, the goal for 2030 was a 20% reduction compared to 2005 levels. Under the net-zero pathway, this target was adjusted to a reduction of 24% $\pm$ 1% by 2030.

Of course, in order to achieve the net-zero carbon pathway, the government has also proposed key reduction strategies for each sector and annual key reduction policies, as shown in Tables 1 and 2.

To achieve the net-zero carbon emission goal, the low-carbon transformation of the power sector is absolutely essential in Taiwan. Looking at Taiwan's greenhouse gas emissions status (based on the 2022 National Greenhouse Gas Inventory), the

**Table 1** Carbon reduction policy planning for various sectors under Taiwan's net-zero carbon emission goal

| Sector | Net-zero pathway policy planning |
|---|---|
| Power | Continuous expansion of renewable energy, development of new energy technologies, energy storage, and grid upgrades |
| Industry | Enhancing energy efficiency, fuel conversion, circular economy, and innovative processes |
| Transport | Changing transportation modes, reducing transportation demand, electrification of vehicles |
| Building | Enhancing building envelope design, building energy efficiency, and appliance energy efficiency standards |
| Other technologies | Carbon capture and storage (CCS) technologies |

Source: Taiwan Environmental Protection Administration [12]. The General Explanation of Taiwan's 2050 Net Zero Emissions. Compiled based on this study

**Table 2** Key carbon reduction strategies for each year under the net-zero carbon emissions target

| Year | Net-zero pathway policy planning |
|---|---|
| 2025 | No new coal-fired power plants<br>35% of urban public buses are electric |
| 2030 | New public buildings are energy efficiency class 1 or nearly zero-emission<br>All urban public buses and official cars are electric<br>30% of car sales are electric<br>35% of scooter sales are electric<br>The manufacturing industry gradually replaces the equipment<br>15% of electricity consumption in the industry is green<br>100% LED lights in commercial buildings<br>60% of air conditioner are operated optimized<br>40 GW of wind and solar power capacity |
| 2035 | 60% of car sales are electric<br>70% of scooter sales are electric<br>Installation of smart meters reaches 100% electric |
| 2040 | 50% of existing buildings are upgraded to building energy efficiency class 1 or nearly zero- emission<br>100% of car sales are electric<br>100% of scooter sales are electric<br>Introduce low-carbon process into industrial demonstration. (Ironmaking using hydrogen energy,<br>Recycle $CO_2$ to synthesize hydrocarbon fuels<br>Installation of CCUS in coal and gas-fired power plants |
| 2050 | 100% of new buildings and over 85% of existing buildings are nearly zero-emission<br>Widely replace equipment<br>In Industry (80–90% in steel industry, 100% in textile industry)<br>Fully adopt low-carbon process<br>Renewable electricity accounts more than 60%<br>Installation of smart substations reaches 100% |

Source: Taiwan Environmental Protection Administration [12]. The General Explanation of Taiwan's 2050 Net Zero Emissions. Compiled based on this study

**Table 3** Proportion of power generation from different types of power plants nationwide

| | 2020 | 2021 | 2022 |
|---|---|---|---|
| Coal-fired | 44.97% | 44.27% | 42% |
| Oil-fired | 1.58% | 1.83% | 1.54% |
| Natural gas | 35.67% | 37.23% | 38.87% |
| Nuclear | 11.23% | 9.55% | 8.24% |
| Renewable energy | 5.42% | 6.02% | 8.28% |
| Pumped hydro | 1.13% | 1.09% | 1.06% |

Source: Ministry of Economic Affairs, Bureau of Energy. Statistics on the Power Generation Proportions of Various Types of Power Plants. Compiled based on this study

country's energy-related emissions in 2020 amounted to 259.385 $MtCO_2e$, accounting for 91% of the total emissions of 285.131 $MtCO_2e$. Among these, electricity generation accounted for 70% of the energy sector emissions, highlighting its significance.

Looking at the proportion of power generation from different types of power plants in Taiwan from 2020 to 2022 (in Table 3), the reduction in coal-fired power generation is relatively small, with only a decrease of about 3% over the three years. Nuclear power generation also decreased by approximately 3%. On the other hand, the combined increase in natural gas and renewable energy sources was only about 6%. As a result, the carbon emissions from power production were unable to significantly decrease during this period.

The main reasons for the changes in Taiwan's power generation mix are as follows:

1. Non-Nuclear Homeland: Due to government policy objectives, Taiwan has no plans to build new nuclear power plants and aims to decommission existing ones as scheduled. Although the European Union officially classifies nuclear energy as green energy in 2022, Taiwan still aims to achieve a non-nuclear homeland by 2025.
2. Limited Development of Renewable Energy: Taiwan is an island with limited land area but a large population, which poses challenges to the development of renewable energy. For example, in the case of solar photovoltaic (PV), large land areas are required for installation, and various regulations related to the environment and construction impose limitations, making it difficult to utilize all available land. The same applies to onshore wind power, which is subject to regulations requiring distance from residential areas and has faced increasing difficulties in recent years due to stricter environmental impact assessment regulations. While there are options such as offshore wind power and geothermal energy, the decreasing number of available options and saturation have led to a slowdown in development in recent years.
3. Decreased Return on Investment for Natural Gas Power Plants: Natural gas is an important low-carbon energy source in Taiwan. It occupies a small land area, enables large-scale power generation, and can replace coal and nuclear power as a baseload energy source. However, the increased costs of power generation

equipment and natural gas prices due to recent events such as the pandemic and conflicts like the Ukraine-Russia war have resulted in higher costs for gas-fired power generation. Additionally, Taiwan's electricity market has not fully liberalized, and electricity prices cannot freely adjust to reflect costs. As a result, the state-owned Taiwan Power Company has faced difficulties in providing appropriate electricity prices that align with cost-effectiveness when tendering for gas-fired power plants. This has hindered the smooth progress of natural gas power plant projects.

In summary, these factors have prevented a significant reduction in the share of coal-fired power generation. In order to achieve long-term emission reduction goals, Taiwan has set specific targets for the power generation sector. Prior to 2020, the government's energy department planned to achieve a coal-fired power generation share of 30%, a natural gas power generation share of 50%, and a renewable energy share of 20% by 2025. Comparing these targets with the current situation, there is still a need to increase natural gas power generation by approximately 11% and increase renewable energy by approximately 12%. If nuclear power is considered as a form of renewable energy and with the gradual development of offshore wind power, there is a possibility of achieving the 20% renewable energy target by 2025. Therefore, the key to achieving Taiwan's 2025 emission reduction goal lies in increasing the capacity of natural gas power plants to replace coal-fired power generation.

For Taiwan's largest power generation company, Taiwan Power Company (Taipower), there are plans for the addition and retirement of coal-fired and gas-fired power plants by 2025 (in Table 4).

**Table 4** Electricity generation capacity addition and retirement plan for various power plants by 2028

| Year | Net-zero pathway policy planning |
|---|---|
| 2021 | New construction: Natural gas power plants 510 MW |
| 2022 | New construction: Natural gas power plants 1124 MW<br>Decommissioning: Natural gas power plants 1100 MW |
| 2023 | New construction: Natural gas power plants 2424 MW<br>Decommissioning: Coal-fired power plants 1000 MW |
| 2024 | New construction: Natural gas power plants 4613 MW<br>Decommissioning: Coal-fired power plants 1200 MW, Natural gas power plants 772 MW |
| 2025 | New construction: Natural gas power plants 1300 MW<br>Decommissioning: Coal-fired power plants 1150 MW |
| 2026 | New construction: Natural gas power plants 1300 MW<br>Decommissioning: Coal-fired power plants 1650 MW |
| 2027 | New construction: Natural gas power plants 1300 MW |
| 2028 | New construction: Natural gas power plants 1300 MW |

Source: Ministry of Economic Affairs, Bureau of Energy. The planned addition and decommissioning of natural gas power plants and coal-fired power plants in Taiwan. Compiled based on this study

In terms of long-term reduction targets, achieving decarbonized electricity is a key focus as it aligns with the net-zero carbon goal. The planned proportions of decarbonized electricity by various types for the year 2050 are as follows: renewable energy accounts for 60% to 70%, hydrogen power generation accounts for 9% to 12%, thermal power generation with carbon capture, utilization, and storage (CCUS) accounts for 20% to 27%, and pumped hydroelectric power accounts for 1%.

Among them, the plan is to further increase the share of renewable energy to 60% to 70%, with a focus on offshore wind power and solar photovoltaic (PV) energy. Offshore wind power aims to achieve installation targets through block development and the development of floating-related technologies. Efforts are also made to support the local industry and operation and maintenance technologies in Taiwan, with the goal of reducing costs as much as possible. In the case of solar PV, the plan involves coordinating with land-use planning, allocating the required land area, and integrating with activities such as aquaculture and agriculture to enhance the value of land multiple use. Additionally, efforts will be made to improve unit efficiency, integrate energy storage and smart grids, and enhance the overall resilience of the power grid to achieve the targets.

Although Taiwan is on a path towards net-zero carbon emissions, it has not completely abandoned thermal power generation as there is still a need for power plants to provide baseload power and maintain power stability and security. However, to meet decarbonization goals, thermal power plants need to develop together with Carbon Capture, Utilization, and Storage (CCUS) technologies, or transition fuels from natural gas to hydrogen (green hydrogen) as much as possible, aiming for a share of 29% to 39%. In terms of technology development, the high cost of carbon capture and the lack of suitable storage sites need to be addressed for CCUS. For hydrogen utilization, a stable hydrogen source is required to expand carbon utilization. In terms of hydrogen energy, based on current technologies, major turbine suppliers like Siemens have stated that their latest turbines are capable of running on 100% hydrogen [11]. The main challenges lie in hydrogen production and storage. To reduce hydrogen production costs, the cost of hydrogen decomposition and renewable energy generation need to be reduced. This allows for the use of excess renewable energy for hydrogen production when there is surplus electricity. As for surplus hydrogen, suitable storage methods and space need to be established, taking into consideration safety considerations although hydrogen storage is similar to natural gas storage.

## 2 Input–Output Model for Low-Carbon Electricity

### 2.1 Taiwan's Input–Output Table

Input–output analysis, originated from the literature by Leontief in 1936 [1], is a method used to analyze the relationship between industries within an economy. It primarily examines the various inputs required by an industry in its production process and identifies which industries utilize the outputs of a particular industry. This analysis is commonly used by countries for policy analysis.

There are several key assumptions in input–output analysis:

1. Single-product assumption: Each industry produces only one type of product.
2. Fixed-coefficient assumption: The proportions of inputs used in production are fixed, indicating a constant level of technology that does not improve over time or with increased production.
3. Fixed-proportions assumption: The input factors used in production maintain a fixed proportion, meaning the ratios of required inputs remain constant.

Looking at a simplified input–output table (in Table 5), the output side includes Z11 to Zn1 as intermediate inputs, and the sum of labor compensation, operating surplus, depreciation of fixed capital, and net indirect taxes represents the initial inputs (also known as GDP). The total inputs (referred to as output or value-added) are the sum of intermediate inputs and initial inputs. On the input side, Z11 to Z1n represent intermediate requirements, and household consumption (C) plus gross fixed capital formation and inventory changes (I) plus government consumption (G) plus exports (X) represent final demand. Subtracting imports (M) from the total demand gives the total output (also known as GDP).

The input–output table, also known as the Interindustry table, is used for conducting input–output analysis. In the case of Taiwan, to perform input–output analysis for Taiwan's economy, it is necessary to refer to the Interindustry tables compiled by Taiwan's statistical agency. The Interindustry tables for Taiwan are compiled once every five years, and the latest available table is the 2016 Interindustry table (completed in 2020). The tables are categorized into three classification systems: 63 sectors, 164 sectors, and 487 sectors. These tables include various forms mentioned in Table 6.

This study aims to analyze the potential economic impacts on Taiwan when transitioning from coal-fired power generation to natural gas. In terms of sector classification in Taiwan, the demand sector to be analyzed is the "Electricity and Steam" industry. This category includes the electricity industry engaged in professional power generation, supply, and distribution activities, as well as the cogeneration industry involved in power and steam production, steam and air conditioning supply, and hot water (including hot springs) and chilled water supply.

Regarding the supply sector, it is necessary to separate the sectors related to natural gas and coal. If we consider the 63 sectors, there is only the "Mineral Products" sector that represents fuel inputs, but it does not distinguish between natural gas and coal. Therefore, we need to drill down to the 164 sectors. Within the 164 sectors, the "Mineral Products" sector is further divided into "Crude Oil and Natural Gas Extraction" and "Sand, Stone, and Other Mineral Products." Although they are not directly labeled as natural gas and coal, the electricity and steam inputs in the "Crude Oil and Natural Gas Extraction" category consist of 100% natural gas, and 99.99% of the "Sand, Stone, and Other Mineral Products" category belongs to coal. Thus, the 164 sectors will be used as the classification basis.

**Table 5** The simplified input–output analysis table

| Output | | Intermediate demand | Final demand | | | | | Less Input | Gross domestic product |
|---|---|---|---|---|---|---|---|---|---|
| Throw in | | Industry | Household consumption | Fixed capital formation | Gross domestic product | Inventory movement | Export | | |
| Intermediate input | Industry | $Z_{11} \sim Z_{1n}$ $\widetilde{Z}_{n1} \sim Z_{nn}$ | | | | | | | |
| Primary investment | Remuneration for work operating surplus depreciation of fixed capital net indirect tax | | | | | | | | |
| Total input | | | | | | | | | |

Source: Taiwan Accounting and General Office, the industrial association table. Compiled based on this study

**Table 6** Various interindustry tables announced by the Taiwan Directorate General of Budget

| Form | Types of industry input–output tables |
|---|---|
| Table 1 | Producer price transaction table |
| Table 2 | Producer price input–output coefficient table |
| Table 3 | Degree of industry interrelation table $(I-A)^{-1}$ |
| Table 4 | Degree of industry interrelation table $[I-(I-\widehat{M})A]^{-1}$ |
| Table 5 | Purchaser price transaction table |
| Table 6 | Domestic product transaction table |
| Table 7 | Domestic product input–output coefficient table |
| Table 8 | Degree of domestic interrelation table |
| Table 9 | Imported goods transaction table (C.I.F. valued) |
| Table 10 | Imported goods input–output coefficient table (C.I.F. valued) |

Source: Taiwan Accounting and General Office, the industrial association table and energy balance table

Regarding the selection of interindustry tables, considering that Taiwan mostly imports fuel internationally, it is more suitable to use Table 1, the producer price transaction table, for evaluation, rather than Table 6, the domestic product transaction table.

Although this study is conducted using the 164 sectors for research and evaluation, due to limitations in the research report, Table 7 only displays the main supply and demand sectors that will be used. Other industry sectors, such as the energy sector, manufacturing sector, construction sector, and service sector, will be presented separately.

## 2.2 *Hybrid Input–Output Table for Electricity*

This study focuses on the analysis of low-carbon electricity and extends it to a hybrid input–output analysis of quantity and quality. Traditional input–output analysis involves matrix calculations using monetary units as the unified measurement. In quantity and quality hybrid input–output analysis, both inputs and outputs can be measured in terms of quantity and quality, allowing for the assessment of their resulting changes. This approach enables the evaluation of the effects of different factors in the production process, not only in terms of the economy but also in resource allocation and industrial development.

Furthermore, this study specifically applies the hybrid input-output analysis to the electricity sector. The use of electricity hybrid input–output analysis was first explored by Ang, B. W. et al to examine the relationship between energy use and environmental impacts [3]. Pérez-López et al. also employed electricity hybrid input–output analysis to assess the energy footprint of electricity production and analyze the influence of different energy sources on the electricity production process [10].

**Table 7** This study establishes an input–output analysis diagram

| | Energy sector | Electricity and Steam | Industrial sector | Residential and commercial sector | Agricultural sector | Final demand | Input | Gross domestic product |
|---|---|---|---|---|---|---|---|---|
| Energy products | | | | | | | | |
| Crude oil and natural oil gas mineral | | | | | | | | |
| Sand, stone and other mineral products | | | | | | | | |
| Industrial product | | | | | | | | |
| Construction product | | | | | | | | |
| Service products | | | | | | | | |
| Original input | | | | | | | | |
| Total input | | | | | | | | |

Source: Taiwan Accounting and General Office, the industrial association and energy balance table, as modified by this study.

To conduct electricity hybrid input–output analysis, an energy-related extension table is required to be merged with the existing industry input–output table, under the same assumptions as the industry input–output table (e.g., the assumption of a single product). After merging, the industry input-output table, originally used for analyzing macroeconomic changes, can be expanded to evaluate the mutual impact between energy input levels and the economy.

For the energy extension table, this study plans to use Taiwan's energy balance table compiled by the Energy Bureau as the merged energy form. The Taiwan energy balance table is produced annually and covers various energy items, including coal and coal products, crude oil and petroleum products, natural gas, biomass, and waste, as well as nuclear and renewable energy. The covered industries include the energy sector, industrial sector, transportation sector, agriculture sector, service sector, and residential sector, which can be corresponded to the 63 sectors in the industry input–output table. Therefore, the energy balance table will primarily be merged into the industry input-output table with 63 sectors. In terms of units, there are two options: original units and heat value units. To standardize the energy units, the energy balance table using heat value units will be used for merging. Additionally, considering that this study focuses on low-carbon energy (specifically, gas power plants), the main energy products imported will be coal and coal products, natural gas, and electricity. The integrated electricity hybrid input–output analysis is presented in the following Table 8.

To satisfy the law of energy conservation, the following equation needs to be fulfilled:

$$a_{mn}X_n = \left(\sum_{i=1}^{n} a_{mi}Z_{in}\right) + f^*_{mn} \tag{1}$$

In Eq. (1), the variables are explained as follows:

$a_{mn}$: The amount of $m$-type energy input (measured in heat units) required by sector $n$ to produce 1 unit of currency (1 monetary unit).
$X_n$: The total output of sector $n$.
$Z_{in}$: The total intermediate input value of product $i$ from sector $i$.
$f^*_{mn}$: The direct consumption of $m$-type energy input by sector $n$ during production.

Since $Z = AX$, where $A$ is the intermediate input matrix, the equation can be rewritten as:

$$a\widehat{X} = aA\widehat{X} + \widehat{F}$$

After rearranging, we obtain:

$$a\widehat{X} - aA\widehat{X} = \widehat{F} \tag{2}$$

**Table 8** This study establishes a diagram for electricity hybrid input–output analysis

| | Energy sector | Electricity and Steam | Industrial sector | Residential and commercial sector | Agricultural sector | Final demand | Input | Gross domestic product |
|---|---|---|---|---|---|---|---|---|
| Energy products | | | | | | | | |
| Crude oil and natural oil gas mineral | | | | | | | | |
| Sand, stone and other mineral products | | | | | | | | |
| Industrial product | | | | | | | | |
| Construction product | | | | | | | | |
| Service products | | | | | | | | |
| Original input | | | | | | | | |
| Total input | | | | | | | | |
| Coal and Coal Products | | | | | | | | |
| Natural gas | | | | | | | | |
| Electricity | | | | | | | | |

Source: Taiwan Accounting and General Office, the industrial association and energy balance table, as modified by this study.

$$a(I-A)\widehat{X}=\widehat{F} \tag{3}$$

$$a(I-A)=\widehat{F}*\left(\widehat{X}\right)^{-1} \tag{4}$$

$$a=\widehat{F}*\left(\widehat{X}\right)^{-1}*(I-A)^{-1} \tag{5}$$

By using Eqs. (2) to (5), we can determine the energy input for each sector in terms of monetary units of production.

## *2.3 Data Collection for Low-Carbon Electricity*

This study will focus on two major analyses. First, an input-output analysis will be conducted to assess the potential economic impact on Taiwan if the fuel source for electricity production shifts from coal to natural gas, with a target of coal-fired power generation accounting for 30% and natural gas accounting for 50% of the total. The changes in intermediate inputs of coal and natural gas in electricity and steam production may have implications for Taiwan's economy.

To achieve this, it is necessary to obtain information such as the annual prices of coal and natural gas, the electricity generation per unit of coal and natural gas, and the electricity generation composition for the respective year from the industry input-output table. Based on statistical data from Taiwan Power Company and the Bureau of Energy, the annual prices of coal and natural gas can be obtained. The electricity generation per unit of coal and natural gas can be estimated using historical data from the Bureau of Energy's energy balance tables, and the electricity generation composition for the specific year can also be obtained from the Bureau of Energy's statistics.

Second, through the process of electricity hybrid input-output analysis and the derived equations mentioned above, an estimation of the coal and natural gas input for electricity and steam industries in terms of monetary units of production will be conducted. This assessment aims to evaluate the differences in power generation efficiency between coal-fired and natural gas power plants from an overall economic perspective.

# 3 Multi-criteria Decision Making

## *3.1 Introduction*

The concept of multi-criteria decision making (MCDM) can be traced back to the early twentieth century when it was first introduced by Koopmans. The fundamental idea behind MCDM is that in the process of management and decision-making, one should not solely consider a single influence. Instead, it is essential to take into account the impacts of various factors such as different objectives, criteria, options, weights, and methods. By quantifying these influences and conducting a balanced comparison and analysis, there is a greater chance of selecting the optimal solution.

MCDM can be applied to various fields, including engineering, environmental, and social issues. To find the best solution to a problem, various methodologies have been developed in succession. The following is a brief description:

1. Weighted Sum Method (WSM): First, identify the criteria necessary for analyzing the decision, then allocate weights to each criterion based on their importance and preferences. Calculate the weighted sum for each option, with the highest-scored decision being chosen as the best solution [13].
2. Technique for Order of Preference by Similarity to Ideal Solution (TOPSIS): Begin by selecting the most ideal decision approach. Then, compare all decisions to the ideal approach, calculating their distances from the ideal solution. Choose the decision with the highest similarity as the best solution [8].
3. Analytic Hierarchy Process (AHP): Similar to WSM but suitable for cases with multiple hierarchical criteria. AHP assigns weights to criteria through pairwise comparisons, providing a clear hierarchy of relationships between criteria and elements. Calculate the weighted sum for each option, with the highest-scored decision being selected as the best solution [2].
4. VlseKriterijumska Kompromisno Rangiranje i Odabir (VIKOR): An enhancement of TOPSIS that maximizes the Majority Rule while minimizing Individual Regret. It ranks and selects the best solution for various strategies after comparison [9].

Additionally, there are other methods that have evolved from MCDM, such as the Elimination and Choice Expressing Reality (ELECTRE) method and the Preference Ranking Organization METHod for Enrichment of Evaluations (PROMETHEE) method.

## 3.2 Research Assumptions and Data Collection

This study will reference the aforementioned research methods to analyze the demand for gas-fired power plants in Taiwan. Given that gas-fired power plants have multifaceted impacts, including environmental, social, and corporate governance factors when viewed from a company's ESG perspective, a comprehensive analysis can be conducted. To further specify, at the national level, various aspects such as economic impact, environmental effects, public health, safety, and landscape impact can be considered. To simplify the analysis, this study will focus on economic impact and air pollution effects as the primary areas of analysis.

Regarding the economic impact, the study will use the simulation results from IO analysis to estimate the economic damage to Taiwan if gas-fired power plants were to account for 50% of power generation.

**Table 9** Air pollution emissions and air pollution fees for each power plant

| | Air pollution emission intensity (kg/kwh) | | |
|---|---|---|---|
| Plant/fee (NTD/kg) | TSP | SOx | NOx |
| Coal power plant | 0.0447 | 0.3417 | 0.4155 |
| Natural gas power plant | 0.0205 | 0.0017 | 0.3446 |
| Air pollution control fee | 60 | 15 | 16 |

Source: Official website of Taiwan Environmental Protection Administration
NTD stands for New Taiwan Dollars, 1 NTD ≒ 0.033 USD

In terms of air pollution effects, the study assumes a transition from coal-fired to gas-fired power plants for Taiwan's baseload electricity generation, resulting in benefits from reduced air pollution and greenhouse gas emissions. To ensure a standardized baseline for comparison, this study plans to monetize the benefits and compare them with the economic impact.

For the air pollution component, this study will collect data on air pollutant emissions intensity from coal-fired and gas-fired power plants in Taiwan in 2016, as well as plan to monetize the cost of air pollution using pollution control fees. The summarized data are presented in Table 9.

In terms of greenhouse gas emissions, in 2016, coal-fired power plants had an electricity emissions factor of approximately 0.925 kilograms of $CO_{2e}$ per kilowatt-hour (kWh), while gas-fired power plants had a factor of 0.415 kilograms of $CO_{2e}$ per kWh. As for the carbon pricing, this study adopts the latest European Union Emissions Trading System (EU ETS) market price as of August 2023, which is approximately 86 euros per metric ton of carbon dioxide (equivalent to around 3000 New Taiwan Dollars per metric ton). Using this information, the study will calculate the benefits arising from the impact of gas-fired power generation on air pollution.

## 4 Results and Discussion

The study first utilizes the electricity hybrid input-output analysis model to evaluate the coal and natural gas usage (measured in terms of heat value) required to produce 1 unit of output value for each sector. The results are shown in Table 10.

From the table above, it can be observed that the coal usage in various sectors in Taiwan is higher than that of natural gas. In the residential and agricultural sectors, where natural gas is more commonly used (such as natural gas for households. However, in the industrial sector, coal usage is significantly higher than that of natural gas. This is primarily due to the fact that Taiwan's industrial sector relies heavily on coal or electricity as the primary sources of energy, with natural gas playing a relatively minor role.

Regarding the focal energy sectors of this study, to generate 1 unit of output value, it requires 1.130 units of coal or 5.956 units of natural gas. The usage of natural gas is approximately 5.27 times higher than that of coal, indicating that the

**Table 10** The heat values of coal and natural gas required for each sector per unit of output

| Sector (measured in terms of heat value) | The amount of coal required to produce 1 unit of output value | The amount of natural gas required to produce 1 unit of output value |
|---|---|---|
| Energy sector | 5.956 | 1.130 |
| Industrial sector | 36.881 | 10.746 |
| Residential and commercial | 4.752 | 2.910 |
| Agricultural sector | 1.255 | 0.398 |

Source: Using Industrial Association and Energy Balance Table From the Taiwan Accounting and General Office, along with the modified Table 8 and Equations (5), this table was estimated.

cost of gas-fired power generation is significantly higher than that of coal-fired power generation.

Therefore, through input–output analysis, this study aims to estimate the potential macroeconomic impacts resulting from changes in intermediate inputs of natural gas (including crude oil and expanded natural gas) and coal (sand, stone, and other expanded products) within the electricity and steam industry.

The proportional changes in intermediate inputs of natural gas and coal will be assumed based on the target of 50% gas-fired power generation and 30% coal-fired power generation in Taiwan by 2025. The calculations will be made in accordance with the intermediate input levels of natural gas and coal from the 2016 industry input–output table, along with the electricity generation proportions of natural gas and coal in that year.

The estimated results of the input–output analysis are presented in the following Table 11.

Under this change, assuming the total electricity generation remains constant and only the intermediate inputs of natural gas and coal in the power and steam industry are altered, the main impact will be on the energy sector. This will result in a reduction in output of approximately 78.4 billion NTD, representing a negative impact of nearly 5% on output. Additionally, the industrial sector, which is heavily reliant on electricity, will experience a decrease in output of approximately 12.1 billion NTD, corresponding to a 0.07% output impact. The impact on the residential and commercial sectors as well as the agricultural sector will be less significant. From a national perspective, the transition in power generation fuels will lead to an overall economic impact of approximately 0.27% in terms of GDP. If we estimate based on the 2016 GDP, it would result in approximately 47.4 billion Taiwan dollars in losses.

In addition to economic losses, this study also estimates the air pollution reduction benefits from transitioning from coal-fired power generation to natural gas. The estimation method is as follows:

Firstly, we need to estimate the increase in natural gas consumption for power generation, using 2016 as the baseline year for comparison.

$$E_1 = E_2 - E_3 \times 30\%$$

**Table 11** The impact on output of each sector under changes in natural gas and coal inputs in the power industry

Intermediate gas inputs will increase by 159%
Intermediate coal input will be reduced by 65%
Unit: million NTD

| Sector | Initial output | Revised output | Change in output | Growth rate or change rate |
|---|---|---|---|---|
| Energy sector | 1,590,782 | 1,512,341 | −78,441 | −4.93% |
| Industrial sector | 18,176,766 | 18,164,712 | −12,054 | −0.07% |
| Residential and commercial sectors | 15,688,375 | 15,680,143 | −8232 | −0.05% |
| Agricultural sector | 631,257 | 631,174 | −83 | −0.01% |
| Total amount | 36,087,180 | 35,988,370 | −98,810 | −0.27% |

Source: Estimated by this study

**Table 12** Estimation results of air pollution and greenhouse gas reduction benefits

| Item | Formula: $E_1$ (Coal Power Plant Emission Intensity − Natural Gas Power Plant Emission Intensity) Control/Trade Fee) | Benefit (billion NTD) |
|---|---|---|
| TSP | $E_1$ (0.0447–0.0205 kg/kwh) 60NTD/kg | 0.6 |
| SOx | $E_1$ (0.3717–0.00017 g/kWh) 15 NTD/kg | 2.14 |
| NOx | $E_1$ (0.4155–0.3446 g/kWh) 16 NTD/kg | 0.47 |
| GHG | $E_1$ (0.925–0.415 $kgCO_{2e}$/kWh) 3000 NTD/$tCO_{2e}$ | 643 |
| Sum | | 646 |

Source: Estimated by this study

where:

$E_1$ = Increase in natural gas consumption for power generation in 2016.

$E_2$ = Actual coal-fired power generation in 2016 (which was 1212 billion kWh).

$E_3$ = Total national power generation in 2016 (which was 2640 billion kWh).

From the calculation, $E_1$ is 420 billion kWh, which would result in a reduction in air pollution and greenhouse gas emissions when transitioning from coal to gas. The estimated total benefit of this reduction can be found in the Table 12, and it amounts to approximately 64.6 billion in losses; it seems to offset the economic losses. However, it is important to note that carbon trading prices have a significant impact on the analysis results. If we use the carbon trading price in mainland China in August 2023, which is 69 RMB per ton (equivalent to about 302 NTD per ton), the reduction benefits will be significantly lower than the economic impact.

## 5 Conclusion

In order to achieve its self-established decarbonization pathway, Taiwan has been actively promoting emission reduction measures in various sectors. In the power generation industry, as part of the non-nuclear homeland vision, there is a need to reduce nuclear and coal-fired power generation while maintaining power security and stability. Natural gas power generation is the only viable option to compensate for the reduction in nuclear and coal-fired power generation. Taiwan aims to achieve a target of 30% coal-fired power generation, 50% natural gas power generation, and 20% renewable energy by 2025.

However, the simulation results in this study reveal that in the energy sector, the output generated by coal and coal products in 2016 was significantly higher compared to natural gas. This is not only due to the high coal-fired power generation volume in that year but also indicates that coal-fired power generation costs are noticeably lower than natural gas. Additionally, coal-fired power generation is more efficient than gas-fired power generation, resulting in these outcomes.

Therefore, in this study, it is assumed that if the intermediate inputs between industries remain unchanged and only the inputs of coal and natural gas in the power generation industry are altered according to the 2025 power generation targets (30% coal-fired, 50% natural gas), there will be a significant impact on the energy sector in terms of output (−4.93%). This will also lead to an overall economic impact of −0.27% on the national level.

This impact is significant for Taiwan, considering that its economic growth rate in 2022 is approximately 2.35%. A single energy policy alone causing a −0.27% impact on energy indicates the severity of the situation. Furthermore, the present study only calculates the impact of changing power generation fuels. However, transitioning from gas-fired to coal-fired power generation involves decommissioning coal plants and constructing new gas plants, which will have a substantial economic impact on Taiwan.

If we consider MCDM, it becomes evident that gas-fired power plants can indeed bring about benefits in terms of air pollution and greenhouse gas reduction. However, these benefits are contingent upon the carbon price. If global carbon pricing gradually increases in the future, there is a possibility that these reduction benefits will offset the economic impact. However, if the carbon pricing does not meet expectations, it may not lead to significant reduction effects.

Of course, economic impact and air pollution benefits are not the only factors to consider when transitioning to gas-fired power plants. If we view Taiwan as a large enterprise from an ESG perspective, in addition to environmental and socio-economic impacts, corporate influence (including corporate transformation) is also crucial. The construction of gas-fired power plants can serve not only for gas-based electricity generation but also for the recently developed gas turbine power generation. According to manufacturers such as Siemens, these power generation units can also utilize pure hydrogen for electricity generation. Therefore, in the context of global net-zero goals, the development of gas-fired power plants may offer even greater benefits.

However, regardless of whether gas-fired power plants will need to be complemented with Carbon Capture, Utilization, and Storage (CCUS) or hydrogen technologies in the future, both of these technologies are currently associated with high costs, which may further contribute to economic impacts.

If CCUS and hydrogen technologies cannot be developed in the future, and considering that natural gas is a low-carbon energy source rather than zero-emission like renewable energy, Taiwan may face challenges in achieving its net-zero carbon emissions target. This could result in gas-fired power plants becoming transitional energy sources, which would not be favorable for Taiwan's energy development.

In summary, this study suggests that Taiwan needs to take a more rigorous approach to energy policy considerations. In addition to evaluating the proportion of different fuel sources in the decarbonization pathway, the economic impact should also be simultaneously taken into account. This will ensure that future energy policies are both feasible and beneficial.

## References

1. Andronov V (1936) The interdependence of industries: a study of the input-output system. Econometrica 4(1):53–64
2. Anderson DR, Sweeney DJ, Williams TA, Camm JD, Cochran JJ (2016) An introduction to management science: Quantitative approaches to decision making 14th, 14 edn. GENGAGE Learning
3. Ang BW, Zhang FQ (2000) A survey of index decomposition analysis in energy and environmental studies. Energy 25(12):1149–1176
4. Cabinet Secretariat, Government of Japan (2020) Japan's Long-Term Strategy under the Paris Agreement
5. Leontief W (1941) The structure of the American economy, 1919–1929: an empirical application of equilibrium analysis. Harvard University Press
6. Ministry of Environment (MOE) (2021) South Korea's Long-Term Strategy for Carbon Neutrality 2050
7. National Development and Reform Commission (NDRC) (2020) China's Nationally Determined Contribution (NDC) and Long-term Strategy for Low Greenhouse Gas Emissions Development
8. Opricovic S, Tzeng GH (2004) Compromise solution by MCDM methods: a comparative analysis of VIKOR and TOPSIS. Eur J Oper Res 156:445–455
9. Opricovic S, Tzeng GH (2007) Extended VIKOR method in comparison with outranking methods. Eur J Oper Res 178:514–529
10. Pérez-López G, Lacal-Arántegui R, Aranda-Usón A (2013) An input-output approach for the energy footprint analysis of electricity production considering technology mixes. Energy Policy 62:500–509
11. SIEMENS (2020) Hydrogen power with Siemens gas turbines Reliable carbon-free power with flexibility
12. Taiwan Environmental Protection Administration (2022) The General Explanation of Taiwan's 2050 Net Zero Emissions
13. Zhang K-S, Han Z-H, Li W-J, Song W-P (2008) Bilevel adaptive weighted sum method for multidisciplinary multi-objective optimization. AIAA J 46:2611–2622

# Part IV
# Circular Production and Management

# Circular Strategies in Southeast Asia: Cambodian Customer Perception of Remanufactured Motorcycle Parts

**R. Casper, J. Vogt Duberg, S. Seang, J. Kurilova-Palisaitiene, E. Sundin, M. Lindahl, and R. Banomyong**

**Abstract** Originating from years of economic instability, a recent positive development of the Cambodian economy is leading to a rising number of registered vehicles, mainly motorcycles. With this rise, the demand for spare parts is also increasing significantly. Differing from other countries, where a share of the spare parts is covered by remanufactured parts, there is no remanufacturing alternative to new parts available in Cambodia. As such, little is known about the Cambodian remanufacturing market. Therefore, this research aims to explore the awareness of motorcycle users in Cambodia of remanufactured parts as an alternative to new spare parts. To fulfil the aim, data from 491 Cambodian motorcycle users were collected and analyzed through structured interviews and online surveys. The main findings conclude that a vast majority of the Cambodian motorcycle users are aware of the remanufacturing alternatives to new spare parts, but in many cases the perception is negative, and the new parts are seen as superior to remanufactured parts.

**Keywords** Circular economy · End-of-use (EoU) · End-of-life (EoL) · Closed-loop supply chain (CLSC) · Sustainability · Automotive · Spare parts · Motorcycle · Remanufacturing · Cambodia

R. Casper (✉)
Chiang Mai University, Faculty of Engineering, Chiang Mai, Thailand

J. Vogt Duberg · J. Kurilova-Palisaitiene · E. Sundin · M. Lindahl
Linköping University, Linköping, Sweden

S. Seang
National University of Management, Phnom Penh, Cambodia

R. Banomyong
Thammasat University, Center of Excellence in Connectivity, Bangkok, Thailand



S. Fukushige et al. (eds.), *EcoDesign for Circular Value Creation: Volume I*,
https://doi.org/10.1007/978-981-97-9068-5_21

# 1 Introduction

Cambodia, a Southeast Asian country with around 17 million inhabitants, is after many years of political and economic instability benefiting from growing wealth: An increase in the national GDP in the last decade and a further forecasted growth of 8% per year [1], raises the purchasing power of the Cambodian people. One effect of this is the significant rise in the number of newly registered motorcycles. Together with tricycles (so-called tuk-tuks), motorcycles represent 85% of all registered vehicles in Cambodia [2, 3]. The latest numbers indicate that there are currently a total of over 4.9 million motorcycles registered in Cambodia [4] leading to high demand for spare parts to satisfy both maintenance and repair activities. As a consequence of the growing number of registered vehicles, this spare part demand is expected to increase further.

In certain regions of the world, especially in North America, Europe and some regions in Asia, a relevant percentage of spare part demands are covered by remanufactured parts. However, in Cambodia, such strategies have as of today no relevance, and it is unknown why this is the case. Therefore, it is of interest to target the Cambodian market for spare parts to identify the attractiveness of remanufactured alternatives. As such, this research aims to explore the awareness of motorcycle users in Cambodia of remanufactured parts as an alternative to new spare parts. Moreover, it investigates the perception of this group of people toward such parts. The aim is achieved by conducting an analysis based on data from structured interviews and online-based surveys.

# 2 Remanufacturing

## *2.1 Background to Remanufacturing*

Through remanufacturing, more environmentally friendly products compared to equivalent new products can be delivered to the market [5]. This is achieved by restoring used, worn-out, or broken products and bringing them back to an *as-good-as-new* condition [6]. As such, the embedded value of the products is preserved which is giving them environmental and economic benefits [7]. For decades these benefits have been utilized by industries by offering remanufactured products to the market [6, 8]. For certain industries, profitableness is the main motivation to perform remanufacturing, while for others, it is the resource-saving capabilities [9]. The latter is a result of the capability to extend the phase of product usage which enables the retention of the material and energy value embedded in them [7]. This leads to lower consumption of resources, which in turn leads to positive environmental contributions [5, 10].

Additionally, remanufacturing can provide economic advantages for the involved parties. For example, it can provide competitive advantages and new product opportunities for companies [11]. However, this is only possible if the remanufacturing system is well designed. To enable remanufacturing systems to operate more efficiently and to facilitate its activities, the realization of circular business models, streamlining of reprocessing operations, and core acquisition policies can be accomplished [12–15]. These activities have been analyzed from a wide perspective ranging from a market-orientated analysis focusing on opportunities for remanufactured parts to a product-orientated analysis identifying measurements to increase the production efficiency [16–18]. Given that remanufacturing is performed efficiently, it has the potential to gain a significant market share [9], and the ability to disconnect the increase of an economy from its negative consequences on the environment [19]. Thereby remanufacturing has the potential to contribute toward sustainable development [20].

However, the success of remanufacturing and its industrial importance relies on the attractiveness of remanufactured products on the market. If the products cannot be sold, there is no incentive to provide remanufactured products. Hence, the attractiveness of offering remanufactured products depends on the customer perception and willingness to pay for them as well as their market viability.

## 2.2 *Customer Perception of Remanufactured Parts*

At the sales moment of a product, the customer has a set of criteria that must be fulfilled, otherwise, the product will not be chosen. It is based on the principle that if the perceived value of a product meets or exceeds the value required for its acquisition, the customer considers the offering and purchases the product [21]. Multiple factors influence how a potential customer perceives the value of an offered product, including its environmental impact, quality specification, performance, durability, and price level [22, 23].

Remanufactured products tend to be compared with new counterparts, meaning that their perceived value must exceed the value of new products. Therefore, there are two alternatives for successful market penetration: either (1) the attributes of the remanufactured products must be attractive, for example, by a new design [24], its environmental contribution [25, 26], additional service offerings [27], or (2) lower pricing [28].

## 2.3 *Motorcycle Remanufacturing in Southeast Asia*

The multifaceted view of the customer perception is investigated in this paper to gain a better understanding of remanufactured motorcycle spare parts in the Cambodian market. Today, the traffic in Cambodia is dominated by motorcycles as the primary mode of transport. Despite this, the market for remanufactured motorcycle spare parts has yet to reach a broad customer base [29]. A broad availability of cores [30] and a large market potential [11] are usually beneficial from a remanufacturing perspective. Therefore, it is likely another factor that has hindered a prosperous Cambodian remanufacturing market. By learning how the market perceives remanufactured parts, remanufacturing offerings could be designed for the Cambodian market, and remanufacturing promotion could target potential hindrances to overcome them [31]. A comparable approach has been performed in the Vietnamese motorcycle market. There is a high concern about the provided quality of remanufactured products as well as uncertainty on how parts would perform in comparison to bootleg parts was present [32]. Such knowledge is valuable when shaping and adapting the marketing strategy regarding these perceptions.

## 3 Research Methodology

### 3.1 Primary Data Collected Using Surveys

As highlighted in the previous section, remanufacturing is a mature industrial concept with market relevance mainly in North America, Europe, and parts of Asia. As demonstrated, today remanufacturing and specifically remanufacturing of motorcycle parts have no market significance in Cambodia. Consequently, no data on remanufactured motorcycle parts in Cambodia is available. In order to fulfil the research aim, primary data has been collected using surveys.

The survey method enables generation of data to explore, for example, the perception, opinion, and knowledge of a defined group in respect of certain phenomena using a specifically designed set of questions gathered and structured in a survey [33]. Surveys can be executed verbally, using paper surveys or online. Advantages are especially their inexpensiveness, flexibility, and fastness [34]. To receive good data and an outcome of high quality, the survey design, described in the following section, is of great importance [35].

### 3.2 Survey Design

Ballou and Markesich [36] stressed that the quality of surveys is strongly dependent on a sufficient selection of participants, an appropriate wording of questions and an adequate qualification of the researchers. To satisfy these requirements, a focus within the design process of this research was set on those features.

The survey was created both in English and Khmer language by experienced researchers using a proven and tested online survey tool. The researchers had a deep knowledge of Khmer language and culture.

The wording of the survey questions was framed in regard to remanufacturing barely being known in Cambodia. Additional explanations on potentially unknown terms were provided within the survey. In relation to this, the participants were asked whether they know the term "remanufacturing." Only participants affirming this received additional and more remanufacturing-specific questions. Participants not having knowledge about remanufacturing were led automatically to a later point of the survey skipping the detailed questions on remanufacturing.

After finalizing the survey design, the survey was validated with ten potential survey participants to check its functionality and understandability. Their feedback was used to improve the survey prior to initializing the field phase described in the following section.

The survey questions are presented in Appendix A.

### 3.3 Sample Size Calculation

This research bases its findings on a selected group of people from the full population. This group is aimed to reflect the demographic characteristics in regard to gender, age, education, and income. To enable a generalization of the findings from the sample to the population, an adequate sample size must be calculated. In this research, the formula designed by Taro Yamane [37] was used to estimate the required sample size; see Eq. (1).

$$\mathrm{n} = \frac{N}{1 + Ne^2} \tag{1}$$

Since there is no data available on the number of Cambodian people using a motorcycle this research study relied on the total population $N$ of registered motorcycles, 4,865,396, in Cambodia [4]. However, as it is infeasible to consider the entire population, a sampling error $e$ must be included representing an acceptable level of inaccuracy [38, 39]. Using a sampling error $e$ of 5% implies a lower bound of 400 valid survey answers required.

### 3.4 Field Phase

In this research study, motorcycles are defined as all two-wheeled vehicles using a combustion engine or electric motor for propulsion. The peer group for the structured interviews and online surveys were all motorcycle users in Cambodia with a minimum age of 15 years.

Participants were approached in two different forms: *Field survey approach:* Participants were approached on the street and interviewed by the researchers using the set of questions from the survey. The participants' answers were immediately recorded by entering them into the online survey using a mobile device.

*Online survey approach:* The participants have been approached by sharing a link to the online survey. The link was shared by URL or QR code on different websites and in various newsgroups, for example from local universities or organizations.

### 3.5 Ethics of Research with Human Participants

This exploration involves human participants. In all phases, the research was conducted respecting the ethics of all participants involved. The method was evaluated and received approval from the Human Research Ethics Committee of the National University of Management (NUM) in Phnom Penh.

## 4 Findings

### 4.1 Valid Survey Answers

During the data collection phase from March to May 2023 557 survey answers were collected. Approximately one-fourth (144 answers) was collected using the structured interviews ("field surveys"), and the remaining 413 were collected from the online surveys.

All participants were older than 15 years and lived in Cambodia. Nevertheless, 66 answers were excluded from the evaluation process, due to not using a motorcycle (23 answers) or due to providing an incomplete survey (43 answers). Given these exclusions, there were 491 valid survey answers remaining. Since the number exceeds the minimum sample size of 400 answers, it is assumed a solid basis for generalization from the sample to the full population was realized. The exclusion procedure is provided in Table 1.

Of the valid survey answers, around 70% were from participants living in the capital Phnom Penh. This reflects the fact that the field phase was conducted there

**Table 1** Overview of received and valid surveys

| Received Surveys | Quantity | % |
|---|---|---|
| **Answers from survey interviews** | 144 | 26% |
| **Answers received online** | 413 | 74% |
| **Total sum of answers from participants** | **557** | **100%** |
| ... not using a motorcycle | 23 | 4% |
| ... outside of focused age group | 0 | 0% |
| ... not living in Cambodia | 0 | 0% |
| ... not completing the full survey | 43 | 8% |
| **Valid received surveys** | **491** | **88%** |

**Table 2** Distribution of valid surveys using key characteristics of the participants and their motorcycles

| **Province within Cambodia** | **Quantity** | **%** |
|---|---|---|
| Urban (Phnom Penh) | 335 | 68% |
| Rural | 136 | 28% |
| **Gender** | **Quantity** | **%** |
| Female | 218 | 44% |
| Male | 253 | 52% |
| **Highest Education** | **Quantity** | **%** |
| High School | 115 | 23% |
| University/College | 305 | 62% |
| PhD | 10 | 2% |
| **Car Ownership** | **Quantity** | **%** |
| Motorcycle age above 10 years | 38 | 8% |
| Motorcycle out of warranty | 357 | 73% |
| Motorcycle privately owned | 420 | 86% |

Remaining value to 100%: participant has skipped question, chosen "other" or has chosen the option "I wish to not answer". Rounding differences are possible

and that the used channels to distribute the surveys were often connected to institutions located in Phnom Penh.

The share of female participants (44%) was slightly lower than the participation rate of male participants (55%). Participants with higher educational backgrounds, such as a university or college degree or PhD (64%), significantly exceeded the country's average in this respect. The reason for this is the chosen distribution channels, which in many cases used university and academic groups to approach participants online.

The key characteristics of the survey participants' motorcycles is presented in Table 2 and summarized as follows: *Types:* Most participants used scooters (57%). Nearly all of them consisted of a combustion engine. Less than 1% had an electric motor for propulsion.

*Brands:* A vast majority of motorcycles were produced by the Japanese manufacturer *Honda* (76%) followed by the *Suzuki* (6%) and *Yamaha* (2%).

*Age & warranty:* Most motorcycles were privately owned (86%) and between 3 and 5 years old (34%). 38 participants (8%) stated that the age of their motorcycle exceeded 10 years. More than two-thirds of motorcycles were outside their warranty period (73%).

## 4.2 Level of Awareness of Remanufacturing

The survey investigated the participants' knowledge of remanufacturing and their awareness of remanufactured parts. The findings based on the received answers are summarized as follows:

- A vast majority of more than two-thirds (72%) state that they have heard about remanufacturing and are aware of remanufactured parts.
- No significant difference between the genders was found (female 71%, male 74%).
- Also, no major difference could be found between the different age groups (<25 years 74%, >25 years 71%) or groups with different educational backgrounds.

## 4.3 Level of Perception of Remanufacturing

Participants stating that they were aware of remanufacturing, obtained additional questions regarding their perception and opinion of remanufactured motorcycle parts. The key findings are depicted in Fig. 1 and described as:

- About half of the answers (44%) stated a rather positive perception when asked about remanufacturing. This group sees remanufactured parts as good as new parts or even as superior to them.
- The remaining half (45%) stated a rather negative perception. Within this group, most answers (39%) stated that remanufactured parts would be inferior when opposed to new parts. Moreover, a minor group (6%) even expressed that remanufactured parts are not an adequate alternative in comparison to new parts.
- More women (49%) than men (41%) had a positive perception of remanufactured parts.
- Younger people (46%) under 26 years of age perceived remanufacturing slightly more positively than older people (42%).

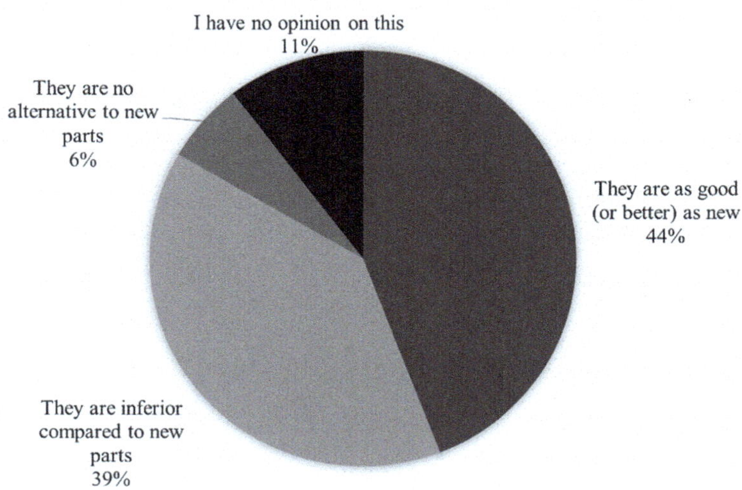


**Fig. 1** Perception of the remanufactured spare part quality and performance

- People with a higher educational background (43%) were slightly more critical about the quality of remanufactured parts than people without an academic background (48%).

## 4.4 Level of Willingness to Purchase Remanufactured Parts

Additionally, the will and intention of Cambodian motorcycle users to purchase remanufactured parts were explored. In the survey two fictive offers for remanufactured parts were proposed to the participants in relation to new counterparts allowing them to describe their preferences.

Offer 1. Focus on the economic advantages of remanufactured parts. The participants were provided with a fictive offer proposing a spare part with identical attributes (quality, performance, warranty, etc.) as a new spare part but to a lower price.

A majority of 56% would likely or most likely opt for such an offer; see Fig. 2. A big group (30%) remains uncertain, and a small group (14%) would rather not be willing to pay for such a "good-as-new" part with pricing advantages.

Offer 2. Focus on the ecological advantages of remanufactured parts. For comparability, a "good-as-new" spare part is included in the fictive offer. Here, ecological advantages were offered at an identical price of a new counterpart.

A vast majority (76%) of people *extremely likely* or *likely* preferred this offer; see Fig. 3. In comparison to Offer 1, the group of uncertain participants decreased (16%) and only a small group (8%) would not consider this it.

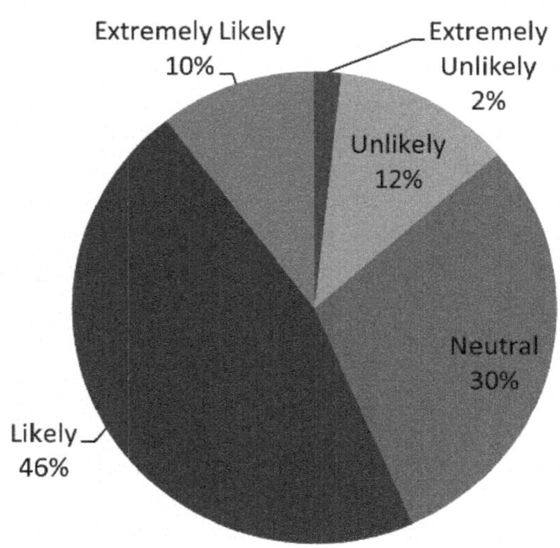


**Fig. 2** Willingness to pay for a "good-as-new part" over new spare part given additional economic advantages

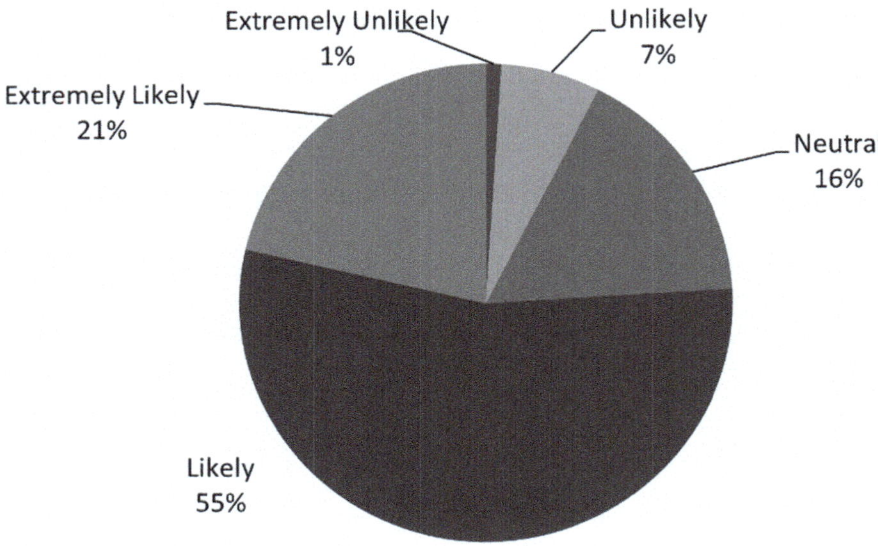


**Fig. 3** Willingness to pay for a "good-as-new part" over a new spare part given additional ecological advantages

### *4.5 Prospects of Cambodian Motorcycle Users on Remanufacturing*

In investigating Cambodian motorcycle users' prospects and expectations of remanufactured parts, the participants were able to define their perception of which criteria remanufactured parts should be evaluated upon. A clear majority (86%) defined the need for a warranty period as an essential criterion. This is followed by identical durability & level of quality (69%), appearance (68%), and functionality (67%) as a new spare part. Comparably, the pricing of a remanufactured part was ranked as lower importance (55%). The full list of the criteria and ranking is found in Table 3.

Furthermore, the pricing of remanufactured parts was explored. A majority (60%) of participants favor or expect a price level of 70 to 90% of the sales price for new counterparts given that they are equivalent in terms of performance, quality, and warranty. See Fig. 4 for the full distribution of participants' answers.

## 5 Discussion

Based on the findings presented in the previous sections, the research aim has been achieved, and recommendations can be given as described in the following section. In an initial step, the research aims are reflected. Afterward, recommendations for relevant groups of stakeholders are provided.

**Table 3** Criteria of remanufactured parts ranked by importance defined by survey participants

| Criteria of part | Agreeing or strongly agreeing |
|---|---|
| The product must have a warranty period | 86% |
| The durability & quality of the product must be the same as a new part | 69% |
| The product must look like a new | 68% |
| The product must not have any functional defects | 67% |
| The product is from a brand with good reputation | 66% |
| The product must be classed as environmentally | 65% |
| The product must perform as well or better than a new spare part | 58% |
| The product must have the latest features | 57% |
| The product must be cheaper than a new part | 55% |

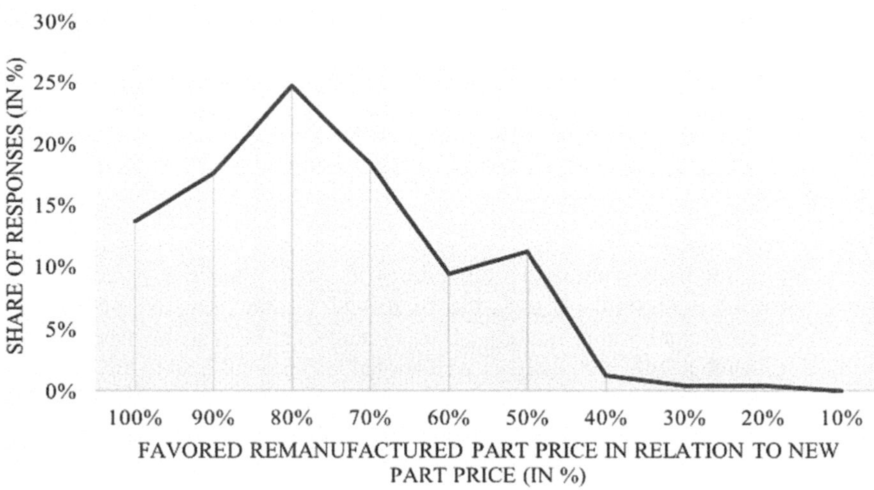


**Fig. 4** Favored price level for remanufactured parts in relation to new spare parts with equivalent performance, quality, and warranty

## 5.1 *Awareness of Motorcycle Users in Cambodia of Remanufactured Parts as an Alternative to New Spare Parts*

A majority—more than two-thirds—of motorcycle users in Cambodia are aware of the term "remanufacturing" and of remanufactured parts. No significant difference was found between genders, age groups or among people with different educational backgrounds.

These findings on the high level of awareness are significant, as there is a disparity between the market relevance and the awareness of remanufactured parts.

As covered in Sect. 2.3, remanufactured products and, especially, remanufactured parts for motorcycles today do not exist in the Cambodian market. Nevertheless, the respondents were aware of these concepts, thus indicating a potential prosperous embracement of remanufacturing in the market. Such trends have been shown in the literature. For example, Andrew-Munot and Ibrahim [40] explored the processes and challenges of remanufacturing and concluded that to establish remanufacturing successfully there has to be an awareness of the market. However, this conclusion varies between markets as Cao et al. [41] showed for the remanufacturing industry in China. There, customers were willing to purchase remanufactured parts and showed active support for remanufacturing even though they displayed low awareness. In general, the trend for the automotive market is that customers are steadily becoming more aware of sustainability issues and potential alternatives to lessen the negative trend such as by implementing remanufacturing [42]. As such, this disparity between the market relevance found in the literature, market awareness found in this research, and remanufacturing establishments opens up the need for further research studies.

### 5.2 Perception of Motorcycle Users in Cambodia of Remanufactured Parts as an Alternative to New Spare Parts

The group of motorcycle users in Cambodia with a rather positive perception of remanufacturing is of similar size as the group with a rather negative perception. Moreover, female and young motorcycle users tend to have a positive perception more often, while people with a higher educational background tend to have a more negative perception. The cause has not been identified, but when Hazen et al. [43] explored the customer perception of remanufactured parts they found that there is a tolerance level of ambiguity which affects customer's perception of remanufactured parts. Similarly, Abbey et al. [44] identified a relation between customer awareness and perception of remanufactured parts which consequently leads to customers having a high environmental consciousness and also viewing remanufactured products more positively.

This phenomenon requires further investigation since a relation between the level of knowledge and perception of remanufactured parts on the Cambodian market might exist and could be exploited to establish remanufacturing. The high awareness of remanufacturing in Cambodia is unexpected as remanufactured parts are not available on the current market. However, after receiving rather negative responses from Cambodian motorcycle users regarding such parts, raises the question of how such a perception may have evolved from.

Other results of the survey show that the financial benefits, for example, a price reduction of remanufactured parts relative to new parts, are perceived as of low importance, while product-related criteria (quality, performance, appearance, durability) are ranked as of higher importance. These criteria to evaluate a product offering, as described by Kabel et al. [22] and Wang et al. [23], are of importance, especially for motorcycle producers, original equipment (OE) part suppliers and workshops.

Additionally, the perceived perception of remanufactured parts is in line with previous explorations on other remanufactured parts in the Cambodian and Thailand markets. On one hand, research on opportunities for remanufactured batteries for electric vehicles (EV) in Thailand concluded that there is a low perception of remanufacturing and that the awareness of sustainable production concepts particularly in Thailand is lower compared to other regions globally [45]. On the other hand, research on opportunities for remanufactured car parts in Cambodia concluded that Cambodian car owners have a negative perception of and view remanufactured parts as inferior to new parts [46].

## *5.3 Recommendations to Stakeholders in Cambodia*

Following the concluding remarks of this research, several sets of recommendations are derived for the most important groups of stakeholders within Cambodia. These groups of stakeholders have been chosen, as these specifically would benefit from remanufacturing:

Policymakers. As described in Sect. 2, remanufacturing provides ecological and economic benefits compared to new parts. Such ecological benefits can support the use of environmentally friendly products, reduce the amount of waste, and improve the national carbon footprint. Moreover, the economic benefits offer the potential to provide work for Cambodian people and give them access to good-as-new parts at a lower price. Therefore, incentives to implement remanufacturing strategies initiated by Cambodian policymakers are recommended. Such incentives could be, for example, tax reductions for remanufactured parts, policies defining end-of-life strategies and defining the legal status of remanufactured parts, e.g., in respect of warranty.

Moreover, policy should improve the perception of remanufactured parts by setting legal frameworks in respect of product quality and warranty for remanufactured parts. A national policy defining this might encourage a local remanufacturing market.

This is in line with research on remanufacturing opportunities for EV parts in Thailand, which concluded that "Thai National standards for remanufactured part processes and testing are not existing" and need to be implemented to ensure market growth of remanufactured parts [45].

Motorcycle producers/original equipment (OE) part suppliers. The market for motorcycles and motorcycle spare parts in Cambodia is huge. Cambodian motorcycle users in this research have expressed a high loyalty toward their original equipment manufacturers (OEMs). Eighty-five of the survey participants stated their intention to purchase an exchange engine (or motor) at their OEM if they would need such a spare part. This market potential could be an ideal basis for offering remanufactured parts providing an economic advantage to the customers, improving the carbon footprint of the motorcycle company, having positive effects on the company's turnover, and preventing customers from using alternative options, for example, parts from scrap yards or from unofficial, non-OE sources. Therefore, in line with Östlin et al. [11] describing the competitive advantages and new product opportunities for companies offering remanufactured parts, it is recommended to check opportunities to add remanufactured parts to the product portfolio.

A significant finding is, that although the survey answers says that Cambodian motorcycle users are aware of remanufacturing, they still make a difference between a "reman part" and "as-good-as-new" part. According to Lund [6] a remanufactured part is per definition "as-good-as-new," the reason for such a differentiation should be investigated more deeply. Marketing campaigns should provide customers with relevant information and stress the "good-as-new" feature and the ecological benefits of remanufactured parts. Following the results of this research, focusing exclusively on pricing advantages should be prevented.

Marketing campaigns should emphasize the environmental and sustainability impact of remanufacturing described by Nasr and Russell [7]. The survey data shows that motorcycle owners attached great importance to these matters. Moreover, such marketing campaigns should respect that young and female participants today have a more positive perception of remanufacturing and might be more open to purchase such spare parts.

As the survey participants highlighted their wish for an adequate warranty period (Table 3), motorcycle manufacturers and part suppliers should provide the same warranty period for remanufactured parts as for new parts.

Motorcycle workshops. Providing remanufactured parts besides new parts can offer additional business opportunities, raise the customer's satisfaction, and prevent customers from using alternatives outside the product program of the workshop ("brand-protection," ref. to Östlin [14]). A great majority of motorcycles in Cambodia are outside their warranty period and a significant amount of them are older than 10 years (Table 2). As such, motorcycle users in a low-income country like Cambodia, might appreciate having alternatives to new spare parts at a lower price. Offering such an alternative might raise the satisfaction with their motorcycle workshop and prevent them from searching for substitutes elsewhere, for example in scrap yards.

Nevertheless, such an extension of the product portfolio of motorcycle workshops would require suitable sourcing opportunities for remanufactured parts within Cambodia or abroad. Therefore, it is recommended to search for sourcing opportunities to adopt remanufactured parts to the existing product portfolio and to request such opportunities from their suppliers.

### 5.4 *Limitations and Outlook*

This research is limited to the perception of motorcycle users in Cambodia. The increasing number of car owners in Cambodia is not respected here. This perspective is explored in parallel research using the same method focusing on Cambodian car owners' perception of remanufactured parts [46].

Furthermore, due to the research design, this exploration has a regional focus on Phnom Penh and focuses on persons with a higher educational background. Additional research on rural areas and on people with different educational backgrounds might be beneficial.

## 6 Conclusion

Based on an adequate number of surveys from Cambodian motorcycle users, three main conclusions are drawn. Firstly, the great majority of Cambodian motorcycle users have knowledge about remanufactured motorcycle parts. Nevertheless, the perception of remanufacturing from every second user is rather negative. Reasons for such a perception need further investigation. The perception changes when the term "remanufacturing" is not used, and instead "good-as-new" parts are offered providing ecological and economic benefits. Secondly, measures to increase the positive perceptions of remanufactured parts could be to stress the identical characteristics (quality, performance, warranty, etc.) of remanufactured parts more and to grant the same warranty conditions as for new parts. Thirdly, the market potential for remanufactured parts in this important market for motorcycle parts is huge. This circumstance is underlined by the fact that in a cost-sensitive market like Cambodia, the participants noted that the price is not of the highest importance when making their purchasing decision. The environmental benefits of remanufactured parts described by Parker et al. (2015) are of significantly higher importance to motorcycle users than the lower purchasing costs.

**Acknowledgments** Thank you to the faculty members and students at the National University of Management (NUM) in Phnom Penh and the American University of Phnom Penh (AUPP) for their support in the field phase by distributing the online survey and by performing survey interviews.

# Appendix A

Survey questions in English and Khmer language.

| Number | Question in English Language | Question in Khmer Language |
|---|---|---|
| 1 | If you would have the chance to consider an alternative to a new spare part for your motorcycle, which criteria would be of high importance to influence your purchase decision? | ប្រសិនបើលោកអ្នកមានឱកាសពិចារណាជម្រើសទៅលើគ្រឿងបន្លាស់ថ្មីសម្រាប់ម៉ូតូរបស់លោកអ្នក តើលក្ខណៈវិនិច្ឆ័យណាមានសារៈសំខាន់ខ្លាំងដែលជះឥទ្ធិពលលើការសម្រេចចិត្តទិញរបស់លោកអ្នក? |
| 2 | If you would need to exchange the engine (or motor) of your scooter/motorcycle, which of these options would you most likely use? | ប្រសិនបើលោកអ្នកចាំបាច់ត្រូវប្តូរម៉ាស៊ីនម៉ូតូរបស់លោកអ្នក តើជំរើសមួយណា លោកអ្នកទំនងជាចង់ប្រើជាងគេបំផុត? |
| | New engine (or motor) from the original manufacturer/OEM (e.g. Honda, Suzuki, Yamaha, etc.) | ម៉ាស៊ីនម៉ូតូថ្មីមកពីរោងចក្រផលិតម៉ាស៊ីនប្រភពដើម (ដូចជារោងចក្រហុងដា ស៊ុយស៊ូគី យ៉ាម៉ាហា។ល។) |
| | New engine (or motor) from an independent manufacturer (e.g. a Chinese manufacturer) | ម៉ាស៊ីនម៉ូតូថ្មីមកពីរោងចក្រផលិតម៉ាស៊ីនឯករាជ្យណាមួយ (ដូចជារោងចក្រផលិតនៅចិន។ល។) |
| | Used engine (or motor) (e.g. from a scrap yard) | ម៉ាស៊ីនម៉ូតូជជុះ (ដូចជាមកពីហាងលក់ម៉ាស៊ីនជជុះ) |
| | Remanufactured/refurbished engine (or motor) | ម៉ាស៊ីនផលិតឡើងវិញឬកែច្នៃឡើងវិញ |
| | I do not know | ខ្ញុំមិនដឹងទេ |
| | I wish not to answer | ខ្ញុំមិនចង់ឆ្លើយ |
| 3 | How likely would you use the following offers: | តើលោកអ្នកទំនងជានឹងប្រើប្រាស់ការផ្តល់ជូនដូចខាងក្រោមយ៉ាងដូចម្តេច? |
| | Offer 1: My motorcycle repair shop offers me an alternative to a new spare part with the same performance, quality, and warranty but at a lower price. | ហាងជួសជុលម៉ូតូរបស់ខ្ញុំផ្តល់ជំរើសជំនួសគ្រឿងបន្លាស់ថ្មីដែលមានដំណើរការ គុណភាព និងការធានាដូចគ្នា ហើយតំលៃទាបជាង |
| | Offer 2: My motorcycle repair shop offers me an alternative to a new spare part with the same performance, quality, warranty, and an identical price but with better sustainability (more environmental-friendly / "green") | ហាងជួសជុលម៉ូតូរបស់ខ្ញុំផ្តល់ជំរើសជំនួសគ្រឿងបន្លាស់ថ្មីដែលមានដំណើរការ គុណភាព ការធានាដូចគ្នា និងតំលៃប្រហាក់ប្រហែលគ្នា ហើយមាននិរន្តរភាពជាង (មិនប៉ះពាល់ដល់បរិស្ថាន) |
| 4 | What price of an alternative to a new spare part would be appropriate for you if it offers the same performance, quality, and warranty as the new spare part? | តើតម្លៃមួយណានៃជំរើសគ្រឿងបន្លាស់ថ្មីដែលសមស្របសម្រាប់លោកអ្នក ប្រសិនបើគ្រឿងបន្លាស់នោះផ្តល់នូវដំណើរការ គុណភាព និងការធានាដូចគ្នានឹងគ្រឿងបន្លាស់ថ្មី? |
| 5 | Have you ever heard of "remanufactured" or "refurbished" parts?<br>Info: Remanufactured parts are undergoing an industrial process in order to make them perform as good or even better than new. They have the same specifications, performance, appearance and warranty as a new part. | តើលោកអ្នកធ្លាប់ឮពាក្យគ្រឿងបន្លាស់ "ផលិតឡើងវិញ" ឬ "កែច្នៃឡើងវិញ" ឬទេ? ព័ត៌មាន៖ ការផលិតឡើងនៃគ្រឿងបន្លាស់ ស្ថិតនៅក្រោមដំណើរការនៃឧស្សាហកម្ម ក្នុងគោលបំណងធ្វើឱ្យគ្រឿងបន្លាស់មានដំណើរការល្អ ឬអាចប្រសើរជាងរបស់ថ្មី។ គ្រឿងបន្លាស់មានមុខងារ ការដំណើរការ រូបរាង និងការធានាដូចគ្នាទៅនឹងគ្រឿងបន្លាស់ថ្មីដែរ។ |
| 6 | What is your perception of "remanufactured" parts? | តើអ្នកយល់ឃើញយ៉ាងណា ចំពោះគ្រឿងបន្លាស់ "ផលិតឡើងវិញ"? |
| 7 | What information should be included when purchasing a remanufactured spare part? Multiple answers are allowed. | តើព័ត៌មានអ្វីខ្លះដែលគេគួរតែបញ្ចូលនៅពេលទិញគ្រឿងបន្លាស់ដែលបានផលិតឡើងវិញ? អាចឆ្លើយលើសពីមួយចម្លើយ។ |
| 8 | Your gender | ភេទរបស់អ្នក |
| 9 | Your age | អាយុរបស់អ្នក |
| 10 | In which Cambodian province are you living at the moment? | តើបច្ចុប្បន្ននេះ លោកអ្នករស់នៅខេត្តឬក្រុងណា? |
| 11 | In which Cambodian province are you born? | តើអ្នកកើតនៅខេត្តឬក្រុងណា? |
| 12 | Education | កម្រិតវប្បធម៌ |
| 13 | Your monthly income | ចំណូលប្រចាំខែរបស់អ្នក |
| 14 | Your motorcycle type | ប្រភេទម៉ូតូរបស់អ្នក |
| 15 | Your motorcycle brand | ម៉ាកម៉ូតូរបស់អ្នក |
| 16 | Your motorcycle's age | អាយុម៉ូតូរបស់លោកអ្នក |
| 17 | Is your motorcycle still in warranty? | តើម៉ូតូរបស់លោកអ្នកកំពុងស្ថិតនៅក្នុងកំឡុងពេលនៃការធានាឬ? |
| 18 | Who is the owner of the motorcycle? | អ្នកណាជាម្ចាស់ម៉ូតូ? |

## References


1. Statista (2022) Cambodia's economy & society – data and analysis. https://www.statista.com/study/48420/cambodia/
2. ASEAN Statistics Division (2023) Total number of registered motorcycles (in thousand). In: ASEANstatsDataPortal. https://data.aseanstats.org/indicator/ASE.TRP.ROD.B.011
3. Kong S (2018) Country presentation Cambodia, DDG, general Department of Land Transport, Ministry of Public Works and Transport, Cambodia. In: Presentation at: Intergovernmental eleventh regional environmentally sustainable transport (EST) forum in Asia 2–5 October 2018, Ulaanbaatar, Mongolia
4. Ministry of Public Works and Transport of Cambodia (MPWT), General Department of Land Transport (2020) Statistic Report of Registered Vehicles as of 25 Aug 2020
5. Sundin E, Lee HM (2011) In what way is remanufacturing good for the environment? In: Proceedings of 7th international symposium on environmentally conscious design and inverse manufacturing (EcoDesign-11). Springer, Kyoto, Japan, pp 551–556
6. Lund RT (1984) Remanufacturing: the experience of the United States and implications for developing countries, UNDP Project Management report number 2, Integrated resource recovery series. The World Bank
7. Nasr NZ, Russell JD (2018) Re-defining value - the manufacturing revolution. Remanufacturing, refurbishment, repair and direct reuse in the circular economy, a report of the international resource panel. United Nations Environment Programme
8. Guide GV, Daniel JRR, Harrison TP, Van Wassenhove LN (2003) The challenge of closed-loop supply chains, informs. J Appl Analy 33:3–6. https://doi.org/10.1287/inte.33.6.3.25182
9. Parker D, Riley K, Robinson S, Symington H, Tewson J, Jansson K, Ramkumar S, Peck D (2015) Remanufacturing market study, Report from the Horizon 2020 project called ERN - European Remanufacturing Network, Grant Agreement No 645984. Accessible from www.remanufacturing.eu
10. Kerr W, Ryan C (2001) Eco-efficiency gains from remanufacturing: a case study of photocopier remanufacturing at Fuji Xerox Australia. J Clean Prod 9:75–81. https://doi.org/10.1016/S0959-6526(00)00032-9
11. Östlin J, Sundin E, Björkman M (2008) Business drivers for remanufacturing. In: Proceedings of the 15th CIRP international conference, Presented at the Life Cycle Engineering Conference. The University of New South Wales, Sydney, Australia, pp 581–586
12. Barquet AP, Rozenfeld H, Forcellini FA (2013) An integrated approach to remanufacturing: model of a remanufacturing system. J Remanufacturing 3. https://doi.org/10.1186/2210-4690-3-1
13. Junior ML, Filho MG (2012) Production planning and control for remanufacturing: literature review and analysis. Prod Plan Control 23:419–435. https://doi.org/10.1080/09537287.2011.561815
14. Östlin J (2008) On remanufacturing systems: analysing and managing material flows and remanufacturing processes. Dissertation, Linköping University, Department of Management and Engineering, Assembly technology, Linköping University, The Institute of Technology
15. Östlin J, Sundin E, Björkman M (2008) Importance of closed-loop supply chain relationships for product remanufacturing. Int J Prod Econ 115:336–348. https://doi.org/10.1016/j.ijpe.2008.02.020
16. Goodall P, Rosamond E, Harding J (2014) A review of the state of the art in tools and techniques used to evaluate remanufacturing feasibility. J Clean Prod 81:1–15. https://doi.org/10.1016/j.jclepro.2014.06.014
17. Priyono A, Ijomah W, Bititci U (2016) Disassembly for remanufacturing: a systematic literature review, new model development and future research needs. J Ind Eng Manag 9:899–932. https://doi.org/10.3926/jiem.2053

18. Russell JD, Nasr NZ (2023) Value-retained vs. impacts avoided: the differentiated contributions of remanufacturing, refurbishment, repair, and reuse within a circular economy. J Remanufacturing 13:25–51. https://doi.org/10.1007/s13243-022-00119-4
19. Gilli M, Marin G, Mazzanti M, Nicolli F (2017) Sustainable development and industrial development: manufacturing environmental performance, technology and consumption/production perspectives. J Environ Econ Policy 6:183–203. https://doi.org/10.1080/21606544.2016.1249413
20. World Commission on Environment and Development (WCED) (1987) Report of the world commission on environment and development: our common future. Oxford University Press
21. Hanemann WM (1991) Willingness to pay and willingness to accept: how much can they differ? Am Econ Rev 81:635–647
22. Kabel D, Elg M, Sundin E (2021, 1954) Factors influencing sustainable purchasing behaviour of remanufactured robotic Lawn mowers. Sustain For 13. https://doi.org/10.3390/su13041954
23. Wang Y, Hazen BT, Mollenkopf DA (2018) Consumer value considerations and adoption of remanufactured products in closed-loop supply chains. Ind Manag Data Syst 118:480–498. https://doi.org/10.1108/IMDS-10-2016-0437
24. Govindan K, Jiménez-Parra B, Rubio S, Vicente-Molina MA (2019) Marketing issues for remanufactured products. J Clean Prod 227:890–899. https://doi.org/10.1016/j.jclepro.2019.03.305
25. Abbey JD, Meloy MG, Blackburn J, Guide VDR Jr (2015) Consumer markets for remanufactured and refurbished products. Calif Manag Rev 57:26–42. https://doi.org/10.1525/cmr.2015.57.4.26
26. Atasu A, Guide VDR Jr, Van Wassenhove LN (2008) Product reuse economics in closed-loop supply chain research. Prod Oper Manag 17:483–496. https://doi.org/10.3401/poms.1080.0051
27. Sakao T, Sundin E (2019) How to improve remanufacturing?—a systematic analysis of practices and theories. J Manuf Sci Eng 141:021004-1–13. https://doi.org/10.1115/1.4041746
28. Sundin E, Sakao T, Lindahl M, Kao C, Joungerious B (2016) Map of remanufacturing business model landscape, report from the horizon 2020 project called ERN - European remanufacturing network, Grant Agreement No 645984. Accessible from www.remanufacturing.eu
29. Saenprasarn L, Ruiz JV, Vivathirun T (2021) ASEAN motorcycle markets and their impact on car markets. White paper, ABeam Consulting
30. Wei S (2015) Core acquisition management in remanufacturing: current status and modeling techniques., Doctoral dissertation. Linköping University, Linköping. https://doi.org/10.3384/diss.diva-113251
31. Kojima M (2022) The diffusion of remanufacturing policies in Asia. Int J Autom Technol 16: 824–830. https://doi.org/10.20965/ijat.2022.p0824
32. Guidat T, Wewer A, Kohl H, Seliger G (2020) Demand-oriented barriers and potentials for remanufacturing in Vietnam. Int J Sustain Manuf 4:336–349. https://doi.org/10.1504/IJSM.2020.107124
33. Nayak MSDP, Narayan KA (2019) Strengths and weaknesses of online surveys. IOSR J Humanit Soc Sci 24(5):31–38
34. Ball HL (2019) Conducting online surveys. J Hum Lact 35(3):413–417. https://doi.org/10.1177/0890334419848734
35. Groves RM, Fowler FJ Jr, Couper MP, Lepkowski JM, Singer E, Tourangeau R (2009) Survey methodology. Wiley
36. Ballou J, Markesich J (2009) Survey data collection methods, Chapter 8. In: Counting working-age people with disabilities: what current data tell us and options for improvement. W.E. Upjohn Institute for Employment Research, pp 265–297. https://doi.org/10.17848/9781441612687.ch8
37. Uakarn C, Chaokromthong K, Sintao N (2021) Sample Size Estimation using Yamane and Cochran and Krejcie and Morgan and Green Formulas and Cohen Statistical Power Analysis by G*Power and Comparisons. APHEIT Int J Interdiscip Soc Sci Technol 10(2)

38. Memon M, Ting H, Hwa C, Ramayah T, Chuah F, Cham TH (2020) Sample size for survey research: review and recommendations. J Appl Struct Equ Model 4(2). https://doi.org/10.47263/JASEM.4(2)01
39. Taherdoost H (2017) Determining sample size, how to calculate survey sample size. Int J Econ Manag Syst 2
40. Andrew-Munot M, Ibrahim RN (2013) Remanufacturing process and its challenges. J Mech Eng Sci 4:488–495. https://doi.org/10.15282/jmes.4.2013.13.0046
41. Cao J, Chen X, Zhang X, Gao Y, Zhang X, Kumar S (2020) Overview of remanufacturing industry in China: government policies, enterprise, and public awareness. J Clean Prod 242: 118450. https://doi.org/10.1016/j.jclepro.2019.118450
42. Subramoniam R, Huisingh D, Chinnam RB (2009) Remanufacturing for the automotive aftermarket-strategic factors: literature review and future research needs. J Clean Prod 17(13): 1163–1174. https://doi.org/10.1016/j.jclepro.2009.03.004
43. Hazen BT, Overstreet RE, Jones-Farmer LA, Field HS (2012) The role of ambiguity tolerance in consumer perception of remanufactured products. Int J Prod Econ 135(2):781–790. https://doi.org/10.1016/j.ijpe.2011.10.011
44. Abbey JD, Meloy MG, Guide VDR Jr, Atalay S (2014) Remanufactured products in closed-loop supply chains for consumer goods. Prod Oper Manag. https://doi.org/10.1111/poms.12238
45. Nanthakorn B, Sundin E, Lindahl M, Banomyong R (2022) Challenges and opportunities regarding circular strategies for electric vehicle components in Thailand. In: 12th international conference on logistics & transport 2022 (ICLT). 17–19 November, Thailand
46. Nanthanon B, Vogt Dolberg J, Sok S, Kurilova-Palisaitiene J, Sundin E, Lindahl M, Banomyong R (2023) Cambodian customers' perception of remanufactured automotive parts. In: International conference of remanufacturing. 27–29 June 2023, Amsterdam

# Information Needs for Establishing Circularity-Focused Collaborations Among Product Manufacturers and Recyclers

**Giovanni Formentini, Rembrandt H. E. M. Koppelaar, Deborah Leone, Siro Dell'Ambrogio, Alessandro Fontana, Fabiana Guggiari, Tina Frangez, Slavko Dvoršak, Janez Matavš, Kristina Obu Vazner, Lucía Herreras Martínez, Eniko Hajosi, and Devarajan Ramanujan**

**Abstract** The realization of a fully circular economy has become a crucial aspect of the European Union's climate aspirations. In this regard, end-of-life product recovery and recycling play a critical role in preserving the economic value embedded in products. Effective information sharing between product manufacturers and recyclers is therefore necessary for identifying existing pain points and opportunities for improving the product design and recovery processes with respect to circularity. However, the common motivations and information needs for establishing collaborations between the aforementioned stakeholders are currently unclear. To this end, we present the methodology and results from a survey-based study performed in the EU-funded CircThread project that investigated (1) common motivations to foster collaboration between recyclers and product manufacturers and (2) desired production information exchange between these stakeholders with regards to product end-

G. Formentini (✉)
Department of Mechanical & Production Engineering, Aarhus University, Aarhus, Denmark
e-mail: giof@mpe.au.dk

R. H. E. M. Koppelaar
EcoWise Ekodenge Ltd., London, UK

D. Leone · S. Dell'Ambrogio · A. Fontana · F. Guggiari
Department of Innovative Technologies, University of Applied Sciences and Arts of Southern Switzerland, Viganello, Switzerland

T. Frangez · S. Dvoršak
Surovina d.o.o. Ul, Maribor, Slovenia

J. Matavš · K. O. Vazner
Gorenje d.o.o, Velenje, Slovenia

L. H. Martínez · E. Hajosi
WEEE-Forum, Brussels, Belgium

D. Ramanujan
Department of Mechanical & Production Engineering, Aarhus University, Aarhus, Denmark

Centre for Digitalisation, Big Data and Data Analytics, Aarhus, Denmark

 
S. Fukushige et al. (eds.), *EcoDesign for Circular Value Creation: Volume I*,
https://doi.org/10.1007/978-981-97-9068-5_22

of-life disassembly. Two separate survey designs were distributed among five recyclers and five product manufacturers to address the above questions. Results show that the main motivation for seeking collaborations was to enable product manufacturers to decrease costs and for recyclers to establish a new source of revenue. However, the lack of product lifecycle information and information-sharing structures poses a significant challenge toward fostering profitable collaborations. The paper ends with a discussion on how the EU CircThread platform and the design of Digital Product Passports could overcome the identified information gaps and reduce the barriers for industrial circular transition.

**Keywords** Circular economy · Disassembly · End of life · Digital product passports

## 1 Introduction

The concept of circular economy (CE) suggests that resources should be retained within the economic system for as long as possible before being discarded [1]. It aims to enable industrial and societal transition from a linear make-use-dispose economy to a circular one, through strategies, including reducing material intensity of production and consumption, designing durable products, and recovery and reuse of end-of-use materials and products. This enables preserving material value and offering a cleaner and more sustainable approach to manufacturing, while also enhancing businesses' resilience against resource scarcity [2]. As the CE concept spans the entire product lifecycle, it necessitates the collaboration of stakeholders throughout the entire value chain of products—from primary material producers to original equipment manufacturers (OEMs), and end-of-use stakeholders such as waste collection agencies and recyclers. Forming these collaborations requires common motivations and incentives for different partners to share product life cycle information and resources, as well as appropriate standards to facilitate the clear and efficient exchange of information. Stakeholder collaboration and information sharing are also vital to improve value chain transparency and are central to the success of ongoing efforts in establishing sustainability and circularity-focused product labels [3] and digital product passports (DPPs) [4]. In this regard, the partners in the EU-funded CircThread project [5] are working together on designing information technology (IT) services to capture, manage and share product life cycle information to enable manufacturers, retailers, consumers, collectors, repairers, recyclers, and reverse logistics operators to work together, by creating an information sharing IT ecosystem. Up to now, CircThread project managed to:

1. successfully conceptualize market opportunities and new business models within the circular economy domain. This involved engaging various stakeholders throughout the entire product life cycle, including consumers, manufacturers, recyclers, and others.
2. identify the information needs of recyclers, specifically focusing on materials declarations to enhance recycling processes.

3. create an information system for circular disassembly efforts, aimed at improving design and reducing disassembly time and resources.
4. monitor regulatory developments related to green claims and purchasing decisions, including consumer labelling requirements.
5. drafted an IT architecture for the Digital Thread, which serves as a foundation for building replicable data spaces and an information exchange platform for circular economy initiatives. This infrastructure can be adopted by other businesses and sectors.

This paper presents results from the CircThread project related to developing new frameworks for circularity-focused design decision making. Specifically, we analyse the aspects of collaboration between OEMs and recyclers and identify the types of disassembly-related product information that could be shared among these stakeholders to support CE.

## 2 Background

The CE strategy is a pioneering concept that challenges all actors of the life cycle of products [6]. Ongoing legislative work related to CE is expected to push industry and consumers to change the way they used to think and work with product life cycle information [7]. CE necessitates altering the manner in which information is organized and disseminated throughout the entire value chain of a product [8]. This gives rise to concerns pertaining to the accessibility of information among various stakeholders, the comparability and authentication of data, the secure identification of users, and the allocation of permits and confidentiality to users. This information will be crucial for effectively managing Electrical and Electronic Equipment (EEE) and their waste (WEEE). According to the directive 2012/19/EU of the European Parliament and of the Council [9] on WEEE, information should be provided to recyclers to facilitate the handling of WEEE waste, including maintenance, upgrade, refurbishment, and recycling. All EU member states are required to provide required measures to ensure that OEMs provide information free of charge within 1 year after any new EEE is placed on the market. This information enables the preparation of WEEE for reuse and the correct and environmentally efficient treatment of WEEE. This information must also be made available to treatment facilities in the form of written instructions or on a CD, via an online service or by other electronic means [9]. Currently, communication among various stakeholders (such as recyclers, collectors, and OEMs) regarding waste management predominantly occurs through email and telephone. OEMs inform recyclers about the specific type and amount of waste, which recyclers obtain by manually reading the information on the product's identification plate. This approach is antiquated and necessitates modernization to accommodate the evolving information demands in the upcoming years. The implementation of the EcoDesign (ESPR) legislation [9] will necessitate the gathering of a greater amount of information throughout the entire lifespan of a product. This includes the compilation of a list of critical raw materials present in each product, which are crucial to the European economy but also pose significant supply-related risks. Presently, there is a lack of regulations mandating transparency of data across the entire supply chain, making it extremely challenging to obtain reliable information. To address such issues, the EU-funded CircThread project [5] seeks to improve communication among stakeholders by establishing a platform that facilitates the collection and sharing of product information throughout the product's entire life cycle, to help stakeholders achieve better communication and meet standards requirements. Achieving this requires finding alternative methods to encode and manage product lifecycle information. With this regard, one of the most ambitious initiatives is the development of a Digital Product Passport [13] framed under the sustainable product initiative. Digital Product Passports are tools that can significantly enhance the end-to-end traceability of a product, complementing information provided in product manuals or labels. It should help consumers make informed choices by improving their access to product information, allow repairers or recyclers to access relevant information, and improve enforcement of legal requirements by authorities. The implementation of a DPP presents some challenges [5], such as (i) difficulty to access and input of information from the different stakeholders identified, e.g., when most OEMs are assemblers, detailed information on the composition of a product may become difficult to access and requires the collaboration of actors upstream, (ii) data comparability and verification, standards are needed to ensure that the data provided on products is comparable and reliable, and (iii) secure user identification, permits and confidentiality assigned to users, confidential data may be stored and be essential for improving circularity, a secure storage and sharing must be in place.

There are also concerns and challenges identified regarding supporting technologies, in particular the availability of a technology that will tolerate the whole life cycle of a product (e.g., e-products like photovoltaic panels may last more than 20 years), currently RFID and QR codes are under discussion in most of the fora about DPPs. The information should be easy to access, read and entered/updated during all steps of the life cycle, and stored in a secure manner for a long period of time.

In this scenario, OEMs and recyclers play a central role in the stakeholder network as they are key components of the supply chain [10]. The former is responsible for producing products, while the latter facilitate the extraction of elements from these products for reintegration into the system. With regard to recyclers, two typical fates exist: material recycling and component harvesting (for eventual reuse and remanufacturing). Material recycling can be considered as a sub-optimal scenario as the value embedded in producing a product/component is lost. It also faces limitations due to material constraints. For instance, recycled polymers exhibit inferior mechanical properties compared to virgin ones and manufacturers often prefer virgin polymers over recycled ones [11]. On the other hand, component harvesting presents challenges in terms of disassembly effort, but it offers better sustainability and product properties; harvested components can be directly used to create new products [12]. To enable the implementation of the component harvesting solution, it is necessary to determine the information that should be exchanged and establish the appropriate means of communication. Currently, there are no existing standards or guidelines for recyclers and OEMs regarding how this collaboration should be conducted. To address this gap, two surveys were created and distributed among recyclers and manufacturers. The survey results assist companies in understanding how the CE paradigm can be applied in practical terms. The remainder of the paper is organized as follows: Sect. 3 outlines the methodology employed for the survey, Sect. 4 presents the findings obtained, and Sect. 5 concludes the paper by discussing future developments.

## 3 Methodology

When addressing the entire product life cycle, there is a wealth of information available. However, it is important to determine the specific information that is necessary to acquire and share to make it valuable for the stakeholders involved. Our focus is on OEMs and recyclers among the available stakeholders. This choice is motivated by the intention to address information gaps identified in Formentini and Ramanujan [12], such as disassembly steps, and potential challenges that may arise when reclaiming products at their end of life (EoL). The proposed *design for circular disassembly* approach requires significantly more time and resources from OEMs and recyclers due to additional information sharing burdens. However, this was justified by the fact that it enables the creation of circular products and results in increased value for both partners. For instance, recyclers can efficiently extract valuable components (in a lesser amount of time), while OEMs can acquire economically valuable components that can be further used as spares or for making new products. However, the specific motivations and information sharing needs for these partners were not explored in their work.

The methodological approach in this work is based on a survey design where the main objective is to assess the perceived value of products at their end of life by OEMs and recyclers. We also wanted to determine what information types (related to product disassembly) would be beneficial to share among the partners. As shown in Fig. 1, the creation of the survey was performed in six distinct phases.

To create the survey, first a brainstorming session with 4 partners: 2 universities, 1 OEM and 1 recycler was performed. The information collected during the brainstorming session allowed identification of macro-level topics of interest. Following the brainstorming session, the survey was drafted. Since the perception of value may be different between the partners, two distinct surveys were created: one for OEMs and one for recyclers. The complete version of the surveys is available at https://doi.org/10.6084/m9.figshare.23608467.

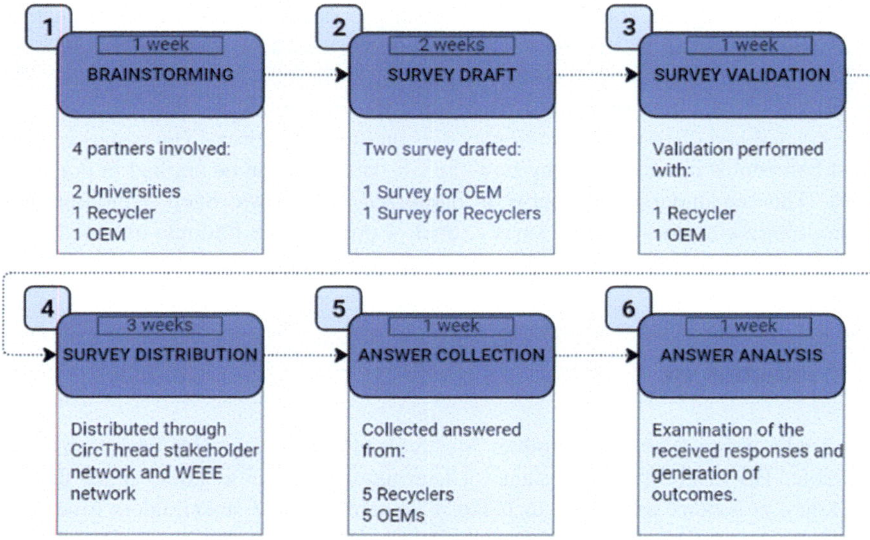


**Fig. 1** Flowchart of the survey creation, dissemination, and results analysis process

Both surveys were divided into three sections, each addressing distinct aspects:

- *Recoverability Aspect.*

This section seeks to understand the essential attributes that make a component or item valuable for the purpose of recovery. Five and seven questions for OEMs and recyclers respectively were framed to assess this aspect.

- *Disassembly Issues.*

This section aims to recognize the common challenges that arise during the process of product disassembly. These challenges can occur either during reparability assessments, where the product is in perfect condition, or at the end-of-life stage, where issues like rusted elements or worn-out parts may be present. For OEMs, this section was evaluated using three specific questions, while recyclers were presented with a matrix-style question. Additional details regarding the matrix question is provided in the subsequent section in this paper.

- *Product Design.*

The purpose of this section is to evaluate the potential influence of product design on end-of-life considerations and the ease of disassembly. OEMs were asked five questions, while recyclers were asked six questions, aiming to identify suggestions for redesign based on end-of-life considerations.

We used a combination of Likert scale questions (rated on a scale of 1–5), open-ended questions, and matrix-style questions in both surveys. This allowed is to gather both qualitative and quantitative information from the respondents. It should

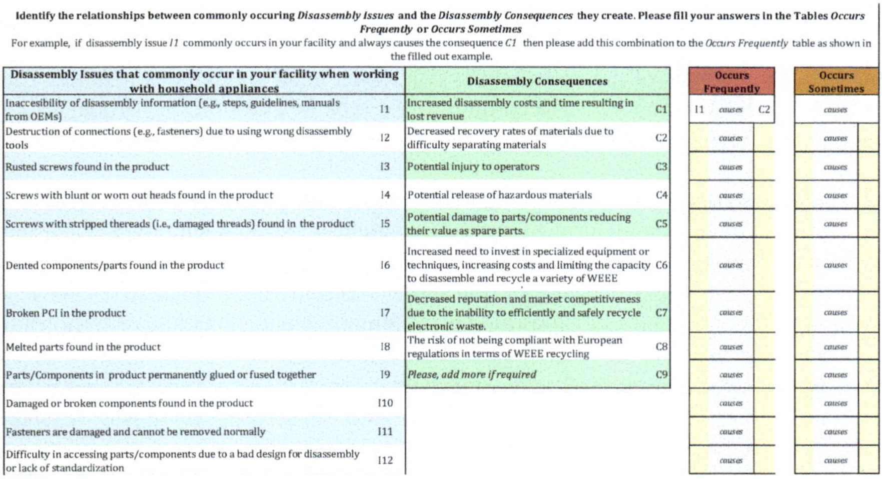

**Identify the relationships between commonly occuring *Disassembly Issues* and the *Disassembly Consequences* they create. Please fill your answers in the Tables *Occurs Frequently* or *Occurs Sometimes***

For example, if disassembly issue *I1* commonly occurs in your facility and always causes the consequence *C1* then please add this combination to the *Occurs Frequently* table as shown in the filled out example.

| Disassembly Issues that commonly occur in your facility when working with household appliances | | Disassembly Consequences | | Occurs Frequently | Occurs Sometimes |
|---|---|---|---|---|---|
| Inaccesibility of disassembly information (e.g., steps, guidelines, manuals from OEMs) | I1 | Increased disassembly costs and time resulting in lost revenue | C1 | I1 *causes* C2 | *causes* |
| Destruction of connections (e.g., fasteners) due to using wrong disassembly tools | I2 | Decreased recovery rates of materials due to difficulty separating materials | C2 | *causes* | *causes* |
| Rusted screws found in the product | I3 | Potential injury to operators | C3 | *causes* | *causes* |
| Screws with blunt or worn out heads found in the product | I4 | Potential release of hazardous materials | C4 | *causes* | *causes* |
| Scrrews with stripped thereads (i.e., damaged threads) found in the product | I5 | Potential damage to parts/components reducing their value as spare parts. | C5 | *causes* | *causes* |
| Dented components/parts found in the product | I6 | Increased need to invest in specialized equipment or techniques, increasing costs and limiting the capacity to disassemble and recycle a variety of WEEE | C6 | *causes* | *causes* |
| Broken PCI in the product | I7 | Decreased reputation and market competitiveness due to the inability to efficiently and safely recycle electronic waste. | C7 | *causes* | *causes* |
| Melted parts found in the product | I8 | The risk of not being compliant with European regulations in terms of WEEE recycling | C8 | *causes* | *causes* |
| Parts/Components in product permanently glued or fused together | I9 | *Please, add more if required* | C9 | *causes* | *causes* |
| Damaged or broken components found in the product | I10 | | | *causes* | *causes* |
| Fasteners are damaged and cannot be removed normally | I11 | | | *causes* | *causes* |
| Difficulty in accessing parts/components due to a bad design for disassembly or lack of standardization | I12 | | | *causes* | *causes* |

**Fig. 2** Matrix-style question to assess disassembly issues and consequences for recyclers. The blue and green columns list predefined entries. The red and brown column indicates occurrence frequencies of issue–consequence pairs. Yellow cells indicate input cells

be noted that the matrix-style question was used to assess aspects on disassembly issues for recyclers only. This question intends to determine the observed frequency of prominent Disassembly issues and the resulting consequences. The frequency levels were categorized into occurring *frequently* and occurring *sometimes*. As shown in Fig. 2, this question consisted of two lists containing predefined disassembly issues and disassembly consequences. The blue column enumerates failures that may occur during a disassembly process, while the green column lists the potential consequences these failures may have on the harvested part. On the right-hand side, the two results columns enable respondents to correlate a disassembly issue with a corresponding disassembly consequence. The red column captures responses for issue–consequence pairs that *frequently* occur, and the brown column for pairs that occur *sometimes*. Involved in survey administration, and a recycler for validation. The purpose was to identify any confusing or unclear questions, assess the length of the survey, and ensure the method to fill the survey was adequately described. After refining the survey structure based on the collected observations, the questions were reworked to create a final version. The validated and fine-tuned version was distributed to the designated stakeholders. In the final stage, a survey draft was shared with partners The surveys were distributed using two methods. Firstly, direct email contact was made with recyclers and OEMs who had expressed interest in the CircThread project and provided their contact information. Secondly, recyclers and OEMs who were members of the WEEE forum a.i.s.b.l. were approached. Participants were given a period of 3 weeks to answer the questions and send their responses via email. In total, we received responses from five original equipment manufacturers (OEMs) and five recyclers. The companies that responded are mainly large enterprises with more than 250 employees and are active in the European market. The analysis of the answers was conducted in an anonymous manner. The analysis of the answers varied according to the type of question. In particular, the Likert scale questions were analysed by comparing the mean value of the ratings for each answer.

For the open-end questions, we manually classified responses for each question and grouped them based on their underlying themes. Significant concepts in the responses were identified and are presented in the paper. Given the limited number of responses, we decided against using other, more involved qualitative analysis methods. Therefore, our analysis provides a preliminary understanding of significant concepts, and may not necessarily be representative. Lastly, the matrix-style questions were analysed based on the occurrence frequencies of issue–consequence pairs. In other words, we counted the number of times a disassembly issue was matched with a disassembly consequence, across the occurs *frequently* and occurs *sometimes* categories.

# 4 Results

The open-ended questions, which mainly focused on recoverability aspects, were manually analysed, and the main findings were summarized. Table 1 shows the three main product characteristics and challenges with regard to taking back elements from the perspective of the surveyed recyclers and OEMs. As shown, only one aspect of take-back was a common motivation for both stakeholders: if the element (product/component) contained critical raw materials (CRMs). All reported take-back related challenges were different.

**Table 1** Recoverability aspects for OEMs and recyclers

| **Elements features to take-back** | **OEMs** | **Recyclers** |
|---|---|---|
| | Contains CRMs | Contains CRMs |
| | Reusable as a spare part | Free from hazardous materials |
| | It can be refurbished | Simple to inspect |
| **Main limitation to elements take-back** | **OEMs** | **Recyclers** |
| | Element/Component might be harmful if not properly checked | Legal framework missing (liability) |
| | Functional obsolescence | Harvested parts have variable levels of quality |
| | Hard to determine the quality of used parts | Virgin material cheaper than recycled material |

**Table 2** OEMs most common disassembly issues

| Disassembly issues | Mean (μ) | Standard deviation (σ) | Coeff. of variation (CV) |
|---|---|---|---|
| Move product around to perform actions | 3.75 | 1.09 | 0.29 |
| Components cannot be recovered | 3.00 | 1.41 | 0.47 |
| Use of different type of screws | 3.00 | 0.71 | 0.24 |
| Disassembly process | 2.75 | 1.30 | 0.47 |
| Rusted Screw | 2.50 | 0.50 | 0.20 |
| Elevate cost of repair | 2.50 | 1.12 | 0.45 |

Table 2 illustrates the mean value of the reported score (1 = Never; 5 = Always), standard deviation, and coefficient of variation for the six highest scoring disassembly issues as reported by the surveyed OEMs. These issues were mainly tackling reparability aspects.

Results show that *Move product around to perform actions* was the most problematic disassembly issue, however its coefficient of variation is moderate ($\mu = 3.75$; CV = 0.29), meaning it is not shared among all companies. On the other hand, the presence of *Rusted screw* ($\mu = 2.50$; CV = 0.20) and the *Use of different type of screws* ($\mu = 3.00$; CV = 0.24), appear to be a more consistent issue for companies since their coefficient of variation is lower. The issue pertaining to *Components cannot be recovered* had the second highest average score, but seemed to be of differing importance to OEMs, based on the relatively higher CV ($\mu = 3.00$; CV = 0.47).

The analysis of disassembly issues reported by recyclers focused on examining the frequency of answers to the matrix-style question. Specifically, we recorded the occurrences where responders reported issues and corresponding consequences as occurring *frequently* and *sometimes*. This approach allowed us to gain an understanding of the disassembly issues that are commonly observed among various recyclers. The results obtained from this analysis are presented in Fig. 3. A total of 4/5 OEMs reported that *Screws with blunt or worn out heads found in the product* (I4) caused *Increased disassembly costs and time resulting in lost revenue* (C1). The second highest reported correlations (3/5) were that *Rusted screws found in the product* (I3) and the *Need to constantly move the appliance on different axis in order to reach components* (I17) caused *Increased disassembly costs and time resulting in lost revenue* (C1) as well as, *Parts/Components in product permanently glued or fused together* (I9) caused *Increased disassembly costs and time resulting in lost revenue* (C1) and *Decreased recovery rates of materials due to difficulty separating materials* (C2).

Results from analysing actions to improve product design reparability and disassembly respectively for OEMs and Recyclers are shown in Tables 3 and 4.

Based on the gathered results, we found that disassembly information is an important consideration for both OEMs ($\mu = 3.50$; CV = 0.14) and recyclers ($\mu = 1.60$; CV = 0.49). Recyclers require that accurate disassembly instructions

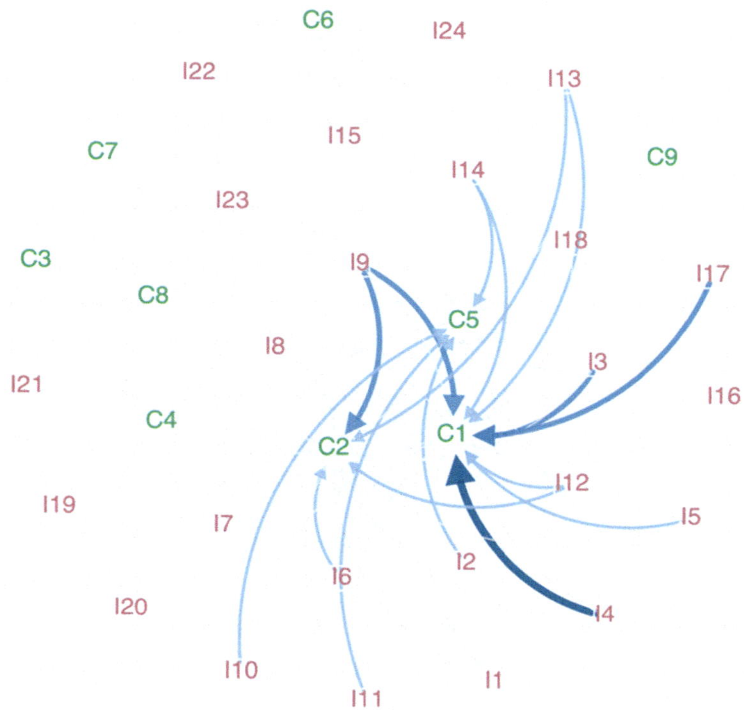


**Fig. 3** Correlations between disassembly issues (I#) and consequences (C#) for the occurs *frequently* category. Thicker arrows with a darker colour represent a greater degree of correlation. The names of the specific issues and consequences are available in the OEM survey accessible at: https://doi.org/10.6084/m9.figshare.23608467

**Table 3** Answers from OEMs for the section Product Design (1, totally disagree; 5, totally agree)

| Disassembly issues | Mean (μ) | Standard deviation (σ) | Coeff. of variation (CV) |
|---|---|---|---|
| Improve reparability by changing fastener material | 3.75 | 1.09 | 0.29 |
| Improve reparability by changing element position | 3.25 | 0.83 | 0.25 |
| Improve reparability by changing disassembly procedure | 3.25 | 0.83 | 0.25 |
| Improve reparability by changing instructions | 3.50 | 0.50 | 0.14 |
| Interest in using used parts that perform like original | 3.25 | 1.48 | 0.45 |
| Interest in buying recovered parts from third parties | 3.00 | 0.71 | 0.24 |
| Interest in buying certified recovered parts | 3.50 | 0.50 | 0.14 |

**Table 4** Answers from recyclers for the section Product Design (1, totally disagree; 5, totally agree)

| Disassembly issues | Mean (μ) | Standard deviation (σ) | Coeff. of variation (CV) |
|---|---|---|---|
| Do not use screws/bolts | 2.40 | 0.80 | 0.33 |
| Do not use clips/connectors - Multiuse (e.g., latch connections) | 2.00 | 1.10 | 0.55 |
| Do not use clips/connectors - Single use (e.g., zip tie) | 2.40 | 1,02 | 0.42 |
| Do not use permanent fix (e.g., glue/welding) | 3.40 | 1.20 | 0.35 |
| Disassembly instructions are available | 1.60 | 0.49 | 0.30 |
| Disassembly instructions, when available are easy to follow | 4.00 | 0.00 | 0.00 |
| Severe variation in product conditions at its EoL | 4.20 | 0.75 | 0.18 |

are available and are easy to follow; currently, the provision of such information (from OEMs) seems to be lacking. Interesting, when instructions are provided, all recyclers agree that they are easy to follow ($\mu = 4.00$; CV $= 0.00$). OEMs stated that enhancing product reparability goes beyond physical modifications such as the used materials, fasteners, and the positioning of elements; it also involves improving the quality of disassembly instructions. It is interesting to notice that all OEMs show interest in the possibility to buy used parts, regardless of the availability of components certifications (without certification—$\mu = 3.00$; CV $= 0.71$; with certifications $\mu = 3.50$; CV $= 0.50$). However, not all OEMs agreed on the possibility to use parts that perform like the original in new products ($\mu = 3.25$; CV $= 1.48$). In the next section, we discuss survey results in relation to the three aspects addressed in the questionnaire: (i) recoverability, (ii) disassembly issues, and (iii) product design.

# 5 Discussion

## *5.1 Recoverability*

OEMs identified various features that make harvesting components from end-of-life products desirable. Specifically, components that contained CRMs and/or electronics were of particular interest. Additionally, components that could be reused as spare parts, especially if the product was no longer available for sale, or for the creation of new products were considered desirable. Furthermore, components with a design life that extended over a long period were often seen as target components for the disassembly process.

## 5.2 Disassembly Issues

When products undergo disassembly, they present various problems such as rusted screws, dented parts, etc. These problems could be present in the product when received by recyclers but can also occur when products are disassembled to extract components at their end of life.

In addition, the disassembly process was often prolonged due to the presence of rusted screws, which made it difficult to remove components efficiently. Interestingly, one of the most time-consuming aspects of disassembly was the need to move the product multiple times on different axes to complete the process, as well as the requirement for various types of disassembly tools (e.g., screwdrivers). These design features contributed to a time-consuming disassembly procedure, which diminished the benefits of such activities, both in terms of economic benefits and circularity (i.e., harvesting of components for reuse and remanufacturing). Furthermore, the survey reveals that OEMs typically do not receive feedback from repairers or recyclers regarding their products. This represents a missed opportunity for OEMs to gain insights into the disassembly performance of their products and identify potential alternative revenue streams, such as selling components with high amounts of precious materials to interested parties. The lack of an effective communication system between these stakeholders represents a significant information gap and negatively impacts the CE potential of a product.

We found recyclers demonstrate a strong understanding of disassembly procedures and failures. Based on the analysis of the open-ended responses, poor product design, including non-optimized disassembly and glued elements, were identified as major issues. This led to increased disassembly costs and time, reduced material recovery rates, potential damage to valuable components, decreased economic value for both OEMs and recyclers, and created the need to invest in new equipment and techniques. Additionally, the absence of information regarding disassembly steps, manuals, and safety precautions prolonged disassembly procedures and increased the risk of damaging target components. This, in turn, reduced the economic value and raises safety concerns for harvesting target components as many Waste Electrical and Electronic Equipment (WEEE) items contain hazardous materials like lead and mercury, which pose risks to both disassembly operators and retaining compliance with European regulations.

## 5.3 Product Design

The survey also covered product design and aimed to understand how OEMs approach product design in relation to disassembly, as well as to gather suggestions from recyclers. According to the open-ended responses from OEMs, the primary focus during product design is on reparability, while aspects such as reuse, refurbishment, recovery, and remanufacturing are often overlooked. This may be attributed to the fact that OEMs typically do not bear responsibility for end-of-life product

**Table 5** Summary of desired information exchange between OEMs and recyclers

| Desired information | Owner | To share with |
|---|---|---|
| Presence of critical raw material | OEMs | Recyclers |
| Components usable as spare parts | OEMs | Recyclers |
| Disassembly issues | Recyclers | OEMs |
| Feedbacks on the disassembly process | Recyclers | OEMs |
| Disassembly steps and manuals | OEMs | Recyclers |
| Required safety precautions | OEMs | Recyclers |
| Presence of hazardous materials | OEMs | Recyclers |
| Redesign suggestions for improving disassembly | OEMs | Recyclers |

management. To enhance reparability, OEMs typically made changes to fastener materials, provided reparability instructions, and adjusted the positioning of elements. However, they did not make alterations to factors such as disassembly tools, element size, or element materials.

Regarding the utilization of recycled materials and reused parts, OEMs expressed interest in procuring recovered parts if they were of equivalent quality to new parts and certified by third parties. Recyclers suggested that modular products be designed, utilizing the same fasteners, and implementing colour coding. They also suggested that product design should prioritize minimizing the need to change product axes during disassembly and avoiding hazardous materials and risky designs, such as sharp edges, to reduce the risk of injuries. Fasteners should be designed for easy disassembly, with multiuse clips being the most preferable option, followed by threaded connections and single-use clips. Non-reversible binding mechanisms such as welding, or glue should be avoided. These suggestions represent design for disassembly principles that are well known in literature [14]; however, there seems to be a significant gap in terms of applying such principles in practice for the product categories such as household appliances and WEEE. Based on the collected responses, it can be inferred that establishing an improved information framework between recyclers and OEMs is crucial to foster a mutual willingness to share information. Table 5 illustrates the specific information that needs to be exchanged between these two entities.

Such information, if present in product labels, manuals, and in DPPs could significantly benefit collaboration between OEMs and recyclers. To facilitate the necessary information exchange, it is essential to establish suitable and interoperable data structures, ensuring ease of access for the relevant partners. It is also important to note that such information (e.g., disassembly issues, disassembly manuals) could contain proprietary elements, making it challenging for OEMs to share them publicly. Therefore, there is a need to establish trustable information technology systems that enable secure sharing of relevant information to the relevant stakeholders. The implementation of the Digital Product Passport (DPP), in conjunction with the ongoing development of the CircThread project, will provide the means to achieve this goal.

# 6 Conclusion and Future Work

The collaboration between stakeholders across the product life cycle plays a vital role in realizing a circular economy. However, there is still uncertainty regarding the specific information that needs to be shared among stakeholders. This research paper begins to address this gap by conducting two surveys—one for recyclers and another for original equipment manufacturers (OEMs)—to understand the necessary product and disassembly-related information for facilitating collaboration between them. The overarching goal is to enable the implementation of solutions that mutually benefit these stakeholders from an economic and circular economy perspective.

We analysed responses from a total of ten companies, consisting of five recyclers and five OEMs. The analysis of the survey results revealed significant gaps exist with regard to information sharing between OEMs and recyclers. The survey also identified the types of information sharing that could establish better collaboration between these stakeholders. In particular, OEMs expressed a willingness to receive feedback on the disassembly process and the challenges faced by recyclers during the product retrieval phase. This willingness arises from the desire to obtain data that is currently unavailable within their own company, with the aim of enhancing product quality and exploring opportunities for component reuse. Recyclers demonstrated a willingness to obtain specific information from OEMs. This includes details regarding the quantity and type of critical raw materials present in the products, identification of hazardous materials, and recommended safety measures to protect operators from potential safety hazards. Additionally, we found a mutual interest from both parties to receive and expand upon disassembly-related information, particularly in the form of comprehensive disassembly guidelines. Sharing such guidelines could enhance the repairability for OEMs and facilitate smoother disassembly processes during the product's end of life for recyclers.

With regard to future work, it is necessary to explore how data sharing can be facilitated not just in terms of the required information technology infrastructure but also from a legal standpoint. This becomes particularly important as recyclers often deal with products from multiple OEMs, posing challenges in maintaining the confidentiality of product data. Multiple OEMs raised concerns about this issue during follow up discussions. An approach suggested by OEMs is that they handle component harvesting activities by themselves (e.g., instituting take-back policies), as it tends to require sensitive information. On the other hand, less data-intensive activities like recycling and disposal could be outsourced to third-party recyclers. However, it remains uncertain whether this approach would prove more advantageous compared to a centralized end-of-life processing model administered by a third party across OEMs. Further research is needed to investigate this question, as well as the provision of relevant information technology systems to support such contexts.

**Acknowledgements** This paper is supported by the European Union's Horizon 2020 research and innovation program under grant agreement No. 958448, project CircThread (Building the Digital

Thread for Circular Economy Product, Resource & Service Management). The findings and opinions stated in this paper reflect the opinion of the authors and not the opinion of the European Commission. The authors would also like to thank the survey respondents for their valuable contributions.

## References

1. Kirchherr J, Reike D, Hekkert M (2017) Conceptualizing the circular economy: an analysis of 114 definitions. Resour Conserv Recycl 127:221–232
2. Baars J et al (2021) Circular economy strategies for electric vehicle batteries reduce reliance on raw materials. Nat Sustain 4(1):71–79
3. Blomsma F, Brennan G (2017) The emergence of circular economy: a new framing around prolonging resource productivity. J Ind Ecol 21(3):603–614
4. Adisorn T, Tholen L, Götz T (2021) Towards a digital product passport fit for contributing to a circular economy. Energies 14(8):2289
5. Commission, E (2021) Building the digital thread for circular economy product. Resour Serv Manag [Online]. 14-11-2022
6. Dias VMR et al (2022) Possibilities for applying the circular economy in the aerospace industry: practices, opportunities and challenges. J Air Transp Manag 102:102227
7. European Commission (2020) Circular economy action plan: for a cleaner and more competitive Europe. Publications Office of the European Union
8. Genovese A et al (2017) Sustainable supply chain management and the transition towards a circular economy: evidence and some applications. Omega 66:344–357
9. Union, E.P.a.t.C.o.t.E. (2012) Directive 2012/19/EU of the European Parliament and the Council of the European Union on waste electrical and electronic equipment
10. COUNCIL, D.E.O.T.E.P.A.O.T. (2009) Directive 2009/125/EC of the European Parliament and of the Council of establishing a framework for the setting of ecodesign requirements for energy-related products
11. Berlin D, Feldmann A, Nuur C (2022) The relatedness of open-and closed-loop supply chains in the context of the circular economy; framing a continuum. Clean Logist Supply Chain 4:100048
12. Karaagac E, Koch T, Archodoulaki V-M (2021) The effect of PP contamination in recycled high-density polyethylene (rPE-HD) from post-consumer bottle waste and their compatibilization with olefin block copolymer (OBC). Waste Manag 119:285–294
13. Formentini G, Ramanujan D (2023) Design for circular disassembly: evaluating the impacts of product end of life status on circularity through the parent-action-child model. J Clean Prod 405: 137009
14. Group, T.B. (2023) Enabling circularity through transparency: introducing the EU digital product passport. World Business Council for Sustainable Development, Switzerland
15. Favi C, Marconi M, Germani M, Mandolini M (2019) A design for disassembly tool oriented to mechatronic product de-manufacturing and recycling. Adv Eng Inform 39:62–79

# Thermal-Based Allocation Method for Multiple Co-product Resource Consumption in Plastic Recycling: A Case Study in Malaysia

**Ming Hui Tan, Meng Soon Chiong, Invidiadoloria Faian Ferolin, Yoon-Young Chun, Kenichiro Tsukahara, and Kiyotaka Tahara**

**Abstract** Plastic recycling plays a vital role in mitigating the environmental impact of plastic waste. Life cycle assessment (LCA) is a powerful tool for evaluating the environmental performance of recycling processes, relying on accurate and transparent life cycle inventory (LCI) data. However, LCI data collection in plastic recycling poses challenges, particularly in regions with limited data availability and diverse co-products. This paper addresses the need for improved LCI data collection by proposing a thermal-based allocation method for electricity consumption in plastic recycling processes. The method considers resin-specific thermal properties during the pelletizing process and offers a more realistic estimation of resource consumption. A case study conducted in Malaysia focused on four resin types—PP, HDPE, LDPE, and LLDPE—demonstrated the effectiveness of the allocation method. The findings revealed that the conventional assumption of equal electricity consumption for all resins resulted in significant overestimation for some resins, leading to inaccurate assessments of environmental impacts. In contrast, the proposed thermal allocation method provided resin-specific estimates, improving the accuracy of electricity consumption estimation and subsequent impact assessments. The results of this study contribute to enhancing the accuracy and transparency of LCI data collection in plastic recycling processes. By considering resin-specific thermal properties, LCA practitioners can obtain more reliable

M. H. Tan
Faculty of Mechanical Engineering, Universiti Teknologi Malaysia, Johor Bahru, Johor, Malaysia

M. S. Chiong (✉)
Faculty of Mechanical Engineering, Universiti Teknologi Malaysia, Johor Bahru, Johor, Malaysia

UTM LoCARtic, Institute for Sustainable Transport, Universiti Teknologi Malaysia, Johor Bahru, Johor, Malaysia
e-mail: chiongms@utm.my

I. F. Ferolin · Y.-Y. Chun · K. Tsukahara · K. Tahara
National Institute of Advanced Industrial Science and Technology, Tsukuba, Ibaraki, Japan

 
S. Fukushige et al. (eds.), *EcoDesign for Circular Value Creation: Volume I*,
https://doi.org/10.1007/978-981-97-9068-5_23

estimates of electricity consumption, leading to improved assessments of environmental impacts. The study highlights the importance of standardized guidelines and approaches in LCA reference documents to ensure efficient and consistent LCI data collection.

**Keywords** Plastic recycling · LCI data · Multiple co-products · Resource allocation · Thermal properties

## 1 Introduction

The sustainable management of plastic waste has emerged as a critical concern for environmental protection worldwide. Plastic recycling plays a crucial role in reducing waste and minimizing the environmental impact of plastic materials. Life cycle assessment (LCA) is a comprehensive methodology used to evaluate the environmental impacts of plastic recycling, encompassing the entire life cycle of a product from raw material extraction to final disposal. To conduct an accurate LCA, reliable life cycle inventory (LCI) data are essential, reflecting the specific conditions of local industries.

However, constructing localized LCI data poses significant challenges, particularly in countries with limited data availability. Many industries in these regions do not separately measure resource consumption by the process but instead rely on lump-sum resource consumption from utility bills. This approach can complicate LCI data collection, especially when a company produces multiple co-products.

Addressing the need for improved resource allocation methods, this research chapter proposes a methodology to effectively allocate resource consumption for multiple co-products in plastic recycling. The proposed allocation method is based on the thermal properties of the plastics undergoing the palletizing process.

With the pressing need to reduce plastic waste and promote sustainable practices, the development of localized LCI data and improved methodologies is crucial. This research aims to enhance the accuracy and transparency of LCI data collection in plastic recycling processes.

## 2 Literature Review

Life cycle inventory (LCI) data collection is a crucial aspect of life cycle assessment (LCA) studies, but it is often time-consuming and data-intensive. LCI data can be categorized into three types [1]:

- The LCI of the subsystem only, i.e., from "gate-to-gate"; these are *processes* like extrusion, cutting, molding. Processes in the Use Phase are also gate-to-gate (e.g., cooking, driving, transport),
- The LCI of a "cradle-to-gate" sub-system, including upstream subsystems of all *inputs* (e.g., materials, fuels, electricity, and heat),
- The LCI of end-of-life subsystems, from "gate-to-grave," ending up in landfill, combustion with or without heat recovery, or stockpiling for recycling.

The absence of specific guidelines in LCA reference documents often leads to challenges and inconsistencies in the LCI phase. Carlson et al. [2] highlighted the importance of "activity" and "flow" as key concepts in LCI data modeling and database design. "Activity" corresponds to the *processes* within the LCI subsystems, while "flow" represents the *inputs* associated with those processes.

*Processes* can be either aggregated or plain, with the possibility of aggregating or disaggregating processes for transparent reporting. Saavedra-Rubio et al. [3] proposed a stepwise guidance for LCI data collection, including planning, data gathering using LCI blocks representing *aggregated processes*, and finalization of LCI blocks.

The challenges in LCI data collection vary for manufacturing and end-of-life processes. For manufacturing, the overwhelming number of *inputs* and *plain processes* in aggregated processes pose difficulties and variabilities in measurement [4]. In the case of end-of-life treatment, the measurement of *inputs* and *outputs* for each product becomes challenging due to the presence of multiple products, which is crucial for data transparency.

In the context of mechanical plastic recycling in Malaysia, Tan et al. [5] demonstrated that facilities may process multiple resin types with similar process flows. However, allocating *inputs* and *outputs* based solely on production volume is inadequate because different plastic resins have distinct physical properties that result in varying resource requirements during the end-of-life process.

Despite the existence of multiple *plain processes* in plastic recycling, most LCI data treat it as a single process, as highlighted by Tan et al. [6]. While this might have a negligible impact on overall impact assessment, it hinders data transparency and the reusability of process data.

Furthermore, Bernstein et al. [7] emphasized the need for accurate quantification of upstream manufacturing inventories through a combination of simulation-based models and data-driven estimates. Applying a similar approach to the end-of-life process can enhance the accuracy and transparency of LCI data.

In summary, LCI data collection is a complex and challenging task in LCA studies. The literature has addressed issues related to *aggregated* and *plain processes*, proposed stepwise guidance for data collection, and highlighted the challenges specific to manufacturing and end-of-life processes. Addressing these challenges is crucial for improving data transparency, accuracy, and reusability in LCI databases.

# 3 Methodology

This chapter aims to develop a thermal-based input allocation method for the plastic recycling process. The study focuses on allocating electricity consumption, identified as the primary input in prior studies [5, 6]. The method is derived based on the physical properties of the resin and is developed through a field study conducted at a plastic recycling facility in Malaysia.

The recycling facility processes four types of resins: PP, HDPE, LDPE, and LLDPE, all following a similar process flow (Fig. 1). The process involves four *plain processes*[1]: (1) crushing + washing, (2) pelletizing, (3) wastewater treatment, and (4) emission scrubbing.

Initially, the incoming sorted plastic waste is crushed into smaller pieces known as flakes and washed. Wastewater generated during the washing process is treated in a closed-loop wastewater treatment plant before being discharged. The washed plastic flakes are then naturally dried[2] and pelletized into recycled plastic pellets. An emission scrubber filters the emissions from the pelletizing process before releasing them into the ambient air.

While there are other *inputs* associated with the *aggregated process* shown in Fig. 1, their discussion and allocation details are not provided here as their quantities for each *plain process* can be readily obtained from the operation records.

The first step involves allocating the total electricity consumption to each *plain process* in the *aggregated process* (Fig. 1). The allocation is based on the assumption that the proportion of electricity consumption for each *plain process* is proportional to the machine power rating or ampere rating. The ampere rating, and consequently the electricity allocation among the four plain processes, is depicted in Fig. 2.

The machine configuration considered in this study is capable of processing 350 metric tonnes of plastic waste per month. Similar relative machine sizes can be assumed for facilities of larger or smaller scales.

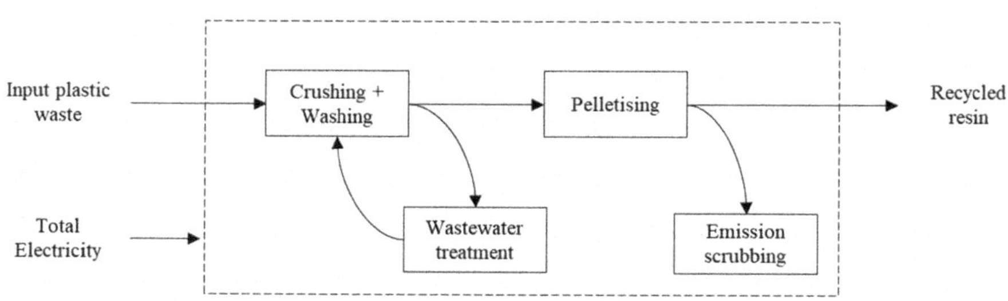


**Fig. 1** *Aggregated process* flow of plastic recycling

[1] That consuming electricity.

[2] Without consuming any energy.

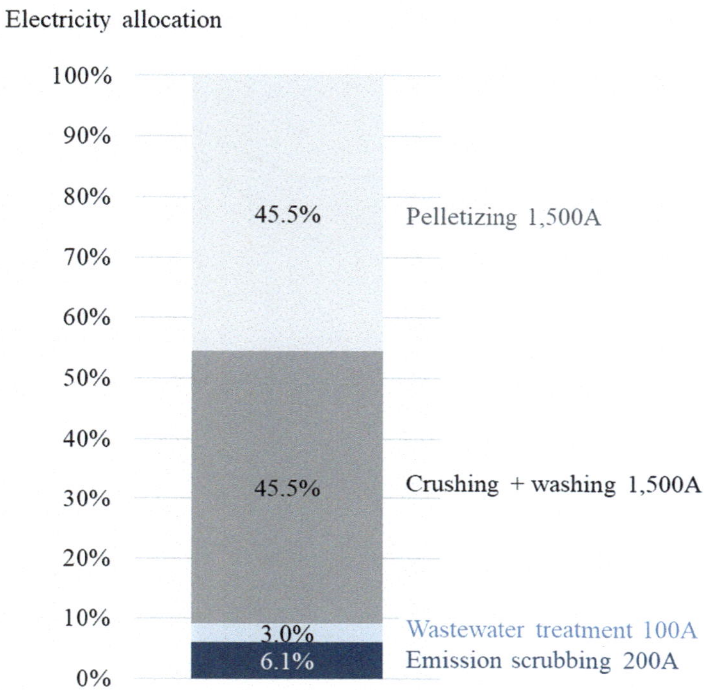


**Fig. 2** Electricity allocation between *plain processes* of plastic recycling

However, for facilities processing multiple resin types, assuming the same electricity consumption for the pelletizing process across all resin types would result in significant prediction errors due to variations in resin thermal properties.

To address this issue, a second-step thermal allocation method is proposed to realistically allocate the electricity consumption for different resins during the pelletizing process. The heat addition during the pelletizing process can be divided into three stages (Fig. 3).

First, the plastic flakes are heated from ambient temperature to their respective melting temperatures, $T_{\text{melting}}$ (Table 1). The heat addition, $q_1$, associated with this heating process can be approximated using the specific heat of the resin in its solid state and the temperature rise (Eq. 1).

$$q_1 = C_{\text{p,solid}} \cdot (T_{\text{melting}} - T_{\text{ambient}}) \tag{1}$$

After reaching $T_{\text{melting}}$, the resin flakes undergo phase change from solid to liquid state. The phase change occurs at a constant temperature, and the heat addition, $q_2$, during this process equals the latent heat of the plastic resins (Table 1), as given in Eq. 2.

$$q_2 = L_{\text{solid}-\text{liquid}} \tag{2}$$

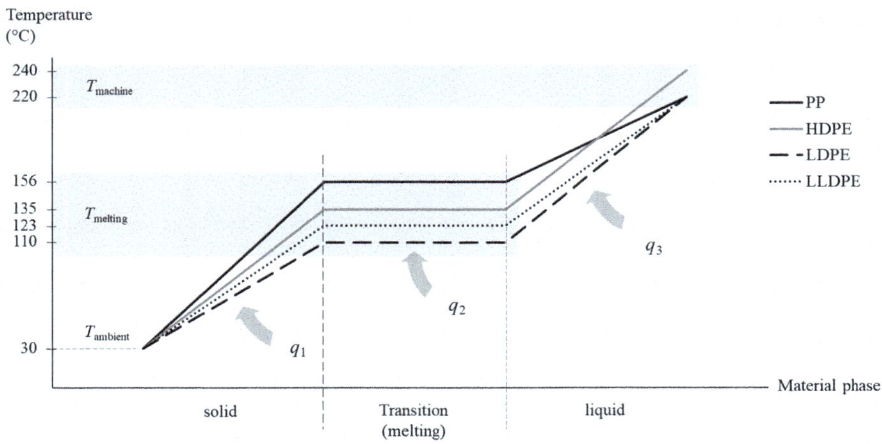


**Fig. 3** Energy flow in the plastic resin during the palletizing process

**Table 1** Thermal properties of plastic resin [8–15]

| Resin types | $C_{p,\text{ solid}}$ (J/kg°C) | $C_{p,\text{ liquid}}$ (J/kg°C) | $L_{\text{solid-liquid}}$ (J/kg) | $T_{\text{melting}}$ (°C) | $T_{\text{machine}}$ (°C) |
|---|---|---|---|---|---|
| PP [8–10] | 2000 | 2000 | 170,000 | 156 | 220 |
| HDPE [11] | 2250 | 2250 | 202,000 | 135 | 240 |
| LDPE [12–14] | 2600 | 2600 | 102,000 | 110 | 220 |
| LLDPE [15] | 5380 | 3240 | 71,750 | 123 | 220 |

In the final stage of heat addition, $q_3$, the melted plastic resins are heated to the machine temperature set point, $T_{\text{machine}}$. The $T_{\text{machine}}$ values in Table 1 were obtained through interviews with the facility manager. The $q_3$ can be correlated using Eq. 3, which considers the specific heat of the resin in its liquid state and the corresponding temperature rise.

$$q_3 = C_{\text{p,liquid}} \cdot \left(T_{\text{machine}} - T_{\text{melting}}\right) \tag{3}$$

Table 2 presents the resulting heat energy required by each resin type during the pelletizing process, considering the sum of the three stages ($q_1$, $q_2$, and $q_3$). Notably, HDPE is not the most energy-intensive resin to be pelletized, despite having the highest $T_{\text{machine}}$. Similarly, PP requires the least energy during pelletizing, even though its $T_{\text{melting}}$ is the highest.

The total heat addition, $q_{\text{total}}$, is the sum of the three stages and is used for electricity allocation according to Eq. 4. Solving Eq. 4 requires expressing the total resins' specific heat, $q_{\text{HDPE}}$, $q_{\text{LDPE}}$, and $q_{\text{LLDPE}}$ in terms of $q_{\text{pp}}$ as shown in

**Table 2** Heat energy required during pelletizing process

| Resin types | $q_1$ (kJ/kg) | $q_2$ (kJ/kg) | $q_3$ (kJ/kg) | $q_{total}$ (kJ/kg) | Solution |
|---|---|---|---|---|---|
| PP | 252.0 | 170.0 | 128.0 | 550.0 | $q_{PP}$ |
| HDPE | 236.3 | 202.0 | 236.3 | 674.6 | $q_{HDPE} = 1.23q_{PP}$ |
| LDPE | 207.7 | 102.0 | 285.6 | 595.3 | $q_{LDPE} = 1.08q_{PP}$ |
| LLDPE | 500.3 | 71.6 | 314.3 | 886.2 | $q_{LLDPE} = 1.61q_{PP}$ |

Table 2. The weight factor used in Eq. 4 is the respective resin's production volume, i.e., $V_{PP}$, $V_{HDPE}$, $V_{LDPE}$, and $V_{LLDPE}$. This enables a more realistic allocation of electricity consumption for pelletizing based on the energy requirements of each resin type.

$$(q_{PP} \cdot V_{PP}) + (q_{HDPE} \cdot V_{HDPE}) + (q_{LDPE} \cdot V_{LDPE}) + (q_{LLDPE} \cdot V_{LLDPE}) \\ = \text{total pelletizing electricity} \tag{4}$$

## 4 Results and Discussion

Figure 4 presents a comparison of the resultant electricity consumption for pelletizing PP, HDPE, LDPE, and LLDPE, calculated using two different methods:

1. Method 1: Assuming all resins have the same electricity consumption for pelletizing.
2. Method 2: Allocating electricity consumption based on the respective thermal properties of the resins.

It is evident that Method 1, which assumes equal electricity consumption for all resins, results in a constant value of 0.486 kWh/kg. In comparison, Method 2, which considers the resin thermal properties, provides more realistic estimates. Method 1 tends to overestimate the pelletizing electricity consumption for most resins, particularly for PP, which exhibits significantly lower energy requirements. The overestimation can reach approximately 25%.

On the other hand, Method 1 underestimates the pelletizing electricity consumption for LLDPE, which has substantially higher energy requirements compared to PP, HDPE, and LDPE. The underestimation can be around 22%.

Consequently, the realistic electricity consumption for each plain process in the aggregated plastic recycling process can be obtained (Fig. 5). This allocation of *inputs* by plain process improves the transparency of *aggregated process* data and promotes the reusability of individual *plain processes*.

Figure 6 compares the overall electricity consumption estimated using Method 2 with a case study conducted by Tan et al. [6] that focused on a recycling facility processing only LLDPE. The study by Tan et al. [6] provided the unit electricity consumption for the *aggregated process*, as it was a single-product operation.

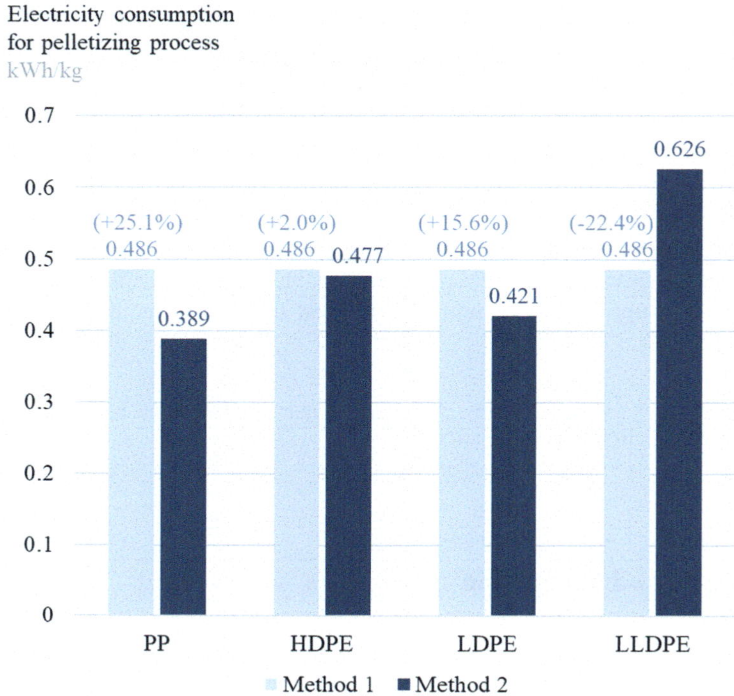


**Fig. 4** Resultant electricity consumption for pelletizing PP, HDPE, LDPE and LLDPE

The overall electricity consumption estimated by Method 2 is approximately 25% higher than the case study by Tan et al. [6]. Several factors contribute to this discrepancy.

Firstly, variations in machine specifications and operating efficiency, which are closely related to maintenance practices, can be challenging to quantify without controlled environments.

Secondly, variations in production duration play a role. The facility studied by Tan et al. [6] had continuous operation due to processing a single resin, while the facility in the current study had intermittent production of recycled PP resin, potentially leading to less efficient processes.

Although still not perfect, the thermal allocation method (Method 2) enables a more scientific-based allocation of pelletizing electricity consumption for different resin types. However, there are still areas for future improvement and investigation, including:

1. Further validation of the methodology on additional facilities to assess its accuracy, particularly in Step 1 regarding the allocation between plain processes.
2. Conducting field measurements, whenever possible, at each machine involved in each plain process to provide ultimate validation.

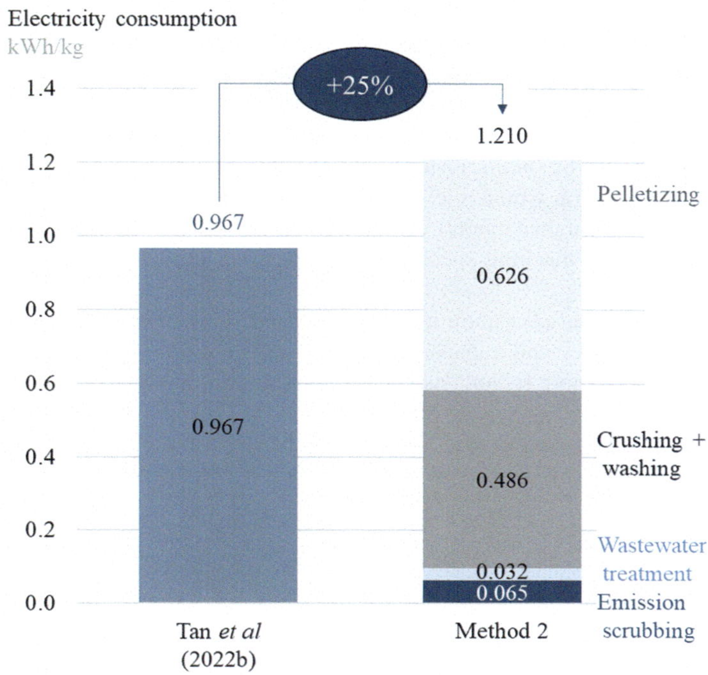


**Fig. 5** *Aggregated process* flow of plastic recycling

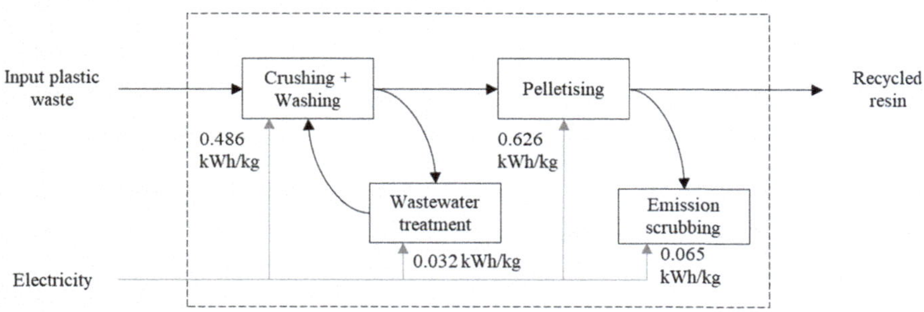


**Fig. 6** Overall electricity consumption for LLDPE recycling process compared to Tan et al. [6]

## 5 Summary

In conclusion, this chapter developed a thermal-based input allocation method for the plastic recycling process, specifically focusing on electricity consumption allocation. By considering the physical properties of different resins and conducting a field study at a plastic recycling facility in Malaysia, the limitations of assuming equal electricity consumption (Method 1) and the importance of considering resin-specific thermal properties (Method 2) were highlighted.

Method 1 resulted in overestimation for most resins and underestimation for high-energy requirement resins, while Method 2 provided more realistic estimates. The allocation of *inputs* by *plain process* improved data transparency and process reusability. Comparison with a relevant case study showed that Method 2 estimated approximately 25% higher overall electricity consumption, which could be attributed to variations in machine specifications, operating efficiency, and production duration.

Overall, the thermal allocation method presented in this research contributes to enhancing the accuracy and transparency of LCI data collection in plastic recycling processes, enabling more reliable assessments of environmental impacts.

**Acknowledgments** This research was funded by the Ministry of Higher Education Malaysia (MoHE)—registered program cost centre:#R.J130000.7851.4L893, and the National Institute of Advanced Industrial Science and Technology (AIST).

## References

1. TU Delft (n.d.) LCI structure. Available online: https://www.ecocostsvalue.com/lca/lci-structure/#av_section_2. Last assessed: 03.07.2023
2. Carlson R, Tillman A-M, Steen B, Lofgren G (1998) LCA data modelling and a database design. Int J LCA 3(2):106–113
3. Saavedra-Rubio K, Thonemann N, Crenna E, Lemoine B, Caliandro P, Laurent A (2022) Stepwise guidance for data collection in the life cycle inventory (LCI) phase: building technology-related LCI blocks. J Clean Prod 366:132903. https://doi.org/10.1016/j.jclepro.2022.132903
4. Glišić M, Sarfraz S, Veluri B, Ramanujan D (2022) A systematic framework for quantifying production system-specific challenges in life cycle inventory data collection. Procedia CIRP 105:210–218. https://doi.org/10.1016/j.procir.2022.02.035
5. Tan MH, Chiong MS, Chun Y-Y, Tsukahara K, Tahara K (2022) An analysis of practices and challenges for plastic recycling industry in Malaysia. Int J Autom Technol 16(6):831–837. https://doi.org/10.20965/ijat.2022.p0831
6. Tan MH, Chiong MS, Chun Y-Y, Tsukahara K, Tahara K (2022) Life cycle inventory of recycled polyethylene in Malaysia. In: Proceedings of 24th cross straits symposium on energy and environmental science and technology. Paper CCS-EEST 24
7. Bernstein WZ, Boettjer T, Ramanujan D (2021) Quantifying life cycle inventories for machining processes at detailed design. Procedia CIRP 98:370–375. https://doi.org/10.1016/j.procir.2021.01.119
8. ECK Plastic Arts (n.d.) Polypropylene (PP). Available online: https://eckplastics.com/polycarbonate-pc/. Last assessed: 04.07.2023
9. Lanyi FJ, Wenzke N, Kaschta J, Schubert DW (2019) On the determination of the enthalpy of fusion of α-crystalline isotactic polypropylene using differential scanning calorimetry, x-ray diffraction, and fourier-transform infrared spectroscopy: an old story revisited. Adv Eng Mater 22(9):1900796. https://doi.org/10.1002/adem.201900796
10. Remichem (n.d.) Polypropylene (PP). Available online: http://remichem.com/polypropylene-pp/. Last assessed: 04.07.2023

11. Woo MW, Wong P, Tang Y, Triacca V, Gloor PE, Hrymak AN, Hamielec AE (1995) Melting behavior and thermal properties of high density polyethylene. Polym Eng Sci 35(2):151–156. https://doi.org/10.1002/pen.760350205
12. Grünschloss E, Escobar MDPN (2017) Improved approximations in thermal properties for single screw extrusion. Materials Science. Available online: https://extrusion.4spe.org/wp-content/uploads/2019/01/2012-06-SPE-Extrusion-Division-Technical-Article.pdf. Last assessed: 04.07.2023
13. Sepe M (2001) The materials analyst, part 45: I can't believe it's not crystalline (part 2). Available online: https://www.plasticstoday.com/materials-analyst-part-45-i-cant-believe-its-not-crystalline-part-2. Last assessed: 04.07.2023
14. Mccord P (2022) Chembook—8.7: LDPE vs HDPE. Available online: http://chembook.org/page.php?chnum=8&sect=7. Last assessed: 04.07.2023
15. Chavan S, Gumtapure V, Perumal DA (2019) Characterization of linear low-density polyethylene with graphene as thermal energy storage material. Mater Res Express 6:065511. https://doi.org/10.1088/2053-1591/ab0e36

# Material Flow Analysis for the Biomass Waste Management of Palm Oil Mills with Regard to Sustainable Palm Oil Production

**Noor Azimah Darus, Ahmad Fariz Mohamed, and Nor Diana Mohd Idris**

**Abstract** Malaysia is one of the largest palm oil producers in the world. In 2022, 5.67 million hectares of oil palm plantations in Malaysia produced 18.4 million tons of crude palm oil (CPO). Empty fruit bunches (EFB), palm kernel shells (PKS), mesocarp fibers (MF), boiler ash, and palm oil mill effluent (POME) are examples of solid and liquid biomass waste produced by the palm oil mill sector in vast amount. These biomass wastes can be used not only as a nutrient-rich resource for the production of organic fertilizer for plantations but also as a feedstock for renewable energy in alternative management. In this study, input–output management for sustainable waste production of two different palm oil mills production capacity in Terengganu is evaluated and tracked using material flow analysis. The results showed that the empty fruit bunches (EFB) were the largest potential solid biomass in terms of quantity and energy generated compared to other waste, which is generated from fresh fruit bunches (FFB) around 22% (3300 ton/month) for mill 1 and 20% (1583 ton/month) for mill 2, respectively The empty fruit bunches were returned to the plantation as additional fertilizer to reduce input costs under conventional management, but there are several alternative management options for empty fruit bunches that provide additional benefits to oil mills. The results also showed that palm oil mills have already self-managed their energy consumption derived from palm kernel shells and mesocarp fiber. All production of 750 ton/month PKS and 2190 ton/month MF from mill 1 and 474 ton/month PKS and 1187 ton/month MF from mill 2 were used for boiler steam generator. Meanwhile, POME is treated before it is applied to land as fertilizer or discharged into rivers. As more and more of the palm oil mill biomass is utilized, a symbiotic relationship develops where

N. A. Darus
Institute for The Environment and Development (LESTARI), Universiti Kebangsaan Malaysia, Bangi, Selangor, Malaysia

Faculty of Plantation and Agrotechnology, Universiti Teknologi MARA Pahang, Bandar Tun Razak, Pahang, Malaysia

A. F. Mohamed (✉) · N. D. M. Idris
Institute for The Environment and Development (LESTARI), Universiti Kebangsaan Malaysia, Bangi, Selangor, Malaysia
e-mail: fariz@ukm.edu.my

 
S. Fukushige et al. (eds.), *EcoDesign for Circular Value Creation: Volume I*,
https://doi.org/10.1007/978-981-97-9068-5_24

"waste" becomes an input to other industries known as by-products. This leads to a zero-waste route for the palm oil industry, especially toward achieving sustainable biomass waste management. In conclusion, this study recommends various options for alternative biomass wastes utilization to contribute to the sustainable palm oil production, which is important to the economic growth and sustainable development of Malaysia.

**Keywords** Biomass · Wastes · Sustainability · Resource · Fertilizer · By-product

## 1 Introduction

Malaysia is the second-largest palm oil producer in the world after Indonesia. According to the 2022 year-end report of the Malaysian Palm Oil Board (MPOB), a total of 5.67 million hectares of land were planted with oil palm trees, which produced an average of 15.49 metric tonnes per hectare of fresh fruit bunches (FFB) for processing in palm oil mills. The total number of palm oil mills in Malaysia in 2022 was reported to be 450, of which 238 were in Peninsular Malaysia, 84 were in Sarawak, and 128 were in Sabah, with crude palm oil (CPO) production of about 18.45 million metric tonnes [1].

CPO production from palm oil mills produces some of the biomass waste from palm oil. The higher the value of FFB processed for CPO production, the more biomass waste is produced. This is because 10% of the total FFB is CPO production, and the remaining 90% is palm oil biomass [2]. It comprises oil palm trunk (OPT), empty fruit bunches (EFBs), palm oil mill effluent (POME), decanter cake (DC), mesocarp fibers (MF), palm kernel shells (PKS), and boiler ash [3]. A palm oil mill wastes around 12–15% MF, 5–7% PKS, and 20–23% EFB based on its capacity. However, this biomass waste needs to be utilized effectively to overcome its disposal problem since environmental concerns today.

However, the presence of this palm oil biomass waste has caused disposal problems in palm oil mills. This will be linked to the issue of environmental pollution and a lack of storage space. It occurs when there is an increase in the production of biomass waste. Irregular biomass waste management systems can lead to unsustainable oil palm mills [4]. For example, the use of POME open ponds leads to the release of methane gas and odor pollution, and the activity of returning EFB to the farm as additional fertilizer is seen as less effective because when EFB rots and rots, it will cause the release of methane gas in the air [5].

The effective use of these wastes can contribute not only to the sustainable production of renewable energy and renewable resources but also to preventing general waste issues such as waste pollution and illegal disposal. Palm oil mills are well known as a surplus energy industry because of the utilization of biomass waste. However, if the method of biomass waste management in palm oil mills is inefficient and less organized, it can increase the level of pollution and greenhouse effect on the environment and thus create sustainability issues [6].

Although palm biomass is widely recognized and used as a source of renewable energy, few electric generating plants actually use it. The National Energy Policy 2022–2040 aims to expand the share of biomass-based energy by developing five Key Economic Growth Activities (KEGAs) that are directly related to the energy sector, such as sustainable mobility, renewable energy (RE), and the green economy. In the present, the Malaysian Sustainable Palm Oil (MSPO) MS2530:2022 standard highlights that the use of renewable energy shall be encouraged and applied where practical. The application will help the Malaysian government achieve about 20 percent renewable energy (RE) capacity in 2025. Following the effort, the Malaysian government is committed to contributing to ASEAN's renewable energy (RE) target of 35% RE capacity by 2025 [7].

The oil palm processing mills, on the other hand, contribute substantially to environmental deterioration on both the input and output sides of their operations. On the input side, crude palm oil (CPO) consumes vast amounts of water and electricity. In contrast, on the output side, processing practices accumulate large quantities of solid waste, wastewater, and air pollution [8]. Therefore, in an effort to develop a sustainable and environmentally friendly palm industry, knowing the flow of input–output and waste management of palm oil biomass must be done in order to meet the requirements of safety, quality, and environmental sustainability. Therefore, the purpose of this study is to evaluate input–output management for sustainable palm oil biomass production by using material flow analysis.

## 2 Methodology

### *2.1 Palm Oil Mills*

The entire palm oil mill operated in Malaysia has a more or less similar production process and palm oil mill biomass management. Their main activity is the processing of fresh fruit bunches (FFB) into crude palm oil (CPO) and palm kernel oil (PKO). In this study, two palm oil mills in Terengganu with production capacities of mill 1 with 40 ton and mill 2 with 25 ton of fresh fruit bunches (FFB) per hour were selected.

Based on material mass balance, each palm fresh fruit bunch (FFB) processed in a palm oil mill produces palm oil and 14–16% of oil palm empty fruit bunches (EFB). In addition, the process produces oil palm midrib (OPM) and palm kernel shell (PKS) of 12–14% and 6.8–7.4%, respectively [6].

### *2.2 System Boundaries*

The system boundary is the palm oil biomass produced in the mill. The flows are focusing on materials including the production of crude palm oil (CPO), kernel,

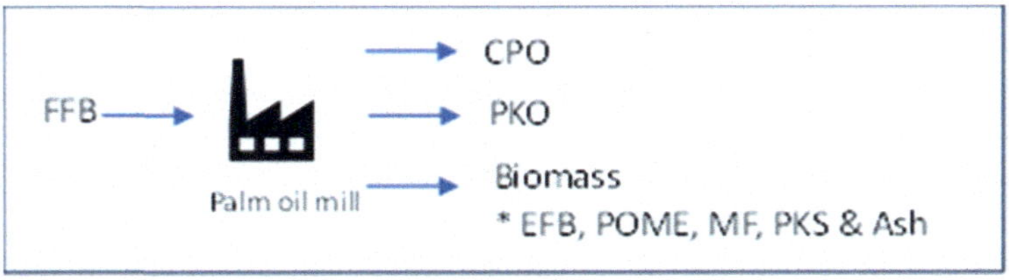


**Fig. 1** System boundary

shell, and other by-products of the mill, including the input of FFB with water and electricity consumption. The mill process flow should be identified to trace the input–output system boundary. Figure 1 shows the system boundary in this study.

### 2.3 *Data Collection*

Information on palm oil biomass processing was gathered from the mill. Flow data was assembled in a variety of ways, including a literature review and a questionnaire survey with staff.

### 2.4 *Input–Output Management and Material Flow Analysis*

Material flow analysis (MFA) is a systematic assessment of the flows and stocks of materials within a system defined in space and time [9]. MFA is a system of assessment that analyses data to find out how well input resources are utilized to create output. Because it is practical to describe input and output in a system based on the fundamental principle of conservation of mass, MFA is a valuable tool to decide resource management, waste management, and environmental management [10]. This study used MFA to apply input–output management to biomass management for environmentally friendly palm oil production.

## 3 Results and Discussion

### 3.1 *Analysis of Biomass Waste*

Theoretically, with increasing FFB processed, there is more palm biomass waste and less energy required. There will be extra handling of waste that influences the cost. Two bar charts below describe the composition of EFB, fiber, and shell in mill 1 and mill 2 as assessed. Most palm oil mills have around 20% EFB, 15% MF, and 6%

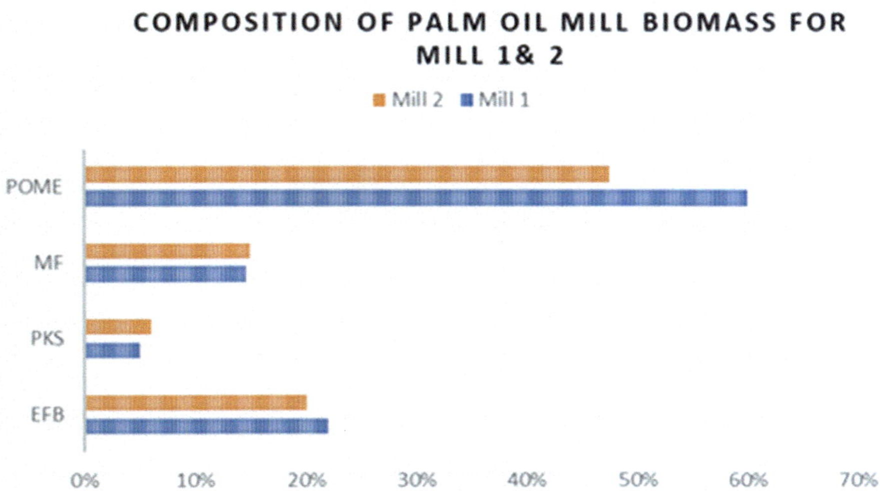


**Fig. 2** Composition (%) of EFB, POME, MF, and PKS

PKS by palm oil mill capacity FFB production (Fig. 2). EFB and POME are the biggest wastes, followed by MF and PKS. Therefore, utilization of MF and PKS as the energy sources from the steam generator has an effect on the amount of energy produced and waste reduced.

## 3.2 *Material Flow in Palm Oil Mill Biomass*

Productivity of CPO varies from company to company due to different factors such as mill capacity, processing conditions, and the quality and quantity of FFB sent to the mill. According to [11], based on the intended capacity, the size of POM is divided into small and large-scale operations. Small scale is defined as 25 tonnes per hectare or less. Additionally, compared to mills located far from the plantation region, the amount of processed FFB differs between integrated mills that continually process FFB from the harvested FFB. A mill that has better management of the biomass produced after processing CPO would have higher sustainable productivity. FFB serves as the major and essential input of the mill's processing since it is the principal raw material for the production of crude palm oil (CPO).

Figures 3 and 4 show the input–output of material flow in the observed palm oil mill with different capacities of production. Biomass waste generation rates were not much different at each palm oil mill. Based on the calculation, both palm oil processes produced solid waste in the forms of EFB, fiber, and shell, which were about 41% and 42% of the total FFB, respectively. From the observation and questionnaire, palm oil mills 1 and 2 have already used the biomass waste MF and PKS to generate electricity and steam for palm oil processing. At the same time, the

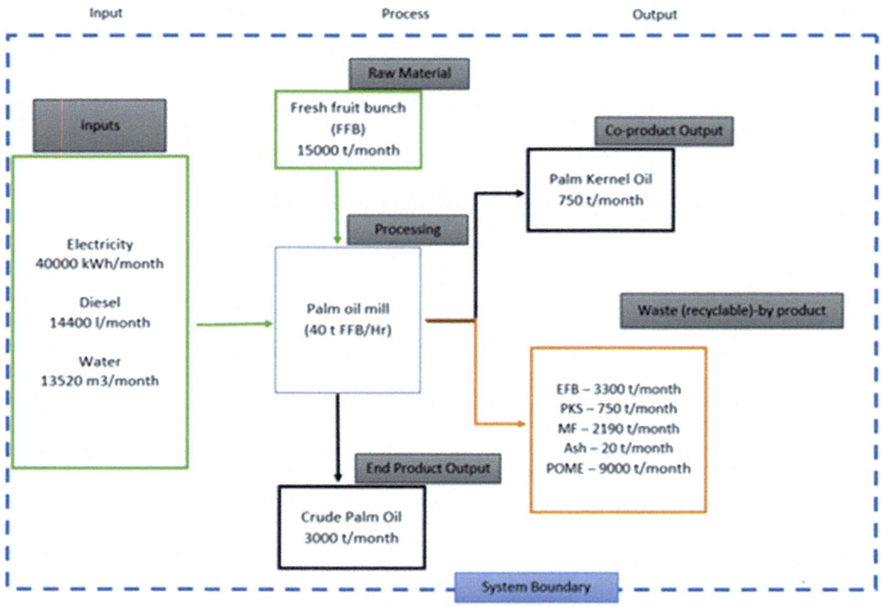


**Fig. 3** The input–output material flow of 40 t FFB/h

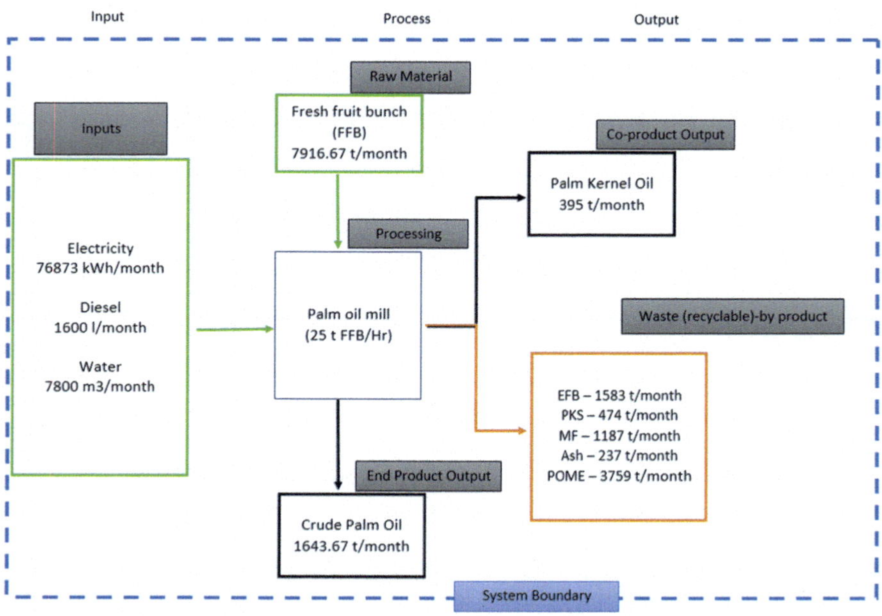


**Fig. 4** The input–output material flow of 25 t FFB/h

**Table 1** Total input–output analysis for mills 1 and 2

| | Mill 1 (ton/month) | Percentage based on input | Mill 2 (ton/month) | Percentage based on input |
|---|---|---|---|---|
| Input (Raw Material)—FFB | 15,000 | | 7916.67 | |
| Output | 3750 | 25% | 2038.67 | 25.75% |
| Waste | 15,260 | 101.73% | 7240 | 91% |

mills sold the waste to a third party as a by-product. During the initial phase, only diesel was used to generate electricity. Table 1 shows the proportion of total input–output analysis in this study. The results show an imbalance of input materials with product output and waste generation during processing. Different hours of operation, mill capacity, year of mill, total operation day, and infrastructure of the mill affect the production of palm oil biomass.

Although Malaysia has a reputation as the world's leading producer and exporter of palm oil, this industry faces some criticism for environmental sustainability issues. Among the issues are palm oil biomass management activities in mills criticized as not sustainable industries and contributors to environmental pollution, such as emissions of methane gas by open lagoon POME treatment and the decay of EFB in factories, air pollution, water pollution, and odor pollution [12, 13]. In common practice, EFB is used as a mulch for soil conditioning and nutrient recycling in oil palm plantations, but such practices may cause increased soil toxicity and increase the cost of transporting EFB from mill to plantation. There is no data provided from mill 1 and mill 2, and the amounts of EFB returned to the plantations as mulch and additional fertilizer make it difficult to ascertain the current status of the EFB management.

Most palm oil mills in Malaysia use an open anaerobic lagoon for POME treatment. The main issues with this system are the emission of methane gas, lagoon maintenance, overloading lagoon storage capacity, and non-compliance in water discharge. In this study, mill A, with a capacity of 40 t FFB/h, has its own close anaerobic digester tank to capture methane gas. But, for mill B, it only has open lagoon treatment for POME. A close anaerobic digester tank is used for biogas production from POME and is also expected to be a renewable energy source for the mill. POME can produce large amounts of biogas through anaerobic digestion methods for self-use in palm oil mills. However, for small mills, the investment cost for using biogas is considered quite high because of the need for good infrastructure and a close digester tank, which is costly at about RM200,000.00 for 1 tank (1800 $m^3$ tank capacity).

### 3.3 Palm Oil Mill Biomass Potential as a Symbiotic Relationship

Palm oil mills produced several biomasses wastes that contained energy, nutrients, and high-value secondary income. In general, palm oil mill production processes in Malaysia are less similar to each other and depend on the capacity of the mill. The abundant supply reflects huge growth potential, especially for the manufacturing and renewable energy sectors. Two types of palm oil mill biomass management are conventional and renewable. In conventional management, the biomass waste is either disposed of in abundance, burned, or discarded, where it will naturally decompose into organic fertilizer. The main energy sources for power generation in palm oil mills appear to be MF and PKS [14]. These carbonaceous minerals are used as fuel in boilers to create steam, which is then used to generate electricity. The oil palm biomass components have different potential uses, such as fuel for electricity generation, biochemical products, fertilizer, and biofuel for renewable palm oil biomass management. This biomass can be re-used for many purposes and converted into useful bio-products such as bio-fuel, biogas, bio-fertilizers, bio-composite, and briquettes, as well as fuels like coal to generate electricity. It was found that the main components of the waste management procedure that have been used effectively are mainly areas that favor industrial players in terms of economics, such as heat generation using biomass waste. In Malaysia, the use of closed anaerobic tanks to capture methane gas is still at a low level. But various incentives have been given by the government to the mill that wants to build up anaerobic closed tanks. Palm oil biomass waste emerges as a potential major contributor to renewable energy as the Malaysian has now shifted from conventional energy sources such as coal, oil, and gas to promoting renewable energy sources to increase energy security [15]. In order to meet the 12th Malaysia plan, thrust on RE, palm oil biomass waste have a huge potential for a concept of recycling waste to RE such as biogas. To date, the adoption of POME for biogas utilization in palm oil mill still faces techno-economic challenges and knowledge gaps that hinder deployment [10]. In addition, Malaysia also encourages palm oil mills to use existing EFB as a source of electricity and create biomass power plants for the zero-waste biomass programmed. However, because of its distinct potential, palm biomass is increasingly receiving greater attention and creating a symbiotic relationship because it can be a by-product of another output.

## 4 Conclusion

This study evaluated the input–output palm oil biomass production by applying material flow analysis to two palm oil mills in Terengganu, Malaysia. The results highlighted the importance of biomass residue utilization to contribute to the sustainability of resilient palm oil production. These palm oil mills already self-managed their energy consumption by burning PKS and MF. A Continuous Emission Monitoring System (CEMS) was installed on their premises to ensure operators comply with clean air emissions. The concept of zero waste will change the way the palm oil industry handles waste materials more efficiently and sustainably. Previously, other waste from palm oil processing was simply thrown away by most industry players. However, there are also efforts by some industry players who process the waste material to be used as fertilizer. Promoting biomass palm oil as renewable energy was in line with the UN Sustainable Development Goals (SDGs) agenda, specifically Goal 12, to ensure sustainable consumption and production. It is vital to ensure that production and consumption are done responsibly and to create a balance between environmental, societal, and economic considerations.

**Acknowledgments** This paper is part of PhD research, funded under Bumiputra Academic Training Scholarship (SLAB) provided by the Ministry of Higher Education Malaysia (MoHE) and Universiti Teknologi MARA (UiTM). The study also supported by fund from Universiti Kebangsaan Malaysia TAP K006974.

## References

1. MPOB Malaysia Palm Oil Board (2022) Overview of the Malaysian Oil Palm Industry 2022. http://bepi.mpob.gov.my/images/overview/Overview_of_Industry_2022.pdf
2. Basiron Y (2007) Palm oil production through sustainable plantations. Eur J Lipid Sci Technol 109(2007):289–295. https://doi.org/10.1002/ejlt.200600223
3. Singh RP, Hakimi IM, Norizan E, Iliyana MS (2013) Composting of waste from palm oil mill: a sustainable waste management practice. Rev Environ Sci Biotechnol 9:331–344
4. Abdullah N, Sulaiman F (2013) The oil palm waste in Malaysia. In: Dlm biomass now sustainable growth and use, pp 75–100. https://www.researchgate.net/publication/259440691_The_Oil_Palm_Wastes_in_Malaysia
5. Er AC, Abd RMN, Katiman R (2011) Palm oil milling wastes and sustainable development. Am J Appl Sci 8(5):436–440
6. Muhammad N, Tjahjono H, Meta R (2014) Analysis of palm biomass as electricity from palm oil mills in North Sumatera. Energy Procedia 47:166–172. https://doi.org/10.1016/j.egypro.2014.01.210
7. Paren Anak Nyawi MPOCC (2023) Biomass palm oil as renewable energy. https://www.mpocc.org.my/mspo-blogs/biomass-palm-oil-as-renewable-energy
8. Jafri NHS, Jimat DN, Azmin NFM, Sulaiman S, Nor YA (2021) The potential of biomass waste in Malaysian palm oil industry: a case study of Boustead Plantation Berhad. In: IOP Conf. Series: Materials Science and Engineering. 1192 (2021) 012028. IOP Publishing. https://doi.org/10.1088/1757-899X/1192/1/012028
9. Brunner PH, Rechberger H (2003) Practical handbook of material flow analysis: advanced methods in resource and waste management series 1. CRC Press LLC, Florida
10. Foong SZY, Chong MF, Ng DKS (2021) Strategies to promote biogas generation and utilisation from palm oil mill effluent. Process Integr Optim Sustain 5:175–191
11. Akhbari A, Kutty PK, Chuen OC, Ibrahim S (2020) A study of palm oil mill processing and environmental assessment of palm oil mill effluent treatment. Environ Eng Res 25(2):212–221
12. Abdullah I, Mahmood WHW, Muhammad HF, Ab Rahman MN (2015) Sustainability in Malaysian palm oil: a review on manufacturing perspective. Pol J Environ Stud 24(4): 1463–1475

13. Oosterveer P (2014) Promoting sustainable palm oil: viewed from a global network and flows perspective. J Clean Prod 107(2014):1–8
14. Yusoff S (2006) Renewable energy from palm oil-innovation of effective utilization of waste. J Clean Prod 14(1):87–93
15. Aghamohammadi N, Reginald SS, Ahmad S, Zinatizaded AA, Li PW, Nik Meriam NS (2016) An investigation of sustainable power generation from oil palm biomass: a case study in Sarawak. Sustainability 8(5):416

# Readiness Status of Smart Waste Collection and Processing Technologies for Plastic Waste Recycling

**Chaoxia Shan, Andante Hadi Pandyaswargo, and Hiroshi Onoda**

**Abstract** Each year, improperly managing the world's plastic waste raises a hidden environmental threat. As a result, efforts on smart technologies for waste collection, such as route optimization, smart bins based on artificial intelligence (AI), and real-time monitoring software, as well as the adoption of the Internet of Things (IoT) system, have become a future consensus for improving plastic waste recycling. In this study, we built a database of smart waste collection and processing technologies for plastic waste recycling. We categorized them into the following categories: monitoring dashboards and applications for waste management, smart bins, sorting robotics, and take-back vending machines. We analyze the data with the Japanese technology readiness assessment (J-TRA) approach. The results demonstrate that the monitoring dashboard and application are relatively more mature than the smart bin and sorting robots. Improving the smart bin readiness level might be the first key step to enable the rest of the technologies to advance into more collective applications.

**Keywords** Technology readiness assessment · AI · IoT · Plastic waste recycling · Smart technologies

## 1 Introduction

Countries worldwide have committed to reducing greenhouse gas (GHG) emissions to tackle climate change and global environmental crises [1]. Plastic production is associated with carbon emissions; in 2019, 3.4% of global GHG emissions accounted for the plastic lifecycle. Meanwhile, China banned plastic waste imports in 2017, and many countries increasingly rethink the route they manage their plastic waste, not only at the end of the chain, such as disposal and recycling but also

C. Shan (✉) · H. Onoda
Waseda University, Graduate School of Environment and Energy Engineering, Shinjuku City, Japan

A. H. Pandyaswargo
Waseda University, Environmental Research Institute, Shinjuku City, Japan



S. Fukushige et al. (eds.), *EcoDesign for Circular Value Creation: Volume I*,
https://doi.org/10.1007/978-981-97-9068-5_25

throughout the design, production, distribution, use, recollection, and transportation chains [2]. Each stage determines how effectively plastics can be retained at their finest quality throughout the material life cycle. Plastic waste should be recycled efficiently to accomplish plastic waste substance utilization significantly and avoid improper disposal such as landfilling. The most common techniques are mechanical recycling (MR), chemical recycling (CR), and feedstock recycling. Mechanical recycling is a physical process that involves cleaning, grinding, and pelletizing, followed by storage as a resource for the polymer-based goods market. The CR and feedstock recycling are distinct from MR, which focuses on the transformation of trash into gaseous or liquid molecules, followed by the manufacturing of new petrochemical products from feedstock [3]. Chemical Recycling is also utilized as a catalytic agent in steel production [4]. Regardless of recycling techniques, first, plastic waste should be effectively collected. The Organization for Economic Co-operation and Development (OECD) [2] reported that the large of plastic waste causes an environmental burden due to poor collection. Improving the effectiveness of collection for recycling is a global priority. Trash generation will continue to increase along with the expansion of the global economy. Recently, partially triggered by the COVID-19 pandemic, single-use hygiene gloves and packaging undoubtedly push the stress on waste collection [5]. Therefore, the development of smart technologies for waste management has been on the rise [6, 7]. How effective humans can wield smart technology to decrease waste mismanagement would majorly solve an environmental load of plastic waste. This study aims to reveal the readiness status of such technologies. The Internet of Things (IoT)-based waste management systems, smart bins, sorting robotics, and take-back vending machines (targeted for plastic bottles) are particularly useful in improving plastic waste recycling efficiency. This study employs the Japanese technology readiness level assessment (J-TRA) method to reveal the readiness of various smart plastic waste recycling technologies. The method uses seven parameters: supply and demand in a certain market, commercialization, technology development status, cost and risk, safety, societal integration, and verification for a sustainable operation.

## 2 Literature Review

### *2.1 The Plastic Waste Recycling Benefits from Smart Technologies Development*

The definition of smart technology varies depending on the context. However, the common elements of smart technologies are well understood. These elements are the IoT, artificial intelligence (AI), big data mining, and machine learning [8]. When we look at smart devices used in the environmental field, such as those based on big data applied for simulation waste flow and projection, many are powered by AI. This technological development has helped humans to achieve higher progress in environmental preservation. For example, smart wastewater monitoring, modeling for air pollution reduction, and better future prediction of solid waste fluctuation have helped to make proactive measures [9]. Furthermore, as plastic waste has contributed to a substantial part in the generation of solid waste during the last 19 years (2000–2019, 460 million tons) [10], there is an immense demand for automated collection and recycling.

Some studies have identified that efficiency is among the major concerns faced by plastic garbage collection and recycling [11–13]. Some studies offered solutions to improve efficiency. For example, appropriate scheduling for urban curbside garbage collection enabled efficiency and allowed existing trucks to deliver waste in huge quantities. Moreover, research has demonstrated that an IoT-enabled optimal navigation system can save travel mileage based on the location of waste bins and that the system will notify the driver of the status of the bins and if they need to be evacuated [14]. Moreover, IoT system-integrated smart bins have become functional to monitor the waste status in real time to inform the waste manager. A smart bin can detect waste entry through an identification sensor [15], and send waste volume status data to support efficient waste collection. An effective smart bin function is supported by a strong big-data construction within the increased sensors [12]. Most of the time, the radio frequency identification devices (RFID) tag server is equipped in a smart bin. The device can support large-scale information and construct wireless sensor networks (WSNs). These technologies are options for improving waste collection effectiveness [16]. On the other hand, a survey suggested that more than 16% of hard plastic waste only had a lifetime of under 1 year, and plastic film made of polyethylene (PP) and polyethylene (PE) is the most suitable type of waste for recycling [17]. Therefore, plastic waste sorting is another important step in plastic recycling.

Utilizing robots and robotic arms to sort plastic waste to find valuable components in mixed waste is a step toward identifying the materials involved [7]. he sorting robots are anticipated to improve efficiency through AI and machine learning advancements. An efficient sorting robot will increase the speed of identifying waste types, grabbing it, and moving it to a specific location, all under the control of software in the web of IoT. The AI-powered sorting robots also rely heavily on waste data. Training for sorting robots based on big data creation is key to producing a highly efficient robot [16]. Raptopoulos et al. [18] mentioned that the capacity to disassemble complicated plastic-based items would become an AI-driven trend in this industry. Smart technologies are anticipated to boost the accuracy of identifying and sorting materials for efficient recycling.

## 2.2 The Future for Smart Plastic Waste Collection and Processing

Some studies predicted the challenges of deploying smart technologies in environmental management. For example, Gutierrez et al. [12] argued that a massive use of sensors and stored big data will result in higher costs to maintain the system. The study pointed to this potential drawback despite the promised improved efficiency. Due to financial constraints, high costs may lead to a selective application of smart technologies. For example, it would only be applied to sources of waste already concentrated in large amounts. Plastic waste mismanaged, such as urban litter, may still need to be collected by street sweepers and concerned citizen groups [13]. However, many areas of smart technology can help with waste collection efficiency improvement that may also reduce costs. For example, efficient routing systems can create an optimized path for collecting waste only from selected filled bins [14]. This way, time and resources from visiting empty bins can be avoided. The direction of research required to inform policymakers in their support of the appropriate smart technologies has been outlined for the following areas: (1) the control instruments, such as law and regulation, to set environmental standards; (2) the market-based instrument, such as a tax incentive and monetary fines, and (3) information-based instruments, such as environmental information disclosure [19].

Although such research is needed by policymakers in developed and developing countries, it is more urgent for developed countries where the incineration rate of plastics is higher than that in developing countries [20]. Moreover, to firmly increase the implementation of plastic waste recycling with smart technologies, it is essential to create a regulatory environment that can create investor enthusiasm [21]. As it has been shown that the market of recycled products is driven by the development of recycling enterprises and technologies, such as pilot practices [22], unappealing recycling will cause disinvestment. For example, if a recycled plastic waste bottle textile is not cheaper than the original nylon, most people would choose the latter [23]. Therefore, it is necessary to take more criteria evaluation of waste technology readiness evaluation, especially the one that strongly considers society [24].

Various stages of waste management can be improved by smart technologies (Fig. 1.) Evaluating which specific technologies need attention for readiness improvement can accelerate the overall system's efficiency and stages. Particularly, the IoT aids the communication between the stages. This research primarily centers on the readiness analysis based on the J-TRL parameters: A: Market, B: Development of D: Verification, E: Safety, F: Commercialization, F: Cost and Risk.

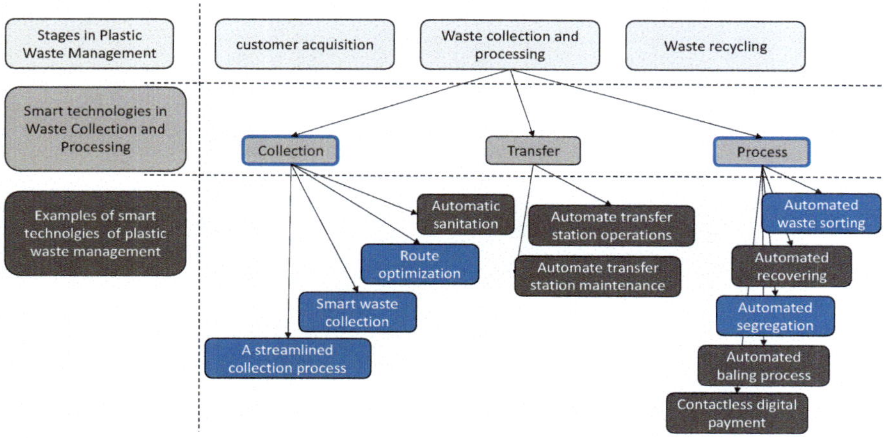


**Fig. 1** Employing smart technologies can improve the collection and processing of plastic waste, technologies covered by the database of this study are indicated by the blue color. (Diagram adapted from Ref. [25])

## 3 Materials and Methods

### *3.1 Resources and Database Building*

In this study, we collect open data from three sources: (1) private enterprises developing intelligent waste management technology, (2) research institutions publicly displaying the development of intelligent waste management technologies, including university laboratories, and (3) The governments' websites where they outline their intentions for adopting regional and national waste management systems, including intelligent technologies.

After gathering and organizing 100 data samples, as seen in Fig. 2, we found that the amount of the products from private enterprises is more than that of others, even though the academic institute is bigger than the government agencies.

Even though the sample size of academic institutions is modest for this research, the frequency where they provide the most recent research advancements and findings is high. Table 1 categorized the collected data based on the smart technology types related to certain waste recycling processes.

According to the collected data, the proportion of monitoring dashboards is greater than other systems. Furthermore, all identified technologies employ IoT, connecting all systems to enable end-user management [26]. The popularity of real-time monitoring and tracking of waste generation and collection status might influence this finding. Similarly, the coevolved AI technologies have enabled trash tracking and the design of the optimal path for truck pickup. IoT has also enabled Smart bin collection systems to improve the collection route and time depending on

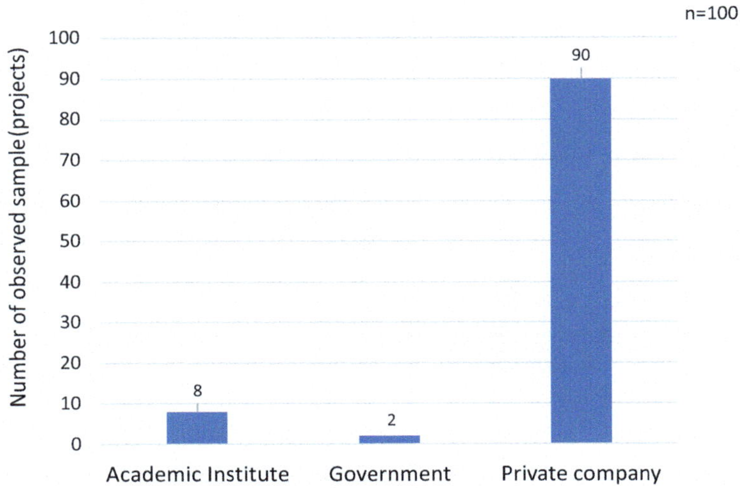


**Fig. 2** The distribution of sample data sources

**Table 1** The descriptive of smart technologies for waste collection

| The elements for smart plastic-waste recycling technology | Smart technologies | N | Components' characteristic |
|---|---|---|---|
| Monitoring + tracking + collection | Monitoring dash-board/application for waste managers | 43 | Hardware + Software |
| Monitoring + collection + classification | Smart bin | 29 | Hardware + Software |
| Sorting + Recycling | Sorting robotics | 24 | Hardware + Software |
| Reward-driven collection | Take-back vending machine | 4 | Hardware + Application |

real-time site conditions [27]. Another type of plastic waste collection smart technology is the cash-back collection vending machine, which typically accepts only specific trash types, such as polyethylene terephthalate (PET) bottles [28].

Sorting robots commonly select and separate certain types of plastic waste from mixed solid waste; some new designs intended to identify the specific polymer-based valuable waste (e.g., polypropylene, polystyrene, polyethylene). An optical sorting sensor usually equips this robot via the robot arm to grab the items and move them to a specified container. Currently, most of these technologies expect to combine AI to increase the motion of speed and accuracy.

## 3.2 The J-TRL Method

In this research, we apply the Japanese technology readiness assessment (J-TRA) method; this method originated from the technology readiness assessment method innovated by the National Aeronautics and Space Administration (NASA). Next, with an adaption that included societal consideration, the present J-TRA method was formed and applied to the Japanese Environmental Ministry (MOE) to evaluate relative technologies development [26].

The J-TRA method covers seven technology readiness evaluation categories in Table 2. The categories are A: Market, B: Technology development, C: Integration, D: verification, E: safety, F: commercialization, and G: cost/risk. This evaluation integrated the internal and external factors into the framework.

The calculation of J-TRA complies with the procedure outlined in [27]. The first step is attained by dividing the status level of the readiness category by the number of levels in that category. Then, the score is normalized as $X_{j,k}$ using Eq. (1). Then, Eq. (2) is employed to gain $N_k$. Finally, the final technology readiness level of the assessed smart technology is the minimum value found among the assessed categories (Eq. (3).

$$X_{j,k} = \frac{\sum_i^n X_i}{n} \tag{1}$$

$$N_k = \frac{X_{j,k}}{\text{Max}_k - \text{Min}_k} \times 100\% \times 8 \tag{2}$$

$$\text{TRL}_j = \text{Min}_{N_k} \tag{3}$$

**Table 2** The J-TRA TRL scoring matrix

| TRL | A | B | C | D | E | F | G |
|---|---|---|---|---|---|---|---|
| 1 | A-1 | B-1 | | D-1 | | | |
| 2 | A-2 | B-2 | | D-2 | | | |
| 3 | A-3 | B-3 | C-1 | D-3 | E-1 | F-1 | G-1 |
| 4 | A-4 | B-4 | C-2 | D-4 | E-2 | F-2 | G-2 |
| 5 | A-5 | B-5 | C-3 | D-5 | E-3 | F-3 | G-3 |
| 6 | A-6 | B-6 | C-4 | D-6 | E-4 | F-4 | G-4 |
| 7 | The equiment and systems have been finalized.<br>Manufacturing and introduction process has been completed. | | | | | | |
| 8 | Manufacturing and introduction process has been completed and is in the stage of mass production products. | | | | | | |

# 4 Results and Discussion

This study's findings are presented in three major sections: (1) the general readiness level of the smart plastic waste management technologies, (2) the changing trend of smart technologies for plastic waste recycling, and (3) the TRL scores calculated using the J-TRL approach.

## *4.1 The Technologies General Readiness Level*

The result of the general readiness level is illustrated in Fig. 3. The readiness levels are categorized into (a) feasibility study, (b) prototypes, (c) pilot-scale, and (d) full commercialization. The results indicated that the take-back vending machine had attained full commercialization compared to the smart bin. The high readiness level of the take-back-vending machine is because the system basis for the machine, such as sensors for PET, has matured. Furthermore, this technology's plastic bottle type of recognition has attained above 90% accuracy [28]. Closed-loop recycling has been proven to be a successful practice for plastic bottles collected in this way.

There are also smart bins, monitoring applications and dashboard, and sorting robotics that have reached full commercialization levels. However, the occurrence of sorting robotics in the gathered database is smaller than in the smart bin.

Conversely, there are more fully commercialized sorting robots than smart bins. This finding may imply that recycling companies have begun to address the efficiency challenge in recycling processes. It also indicates that waste management practices have shifted from the traditional focus of improving waste collection toward improving waste recycling and material recovery through best practice technologies [29].

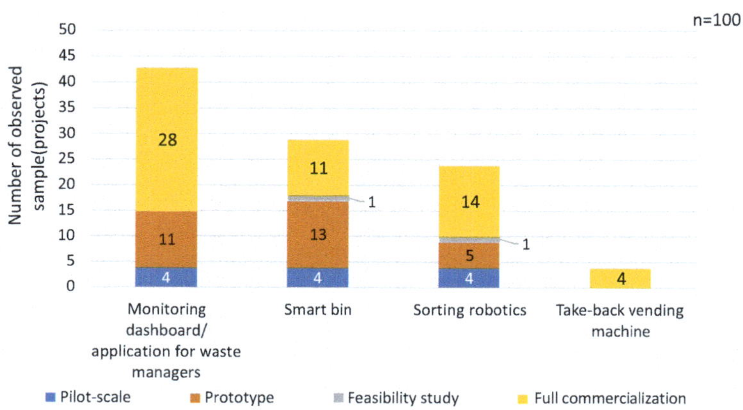


**Fig. 3** The general readiness level smart technologies for plastic waste in certain types

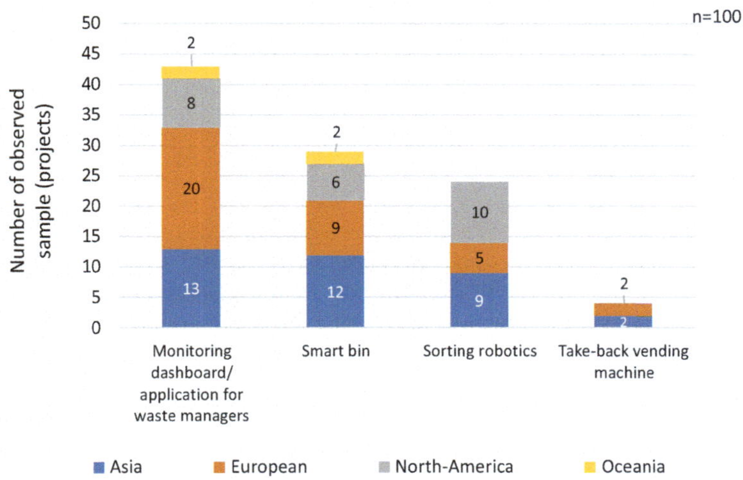


**Fig. 4** The general readiness level smart technologies for plastic waste recycling in different regional

Figure 4 presents the distribution of smart waste technologies for plastic recycling. The development of smart bins is balanced in the four referred regions (Asia, Europe, North America, and Oceania). This balance is a positive sign that efforts on collection have been made across the regions.

Moreover, a substantial part of development can be observed in monitoring dashboards and apps. We found that the European region takes the biggest share of the monitoring system for waste managers. It is not surprising since the European Investment Bank made several recommendations for product design incentives, including a minimum percentage of post-consumer recycled content (PCR) in all new products [30]. While, in general, Europe is known as the front end of waste disposal technology developers, it is interesting to observe that there are fewer technology development projects on sorting robots in Europe compared to North America and Asia. The diversity of technological focus shown by these regions may generate different streams of profits [31]. or take-back vending machines, there is a similar share in Europe and Asia, which may reflect the lifestyle of people in these two regions regarding the consumption of plastic bottled beverages [32].

## *4.2 Trends in Smart Plastic Waste Recycling Technologies Over the Years*

Figure 5 shows that the development of smart plastic waste technologies started with monitoring dashboards/applications in 1999 and experienced explosive growth in 2017. Two other technologies, the smart bin and sorting robots, experienced a

**Fig. 5** The smart technologies for plastic waste recycling evolved during time

similar significant increase in 2017. The increase rates of the smart bin and sorting robots are also similar. This observation revealed that the progress and integration of elemental technologies, such as IoT, data mining, and data analytics, may have surged during this period [33]. Eventually, the market saturation for these technologies may cause a gradual decline; however, this does not imply that the technologies will cease to advance, as the software has yet to embody the number of entities and will focus on updating with new intelligent abilities.

Furthermore, this may be the beginning of radically innovative products. For example, the increased trend for sorting robots had only started in 2013 [34]. Figure 5 also shows that although the growth of technologies peaked in 2017, it then entered a fluctuating phase from 2019 to 2023. It must be noted that the fluctuation happened during the COVID-19 pandemic period. The pandemic might have increased the demand for smart bins. An array of contactless and new collection methods may continue to be developed to respond to this new demand.

## 4.3 *J-TRL Assessment Results*

The results of J-TRL are presented in Fig. 6. The radar chart shows the readiness levels of each J-TRA parameter of the four smart technologies for plastic waste recycling.

It can be observed that all technologies are above the average 4-point level for the technology development category. This finding is a sign that technology-wise, the development of smart technologies has already taken pace. The conventional

**Fig. 6** J-TRL result of smart technologies for plastic waste recycling

recycling phase may have been gradually left as people aim to increase collection, dismantling, and sorting efficiency. The monitoring dashboard and apps have also attained similar commercialization readiness to two other technologies (smart bin and sorting robotics). This is mainly because an IoT system usually powers the monitoring app and dashboard. An IoT system usually relies on big data construction. The availability of data sources for the smart bin and sorting robotics will continue to advance AI technologies [24]. Moreover, a good performance of these new smart technologies will determine how these technologies integrate into society [35]. Adopting the take-back vending machine experience has demonstrated this success in several societies.

Observing results in this study, we argue that the smart bin, which would be the smart technology type that people would encounter for their waste collection, would be a key component in plastic waste collection and processing. Improving its readiness level would have a pronounced effect on effectively expanding the market and increasing the level of business of smart waste technologies. A high readiness smart bin will also induce cost savings and reduce risks. For the overall waste management stages, improving the IoT system will improve communication between collection and processing, and eventually, a smooth collective implementation of smart waste management technologies.

# 5 Summary

The J-TRL readiness level of monitoring dashboards and apps is higher than that of the smart bin in six parameters (market, cost and risk, technology, safety, verification, integration, and commercialization). Some uncertainty was identified in the speed of smart bins' commercialization. Predictably, the delay in the technology development of smart bins is caused by the technology's integration of software and hardware requirements. The "cost and risk" index of fourth-generation technologies is usually still unattractive due to the high cost of operation for software that supports IoT and AI. There might also be a lack of short-term interest in these technologies, such as waste sorting, relative to manual labor.

Moreover, there is a need for employee training to control the smart bin by means such as the vision system if they adopt these new smart technologies. As a result, the integration is slightly delayed than the monitoring dashboards and apps. The monitoring dashboards and apps are more ready regarding cost and risk. It was observed that this technology is mostly utilized for the garbage collection phase to monitor the waste status in the bin and notify the driver when the curbside waste bin has been emptied. Finally, the take-back vending machine scores highly because all the vending machines in the sample are ready for full commercialization.

The readiness level of sorting robots indicates they are close to the smart bin, but the readiness of cost and risk is higher than that of the smart bin. Depending on the location's infrastructure, the smart bin will be placed in a curbside or certain transportation spot; this poses some variation in operational risks. The general readiness result of all the observed sample data suggests that smart bins are still in the pilot stage (14%) and prototype (45%). Consequently, we discovered that sorting robots comprise 58% of fully commercialized smart technologies, whereas 38% consist of smart bins. Most technologies performed poorly on the "cost and risk" index. These four smart technology types are not ready regarding the "cost and risk" because the operating cost depends on the electricity supply, which faces a sustainable supply. However, there is hope for a price decline should the growth of electricity supply from sustainable sources continue in a positive direction.

**Acknowledgments** This research is supported by the Japan Society for the Promotion of Science under project number: 23 K11548 titled "Improving the social acceptance of smart technologies: A people-centric approach for sustainable smart waste management and smart mobility technologies" and the Environmental Restoration and Conservation Agency under project number: JPMEERF20213G01 titled "Development of Contactless Garbage Collection Systems and Scenarios Construction for Social Implementation."

## References


1. UNEP (2023) Emissions gap report 2022. UNEP. https://www.unep.org/resources/emissions-gap-report-2022. Accessed 14 Jun 2023
2. OECD (2022) Plastic pollution is growing relentlessly as waste management and recycling fall short, says OECD. OECD. https://www.oecd.org/newsroom/plastic-pollution-is-growing-relentlessly-as-waste-management-and-recycling-fall-short.htm. Accessed 5 Mar 2023
3. Ragaert K, Delva L, Van Geem K (2017) Mechanical and chemical recycling of solid plastic waste. Waste Manag 69:24–58. https://doi.org/10.1016/j.wasman.2017.07.044
4. Nomura S (2020) Blast furnace feedstock and coke oven chemical feedstock. In: Letcher TM (ed) Plastic waste and recycling. Elsevier, pp 321–358
5. Leal Filho W, Salvia AL, Minhas A, Paço A, Dias-Ferreira C (2021) The COVID-19 pandemic and single-use plastic waste in households: a preliminary study. Sci Total Environ 793:148571. https://doi.org/10.1016/J.SCITOTENV.2021.148571
6. Ebner N, Iacovidou E (2021) The challenges of Covid-19 pandemic on improving plastic waste recycling rates. Sustain Prod Consum 28:726–735. https://doi.org/10.1016/J.SPC.2021.07.001
7. Onoda H (2020) Smart approaches to waste management for post-COVID-19 smart cities in Japan. IET Smart Cities 2:89–94. https://doi.org/10.1049/IET-SMC.2020.0051
8. Shah SFA, Iqbal M, Aziz Z, Rana TA, Khalid A, Cheah YN, Arif M (2022) The role of machine learning and the internet of things in smart buildings for energy efficiency. Appl Sci 12:7882. https://doi.org/10.3390/APP12157882
9. Ye Z, Yang J, Zhong N, Tu X, Jia J, Wang J (2020) Tackling environmental challenges in pollution controls using artificial intelligence: a review. Sci Total Environ 699:134279. https://doi.org/10.1016/J.SCITOTENV.2019.134279
10. OECD (2022) Plastic leakage and greenhouse gas emissions are increasing. Organisation for Economic Co-operation and Development. https://www.oecd.org/environment/plastics/increased-plastic-leakage-and-greenhouse-gas-emissions.htm. Accessed 13 Jun 2023
11. Kaza S, Yao LC, Bhada-Tata P, Van Woerden F (2018) What a waste 2.0. A global snapshot of solid waste management to 2050. https://doi.org/10.1596/978-1-4648-1329-0
12. Gutierrez JM, Jensen M, Henius M, Riaz T (2015) Smart waste collection system based on location intelligence. Procedia Comput Sci 61:120–127. https://doi.org/10.1016/J.PROCS.2015.09.170
13. Lebreton L, Andrady A (2019) Future scenarios of global plastic waste generation and disposal. Palgrave Commun 5:1–11. https://doi.org/10.1057/s41599-018-0212-7
14. Haque KF, Zabin R, Yelamarthi K, Yanambaka P, Abdelgawad A (2020) An IoT based efficient waste collection system with smart bins. In: IEEE world forum on internet of things, WF-IoT 2020 - symposium proceedings
15. Baras N, Ziouzios D, Dasygenis M, Tsanaktsidis C (2020) A cloud based smart recycling bin for waste classification. In: 2020 9th international conference on modern circuits and systems technologies, MOCAST 2020
16. Sarc R, Curtis A, Kandlbauer L, Khodier K, Lorber KE, Pomberger R (2019) Digitalisation and intelligent robotics in value chain of circular economy oriented waste management–a review. Waste Manag (Elmsford) 95:476–492. https://doi.org/10.1016/j.wasman.2019.06.035
17. Faraca G, Astrup T (2019) Plastic waste from recycling centres: characterisation and evaluation of plastic recyclability. Waste Manag 95:388–398. https://doi.org/10.1016/J.WASMAN.2019.06.038
18. Raptopoulos F, Koskinopoulou M, Maniadakis M (2020) Robotic pick-and-toss facilitates urban waste sorting. In: IEEE international conference on automation science and engineering 2020-August, pp 1149–1154. https://doi.org/10.1109/CASE48305.2020.9216746
19. Liao Z (2018) Content analysis of China's environmental policy instruments on promoting firms' environmental innovation. Environ Sci Pol 88:46–51. https://doi.org/10.1016/J.ENVSCI.2018.06.013

20. Wen Z, Xie Y, Chen M, Dinga CD (2021) China's plastic import ban increases prospects of environmental impact mitigation of plastic waste trade flow worldwide. Nat Commun 12(1): 425. https://doi.org/10.1038/s41467-020-20741-9
21. Zhang A, Venkatesh VG, Liu Y, Wan M, Qu T, Huisingh D (2019) Barriers to smart waste management for a circular economy in China. J Clean Prod 240:118198. https://doi.org/10.1016/J.JCLEPRO.2019.118198
22. Alcayaga A, Wiener M, Hansen EG (2019) Towards a framework of smart-circular systems: an integrative literature review. J Clean Prod 221:622–634. https://doi.org/10.1016/J.JCLEPRO.2019.02.085
23. Subramanian K, Chopra SS, Cakin E, Li X, Lin CSK (2020) Environmental life cycle assessment of textile bio-recycling—valorizing cotton-polyester textile waste to pet fiber and glucose syrup. Resour Conserv Recycl 161:104989. https://doi.org/10.1016/J.RESCONREC.2020.104989
24. Fatimah YA, Govindan K, Murniningsih R, Setiawan A (2020) Industry 4.0 based sustainable circular economy approach for smart waste management system to achieve sustainable development goals: a case study of Indonesia. J Clean Prod 269:122263. https://doi.org/10.1016/J.JCLEPRO.2020.122263
25. Akihisa O, Hadi PA, Ryota T, Hiroshi O (2023) Demonstration of a contactless waste collection system: a Japanese case study. IET Smart Cities 5:303–316. https://doi.org/10.1049/smc2.12065
26. Ihara I, Andante Hadi P, Onoda H (2018) Development and the effectiveness of the J-TRA: a methodology to assess energy technology R&D programs in Japan. In: EcoDePS 2018 proceedings. Japan, pp 109–117
27. Pandyaswargo AH, Maghfiroh MFN, Onoda H (2023) Global distribution and readiness status of artificial intelligence application on mobility projects. Energy Rep 9:720–727. https://doi.org/10.1016/J.EGYR.2022.11.070
28. Korucu MK, Kaplan Ö, Büyük O, Güllü MK (2016) An investigation of the usability of sound recognition for source separation of packaging wastes in reverse vending machines. Waste Manag 56:46–52. https://doi.org/10.1016/J.WASMAN.2016.06.030
29. Esmaeilian B, Wang B, Lewis K, Duarte F, Ratti C, Behdad S (2018) The future of waste management in smart and sustainable cities: a review and concept paper. Waste Manag 81:177–195. https://doi.org/10.1016/J.WASMAN.2018.09.047
30. OECD (2022) Global plastics outlook. OECD, Paris, France
31. Lim J, Ahn Y, Kim J (2023) Optimal sorting and recycling of plastic waste as a renewable energy resource considering economic feasibility and environmental pollution. Process Saf Environ Prot 169:685–696. https://doi.org/10.1016/J.PSEP.2022.11.027
32. Pavan VKS, Akhil VSSS, Kumar PNS, Priya BK (2021) Implementation of reverse vending machine technique for dry waste and e-waste. In: 2021 5th international conference on electronics, materials engineering and nano-technology, IEMENTech 2021. https://doi.org/10.1109/IEMENTECH53263.2021.9614824
33. Zhou J, Jiang P, Yang J, Liu X (2021) Designing a smart incentive-based recycling system for household recyclable waste. Waste Manag 123:142–153. https://doi.org/10.1016/J.WASMAN.2021.01.030
34. Furman J, Seamans R (2019) AI and the economy. Innov Policy Econ 19:161–191. https://doi.org/10.1086/699936/ASSET/IMAGES/LARGE/FG9.JPEG
35. Nižetić S, Djilali N, Papadopoulos A, Rodrigues JJPC (2019) Smart technologies for promotion of energy efficiency, utilization of sustainable resources and waste management. J Clean Prod 231:565–591. https://doi.org/10.1016/J.JCLEPRO.2019.04.397

# Research Framework for Urban Industrial Symbiosis

**Philipp Grimmel, Katja Knecht, Grace Abou Jaoude, Mark Mennenga, Olaf Mumm, Christoph Herrmann, and Vanessa Miriam Carlow**

**Abstract** This chapter proposes a novel framework for the development of computational and data-enhanced tools to support urban industrial symbiosis integrating perspectives from engineering, urban planning, and design. The framework serves as a basis for the conceptual development of a research demonstrator that supports the identification, analysis, and design of the interactions between different urban (sub)-systems. The framework aims to facilitate synergetic exchanges between different urban systems to increase resource efficiency, reduce waste, and provide societal, environmental, and economic benefits.

**Keywords** Urban Industrial Symbiosis · Tool Development · GIS · ICT

## 1 Introduction

By 2050, roughly 7 billion people (68% of the world population) are expected to live and work in urban regions [1]. The rising urban population and ongoing urbanization processes will be accompanied by a growing demand for resources, an inherent increase of greenhouse gas emissions, and magnified challenges related to urban resource management [2]. Considering the direction provided by recent policy initiatives, such as the EU Green Deal [3] and EU Circular Economy Action Plan [4], there is an imperative to promote the circular economy and the broadening of industrial symbiosis (IS) activities emphasizing the need for research in this domain.

---

Philipp Grimmel and Katja Knecht contributed equally to this work.

P. Grimmel (✉) · M. Mennenga · C. Herrmann
Chair of Sustainable Manufacturing and Life Cycle Engineering, Institute of Machine Tools and Production Technology (IWF), Technische Universität Braunschweig, Braunschweig, Germany
e-mail: p.grimmel@tu-braunschweig.de

K. Knecht · G. Abou Jaoude · O. Mumm · V. M. Carlow
Institute for Sustainable Urbanism (ISU), Technische Universität Braunschweig, Braunschweig, Germany

 
S. Fukushige et al. (eds.), *EcoDesign for Circular Value Creation: Volume I*,
https://doi.org/10.1007/978-981-97-9068-5_26

These policy initiatives advocate for the reuse of resources to prolong their life cycle and facilitate waste-to-resource exchanges not only between industries but also by considering other urban stakeholders.

The concept of urban industrial symbiosis (UrIS) is a promising strategy for linking sustainable urban development and sustainable production through collaboration and exchange of resources of industrial and non-industrial actors [5]. However, one of the key challenges of UrIS is to understand and develop appropriate methods and tools to organize the interactions between different urban (sub)systems, leading to increased resource efficiency, reduced waste, reduced transport, and improved societal, environmental, and economic benefits [6]. This includes understanding potential synergies and relations between different land uses in cities and how urban (sub)systems and their governance can be optimized to lower or eliminate negative environmental, societal, and economic impacts. Previous approaches for promoting the establishment of synergies have primarily focused on solutions between industrial partners in close vicinity, such as in eco-industrial parks (EIP), and provided benefits for resource matching, thereby facilitating the identification of possible symbiosis partners without considering the potential of UrIS [7]. Given the complexity of interactions, networks, and processes, holistic frameworks that consider synergetic exchanges between different urban components and transcend disciplinary boundaries are necessary for system optimization [8]. Furthermore, the development of processes and strategies and their implementation for UrIS require new methods and tools to link the different systems, functions, and flows.

To this purpose, we propose a framework for the development of computational and data-enhanced tools to support UrIS that integrates perspectives from engineering, urban planning, and design. The framework is intended to provide a research agenda supporting the development of diverse interdisciplinary research activities in the field and lead to a research demonstrator. As a sandbox and collaborative research environment, the demonstrator will allow to test newly developed methods and tools that promote the identification, establishment, and management of symbiosis activities and relationships between different stakeholders in urban areas and thus further an interdisciplinary research agenda in the field of UrIS [9].

This chapter is structured as follows. First, we introduce the theoretical and applied concepts underlying UrIS (Sect. 2), which is followed by an overview and discussion of functions, opportunities, and shortcomings of existing computational tools and platforms in the domains of IS (Sect. 3.1), urban digital twins, and geospatial technologies (Sect. 3.2). Based on this, we identify essential functionalities and requirements for the development of computational tools to promote UrIS and discuss how to implement and deploy these solutions in the scope of a demonstrator (Sect. 4).

## 2 Concepts and Approaches Underlying Urban Industrial Symbiosis

In recent years, circular economy (CE) has gained momentum as solutions to transition from the traditional linear model to closed-loop production. CE offers comprehensive solutions that are not limited to waste management and recycling, reuse, and recovery of materials. The concept extends to the entire life cycle of industrial processes and their interaction with the environment at various scales [7]. Fostering the CE at different scales requires shifts at the city or regional levels, industrial level as well as the consumer level. IS supports the CE through synergetic exchanges and networks across industry sectors [7]. IS typically is defined as a collective process or collaborative approach where separate industries exchange by-products [8, 12]. From a systems perspective, IS is perceived as a process where the interactions of different firms generate an outcome over time [13]. The concept widely proliferated in literature and industrial practice with the success of the Kalundborg EIP (Denmark) as a prominent example of IS [14].

UrIS further expands and generalizes the concept of IS by incorporating additional stakeholders as well as a wider variety of interactions and exchanges within a broader geographic scope. Based on Chertow's (2000) definition of IS [12], Urban Symbiosis (UrS) can be defined as the engagement of traditionally separate entities in an urban environment (e.g., companies, industries, public or private institutions, households) in a collective approach fostering the exchange of physical and/or immaterial resources (e.g., knowledge, expertise, and/or information) [15–17]. If one of the symbiosis partners is an industry, e.g., using by-products from the urban environment as resources for industrial operations [5] or providing industrial waste as input to urban processes [18], this can be referred to as Urban Industrial Symbiosis (UrIS).

Hitherto, geographic proximity between symbiosis partners has played an important role in the development of EIPs or clusters. In particular relevant, if resource exchanges require specific infrastructures, or if infrastructure sharing is part of a synergetic relationship [11]. In scenarios like these, geographic proximity helps for instance to reduce both transport costs and mental distance. In cities urban infrastructure can expand the range of viable synergies [19]. Economic benefits of expanding the geographic search space can result in higher volumes and a diversity of tradable resources as well as in an increase and heterogeneity of potential partners, which reduces economic risks [7]. Nevertheless, geographic proximity can support early phases of the symbiosis process by promoting the rapid development of local networks and trust among partners through participation [19]. Furthermore, for the implementation of a symbiosis, the geographic context is of importance to industrial and commercial partners with respect to the local availability of skilled labor, relevant infrastructures, e.g., for services, and transport, access to finance and capital, as well as joint general regulatory frameworks and conditions due to local policies and environmental standards [19, 20].

The larger and more diverse the number of available resources and partners, the higher is the potential of synergy opportunities for the urban environment. However, this higher potential both increases the system complexity and creates a need to support actors in finding and pursuing synergy opportunities [10]. Digital platforms and computational tools have the potential to facilitate these processes [10].

# 3 Potentials and Short-Comings of Existing Computational Tools and Platforms for the Promotion of UrIS

In recent years, the increased development of digital and, in particular, web-based platforms and tools was witnessed to help to further expand networks and diversify symbiosis opportunities [21]. Although these efforts have primarily targeted IS, their analysis can serve as a starting point for the identification of relevant functionalities, opportunities, and shortcomings in preparation for the development of a research framework. In combination with an understanding of the potentials of existing geospatial technologies and urban digital twins, these insights can holistically inform its conceptual development.

## *3.1 Analysis of Existing IS/UrIS Tools and Platforms*

Existing reviews on IS tools and platforms have investigated their contribution to and their role in different stages of the IS creation process, either across the different stages using IS process phases as basis for categorization [6, 21, 22], in specific phases, such as for the identification of synergy potentials [23], or toward an integration of different waste streams in IS process design [24]. We present a synthesized overview of the creation stages, including a description of functions, a brief evaluation, and examples of existing IS tools that fall into respective categories in Table 1, for which the harmonized framework of IS creation processes by Yeo et al. (2019) [21] serves as a basis. It distinguishes the following phases: (a) pre-assessment, (b) engagement of companies, (c) identification of synergy opportunities, (d) business feasibility, (e) implementation of transactions, (f) documentation and reinforcement [21]. While stages (a) to (c) are important for identifying opportunities for new IS, steps (d) and (e) focus on the operational phase, the establishment and maintenance of a symbiosis [21]. Phase (f) contributes to a central knowledge base, which can be harnessed for the advancement of IS tools [21]. While the framework overlaps with the IS development phases defined by Grant et al. (2010) [6] and Maqbool et al. (2019) [22], who use a revised version of Grant et al.'s classification system and evaluation criteria [6], for stages c) to f), it extends these by adding the early stages of a) pre-assessment and b) engagement of stakeholders.

**Table 1** Synthesized categorization of existing digital platforms and tools according to the framework of IS creation steps

| IS Creation Steps (Yeo et al. 2019) | (a) Pre-Assessment | (b) Engagement of Stakeholders | (c) Identification of Synergy Opportunities | (d) Feasibility Assessment | (e) Implementation of Transactions | (f) Documentation and Reinforcement |
|---|---|---|---|---|---|---|
| Functions | Assessment of the local context and conditions:<br>• Potential network partners<br>• Potential resource, process, and infrastructure availability<br>• Needs and requirements<br>• Existing policies and regulations | Building and growing of IS networks:<br>• Creating awareness<br>• Generating interest<br>• Establishing trust | Matching of potential synergy partners based on:<br>• Free market supply and demand<br>• Characteristics of resource streams, matching of inputs and outputs of processes<br>• Network design and optimization<br>• Industry codes | Evaluation and validation of synergy opportunities with respect to:<br>• Economic feasibility (costs-benefits, risks)<br>• Legal and regulatory compliance<br>• Environmental, social, technical, and spatial assessment criteria | Establishing and maintaining a symbiosis:<br>• Handling of flows of resources<br>• Monitoring<br>• Performance assessment<br>• Eliminating barriers and challenges in the process of IS realization<br>• Commercialization and adaptive management | Creation of a shared IS knowledge base that can inform creation of future IS:<br>• Capturing and sharing information and experiences on successful symbiosis<br>• Documentation of historical case studies<br>• Development of standards |
| IS/UrIS Tool Examples | • SymbioGIS [27] LoopLocal<br>• SWAN [30] | • Predominantly analog, e.g., networking events, calls, mailings, workshops, or via existing digital channels, e.g., emails or social media platforms [21] | • e-Symbiosis platform [28]<br>• SWAN [30]<br>• SymbioGIS [26]<br>• SymbioSyS [31]<br>• SYNERGie® 4.0 [29] | • SWAN [30]<br>• SymbioGIS [26]<br>• SYNERGie® 4.0 [29] | • SYNERGie® 4.0 [29] | • SYNERGie® 4.0 [29] |

(continued)

**Table 1** (continued)

| IS Creation Steps (Yeo et al. 2019) | (a) Pre-Assessment | (b) Engagement of Stakeholders | (c) Identification of Synergy Opportunities | (d) Feasibility Assessment | (e) Implementation of Transactions | (f) Documentation and Reinforcement |
|---|---|---|---|---|---|---|
| Evaluation | • Data-driven<br>• Requires respective geospatial, statistical, and resource-related datasets [21] | • No specific IS tool exists<br>• Usage of predominantly analog methods and tools, e.g., networking events, calls, mailings, emails, workshops [21] | • Majority of existing developments focus this phase [22]<br>• Logic-driven stage with different approaches to matchmaking [21] | • Requires assessment criteria, metrics, indicators, analysis methods, and models, as well as simulations for performance assessment<br>• Supports decision-making [21] | • Requires real-time handling of data on resource flows and transactions for (predictive) performance assessment [21, 22]<br>• Supports decision-making and requires integration into existing project management workflows [21] | • Text-based repositories, wikis [21]) |

### Pre-assessment

In the preliminary assessment stage, the local context and conditions are evaluated in respect to available resources, infrastructures, networks, stakeholders, and existing processes to pre-determine the general prospects for symbiosis in the region [21]. This stage is dependent on the availability of geospatial, statistical, and resource-related datasets, which can be used to estimate probable waste flows [25] or analyze input and output flows of regions [26]. In combination with Geospatial Information Systems (GIS) the available information can be brought and evaluated in its spatial context, e.g., in form of map-based visualizations of datasets (e.g., SymbioGIS, [26]) or heat maps visualizing potentials for development of IS/UrIS activities (e.g., LoopLocal, [24]).

### Engagement of Stakeholders

Although the engagement of businesses—or in the context of UrIS more generally stakeholders—is a key activity in preparation of symbiotic exchanges and collaboration, this phase has not yet been specifically addressed in IS tool development [21]. Activities in this phase therefore revert to traditional, nonspecialized, and often analog methods and tools for communication and engagement, such as networking events, calls, mailings, emails, workshops, or the use of existing social media platforms [21]

### Identification of Synergy Opportunities

The majority of tools and platforms that have been reviewed in the past have been built to support the identification of synergy opportunities and match-making [22]. Two main general approaches can be distinguished: (i) matching based on free market supply and demand approaches, e.g., in the form of e-marketplaces such as e-Symbiosis [28] or (ii) based on input-output streams of processes considering resource characteristics, which can be further enhanced by incorporating network design and optimization, also leading to emergent behavior [18]. Grant et al. (2010) [6] distinguish a third approach to match-making, namely based on industry codes. Notable examples of e-tools and platforms enabling the matching based on process input-output streams are the SYNERGie® 4.0 platform by International Synergies Limited [26], SymbioSyS [28], the SWAN platform [27], and SymbioGIS [23]. Although match-making results are commonly presented in table form, distance calculated based on location-information has been included as a filter category for sorting matches, e.g., in the SYNERGie® 4.0 Platform [26].

### Feasibility Assessment

Although defined as its own stage and category, feasibility assessment is closely interwoven with the identification of potential synergies. In this step, the economic, legal and regulatory, environmental, social, technical, and spatial performance of synergy opportunities is assessed to determine their feasibility and viability based on a varying set of criteria, metrics, indicators, analysis methods and models, as well as simulations, to support decision-making [6, 21, 22]. Platforms such as SWAN [30], SYNERGie® 4.0 [29], and SymbioGIS [26] support this phase. Whereas a main focus of assessment in this phase lies on determining economic feasibility, e.g., using cost-benefit analysis, SWAN also assesses potential synergies with respect to their technical viability [30]. Furthermore, the platforms SymbioSyS and SymbioGIS take the geospatial context into consideration, both for synergy identification and assessment, by employing georeferencing to map opportunities as well as GIS to analyze distances, logistics and, in the case of SymbioSyS, to propose sites for facilities and infrastructure planning [26, 31].

### Implementation of Transactions

Few existing tools support putting a synergy into action and maintaining it, which also includes the removal of barriers, commercialization, and adaptive management, as well as the ongoing monitoring of synergies that are in place [6, 21]. A potential reason for this is that this development step requires a close integration into existing business activities and therefore individual solutions for stakeholders [21]. An exception is the SYNERGie® 4.0 platform, which also provides a management system for IS as well as reporting functionalities [29]. Looking ahead, the integration of Internet-of-Things (IoT) and digital twin technologies in IS/UrIS platforms can support real-time data integration and thus the monitoring and performance assessment of ongoing symbiosis activities further to support management decisions [32].

### Documentation and Reinforcement

Documentation and sharing of successful symbiosis projects and case studies contribute to the creation of a knowledge base to further advance the field. Yeo et al. (2019) identify the existence of several localized initiatives and data bases that support this step [21].

### Challenges and Opportunities

While these various platforms and tools have been introduced to support different IS creation steps, the application of these digital tools for IS has remained limited in industrial practice [21]. Challenges often revolve around data availability, company reluctance to share information, the costs associated with utilizing such tools as well as an unequal share of benefits [7, 23, 33]. Also, the perspective of users, their capabilities, and needs are often neglected in the development of such tools and platforms, in which the focus is often placed on functional rather than social and design aspects or business integration [22, 34]. These aspects have led to the discontinuation of a large number of previously developed IS tools and applications in the past, although their potentials for enhancing different IS creation steps have been shown [23, 34].

With the exception of platforms such as SYNERGie® 4.0, existing tools and platforms tend to focus on supporting only one or a low number of symbiosis creation steps, in particular neglecting the operational phases of synergy development and stakeholder engagement. Furthermore, they are not built for interoperability and cannot be easily combined or integrated into other existing applications [34]. A platform ecosystem perspective and modular, extensible approach to the development of applications for UrIS would not only allow to combine efforts but also provide the opportunity for stakeholders to customize their platform [34, 35].

While most platforms or tools aim to support decision-making and expand IS processes, few have focused on or extended their application to synergetic exchanges between industry and other urban functions and thus did not consider these spatiotemporal potentials. Also, with exception of SymbioSyS and SymbioGIS, the integration of both geospatial data and data on material and energy flows to identify, assess, and visualize synergistic exchanges remains limited in literature and in industrial practice [27, 31].

## *3.2 Potentials of Urban Digital Twin and Geospatial Technologies*

Spatial data about cities has long been used to analyze and manage services and infrastructures, predict latent urban challenges, plan and devise solutions and policies. Accordingly, a variety of 3D geospatial tools and platforms that can rearrange and visualize data onto a three-dimensional landscape have been widely used in urban planning [36]. Examples of these include ESRI's ArcGIS velocity and CityEngine [37], Geodesign Hub [38], and Virtual City Systems [39]. These spatial software environments enable planners to visualize often complex data sets not only through three-dimensional landscapes but also through dashboards that use interactive graphics to represent information. Beyond data visualization, the tools enable planners to conduct spatial analysis, statistical modelling as well as predict and simulate various scenarios. In recent years, furthermore, specific urban digital twin platforms have been developed, which support the integration and linkage of semantic and geospatial properties of urban systems [40, 41]. Examples include the 3DEXPERIENCE platform from Dassault Systèmes, which has formed, for example, the basis for the Digital City Twin development for Singapore, Virtual

Singapore [42]; another example are the iTWIN platform and OpenCities Planner from Bentley Systems (2023) [43]. What distinguishes urban digital twins from other applications used in the field, such as regular 3D models and building information modeling (BIM), is a continuous, mutual interaction between real-world systems and their virtual twin enabled by IoT and artificial intelligence (AI) [41].

Shahat et al. (2021) identified potentials and use cases for urban digital twins in the areas of (a) data management, (b) visualization, (c) situational awareness, d) planning and prediction, as well as e) integration and collaboration [41]. This can provide relevant insights and complement the criteria established in IS literature for the conceptual development of a research framework and identification of potential functionalities to address in a research demonstrator for UrIS (see Table 2 for a summary and overview).

### Data Management

Data management is considered key, as such a platform is required to integrate, manage, and process large amounts of heterogeneous data sets from varied sources [41]. Essential requirements for a platform are data processing, ensuring interoperability and the integration and fusion of different software [41]. However, pivotal is the access to and with it the sharing and exchange of data sets, which includes the development of an open-data or data-sharing approach [44].

### Visualization

Visualization is another important aspect, as it supports the experience and understanding of the urban environment and the underlying interdependencies [41]. In the context of digital urban twins, this includes real-time representations of the city in 3D, e.g., with support by VR and AR applications, but also how to incorporate, represent, and visualize additional datasets, multi-spatial and temporal scales, behavior models using agent-based modelling, or results from network analysis in the model or map-based representations [41]. Additional functionalities to consider include user navigation and the development of personalized information systems, which support the usability and user experience of the platform [41].

### Situational Awareness

Situational awareness ensures the connection of the digital twin to its physical urban systems by collecting, monitoring, and managing data, e.g., using sensors and IoT devices, as well as constitutes the enabling property for urban planning and governance to understand the state and performance of the urban systems at any time by, e.g., using data analysis [41, 45].

**Table 2** Overview of functions and potentials of urban digital twins according to the themes identified by Shahat et al.

| Characteristics of Urban Digital Twins | (a) Data Management | (b) Visualization | (c) Situational Awareness | (d) Planning and Prediction | (e) Integration and Collaboration |
|---|---|---|---|---|---|
| Functions | • Integration, management, and processing of large amounts of heterogeneous data sets from varied sources<br>• Interoperability, integration, and fusion of different software | • Visualization of datasets, e.g., on multi-spatial and temporal scales, behavior models, and network dynamics<br>• Support of an understanding of the urban environment and of a 3D real-time experience, e.g., by using AR/VR<br>• User navigation<br>• Provision of personalized information systems | • Collection, monitoring, and management of historical and real-time data sets, e.g., by employing IoT | • Evaluation of Historical and real-time data sets by pattern identification, detecting changes, and predicting future performance | • Integration of urban (sub)systems and processes, as layers or coupled models<br>• Provision of collaborative environment through data sharing and engagement practices |
| Potentials | • Access to, sharing, and exchange of data sets with an open data or data-sharing approach | • Can depict the status quo but also used for scenario planning and the visualization of planning alternatives | • Data analytics and spatial analysis of collected data enables planning and governance | • Support of decision-making processes<br>• Sandbox for development and testing of scenarios | • Multiple domains integration<br>• Stakeholder participation and engagement<br>• Open platform |

**Planning and Prediction**

The data captured and incorporated in the urban digital twin can be further employed for the purpose of planning and prediction to support decision-making processes. An analysis of historical data sets allows, for example, to identify patterns and learn from the environment or to detect changes and predict likely future performances [41, 44]. Furthermore, the urban digital twin can provide a sandbox for the development and testing of scenarios employing, for example, simulation techniques [41, 44].

**Integration and Collaboration**

Lastly, important aspects to consider are the integration of different layers and models to sufficiently address the complexity of the city and portray the interdependencies and relationships between different domains and elements as well as providing an environment that supports the collaboration and participation of different stakeholders and the engagement of citizens [41]

# 4 Research Framework to Promote UrIS

Based on the findings of Sect. 3, we propose a synthesized, holistic framework for research and the development of applications promoting UrIS by supporting the identification, establishment, and management of symbiosis activities and relationships between different stakeholders in urban areas.

Integrating functionalities of urban digital twins and geospatial technologies commonly used in urban planning as discussed in Sect. 3.2 can enhance the development of applications to better address the specific characteristics and requirements of UrIS as discussed in Sect. 3.1.

## *4.1 Conceptual Framework*

The analysis of existing tools and platforms has revealed shortcomings. These concern foremost the integration of geospatial information and respective visual representations, the engagement of stakeholders, user-centricity, modularity, as well as the support of all phases of the symbiosis creation processes described in Sect. 3.1. An analysis of the potentials of existing geospatial and urban digital twin technologies has in turn brought to light the importance of visualization techniques and data management, the integration of different applications and systems as well as of situational awareness, and the relevance of these tools for decision-making, planning, and prediction. The proposed framework brings these aspects together

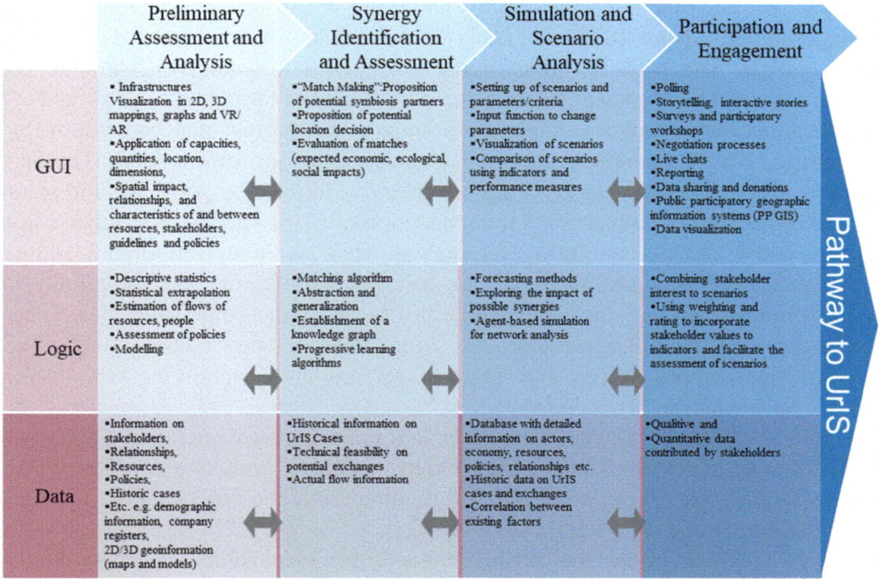


**Fig. 1** Research framework for UrIS

and establishes the requirements through the following key functions: (1) Preliminary Assessment and Analysis, (2) Synergy Identification and Assessment, (3) Simulation and Scenario Analysis, and (4) Stakeholder Engagement. These key functions are conceptualized based on the IS creation steps established by Yeo et al. (2019) [21]. The framework is visualized in Fig. 1.

The key function "Preliminary Assessment" integrates "Pre-Assessment" of IS (Table 1, column a) with the "Situational Awareness" (Table 2, column c) in tools for urban digital twins. Within this assessment contextual conditions, urban and industrial landscape and stakeholders are analyzed to provide insights on potential prospects for UrIS.

The key function "Synergy Identification and Assessment" integrates "Identification of Synergy Opportunities" and "Feasibility Assessment" in IS tools (Table 1, columns c and d) and considers "Situational Awareness" in tools for urban digital twins (Table 2, column c and d). This function also includes the assessment of the economic, environmental, and social implications and the impact of each potential synergy.

The key function "Simulation and scenario analysis" integrates "Feasibility Assessment" in IS tools with "Planning and Prediction" (Table 2, column d) in tools for urban digital twins". It allows users to explore different symbiotic configurations and assess their potential outcomes and impacts under various scenarios. This involves utilizing simulation techniques and scenario analysis to gain insights into the performance and viability of different symbiotic relationships. The scenario development functionality enables users to compare a variety of UrIS scenarios: a reference scenario as well as alternative scenarios [46].

Lastly, the key function "Participation and Engagement" integrates "Engagement of Stakeholders" from IS Tool (Table 1, column b) and the "Integration and Collaboration" from tools for urban digital twins (Table 2, column e). The function aims to involve and empower stakeholders throughout the process.

This function promotes collaboration, knowledge sharing, and co-creation by facilitating effective communication and interaction among stakeholders. The discussions and collaborative approaches are intended to develop place-specific solutions to promote and facilitate the implementation of UrIS. The aim is to engage not only industry partners but also government agencies, nonprofit organizations, nongovernmental organizations (NGOs), research institutions, and the public. It seeks to develop a common understanding of UrIS. While contrasting viewpoints may emerge, potential implementations within this function could consider the stakeholders' interests, values, and tensions and rigorously support the 'Simulation and Scenario Analysis' function.

The three application layers data, logic, and graphical user interface (GUI) stem from software development and intersect the functions to complete the framework. In the cells created by the intersections, the requirements, and functionalities for each function within each layer will be addressed.

The data layer addresses the considerations of data management raised by digital twin applications (Table 2, column a) and IS tools (Table 1, column f). The information and data that are required to support the key functions include incorporating relevant data related to urban and industrial systems, such as infrastructure, resource flows, and socioeconomic factors.

The logic layer describes the models and methods that will be implemented and tested to support each of the four functions. This will involve the development and deployment of advanced algorithms, analytical techniques, and simulation models used in IS/UrIS. The logic layer connects the data layer in the backend with the user-facing GUI layer.

The GUI layer describes, which functionalities and interaction opportunities platform users can access and how visualizations and interactive features allow the user to efficiently and intuitively apply the software and comprehend the results. The integration of geospatial, 3D representations, and interactive dashboards as implemented in urban digital twins (Table 2, column b) allows users to explore and experience the spatial characteristics of symbiotic scenarios, data sets, simulations, and assessments.

## 4.2 Requirements and Functionalities

Considering the existing tools on IS, urban digital twin and geospatial technologies, the requirements for the development of computational tools promoting UrIS have been identified as follows.

### Preliminary Assessment and Analysis

The preliminary assessment and analysis incorporate information on stakeholder details, resource availability, 2D and 3D geospatial information, policies, statistical information, and existing UrIS cases. It integrates diverse datasets to provide a more holistic view of the urban landscape, facilitating informed decision-making and promoting sustainable resource management.

The logic layer employs statistical extrapolation, estimation of material and people flows, and policies. These techniques enable the tool to analyze stakeholders, relationships, resources, existing cases, and regulations. Thereby valuable insights into the existing urban ecosystem are generated, identifying potential synergies for further exploration. Areas and stakeholders with relevance for establishing UrIS can be identified.

The GUI component provides users with visualizations of capacities, quantities, and characteristics of resources, infrastructure, and spatial impact, e.g., visualizing hotspots and development potentials. It furthermore allows for the exploration and evaluation of guidelines and regulations that influence UrIS thereby empowering users to not only grasp the comprehensive landscape of available resources but also understand policy implications.

### Synergy Identification and Assessment

The data layer encompasses existing UrIS cases, technical feasibility studies, actual flow information, prices, and impact information for the flows. It utilizes detailed information on actors, economy, resources, regulations, zoning codes, and relationships. The data layer also incorporates the information that has been generated in the preliminary assessment and analysis.

The logic layer employs matching algorithms, abstraction, and generalization techniques to identify suitable matches for symbiotic partnerships. By assessing existing UrIS cases and technical feasibility, the tool provides a foundation for evaluating potential exchanges. It abstracts and generalizes information to identify viable matches across various industries and urban stakeholders, promoting resource optimization and waste reduction. For the evaluation of economic and environmental impact life cycle costing, environmental and social life cycle analysis need to be implemented and analyzed with progressive learning algorithms.

The GUI component offers suggestions for potential symbiosis partners and locations for the establishment of a new business, where a high symbiosis interaction can be performed. It assists users in identifying and evaluating possible matches for resource exchange and sharing of infrastructure, enabling them to visualize the benefits of these symbiotic relationships. Moreover, the economic, environmental, and social benefits of single synergy matches are presented.

### Scenario Analysis

The data layer consists of a comprehensive database encompassing detailed information on actors, economy, resources, regulations, land uses, and other relevant factors. It also includes historic data typically used in forecasting techniques to simulate different scenarios [47]. The information on synergy identification and assessment is considered as an input for the scenarios. Additionally, information on the correlation between the factors need to be provided. This includes information on the feasibility to perform several symbioses and the influence of the KPI of the simulation model. This data serves as the foundation for forecasting and simulation, providing the necessary information to assess the potential benefits and risks of different scenarios.

The logic layer employs forecasting techniques and complex models to simulate different scenarios and assess the impact of potential synergies. It utilizes simulations, e.g., agent-based or system dynamics for network analysis, enabling the tool to predict the outcomes of different scenarios. Given the complexity of forecasting models and to narrow uncertainty and reduce the risk of inadequate forecasts, sensitivity analysis, which is also used to calibrate models [48], is employed. The logic layer analyzes and assesses different databases to generate alternative scenarios and their consequences.

The GUI application layer supports collaborative sketch planning and facilitates the development of scenarios. In alignment with other tools that have been used to model future development patterns and urban dynamics [49, 50], the GUI layer presents an interactive human–computer work environment for designing and visualizing scenarios. It allows users to input and modify parameters, select a series of performance indicators, and compare their implications and outcomes.

### Participation and Engagement

The data layer includes qualitative data collected from stakeholders capturing their opinions, concerns, values, and aspirations. This data enriches the decision-making process by incorporating the perspectives of residents, businesses, and other stakeholders. By integrating this qualitative information with quantitative datasets, the tool creates a comprehensive understanding of the urban ecosystem and promotes collaborative decision-making.

The logic layer facilitates the integration of stakeholders' perspective through the weighting and rating of values as well as performance indicators selected in the scenario analysis functionality. It ensures that data collected from stakeholders is entered accurately and validated for consistency. The layer combines stakeholder interests to create scenarios that align with the aspirations of the community, promoting inclusive urban development.

The GUI component offers functionalities such as resident polling, negotiation processes, live chats, reporting, and public participation functionalities to contribute individual interests and information. This interface promotes active participation of stakeholders, fostering collaboration, and ensuring that diverse perspectives are considered during the decision-making process.

### *4.3 Implications for Research and Demonstration*

Employing this conceptual framework and analyzing it through the three key levels aims to provide a holistic guideline for research and the development of applications that promote UrIS by implementing the potentials of urban digital twin and geospatial technologies (Sect. 3.2) into the tools for IS (Sect. 3.1).

In the analysis, data availability from companies and urban areas regarding, e.g., waste volume and material usage is crucial for effective UrIS. Overcoming the challenge of obtaining this data requires addressing the concerns and incentives of companies. Motivating potential users to share data can be achieved by demonstrating the benefits of participation, such as improved resource efficiency, cost savings, and enhanced competitiveness. Different strategies, such as legislative measures, can enforce data reporting by companies and other stakeholders, ensuring that crucial information for UrIS is available. Alternatively, leveraging publicly available data sources can provide a foundation for tools, enabling preliminary assessments and analyses.

Creating mutual benefits and fairly sharing resulting profits are essential considerations for successful UrIS. The research framework should facilitate the identification of symbiotic relationships that bring advantages to all participating parties. Clear mechanisms and agreements can be established to ensure that the benefits derived from symbiotic exchanges are distributed equitably among stakeholders.

Data security and privacy are paramount in the development of a platform for UrIS and should therefore also be addressed in research activities. The exchange of data should adhere to the highest standards of security and privacy to protect the interests of participating organizations and individuals. Incorporating robust data governance frameworks, encryption techniques, and strict access controls can help address these concerns and build trust among stakeholders [50].

The proposed research framework for UrIS provides benefits to different stakeholder groups including citizens, city planners, and industry.

## 5 Conclusion and Outlook

With rapid urbanization processes and their associated impacts on resource consumption and the environment, UrIS appears as a potential contributor to improve resource efficiencies by promoting resource exchanges among different actors. While various tools have been employed to facilitate synergetic exchanges, most of them have focused on industrial firms and have failed to address the spatiotemporal dimension of cities or failed to consider nonindustrial partners. Thus, this study proposed a conceptual framework for research and a pathway outlining the development and demonstration of applications promoting UrIS. It encompasses four key functionalities: preliminary assessment and analysis, synergy identification, scenario analysis, and participation and engagement. The presented concept integrates the functionalities of existing applications for IS and of urban digital twins and geospatial technologies.

These functionalities are essential for facilitating sustainable resource management, optimizing resource utilization, and fostering symbiotic relationships among industries.

The need for research arises from several factors. Firstly, as previous tools on IS face challenges and discontinuation, there is a gap in effectively promoting symbiotic relationships and resource efficiency. The concept addresses these challenges by focusing on data availability, company reluctance, and costs, providing mechanisms to motivate data sharing, create mutual benefits, and enhance data security.

Furthermore, the research framework responds to the growing importance of sustainable urban development, circular economy principles, and resource efficiency.

To enhance the success of an UrIS tool, it should be positioned within the broader context of the relevance of productive cities. By emphasizing the importance of sustainable resource management and the principles of the circular economy, the potential benefits of UrIS tools can be underscored. Additionally, spatial analysis can foster a deeper understanding of the tools value and demonstrate how its adoption can be promoted. The proposed research framework aligns with the research agenda set out by, for example, the European Union in order to achieve the sustainable development goals, environmental targets, and foster economic resilience and thus can garner support and interest from stakeholders, including residents, policymakers, urban planners, and industry leaders.

**Acknowledgments** The authors thank the BMBF—Federal Ministry of Education and Research for supporting the project ISUrF-Hub (01DP17049) and the collaborators within the project.

## References

1. UN-Habitat (2022) World Cities Report 2022. UN-Habitat, Nairobi
2. United Nations. Department of Economic and Social Affairs (2022) World Population Prospects 2022
3. European Commission (2023) A European Green Deal: Striving to be the first climate-neutral continent. Retrieved 12/09/2023 from https://commission.europa.eu/strategy-and-policy/priorities-2019-2024/european-green-deal_en
4. European Commission (2023) Circular economy action plan. The EU's new circular action plan paves the way for a cleaner and more competitive Europe. Online: retrieved 12/09/2023 from https://environment.ec.europa.eu/strategy/circular-economy-action-plan_en
5. Van Berkel R, Fujita T, Hashimoto S, Geng Y (2009) Industrial and urban symbiosis in Japan: analysis of the eco-town program 1997–2006. J Environ Manag 90(3):1544–1556

6. Grant G, Seager TP, Massard G, Nies L (2010) Information and communication technology for industrial symbiosis. J Ind Ecol 14(5):740–753
7. Fraccascia L, Yazan D (2018) The role of online information-sharing platforms on the performance of industrial symbiosis networks. Resour Conserv Recycl 136:473–485
8. Dong L, Liang H, Zhang L, Zhaowen L, Zhiqiu G, Mingming H (2017) Highlighting regional eco-industrial development: life cycle benefits of an urban industrial symbiosis and implications in China. Ecol Model 361:164–176
9. Gopinath V, Johansen K, Derelöv M (2018) Demonstrators to support research in industrial safety—a methodology. Proc Manuf 17:246–253
10. Ghisellini P, Cialani C, Ulgiati S (2016) A review on circular economy: the expected transition to a balanced interplay of environmental and economic systems. J Clean Prod 114:11–32
11. Di Baldassarre G, Sivapalan M, Rusca M, Cudennec C, Garcia M, Kreibich H, Blöschl G (2019) Sociohydrology: scientific challenges in addressing the sustainable development goals. Water Resour Res 55(8):6327–6355
12. Chertow M (2000) Industrial symbiosis: literature and taxonomy. Ann Rev Energy Environ:313–337
13. Jiao W, Boons F (2014) Toward a research agenda for policy intervention and facilitation to enhance industrial symbiosis. J Clean Prod 67:14–25
14. Ehrenfeld J, Gertler N (1997) Industrial ecology in practice: the evolution of interdependence at Kalundborg. J Ind Ecol 1(1):67–79
15. Roy M (2021) Sharing economy and sustainability. In: Sustainable development strategies, pp 167–188
16. Kosmol L (2019) Sharing is caring—Information and knowledge in industrial symbiosis: a systematic review. In: 21st IEEE 2019, pp 21–30.
17. Mehdipanah R, Marra G, Melis G, Gelormino E (2018) Urban renewal, gentrification and health equity: a realist perspective. Eur J Public Health 28(2):243–248
18. Fraccascia L, Giannoccaro I (2020) What, where, and how measuring industrial symbiosis: a reasoned taxonomy of relevant indicators. In: Resources, conservation and recycling, p 157
19. Lombardi DR, Laybourn P (2012) Redefining industrial symbiosis. J Ind Ecol 16(1):28–37
20. Lombardi D, Laybourn P (2015) National industrial symbiosis programme connecting industry creating opportunity. In: Proceedings of conferences promoted by ENEA.
21. Yeo Z, Masi D, Low J, Ng Y, Tan P, Barnes S (2019) Tools for promoting industrial symbiosis: a systematic review. J Ind Ecol 23(5)
22. Maqbool A, Mendez Alva F, van Eetvelde G (2019) An assessment of European information technology tools to support industrial symbiosis. Sustainability 11(1):131
23. Silva MG, de Carvalho TS, Castagna, AG, do Rocio Strauhs F, Piekarski CM (2022) The role of online platforms to enable the process of industrial symbiosis: an analysis of tools available in the market. Cleaner Production Letters 3
24. Lawal M, Wan Alwi S, Manan Z, Ho W (2021) Industrial symbiosis tools—a review. J Clean Prod 280
25. Song B, Yeo Z, Kohls P, Herrmann C (2017) Industrial symbiosis: exploring big-data approach for waste stream discovery. Proc CIRP 61:353–358
26. Chen P, Ma H (2015) Using an industrial waste account to facilitate national level industrial symbioses by uncovering the waste exchange potential. J Ind Ecol 19(6):950–962
27. Aid G, Brandt N, Lysenkova M, Smedberg N (2015) Looplocal–a heuristic visualization tool to support the strategic facilitation of industrial symbiosis. J Clean Prod 98:328–335
28. Cecelja F, Raafat T, Trokanas N, Innes S, Smith M, Yang A et al (2015) E-Symbiosis: technology-enabled support for industrial Symbiosis targeting small and medium enterprises and innovation. J Clean Prod 98:336–352
29. Laybourn P (2019) Improving resource efficiency through industrial Symbiosis: a brief world tour. Hg. V. International Synergies Limited

30. Angelis-Dimakis A, Arampatzis G, Pieri T, Solomou K, Dedousis P, Apostolopoulos G (2021) SWAN platform: a web-based tool to support the development of industrial solid waste reuse business models. Waste Manag Res 39(3):489–498
31. Alvarez R, Ruiz-Puente C (2017) Development of the tool symbiosys to support the transition towards a circular economy based on industrial symbiosis strategies. Waste Biomass Valoriz 8: 1521–1530
32. Holgado M, Holgado M, Evans S, Benedetti M, Dubois M, Li Y, Morgan D, Ferrera E, Rossini R, Baptista AJ, Lourenço E, Silva EJ (2018) MAESTRI toolkit for industrial symbiosis: overview, lessons learnt and implications. In: Sustainable design and manufacturing 2018: Proceedings of the 5th SDM.
33. Bin S, Zhiquan Y, Low J, Jiewei D, Kurle D, Cerdas F, Herrmann C (2015) A big data analytics approach to develop industrial symbioses in large cities. Proc CIRP 29:450–455
34. Benedict M, Kosmol L, Esswein W (2018) Designing industrial Symbiosis platforms—from platform ecosystems to industrial ecosystems. In: Tanabu M, Senoo D (eds) 22 Pacific Asia conference on information Systems. PACIS 2018, Yokohama
35. Kosmol L, Leyh C (2020) A vision for industrial Symbiosis: build your platform (ecosystem). In: Twenty-eighth ECIS 2020, Marakesh
36. Dawkins O, Kitchin R, Young G, Zawadzki T (2021) City dashboards and 3D geospatial Technologies for Urban Planning and Management. In: Alasdair Rae und Cecilia Wong, pp 83–101.
37. ESRI ArcGIS Velocity: Analyze real-time and big data. Retrieved 20/06/2023 from: www.esri.com/en-us/arcgis/products/arcgis-velocity/overview.
38. Steinitz C. A framework for geodesign: Changing geography by design: 92373-8100.
39. Virtual City Systems (2023) Virtual City Systems: digital views. Real perspectives: Retrieved 19/06/2023 from https://vc.systems/en/
40. Binyu L, Janssen P, Stoter J, Biljecki F (2023) Challenges of urban digital twins: a systematic review and a Delphi expert survey. Autom Constr:104716
41. Shahat E, Hyun CT, Yeom C (2021) City digital twin potentials: a review and research agenda. Sustainability:3386
42. Dassault Systèmes (2016) Virtual Singapore Online. Retrieved 15/06/2023 from: https://www.3ds.com/insights/customer-stories/virtual-singapore
43. Bentley Systems (2023) iTwin Platform. Retrieved 14/06/2023 from: https ://www.bentley.com/en/products/product-line/digital-twins/itwin-platform
44. Fuller A, Fan Z, Day C, Barlow C (2020) Digital twin: enabling technologies, challenges and open research. IEEE Access 8:108952–108971
45. Ruohomäki T, Kesäniemi O, Airaksinen E, Martikka M, Huuska P, Suomisto J (2018) Smart city platform enabling digital twin. In: 2018 International Conference on Intelligent Systems, pp 155–161
46. Castelli G, Cesta A, Diez M, Padula M, Ravazzani P, Rinaldi G, Savazzi S (2019) Urban intelligence: a modular, fully integrated, and evolving model for cities digital twinning. In: IEEE 16th International Conference on Smart Cities: Improving Quality of Life Using ICT & IoT and AI, Charlotte, pp 033–037
47. Aissani L et al (2019) Life cycle assessment of industrial symbiosis: a critical review of relevant reference scenarios. J Ind Ecol 23(4):972–985
48. Geertman S, Stillwell J (eds) (2009) Planning support systems best practice and new methods, vol 95. Springer Science & Business Media
49. Avin U, Goodspeed R (2020) Using exploratory scenarios in planning practice: a spectrum of approaches. J Am Plan Assoc 86(4):403–416
50. Kosmol L, Leyh C (2021) A conceptual design for digital industrial symbiosis ecosystems. In: The next wave of sociotechnical design, Bd. 12807. Springer International Publishing (Lecture Notes in Computer Science), Cham, pp 362–374

# Methodology to Model the Environmental Impact of the Overall Maintenance of Aircraft Engines

**Michel Nocture, Maud Lemagnen, Bertrand Laratte, and Sarah Atroun**

**Abstract** The majority of aircraft engine life cycle assessment (LCA), and more generally aeronautical products, focuses on the use, the extraction of raw materials, and the manufacturing phases. This focus seems logical since they represent a priori the most significant sources of impacts on the environment. The maintenance phase is usually overlooked and roughly modeled. However, it is precisely relevant as aircraft engines are high-value products, and airlines strive to maintain their engines to last the longest as possible.

The maintenance phase can be considered as an additional life cycle for the product. It implies a vast diversity of swapped parts and repairs. The detailed LCA modeling would be an intricate analysis that would require huge work of inventory. Nevertheless, one has to tackle this objective, as LCA tends to be a reference methodology not only for environmental disclosure but also for eco-design activities for the aeronautical sector.

This chapter proposes a methodology to model the overall maintenance of an aircraft engine throughout its whole life for LCA purposes. This model applies to the whole engine or parts of it. A generic flow diagram of maintenance in workshops has been built based on industrial experience. Then the simplifications of the model, required to make it viable, are introduced. The feasibility is evaluated with a use case on the repair of a specific engine part. Finally, this chapter proposes a discussion about the limits related to the assumptions, the simplifications, and the areas of improvement.

**Keywords** Life Cycle Assessment · Life Cycle Inventory · Maintenance · Aircraft Engines · Model · Environmental Footprint · Flow Diagram

M. Nocture (✉) · M. Lemagnen · S. Atroun
Safran Aircraft Engines, Moissy-Cramayel, France
e-mail: michel.nocture@safrangroup.com

B. Laratte
Arts et Metiers Institute of Technology, University of Bordeaux, Bordeaux, France

 
S. Fukushige et al. (eds.), *EcoDesign for Circular Value Creation: Volume I*,
https://doi.org/10.1007/978-981-97-9068-5_27

## Nomenclature

| | |
|---|---|
| AMM | Aircraft Manual Maintenance |
| EF | Environmental Footprint |
| EoL | End-of-Life |
| ESM | Engine Shop Manual |
| HPT | High-Pressure Turbine |
| LCA | Life Cycle Assessment |
| LCI | Life Cycle Inventory |
| LCIA | Life Cycle Impact Assessment |
| MRO | Maintenance, Repair, and Overhaul |
| OEM | Original Equipment Manufacturers |
| KCP | Key Changed Parts |
| KRP | Key Repaired Parts |
| PPE | Personal Protective Equipment |
| PEF | Product Environmental Footprint |
| ESV | Engine Shop Visit |
| USM | Used Serviceable Material |

## 1 Introduction

Pressure has been increasing in recent years on the aeronautical sector to improve its environmental performance [1, 2]. The first step of this eco-design process is to evaluate the current status to spot areas of improvement. Life cycle assessments (LCA) [3] enable one to perform such analysis from cradle to grave: every phase of a product's lifetime is taken into account. LCAs consider several environmental impact indicators preventing pollution transfer: between two phases of the life cycle or from one environmental impact indicator (e.g., climate change) to another (e.g., ocean acidification).

Previous environmental studies of aeronautical products, and more precisely aircraft engines, focus their analyses on the exploitation phase, when the major impact on climate change is occurring (from 92% [4] to 99% [5, 6]). Other studies strive to try to take into account the extraction and manufacturing phases [7, 8] with different levels of fidelity and complexity. The maintenance and end-of-life (EoL) phases are usually overlooked and roughly modeled [9, 10] if not neglected [5].

However, if the aeronautical sector is asked to optimize its overall operations it is necessary to also have a way to consider the maintenance phase. This is relevant to tackle this topic not only for environmental disclosure but also for eco-design activities: to anticipate engine future degradations and failures as early as the engine design phase. LCA of the maintenance phase is not a trivial work as there is a large diversity of maintained parts, of repair operations, and of logistic flows for each visit to the maintenance workshop, also called Engine Shop Visit (ESV). The detailed and exhaustive LCA modeling would be an intricate model requiring huge work of inventory. The results would be too complex to be analyzed.

The contribution of this study is to fill the lacks in methodologies available in the literature to model the environmental impact of aircraft engines' maintenance. Some studies, mainly on aircrafts, seek to take into account the maintenance phase but with a low level of precision. Rahn et al. [11] estimate the environmental impacts based on costs. They deepen the analysis on on-wing aircraft maintenance (tire and brake change or engine wash) [12] only. The novel methodology proposed in this chapter focuses on the engine maintenance done in workshops, in opposition to line maintenance performed while the engine remains hung to the aircraft. This model can be adapted to discrete event simulation (DES) [13] methods that are able to grasp the effects of occasional events throughout the systems' life. Coupling such models with aircraft emission models enables to optimize the aircraft design and maintenance planning [14] in addition to traditional financial features [15]. Then the results of operation optimizations can be integrated into the design phase of future products as Johanning et al. [5] or Fabre et al. [16] proposed.

This chapter offers a methodology to evaluate the environmental footprint of the maintenance phase of aircraft engines. First, the methodology of data collection based on a generic flow diagram of the Engine Shop Visit is described. Second, the methodology of the engine's overall maintenance model and the simplifications are detailed. Then, a use case of data collection on the repair of a specific part is presented. Finally, a discussion is proposed on the limits of the methodology and on future works.

## 2 Methodology for Data Collection Based on a Flow Diagram

### 2.1 *Objective of the Generic Flow Diagram*

The generic flow diagram of the maintenance in workshops aims at describing the different phases an engine undergoes during an engine shop visit (ESV). It is especially useful to prepare data collection and to build the LCA model. In this diagram, the different input and output flows are detailed without being quantified. The scope covered in this chapter goes from the removal of aircraft engines to their return to service after maintenance.

Six phases are identified and detailed in the next paragraphs.

- Engine removal from the aircraft
- Engine disassembly
- Part repair and/or part change
- Engine reassembly
- Engine performance tests
- Engine reinstallation on the aircraft

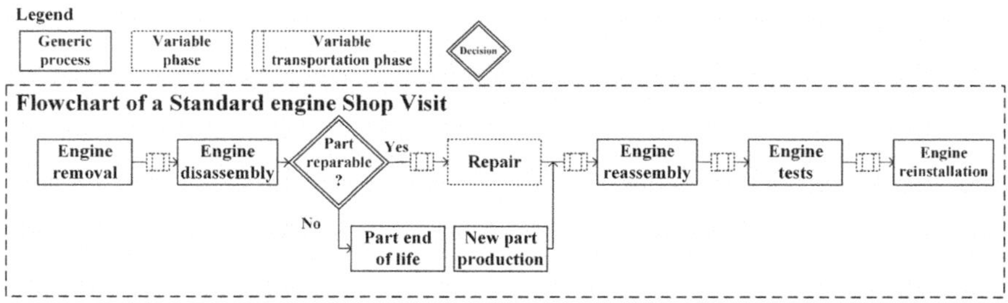


**Fig. 1** Simplified flowchart of aircraft engine maintenance

In addition, there are logistic phases between all these steps to transport the engine and the maintained parts.

Figure 1 represents the flowchart of the engine shop visit. The input and output flows are not illustrated for readability reasons.

This simplified flowchart is meant to be applied to any parts that are maintained on an aircraft engine. It is based on the experience of Safran, an aircraft engine manufacturer performing maintenance actions.

Unlike cradle-to-gate Life Cycle Inventory (LCI), the majority of the input flows at each operation of the engine maintenance is also an output flow as a waste. For instance, the acid used to clean a part is eventually removed from this part and becomes a waste to be treated. The material that is added to the part represents a small fraction of the total flows. This is not true for scrapped parts for which new parts need to be manufactured.

## *2.2 Description of the Flow Diagram*

From this flowchart, we can split the phases into two groups: constant phases and variable phases. The next paragraphs describe the operations done during each of these phases and their major input and output flows.

### Constant Phases

These phases are considered constant regardless of the parts that are maintained. It enables to perform only one data collection that can be referred to for any part.

**Engine Removal**

The engine removal phase is the operation of removing the engine from the aircraft before packing it and shipping it to the first engine maintenance workshop. Field technicians drain the engine fluids and protect the inside of the engine.

### Engine Disassembly and Reassembly

In a maintenance workshop, the engine is disassembled to isolate the parts that are going to be maintained. The surfaces of the engine are cleaned for inspection purposes. Each engine has its specific disassembly level depending on the engine state and what the airline operating the engine asked for. The disassembly is an iterative process, as inspections can bring out parts that need maintenance while it was not planned. After all the parts are maintained, the inverse operation is done: the parts are gathered and the engine is reassembled. The part repairs or part productions are usually done in other workshops.

The maintenance model proposed here considers an average engine disassembly and reassembly. The major flows of this phase are new screws and bolts, oil and lubricants, paper sheets for traceability, and personal protective equipment (PPE). This phase does not extensively use energy (electricity and heat).

### Engine Testing

After the parts of the engine have been maintained and the engine reassembled, the engine functionality and performance (margins compared to certification limits) are verified in a test bench. The engine runs at different operating speeds for approximately 1–3 hours while measurements are checked. The engine consumes kerosene and oil during the performance test.

### Return in Operation

After the engine is maintained and tested, it is reinstalled on the aircraft to go back into operation. It is the inverse operation of the engine removal.

## Variable Phases

### Part Repair or Part Change

The part repair or the part replacement is the phase that can vary the most depending on the part under study. The repair of a static pipe operating in low temperature conditions will not be comparable to the repair of a blade operating at high speed, temperature, and pressure. The mass of the part is not sufficient to have a decent estimation of the environmental footprint. To have an adequate model, it must describe the repair processes (cleaning, machining, material addition, coating, etc.) that are effectively done.

The LCA of the replaced part can be broken down in two LCAs: the production of the new part and the end of life of the rejected one. **This chapter focuses mainly on part repair.**

Many input and output flows take place in this phase: electricity, gases, acids, water, metals, pastes, PPE, etc.

### Transports

Engines and parts are usually shipped by plane or trucks, depending on the distance and the contract deadlines. Traveled distances vary significantly between the different workshops. Some parts can cross oceans to be repaired. No specific effort was made by maintenance, repair, and overhaul (MRO) companies to optimize the maintenance logistics.

### *2.3 Limits of the Flow Diagram*

This simplified flow diagram cannot represent exhaustively the aircraft engine maintenance.

It does not take into account the nonconformities occurring in the process. These nonconformities can be identified at any step of the engine maintenance: quality checks are done at every step and the results are compiled in traceability documentation.

The flow diagram describes the "standard" process, as it is designed by the original equipment manufacturers (OEM). Aside from this standard process, there are broker companies. These companies are specialized in buying old engines to retrieve the serviceable parts and sell them to maintenance workshops. These parts can be counted as recycled ones in the EoL phase of a *cradle-to-grave* aircraft engine LCA. For simplicity and boundary reasons, this broker market is not considered in this study and all changed parts are considered as newly produced ones.

## 3 Methodology for the Aircraft Engine Maintenance Environmental Footprint Model

### *3.1 Modeling Principle*

The objective of this model is to have a simple yet comprehensive model limiting the LCI work to just what is necessary.

The LCI on the constant phase is done once and is applied for all the ESV an engine can undergo.

The variable phases are treated separately. The model philosophy is to focus the analysis on the most frequently repaired or changed parts. These parts are called *key repaired parts* (KRP) and *key changed parts* (KCP). This model aims at describing accurately the processes of the variable phases (transports, repair, or replacement) done on the KRP and KCP.

The repair and production operations are clustered to limit the LCI work. Only one LCI per cluster is sufficient to characterize all the repairs in the same cluster.

Figure 2 represents a high-level flowchart of the model structure for the part repair and part replacement. The indicated numbers refer to the paragraphs of Sect. 3.2.

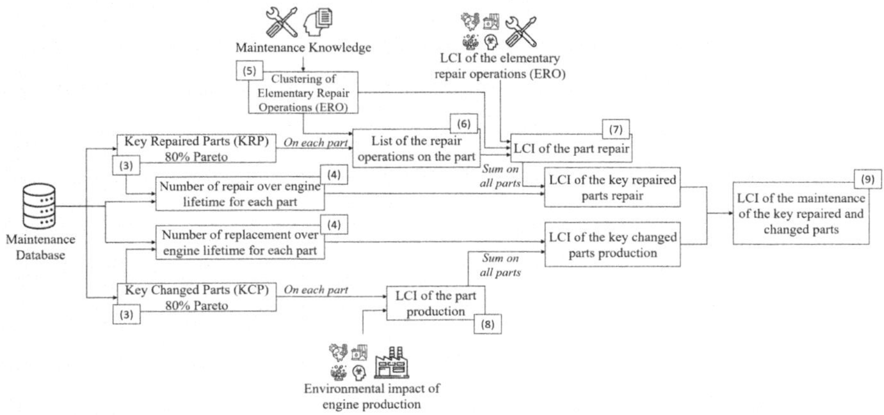


**Fig. 2** Flowchart of the model for part repair and part replacement

## 3.2 *Description of the Different Model Elements*

This section describes the assumptions of each element of the model.

### Constant Phases

The constant phases are assumed to be the same every time an engine undergoes a shop visit. The study of the use case in Sect. 4 describes how the LCI has been done for these phases. They affect the whole engine and occur once for each ESV. One must estimate the number of ESV for one engine to know the input and output flows of these phases over the engine's lifetime. This is estimated with information from databases owned by the companies doing the maintenance, repair and overhaul (MRO) such as OEMs, airlines, or companies specialized in this sector. From Safran's experience, an aircraft engine usually goes from two to four major ESV during its lifetime.

The LCI of the constant phases over the engine lifetime can be expressed as follows:

$$\mathrm{LCI}_{\mathrm{cst\ phases}} = N_{\mathrm{esvol}} * \left(\mathrm{LCI}_{\mathrm{removal}} + \mathrm{LCI}_{\mathrm{disassembly}} + \mathrm{LCI}_{\mathrm{reassembly}} + \mathrm{LCI}_{\mathrm{testing}} + \mathrm{LCI}_{\mathrm{reinstallation}}\right)$$

With $N_{\mathrm{esvol}}$ the number of shop visits an engine undergoes over its lifetime. $\mathrm{LCI}_{\mathrm{xx}}$ represents the LCI of one occurrence of an engine maintenance constant phase.

### Transport Phases

The logistic phases are hard to describe as they depend on several factors: where the engine is operated, the location of the closest maintenance workshop, the abilities of the workshops, etc. It is individually defined each time an engine goes into ESV. To tackle this complex problem, we have chosen a logistic scenario based on our experience. In addition, a sensitivity analysis is done to highlight the weight of transport extension or reduction on the overall environmental footprint. This is the same philosophy as Parolin et al. in their LCA tool for aircraft design [17] to spot the most impactful phases.

The LCI transport calculation is weighted by the mass of the transported materials and the traveled distances. It is expressed as follows:

$$\mathrm{LCI}_{\mathrm{transports}} = \sum_{\substack{i \text{ in engine} \\ \text{or parts}}} d_i^{\mathrm{plane}} * \mathrm{LCI}_{1\mathrm{kg.km}}^{\mathrm{plane}} * m_i + \sum_{\substack{i \text{ in engine} \\ \text{or parts}}} d_i^{\mathrm{truck}} * \mathrm{LCI}_{1\mathrm{kg.km}}^{\mathrm{truck}} * m_i$$

$m_i$ is the transported mass (whole engine or parts of it), $d_i^{\mathrm{plane}}$ or $d_i^{\mathrm{truck}}$ the traveled distance by the parts or engine. $\mathrm{LCI}_{1\mathrm{kg.km}}^{\mathrm{plane}}$ and $\mathrm{LCI}_{1\mathrm{kg.km}}^{\mathrm{truck}}$ are the LCI for the transport of 1 kg over 1 km by truck or plane.

### Pareto of the Most Repaired and Replaced Parts

As mentioned earlier, the model focuses on key parts (KRP and KCP) to limit the LCI work. These parts are identified with a Pareto analysis on the number of times the parts that are repaired or changed. For simplicity reasons the Pareto analysis is done on the number of occurrences while it is usually done on mass in other LCA works. It can be a tall order to have the complete nomenclature (with the mass) of all the maintained parts.

The KRP and KCP are the parts that represent 80% of part repairs and 80% of part changes in number.

This Pareto analysis results in two lists of parts (each one containing 15–20 elements): one for the most repaired parts and one for the most changed parts. The model focuses mainly on these parts.

### Estimation of the Number of Repairs and Replacements Over One Engine Life

Estimating the number of times each KRP and KCP are repaired or changed is not trivial. It is not possible to simply track engines during their entire lifetime as they can last several decades and they can change ownership several times. The followed methodology here is to use MRO company databases on the available (thus limited) period. The perimeter of the LCA is limited to the maintenance tracked by the MRO companies in their databases.

The number of times a part $\mathrm{KRP}_i$ (respectively $\mathrm{KCP}_i$) is repaired $N_{\mathrm{reol}}$ (respectively changed $N_{\mathrm{ceol}}$) are calculated as follows:

$$N_{\text{reol}}^{\text{KRP}_i} = N_{\text{rdb}}^{\text{KRP}_i} * \frac{N_{\text{efol}}}{N_{\text{edb}} * N_{\text{efdb}}}$$

$$N_{\text{ceol}}^{\text{KCP}_i} = N_{\text{cdb}}^{\text{KCP}_i} * \frac{N_{\text{efol}}}{N_{\text{edb}} * N_{\text{efdb}}}$$

$N_{\text{reol}}^{\text{KRP}_i}, N_{\text{ceol}}^{\text{KRP}_i}$ are the number of part repairs and part changes for one engine over its lifetime. $N_{\text{rdb}}^{\text{KRP}_i}, N_{\text{cdb}}^{\text{KRP}_i}$ are the number of repairs or changes recorded in the databases for the part $\text{KRP}_i$ or $\text{KCP}_i$. $N_{\text{efol}}$ is the hypothesis of the number of flights throughout an engine lifetime (around 30,000). A sensitivity analysis can challenge this hypothesis. $N_{\text{edb}}$ refers to the number of engines in the databases and $N_{\text{efdb}}$ refers to the number of flights these engines have done during the period of time covered by the database.

## Groups of Repair Operations

Almost every part has its specific repair procedure with its specifications. A clustering of the different elementary repair operations is done to limit the LCI work and to simplify the repair descriptions. This clustering is based on the OEM maintenance experience and regroups the common operations done on the different parts.

It results in 45 *Elementary Repair Operations* (ERO) segregated into six major groups: cleaning, inspection, stripping, reconstruction, protection, and resizing. These operations are listed in Fig. 3.

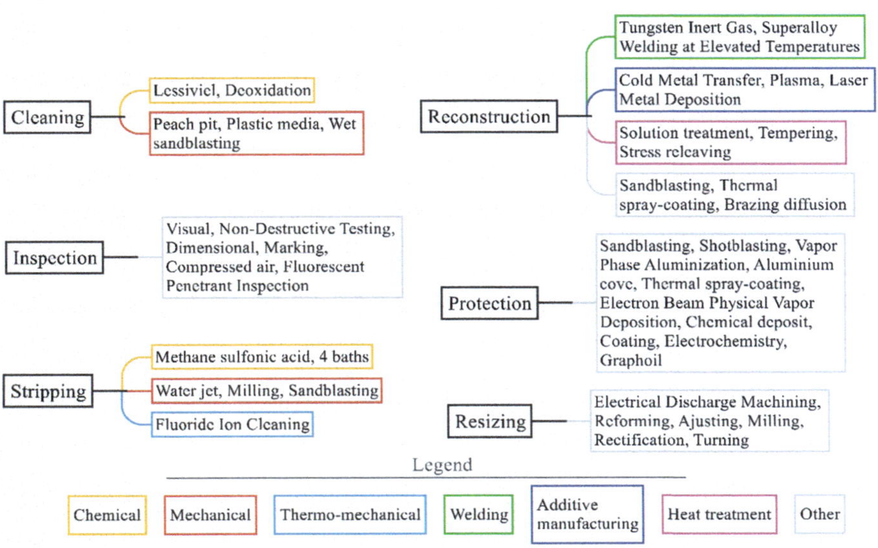


**Fig. 3** Clustering of the elementary repair operations

A unitary metric must be defined for each ERO to quantify "how much" of each ERO is done on each KRP. This metric is based on the repair type and the physical flows involved: what is the driving metric (mass, surface, volume, etc.) for the elementary repair operation?

### List of the Repair Operations on Each Part

The list of the elementary repair operations is identified for each KRP. The matching of the KRP with their EROs is based on maintenance documentation, such as the Engine Shop Manual (ESM), and on workshop experience. The suitable repair metric (mass, surface, volume, etc.) is also associated with each KRP-ERO couple.

### LCI of the Elementary Repair Operations (ERO)

One main requirement of the model is to have the LCI of all the unitary EROs available. This can be summarized as an LCI database for part repairs. For each ERO, the input and output flows have to be estimated for a **unitary repair** (e.g., sandblasting 10 $cm^2$, machining 100 g of material). This represents an important work of data collection. A normalization analysis also needs to be done to convert the flows of a part repair to a unitary repair one. The LCI repair database construction is the most time-consuming work but it can be reused for other studies to do quick LCA on any repair.

This database is then used to combine input and output flows (energy, chemical, emissions, wastes, etc.) for each ERO of the different KRPs.

It is key to estimate the uncertainty during the unitary LCI of each ERO as it is the base of the calculations. The uncertainty is split into two causes: the closeness of the data compared to the measurement and the uncertainty due to the measurement itself. The first uncertainty factor is estimated with the Weidema-Wesnaes [18] matrix which takes into account different factors: reliability, completeness, temporal correlation, geographical correlation, sample size, and technological correlation. The second uncertainty factor is estimated with the Frischknecht [19] method. The uncertainty depends on the flow typology: for instance, electricity measurements are more accurate than gas emission ones.

Once the ERO LCI database is available, the LCI of the repair of the part $KRP_i$ is calculated as follows:

$$\mathrm{LCI}_{\mathrm{KRP}_i} = \sum\nolimits_{k \text{ in } \mathrm{ERO}_{\mathrm{KRP}_i}} \mathrm{LCI}_{\mathrm{ERO}^k}^{\mathrm{unit}} * \mathrm{ERO}_{\mathrm{KRP}_i}^{k,\mathrm{metric}}$$

With $\mathrm{LCI}_{\mathrm{KRP}_i}$ the LCI for the repair of the $i^{\text{th}}$ KRP, $\mathrm{ERO}_{\mathrm{KRP}_i}$ the list of EROs for the $i^{\text{th}}$ KRP, $\mathrm{ERO}_{\mathrm{KRP}_i}^{k,\mathrm{metric}}$ the metric of the $k^{\text{th}}$ ERO for the $i^{\text{th}}$ KRP, and $\mathrm{LCI}_{\mathrm{ERO}^k}^{\mathrm{unit}}$ the unitary LCI for the $k^{\text{th}}$ ERO.

The LCI of all the KRP over the engine lifetime is calculated as follows:

$$\mathrm{LCI}_{\mathrm{repair}} = \sum\nolimits_{i \text{ in KRP}} \mathrm{LCI}_{\mathrm{KRP}_i} * N_{\mathrm{reol}}^{\mathrm{KRP}_i}$$

$\mathrm{LCI}_{\mathrm{repair}}$ is the LCI for the engine repair over its lifetime and KRP is the list of the key repaired parts. $\mathrm{LCI}_{\mathrm{KRP}_i}$ is the LCI of the repair of the $i^{\mathrm{th}}$ KRP and $\mathrm{N}_{\mathrm{reol}}^{\mathrm{KRP}_i}$ is the number of repairs for one engine over its lifetime for the $i^{\mathrm{th}}$ KRP.

## LCI of the Production of the Replaced Parts

Similar work needs to be done on the *key changed parts* (KCP) with an LCA of their material extraction and manufacturing processes. However, this chapter does not detail how the manufacturing processes should be clustered and evaluated. One assumes that the LCI work for the production of the KCP is available. Other studies [6–8] tackle the LCA of the production phase (cradle to gate).

Similarly, to the part repairs, the LCI of the new part productions over the engine lifetime is expressed as follows:

$$\mathrm{LCI}_{\mathrm{part\ prod}} = \sum\nolimits_{i \text{ in KCP}} \mathrm{LCI}_{\mathrm{KCP}_i} * N_{\mathrm{ceol}}^{\mathrm{KCP}_i}$$

With $\mathrm{LCI}_{\mathrm{part\ prod}}$ the LCI for the production of the changed parts over the engine's lifetime, KCP the list of the key changed parts, $\mathrm{LCI}_{\mathrm{KCP}_i}$ the LCI of the part production of the $i^{\mathrm{th}}$ KRP, and $\mathrm{N}_{\mathrm{ceol}}^{\mathrm{KCP}_i}$ the number of part changes for one engine over its lifetime for the $i^{\mathrm{th}}$ KRP.

As new parts replace the rejected ones, the EoL of the latter must be taken into account. For simplicity reasons, it is assumed that metallic parts are recycled as common metal and the other parts are thrown away.

The LCI calculation of the rejected parts EoL is expressed as follows:

$$\mathrm{LCI}_{\mathrm{parts\ EoL}} = \sum\nolimits_{i \text{ in KCP}} N_{\mathrm{ceol}}^{\mathrm{KCP}_i} \left( m_{\mathrm{metal}}^{\mathrm{KCP}_i} * \mathrm{LCI}_{\mathrm{metal}}^{\mathrm{EoL}} + m_{\mathrm{other}}^{\mathrm{KCP}_i} * \mathrm{LCI}_{\mathrm{other}}^{\mathrm{EoL}} \right)$$

With $\mathrm{LCI}_{\mathrm{part\ EoL}}$ the LCI for the EoL of all the rejected parts over the engine's lifetime, KCP the list of the KCP key changed parts. $m_{\mathrm{metal}}^{\mathrm{KCP}_i}$ and $m_{\mathrm{other}}^{KCP_i}$ are the mass of the metal or other material for the $i^{\mathrm{th}}$ KCP, and $N_{\mathrm{ceol}}^{\mathrm{KCP}_i}$ the number of part changes for one engine over its lifetime for the $i^{\mathrm{th}}$ KCP. $\mathrm{LCI}_{\mathrm{metal}}^{\mathrm{EoL}}$ and $\mathrm{LCI}_{\mathrm{other}}^{\mathrm{EoL}}$ are the LCI of the EoL for 1 kg of metal or other material.

## LCI of the Overall Engine Maintenance

The LCI of the constant phases over the engine lifetime is expressed as follows:

$$\mathrm{LCI}_{\mathrm{cst\ phases}} = N_{\mathrm{ESV}} * \left( \mathrm{LCI}_{\mathrm{removal}} + \mathrm{LCI}_{\mathrm{disassembly}} + \mathrm{LCI}_{\mathrm{reassembly}} + \mathrm{LCI}_{\mathrm{testing}} + \mathrm{LCI}_{\mathrm{reinstallation}} \right)$$

With $N_{ESV}$ the number of shop visits over one engine lifetime and $LCI_{xx}$ LCI of one occurrence of an engine maintenance constant phase.

The overall engine maintenance LCI is the sum of the different phase LCIs. It is expressed as follows:

$$LCI_{engine\ maintenance} = LCI_{cst\ phases} + LCI_{transports} + LCI_{part\ prod} + LCI_{part\ EOL} + LCI_{repair}$$

**Environmental Footprint Calculation from the LCI**

The LCI of all the key repaired and key changed parts (KRP and KCP are used for the Life Cycle Impact Assessment (LCIA). All the flows are converted into midpoint or endpoint environmental footprint indicators. The users can choose the most suitable models given the objectives of their studies and the area of work: EF3.1 [20], LIME [21], IMPACT World + [22], etc.

The model proposed in this chapter aims at being as modular as possible. It is designed to be compatible with any database and any impact indicator model. That is why the environmental footprint (EF) calculation is the last step of the model. Before that, the variables manipulated are physical flows estimated with the LCI methodology.

Among the different midpoint impact indicators, environmental studies of the aviation sector examine climate change, summer smog formation, resource depletion, and acidification [23]. The indicators that stand out in the maintenance LCA may be different as the processes occurring during this phase differ largely from the exploitation phase (mainly burning kerosene).

## 3.3 *Extension to the Parts Out of the Pareto*

The LCI calculation detailed in section "Environmental Footprint Calculation from the LCI" is valid for the parts representing 80% of the most repaired and changed parts. The extension to the lacking 20% of the repaired part is done by multiplying the result by 1.25 (i.e., $\frac{1}{0.8}$). Unlike classic LCA, no weighting by mass is recommended here. The mass of a part is not representative of the EF of its repairs. Small defect repairs on large and heavy parts, such as fan blades, can have a much lower EF than small parts such as high-pressure stator vanes that require chemical and heat treatments. The extension to the lacking 20% of the changed parts is done the same way for simplicity reasons. It can be a tall order to access the full nomenclature of an engine, including the mass of all parts.

# 4 Data Collection Use Case on a Specific Engine Part Repair

A use case has been carried out on the repair of a specific part of an aircraft engine. The studied part is a high-pressure turbine (HPT) shroud. The choice of this part has been driven by several factors. Despite the cutting-edge materials and technologies used for its durability, the part degrades over time as it undergoes high pressures and high temperatures. This HPT efficiency decreases over time and the engine fuel consumption increases, leading to higher costs and higher greenhouse gas emissions during aircraft operations. As a result, this part is usually repaired in Safran Aircraft Engine maintenance workshops. Given the complexity of the part, the repair operations involve heavy industrial equipment with significant flows (furnaces, automatic milling machines, coatings, acids, etc.).

This use case was carried out before building the complete model of the engine's overall maintenance, including the substantial workload of collecting data for each elementary repair operation (ERO). The objective of this use case is to prove the feasibility and validity of such LCI and LCIA work.

## *4.1 Data Collection for the Different Phases*

This section describes data availability, the way it is collected, the estimation methods of the missing data or why some flows are neglected. It covers the different phases of the generic flow diagram described in Sect. 2.2. Here the production of a new part is not covered as we assume that its LCA is available.

### Engine Removal and Reinstallation

Airlines often do the operation of removing the engine from the aircraft or reinstalling it after maintenance on airport tarmacs. It is difficult to collect data directly during these operations. Authorizations are required to access operation zones. The Aircraft Manual Maintenance (AMM) describes the procedure of removing the engine. It indicates that few physical flows (mainly energy to lift the engine with a crane and some duct tapes) are involved in this operation. Going on the field to do the LCI would be a lot of effort for a few EF. The different flows are estimated from the specifications and the experience of the team who designed the engine removal procedure at Safran Aircraft Engines.

### Engine Disassembly and Reassembly

The difficulty with this phase is that all the engines are not equally disassembled. Some engines have almost all their parts disassembled while others have only some parts removed as their maintenance plan is strictly specified. The plant under study for the LCI is fully dedicated to engine disassembly and reassembly. The LCI of this phase is done by quantifying the input and output flows at the plant level. Precise data covering a year of exploitation is available from the facilities management and the purchase teams. The result is an estimation of the flows for an average engine disassembly.

### Part Repair

The repair phase is the phase with the highest expected EF as several complex industrial processes are involved. Data collection was realized directly in the plant where the HPT shroud is repaired among several other parts. Electrical measurements were made on the most consuming machines (furnaces and heated acid baths). Process specifications detailed in the ESM indicate the consumption of chemicals. Experienced operators estimated the consumables quantities (welding wires, brazing powder and milling heads). Other consumables, such as PPE, are estimated from the purchase team information. Wastes information could only be available at the site level. Allocation calculation is done when data is only available for processes combining different engine parts (furnaces and wastes). The allocation is done with the number of parts weighted by their mass or their surface depending on the ERO. HPT Shroud data collection is done to be as close as possible to the real operational processes.

### Engine Test After Maintenance

Engine tests are done in a close test bench where the engine runs at different operating points. The engines consume mainly kerosene and oil. The test bench systems track directly the fuel consumption guaranteeing good data quality and completeness. The main flow of the test bench itself is electricity. The electricity consumption is collected at the building level.

### Transports

As said in the methodology description in Sect. 3, the transport phases can differ a lot given the engine maintenance scenario. In this case study, the selected scenario is the one where the maintenance workshops are the one where the LCI is done. The engines are transported by trucks (one engine per truck) for short distances and by airplanes for long distances. Table 1 summarizes the transport modes and distances for the selected scenario. One can see that the transport from the engine disassembly plant to the repair plant is the longest as it crosses the Atlantic Ocean (from France to Mexico) but only the repaired part is shipped for these trips as the engine is disassembled.

**Table 1** Summary of transport modes and distances

| | Distance traveled | |
|---|---|---|
| | by truck (km) | by plane (km) |
| Engine removal → engine disassembly | 64 | |
| Engine disassembly → part repair | 64 | 9373 |
| Part repair → engine reassembly | 64 | 9373 |
| Engine reassembly → engine testing | 58 | |
| Engine testing → engine reinstallation | 60 | |

The EF associated with the transport is calculated using the Product Environmental Footprint (PEF) methodology [20].

The impact of other logistic scenarios is to be studied through sensitivity analysis.

# 5 Conclusion and Future Works

## *5.1 Conclusion*

This chapter provides a methodology to model the environmental footprint (EF) of the overall aircraft engine maintenance. The objective is to reach a balance between representativeness of the maintenance complexity and simplicity to keep it understandable and feasible in a reasonable amount of time. This compromise is reached by segregating constant phases—engine removal, disassembly, re-assembly, testing, and reinstallation—from the variable ones—part repair, part production, end of life of parts, and transports—on which a more detailed analysis is required. The production of new parts is not detailed in this chapter. We rely on other works to detail the methodology. It is assumed that their cradle-to-gate LCAs are available. The proposed model advocates a Pareto analysis based on the repaired and changed parts to limit the data collection work to a limited amount of parts: the *key repaired parts* (KRP) and the *key changed parts* (KCP). The experience of MRO companies is used to estimate the repair (respectively change) number of each KRP (respectively KCP) over engine life.

The model proposes a clustering of the different *elementary repair operations* (ERO). This clustering is then used to describe all the repairs done on each KRP. For each ERO a metric is defined to quantify the repair. The model requires to have an LCI database for each unitary ERO. Establishing this database is the most demanding data collection work of the model. This chapter describes the calculations to model the EF of an aircraft engine maintenance, detailing each phase and the combination of their results.

Data collection of a specific engine part repair, an HPT Shroud, is detailed in a use case study. The chapter describes the methodology to collect the data given the flow typology and the data accessibility. Some flows are directly measured, some are estimated from plant-level data, and others are estimated from ESM specifications.

The limits of this model are outlined. This model does not take into account the broker market. This alternative market can be integrated into a cradle-to-grave aircraft engine LCA which takes into account the phase. The difficulty to build the model described in this chapter is data accessibility. It necessitates three types of information: (1) the operational experience of the most often maintained parts, (2) the repair operations on these parts, and (3) the inventory of input and output flows for each elementary repair. The traceability of the key maintained parts is a difficult task, even for OEMs that maintain engines. Rollinck et al. propose to implement blockchain technology to ensure data quality and accessibility [24]. The information of (2) and (3) usually belong to MRO companies. They are in the best position to carry out such analysis. Building the LCI database with each elementary repair operation is the main workload to address to implement the proposed model. When feasible, it is done with measurements directly in maintenance workshops like in the use case described in this chapter. When the flows are not measurable or estimated in maintenance workshops, the LCIs rely on available literature on the industrial processes (machining [25], welding [26], Electro Discharge Machining [27], etc.).

Finally, this model is static as it assumes the entire environmental impacts are generated instantaneously while the engine maintenance occurs over the whole engine lifetime covering several decades. This model could be adapted to Dynamic Life Cycle Assessments (DLCA) [28] methodology to take into account the lifetime of each pollutant and the temporality of their emissions.

## 5.2 Future Works

The next step is to assess the impacts (LCIA) of the HPT Shroud with the LCI already done. The objective is to compare the different phases to spot the most impactful ones on the different environmental footprint indicators.

After validation of the model feasibility with the HPT Shroud use case, we will start to build the LCI repair database on all the elementary repair operations. This database will enable us to model aircraft engines' maintenance throughout their entire lifetime. We will be able to compare different maintenance scenarios and identify optimums. Eventually, the results of this comparison will be integrated into the design choices of future aircraft engines.

**Acknowledgments** The authors would like to thank César Sanchez and Jean-Luc Duparcq for sharing their vast knowledge of aircraft engine maintenance processes.

This work is supported by the Directorate General for Civil Aviation (France) under the *Plan $CO^2$* project.

## References


1. Barke A, Thies C, Melo SP, Cerdas F, Herrmann C, Spengler TS (2020) Socioeconomic life cycle assessment of future aircraft systems. Procedia CIRP 90:262–267
2. Gössling S, Hanna P, Higham J, Cohen S, Hopkins D (2019) Can we fly less? Evaluating the 'necessity' of air travel. J Air Transport Manag 81:1–22
3. International Standard Organization (2006) ISO 14040: 2006 Environmental management Life cycle assessment Principles and framework
4. Krieg H, Ilg R, Brethauer L (2012) Environmental impact assessment of aircraft operation: a key for greening the aviation sector. J Aerosp Sci Technol Syst 91:73–78
5. Johanning A, Scholz D (2013) A first step towards the integration of life cycle assessment into conceptual aircraft design. In: Proceedings of the German aerospace congress, DLRK, Stuttgart, Germany
6. Howe S, Kolios A, Brennan F (2013) Environmental life cycle assessment of commercial passenger jet airliners. Transp Res Part D: Transp Environ 19:34–41
7. Musacchio A, Serra A, Cencioni L, Colantoni S, Bartocci P, Fantozzi F (2021) Decarbonizing materials and machining for the gas turbines sector through life cycle assessment. In: Proceedings of ASME turbo expo, turbomachinery technical conference and exposition
8. Musacchio A, Serra A, Cencioni L, Corona A (2020) Defining key environmental perfomance factors (KEPF) for gas turbines eco-design and production through life cycle assessment. In: Proceedings of ASME turbo expo, turbomachinery technical conference and exposition
9. Chester MV (2008) Life-cycle environmental inventory of passenger transportation in the United States. Master's Thesis,. University of California, Berkeley, CA
10. Lopes JV (2010) Life cycle assessment of the Airbus A330–200 aircraft. Master's Thesis,. Universidade Técnica de Lisboa, Lisbon, Portugal
11. Rahn A, Wicke K, Wende G (2022) Using discrete-event. Simulation for a holistic aircraft life cycle assessment. Sustainability 2022(14):10598
12. Rahn A, Pohya AA, Wehrspohn J, Wicke K (2021) A modular framework for life cycle assessment of aircraft maintenance. In: AIAA aviation 2021 forum. American Institute of Aeronautics and Astronautics, Reston, VA
13. Choi BK, Kang D (2013) Modeling and simulation of discrete-event systems. Wiley, Hoboken, NJ
14. Pohya AA, Wehrspohn J, Meissner R, Wicke K (2021) A modular framework for the life cycle based evaluation of aircraft technologies, maintenance strategies, and operational decision making using discrete event simulation. Aerospace 8
15. Meissner R, Rahn A, Wicke K (2021) Developing prescriptive maintenance strategies in the aviation industry based on a discrete-event simulation framework for post-prognostics decision making. Reliab Eng Syst Saf 214:107812
16. Fabre A, Planès T, Delbecq S, Budinger V, Lafforgue G (2022) Life cycle assessment models for overall aircraft design. In: Proceedings of the AIAA SciTech 2022 Forum, San Diego, CA. American Institute of Aeronautics and Astronautics, Reston, VA
17. Parolin G, Borges AT, Santos LC, Borille AV (2021) A tool for aircraft eco-design based on streamlined life cycle assessment and uncertainty analysis. Procedia CIRP 98:565–570
18. Weidema BP, Wesnaes MS (1996) Data quality management for life cycle inventories: an exemple of using data quality indicators. J Clean Prod 4
19. Frischknecht Jungbluth R, N., Althaus, HJ. et al (2005) The ecoinvent database: overview and methodological framework (7 pp). Int J Life Cycle Assess 10(1):3–9
20. European Commission, Product Environmental Footprint (PEF) and Organisation Environmental Footprint (EF) (2021) [Online]. Available on: https://eur-lex.europa.eu/legal-
21. Inaba A, Itsubo N (2019) Preface. Int J Life Cycle Assess 23(12):2271–2275
22. Bulle C et al (2019) IMPACT World+: a globally regionalized life cycle impact assessment method. Int J Life Cycle Assess 24(9):1653–1674

23. Keiser D, Schnoor LH, Pupkes B, Freitag M (2023) Life cycle assessment in aviation: a systematic literature review of applications, methodological approaches and challenges. J Air Transp Manag 110:102418
24. Rolinck M, Gellrich S, Bode C, Mennenga M, Cerdas F, Friedrichs J, Herrmann C (2021) A concept for blockchain-based LCA and its application in the context of aircraft MRO. Procedia CIRP 98:394–399
25. Denkena B, Dittrich MA, Jacob S (2016) Energy efficiency in machining of aircraft components. Procedia CIRP 2016(48):479–482
26. Sarkar S, Ahmed M, Chowdhury MA, Melton G (2022) Life cycle assessment (LCA) results of MIG and TIG welding technologies using the IMPACT 2002+ methodology. Int J Mech Eng Robot Res 11:564–568
27. Gamage JR, DeSilva AK, Harrison C, Harrison D (2016) Ascertaining life cycle inventory data for electrical discharge machining. Procedia CIRP 41:908–913
28. Su S, Li X, Zhu C, Lu Y, Lee HW (2021) Dynamic life cycle assessment: a review of research for temporal variations in life cycle assessment studies. Environ Eng Sci 38(11):1013–1026

# Part V
# Digital Technologies for Sustainability

# Digital Twins for Real-Time Life Cycle Assessment of Products

**Theresa Riedelsheimer, Anne Seegrün, and Kai Lindow**

**Abstract** The reduction of the global environmental impact has been established as an overall responsibility of our time. In this context, the regulatory requirements, specifically the EU proposal for a new Ecodesign for Sustainable Products Regulation (ESPR) and Digital Product Passports, are becoming stricter. Companies that want to place physical goods on the EU market need to prepare for more specific reporting and wider scopes of monitoring their environmental impact. In this context, Digital Twins enable the representation and analysis of data and information of specific assets, e.g., products. This chapter focuses on applying the concept of Digital Twins of products to address the new regulatory requirements and depict an accurate environmental impact of an individual product along its lifecycle. First, the authors present relevant personas and user stories as well as the concept at a technical level, focusing on the underlying functional diagram, system architecture, IT infrastructure, and data modeling. Finally, insights from industrial applications of the concept within the automotive industry and the production of an orthosis with bio-based material are discussed. It is shown that the industrial application still faces challenges with regard to standardized cross-company data exchange as well as the interconnection of heterogeneous IT systems for data provisioning. At the same time, the potential of integrating Life Cycle Assessment (LCA) in the creation of Digital Twins for monitoring the environmental impact of an individual asset is emphasized.

**Keywords** Digital Twin · Life Cycle Assessment · Environmental impact · Product life cycle

---

T. Riedelsheimer (✉) · A. Seegrün · K. Lindow
Division Virtual Product Creation, Fraunhofer Institute for Production Systems and Design Technology IPK, Berlin, Germany
e-mail: theresa.riedelsheimer@ipk.fraunhofer.de

 
S. Fukushige et al. (eds.), *EcoDesign for Circular Value Creation: Volume I*,
https://doi.org/10.1007/978-981-97-9068-5_28

## 1 Introduction

The reduction of the global environmental impact has been established as an overall responsibility of our time, underscored not least by initiatives such as the Sustainable Development Goals (SDG) and the Paris Climate Agreement from 2016. The implementation of appropriate countermeasures against the climate crisis with resource scarcity and the destruction of our planet need to be addressed on several levels: on a societal, legislative, regulatory, organizational and technological level. One specific lever lies with manufacturing companies being responsible for the development, production, marketing, operation, and disposal of products. In this context, the requirements for different industries and countries are becoming stricter. Important EU-wide drivers include the following:

- The European Green Deal (2019) [1] and the new Circular Economy (CE) action plan (2020) [2].
- The European Sustainable Products Initiative [3], which involves an update of the Ecodesign Directive 2009/125/EC in 2022 to expand its scope to cover more product types with the Proposal for Ecodesign for Sustainable Products Regulation (ESPR) [4], planned for adoption in 2025 with the first products expected in 2026,
- Digital Product Passports with a first very specific regulation on battery passports [5], planned for 2027,
- The Corporate Sustainability Reporting Directive (CSRD) (2022), planned for 2024 [6].

Consequently, companies need to prepare for more specific reporting and wider scopes of monitoring their environmental impact. One implicit objective is a higher primary data share of sustainability assessments, e.g., via Life Cycle Assessments (LCA), of individual products along the lifecycle and the implementation of more effective measurements to monitor and reduce the environmental impact. The necessary cross-company data exchange and product-specific monitoring demands for standardized technical solutions and interfaces. In this context, Digital Twins (DT) offer a technological approach for the representation and analysis of data and information of specific assets, e.g., products. Consequently, DTs could benefit the LCA execution to provide more product-individual LCA results along the product lifecycle. However, it has been mainly applied for the optimization of product quality and economical process efficiency. Therefore, the question arises, whether and how this concept can be applied to foster sustainability, i.e., to reduce the environmental impact. This chapter addresses the following research questions:

[RQ I] How can a DT of products support companies to increase their primary data share of product-specific LCAs and to monitor and reduce the product-individual environmental impact?

[RQ II] Which core functionalities and system elements need to be considered to enable the establishment of DTs for LCA?

To answer these research questions, the following approach was chosen:

I. Conducting an industrial survey to identify the industrial needs,
II. Researching the state of the art of literature and regulatory demands,
III. Development of the technological concept, following the approach of the DT-V-Model [7] in a user-centered approach and,
IV. Validation within industrial use cases in the automotive industry.

## 2 State of the Art

### 2.1 Digital Twins

The topic of DTs has been discussed in the literature for over two decades and has reached the level of industrial applications. There are multiple perspectives on DTs (see extensive literature review in [8] and [9]), which evolved over time from the focus on simulation (e.g., [10, 11]) to the integration of additional value creation (e.g., feedback to design [12, 13]). Most DT concepts can also be differentiated with regard to the depicted system, such as production lines, products in use, or products during development, and as a newer focus, product reuse, recycling, or other CE strategies.

At the same time, the standardization of technological enablers has evolved. Important concepts are the Asset Administration Shell (AAS) [14], standardized data models as published by the IDTA, as well as open-source applications and interfaces within the production environment, such as Eclipse BaSyx [15].

As this research is based upon the concept of DTs from an engineering perspective with a focus on products, it applies the definition by Stark et al. [16]. Here, DTs are "a digital representation of an active unique product [...] or unique product-service system (a system consisting of a product and a related service)". A DT consists of the interlinkage of digital master data, which are models and information from the planning phase, and instance-specific data (the so-called digital shadow) from production, usage, or end-of-life (EoL). The most added value is created within the DT core, which is responsible for the calculation, simulation, or interlinkage of data and information of the product or asset in focus. Within this chapter, DTs are described with regard to products or product systems as assets.

### 2.2 LCA

LCA is a standardized method used to calculate the environmental impact across more than 15 impact categories (e.g., Greenhouse Gas Emissions) and is defined within the ISO standard 14040/44. It entails the basic process steps:

1. Define goal and scope.
2. Collect data as part of the Life Cycle Inventory (LCI).
3. Assess impact in the Life Cycle Impact Assessment (LCIA) (e.g., in $CO_2$ equivalents).
4. Interpret, identify optimization potential, and optimize.

The results can be presented in different ways, e.g., as part of a standardized reporting document. Here, the Environmental Product Declaration (EPD) is used to present the environmental and human health impact of a product specifying the ISO 14025 standard (Environmental Product Declarations Type III). It can be created for a company-specific product or for average products and must be verified by an independent party. In the future, certain information and impact categories, such as the Product Carbon Footprint (PCF), need to be provided with almost every product that is placed on the market—in a so-called digital product passport [5]. One established accounting and reporting standard is the Greenhouse Gas (GHG) Protocol, which is part of an initiative that was launched in 1998 [17]. It defines different scopes for setting the boundaries of sustainability reporting: scope 1 with direct emissions owned or controlled by the reporting company, to indirect emissions in scope 2 and 3 along the lifecycle.

## *2.3 Digital Twins for LCA*

The main incentive of a DT for LCA is to use the DT concept to enable product-specific LCA and to monitor and potentially reduce the environmental impact of a product. An implicit objective is to increase the share of primary data from the product-specific supply chain, production sites, company environment, product usage, and EoL instead of relying on accessible databases containing average data and information.

The potential of such a concept has been fostered by the initiatives and regulations on EU level. Consequently, the industrial sector must now explore viable approaches to meet these new requirements. As a first step, to understand the industrial perspective on this concept, an industrial study with 26 companies was conducted in 2019. The results showed that 63% of the interviewed companies saw a major potential for future use of DTs to perform sustainability assessments [18]. Specifically, the aspects of environmental impact monitoring and the reduction of energy and resource consumption were put in focus.

This chapter builds upon the previously presented research on the challenges within Life Cycle Sustainability Assessment (LCSA) and the basic concept of a DT [19], the role of DTs for Product Lifecycle Management (PLM) [9], the presented lifecycle of DTs in the context of Biological Transformation [20] as well as the methodical approach to develop such a concept, namely the DT-V-Model [7] and its validation for different use cases [21]. Other authors presented similar concepts of a combination of DTs and LCA (e.g., [22–24]), whereas [23, 24] only focus on solutions for the manufacturing context. None of the identified literature details the design of the DT, its functions, and underlying IT systems, which would be necessary as a blueprint for further implementation. This chapter takes an additional step by providing the development results on a technological level.

Recent research also considers aspects such as cross-company data exchange and CE [9, 25, 26], but does not specifically focus on LCA as an approach.

In general, the published research lacks a detailed description of how to integrate sustainability assessment into the design of the DT, covering not only conceptual and functional aspects but also technological considerations.

## 3 Concept Development

To develop the concept, a user-centered approach was executed in accordance with the DT-V-model (see [7, 21]).

### *3.1 Personas*

Firstly, the main personas were described and validated with industrial partners from the automotive value chain. Therefore, interviews to understand the current and future roles in the context of engineering and sustainability were conducted in 2020 with representatives from the automotive value chain and a home appliances manufacturer ($n = 9$). Their roles varied from product development, research, and innovation to sustainability management. As a result, 23 user stories for 6 roles were identified. Additionally, within the research project BioFusion 4.0, two workshops were conducted to further develop the personas with nine participants from research and industry. For the most relevant personas during the Beginning-Of-Life (BoL) and Mid-of-Life (MoL), user stories (see Fig. 1) were defined to derive functional and nonfunctional requirements. While this chapter does not specifically focus on the EoL phase, it is important to note that DTs hold significant potential for CE and the EoL phase.

The most relevant roles that are potential users of the DT for LCA—besides regulatory bodies or independent certification institutes—are the sustainability manager (SM), the product developer (PD), and the production planner (PP). Their main responsibilities and tasks are depicted in Fig. 1. Additionally, several individual personas have been developed to create a robust concept that meets the needs of a diverse spectrum of future users. These personas later serve as the basis for creating test cases and conducting user testing.

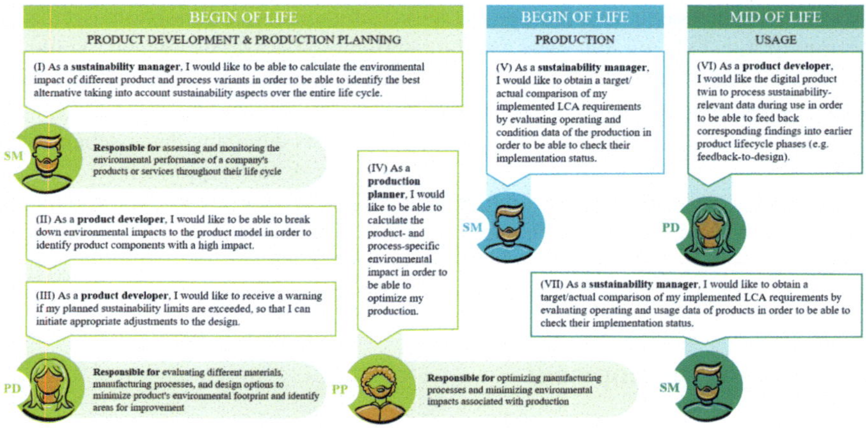


**Fig. 1** Personas and User Stories for the Digital Twin for Life Cycle Assessment

## 3.2 User Stories

The respective user stories are situated within the primary lifecycle phases of BoL, encompassing product development, production planning as well as the production phase itself, and MoL concerning product usage. The EoL phase is not depicted in Fig. 1, but still provides relevant personas from a DT perspective in the context of CE, for example to enable the reuse of products and components.

In BoL, the user stories cover aspects such as monitoring the environmental impact for product and process variants (I), component-specific environmental impact analysis (II), design decision support with regard to prior defined thresholds, and the provision of system triggered warnings (III). Additionally, the production planner requires both product-specific and process-specific breakdowns of the LCA results to optimize the respective production process decisions during planning (IV). During the production itself, one relevant user story is focused on the sustainability manager demanding the monitoring of the target achievement with regard to the as-planned environmental impact (V).

During usage, the main user stories describe feedback to design (VI) and monitoring (VII).

During EoL and the respective downstream processes, consideration of different CE strategies is necessary. The decision support for optimizing the environmental impact across different lifecycles (e.g., reuse or repurpose) demands for a breakdown of the environmental impact to individual components. For each CE strategy, not only sustainability managers but also specific roles in dismantling, refurbishment, or recycling require the provisioning of more accurate data on the possible and most environmental-friendly CE strategy. To meet this demand, dedicated decision support systems are developed (see [25–27]). This chapter, however, focuses on the lifecycle model from cradle to usage.

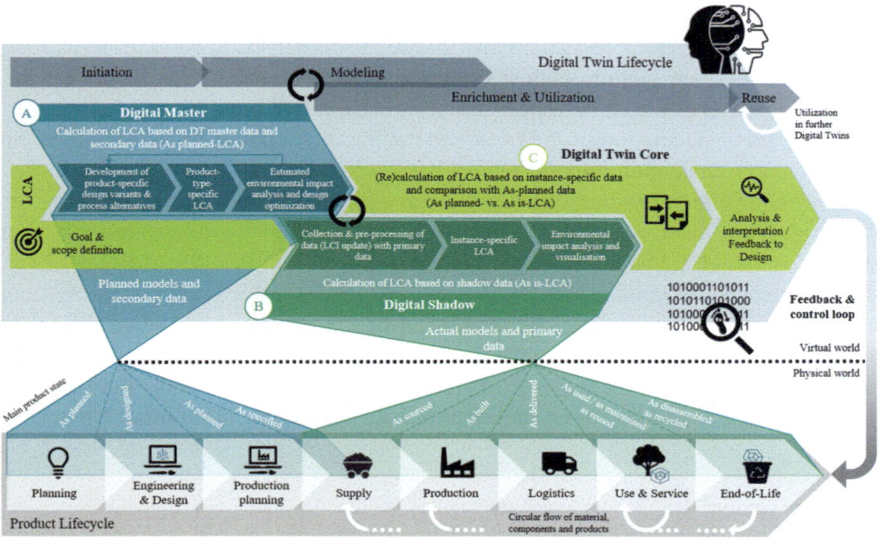


**Fig. 2** Concept of the Digital Twin for Life Cycle Assessment—Overview

## 3.3 *Concept*

For the identified user stories, a requirements list was derived and the basic concept was developed to address them. Figure 2 illustrates the core concept behind the integration of LCA with DTs.

The DT lifecycle, using data and models in the virtual world, is combined with the product lifecycle, representing the physical world. Every DT needs to be developed in alignment with the respective product lifecycle and the relevant data sources along the value stream. The main generic phases of the DT lifecycle include the following:

- *Initiation*, which includes the scoping of a DT based on the main objective, requirements, and data needs of the data consumer side (personas and IT systems), as well as the identification of core functionalities and components.
- *Modeling* the relevant data, information, and models, along with corresponding testing, in alignment with the digital master phase.
- *Enrichment and Utilization*, which starts with the integration of dynamic and instance-specific data during the digital shadow phase.
- The *reuse* in further DTs, for future product generations and to update LCA databases.

The modeling phase can also occur during a later product lifecycle phase, if the DT is developed after a product is already in use (see also development scenarios in [7]).

The digital master phase (A) is executed in parallel with the goal and scope definition of the LCA process and transitions to the digital shadow phase (B) as soon as the first instance-specific data is collected. The DT core (C) is an underlying element of the DT, which is responsible for decision-making and decision support regarding the recalculation of the LCA and the identification of levers to reduce the calculated environmental impact.

The main generic product states, which are relevant for the DT update along the lifecycle, have been identified in alignment with the main milestones during a product's lifecycle. These states are especially relevant for the different LCA execution steps. As soon as one of the depicted main product states is reached, an update of the LCA or at least the evaluation of the necessity to update the LCA, needs to be triggered. It is recommended to define the goal and scope of the LCA early on and comply with requested system boundaries set by regulations or standards (e.g. PCF) while ensuring uniform functional units for all product types.

*Goal and Scope*

The goal and scope need to be defined at the beginning of each LCA. In the context of DTs, the repeatability of the LCA and the comparability of the LCA results along the lifecycle must be ensured. As a result, the functional unit is defined only once, serving as the reference value for assessing the product system's benefits to which the input and output flows from the life cycle inventory can be correlated. Additionally, the system boundaries need to be consistent over the execution of the LCA along the lifecycle. For a comprehensive LCA, it is advisable to include Scope 3 emissions and place particular emphasis on material extraction up to the EoL phase, and possibly even consider reuse scenarios. The DT can support different objectives for the LCA, which can range from product-specific monitoring to reporting and identification of improvement potential. Due to the very characteristic of a DT, which captures instance-specific, dynamic, and even real-time monitoring data (e.g., from production phase), the effort of establishing such a DT must be outweighed by its added value. Therefore, its application should not only focus on reporting but also on the optimization of the product instance or future product generations. It shows its highest potential with complex mechatronic products of high value and with a long lifetime.

(A) *As planned-LCA as part of the digital master*

In the early phase of product development, the data basis is usually not sufficient for a full and accurate LCA. Here, a preliminary sustainability assessment should be conducted for different product design variants and corresponding process alternatives. As the product characteristics mature, the so-called As-Planned LCA can be performed based on planning data and assumptions. Its results and relevant metadata are then stored within the digital master (see A in Fig. 2)—or made traceable by assigning a unique product ID. This facilitates an approximate environmental impact analysis and supports design optimization. Within the DT, the following main product states should be traceable: as planned during product planning, as designed, as planned during production planning, and as specified during sourcing, i.e., provided to the suppliers.

A crucial decision lies in choosing between utilizing primary data obtained from the company's specific supply chain and secondary data sourced from databases. Conducting an early hot spot and data quality analysis can help pinpoint the data points where primary data should be prioritized. Additionally, when using primary data from the supply chain, the role of integrating data from other companies needs to be considered. This poses the challenge of requesting and integrating data from multiple heterogeneous IT systems, data models, types, and formats as well as different interfaces. Addressing this challenge demands standardized data exchange, data models, and interfaces.

(B) *As-is LCA as part of the digital shadow*

With the start of the supply and the production phase, the product batches or even instances start to differ from one another, e.g., depending on the production site or logistics. In the use phase, at the latest, each individual product has its distinct characteristics of usage patterns, performance, and wear—sometimes referred to as the digital shadow. This generates an individual environmental impact, which is captured along consecutive lifecycle phases via unique identifiers within the digital shadow (see B in Fig. 2). In this phase, the LCA is based on dynamic and primary data of the individual product—as long as the effort is appropriate. That is for either meeting regulatory requirements or for the estimated reduced environmental impact (see also scope 4 [17]). First, it is necessary to collect and preprocess the dynamic data and ultimately update the LCI as step 2 of the LCA. The instance-specific LCA is then executed with the updated LCIA. Within the respective phase of the product lifecycle, the environmental impact can be analyzed automatically or manually with an appropriate visualization. Consequently, the relevant data points from the specific phase need to be defined to identify the primary data sources. This refers to:

- *As sourced*: Data from the procurement process, supplier-specific data (usually part of Scope 3 emissions) on the actual material, and components as well as their environmental impact,
- *As built* (usually refers to Scope 1 and 2 emissions): The environmental impact that is caused by the production processes and assembly or other direct processes,
- *As delivered*: The environmental impact of the chosen ways of transport and the actual delivery processes to downstream stakeholders,
- *As used, as maintained, and as reused*: All environmental impact that is caused during the usage phase and the update with changes in the product characteristic, e.g., due to the exchange of components and repair activities,
- And *as disassembled and as recycled*: Environmental impact caused by disassembly, recycling, or waste processes.

The trigger events, triggering IT systems, and frequency need to be defined during the design phase of the DT as well as the quality and granularity of the data used.

(C) *Analysis and optimization within the Digital Twin Core*

The third main phase can be defined as the DT core calculations and assessment. By linking master models and shadow data within the DT Core, lifecycle information from the planning phase and the actual instance-specific lifecycle can be correlated. This includes a recalculation of the LCA results based on instance-specific data, as well as the analysis and interpretation to identify optimization potential. Ultimately, the comparison between "as-planned LCA" versus the "as-is LCA" (see C in Fig. 2) may enable the provision of insights for future product optimizations (feedback to design). However, analysis and interpretation within the DT Core can also be used to execute control commands via direct feedback to the physical system, for instance, to optimize product behavior or machine processes with regard to $CO_2$ emissions.

The design of the visualization environment, as well as interfaces, needs to cater to both data-consuming users and IT systems. This is typically achieved through a user-centered approach based on stakeholder analysis and persona development (see Sect. 3.1). In this context, the main objective is also defined within the goal definition of the LCA and is usually focused on reporting, decision support, or automatic optimization. Therefore, it is crucial to define the appropriate degree of autonomy for DTs.

# 4 Technological Perspective

To detail the technological concept, the elements of a requirements list, use case diagram, functional diagram, and system architecture are developed. Here, the use case of a DT for LCA based on live production data was chosen as an example for the primary digital shadow phase.

## *4.1 Functional Diagram*

Figure 3 shows the operational view of the core functionalities in UML notation, which were derived from the personas, user stories, and requirements. The requirements include the functional and non-functional aspects but also specifically the data demands and necessary data points.

The operational process starts with setting the goal and scope (LCA-phase 1). As part of the digital master data generation, the development process and production

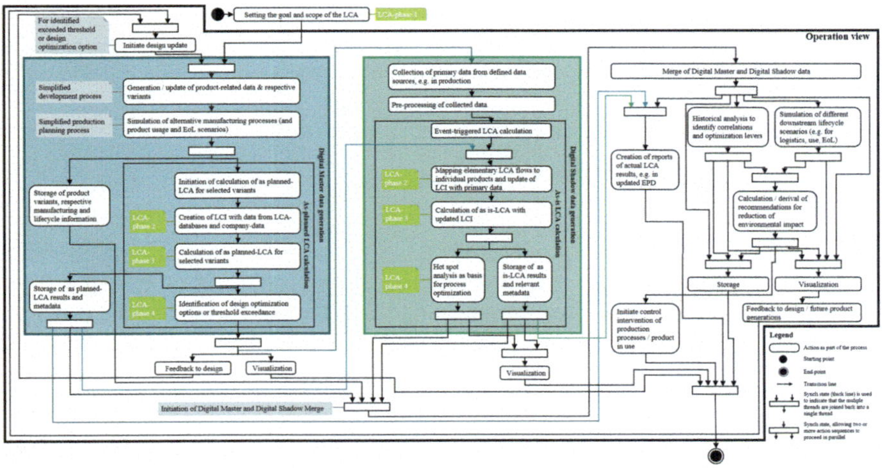


**Fig. 3** Functional diagram of the Digital Twin for Life Cycle Assessment

planning process are executed. The resulting product variants, along with associated manufacturing and lifecycle information, are then stored. To store the as-planned LCA results and metadata, the three main LCA phases are performed. The main output of this phase is the visualization of the results and the identification of design optimization options or threshold exceedances, which initiates feedback to the design process.

Digital shadow data generation is initiated at the start of a product's individual lifecycle. This process involves the collection of primary data from defined data sources; for instance, in production, the preprocessing of the collected data and the event-driven LCA calculation. Again, the three main phases of the LCA are carried out, and for those aspects where primary data is not available, the planning data is used. The main results in this stage are a hot spot analysis as a basis for process optimization, the storage of the as-is LCA results, and relevant metadata, and their visualization.

On this basis, the digital master data and digital shadow data can be merged to fully exploit the potential of the DT analysis. Three major streams are then initiated:

- Creation of reports of actual LCA results, for instance, in an updated EPD.
- Historical analysis to identify correlations and optimization levers of the digital shadow data.
- The simulation of different future downstream lifecycle scenarios depending on the current product state (e.g., for logistics, use, and EoL).

This is followed by a calculation and recommendations on how to reduce the environmental impact thus initiating design feedback and optimization of future product generations. In addition, in situ optimization is initiated, which involves direct control of the represented system through intervention in production processes

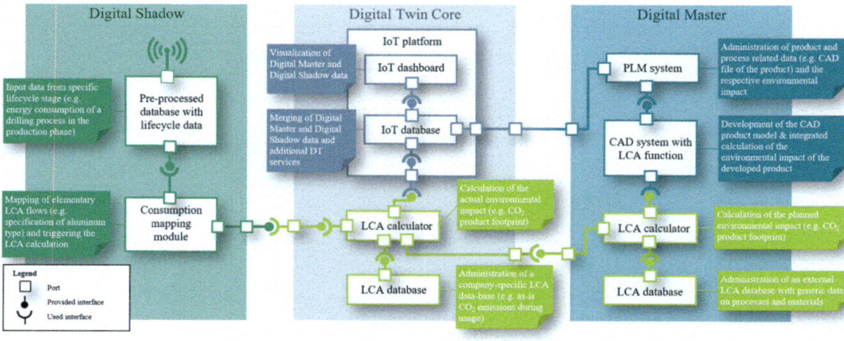


**Fig. 4** Digital Twin for Life Cycle Assessment system architecture—Overview

and products in use. The degree of intervention is highly dependent on the chosen level of autonomy. From the depicted functional perspective, the system-focused architecture is derived.

## 4.2 Solution Architecture

Figure 4 provides an overview of the selected solution-neutral system architecture to implement a DT for LCA. The system architecture depicts three main systems with their respective system elements: digital master, digital shadow, and the DT core.

### *Digital Master*

The digital master is responsible for the management of all planning data and models. The main systems are:

- *PLM system*: Administration of product- and process-related data (e.g., computer-aided design (CAD) file of the product) and the respective environmental impact,
- *CAD system with LCA function*: The management and development of the CAD product model with an integrated planning of the environmental impact,
- *LCA calculator*: the creation of the LCA model and the calculation of the environmental impact of the developed product (e.g., planned $CO_2$ footprint [equivalents] of the product)
- *and LCA database*: The administration of an external LCA database with generic data on processes and materials.

*Digital Shadow*

The specific systems that are part of the digital shadow are highly dependent on the respective lifecycle phase and defined product states in the concept phase. In its basic configuration, the two essential system elements are:

- *Preprocessed database with lifecycle data*: Input data from specific lifecycle stages (e.g., energy consumption of a drilling process in the production phase) and
- *Consumption mapping module*: Mapping of elementary LCA flows (e.g., specification of aluminum type) and triggering the LCA calculation.

*Digital Twin Core*

The main involved systems are the Internet of Things (IoT) platform, the IoT database, the LCA calculator, and the LCA database. To allow for the management of live IoT data, an IoT platform was chosen. These systems execute the following main functions:

- *IoT platform*: Visualization of digital master and digital shadow data (*IoT dashboard*) and merging of digital master and digital shadow data as well as the management of and calculation for additional DT services, such as the recommendation of optimization measures and initiating product control (*IoT database)*.
- *LCA calculator*: Calculation of the current environmental impact (e.g., $CO_2$ footprint of the individual product) with access to the LCA model of the planning phase and the live data, e.g., from production.
- *LCA database*: Administration of a company-specific LCA database (e.g., actual $CO_2$ emissions of the processes along the product lifecycle).

The presented solution architecture can be applied to different use cases with similar requirements, i.e., focusing on the product as an asset and including static planning as well as dynamic live data from the lifecycle. However, given potential variations in the provisioning of DT analysis results, a full integration into other IT systems, such as production control, may be better suited than relying solely on an IoT platform as the DT core system.

## 5 Validation

Within three use cases, the presented concept underwent validation as part of a research project at various levels. The primary goal for all use cases was to monitor the environmental impact of individual products and conduct a hot spot analysis to identify the primary sources of $CO_2$ emissions.

*Use Case 1: Line production of engine parts for automotive vehicles*

- Creation of a DT of a fully automated line production for a combustion engine part with an industrial application partner (OEM).
- As-planned LCA: LCA model for the produced part from raw material to recycling.
- Digital shadow: Data capturing from energy management and Manufacturing Execution System (MES).
- As-is LCA: Real-time quantification of PCF and 15 LCA impact categories per OK part.

*Use Case 2: Line assembly of e-car components*

- Creation of a DT of an experimental assembly line for an e-car assembly with an industrial application partner (OEM).
- As-planned LCA: LCA model for the produced parts from raw material to recycling.
- Digital shadow: Data capturing from smart assets, using events from production to trigger the LCA update after each assembly step.
- As-is LCA: Real-time quantification of carbon footprint and 15 LCA impact categories per assembly step.

*Use Case 3: 3D printing of an orthosis as assembly support*

- Creation of a DT of the orthosis and the 3D printing process of individual product variants using bio-based material.
- As-planned LCA: LCA model for the orthosis from raw material to recycling.
- Digital shadow: Data capturing from 3D printing process and energy monitoring system as well as EoL analysis.
- As-is LCA: Real-time quantification of carbon footprint and 15 LCA impact categories per OK part with a specific focus on the comparison of different CE strategies.

The resulting solution-specific IT system architecture is currently implemented for each use case in the industrial (use case 1 and 2) and laboratory environment (use case 3). For use case 3, open-source-based software was chosen, whereas use case 1 and 2 were implemented within the IT system architecture of the application partner and with proprietary software for PLM and CAD. For all use cases, the LCA calculation software OpenLCA © and the EcoInvent © database were used. For each of the use cases, a qualitative validation of the development process, the establishment effort, and added value was performed. The main outcomes of the validation within the use cases are described below.

### 5.1 Concept and Approach in General

In general, the concept is considered relevant and regulatory requirements encourage its implementation within the industry. The primary challenges emerge when implementing it in global supply chains to cover Scope 3 emissions. It is essential to consider and ensure compliance with continuouslly changing regulatory requirements, including future ones. Two significant challenges were identified for the conceptualization of generating and collecting digital shadow data: defining real-time within this context and determining the optimal level of updating an LCA based on dynamic data from the product's life, such as through sensors.

### 5.2 Development Approach

Due to the primary effort involved in data collection and provisioning, and the benefits being reaped by data consumers, cross-domain collaboration is essential. Such collaboration could be between production personnel and sustainability management. The data-providing stakeholders and IT systems do not usually benefit directly from the DT for LCA services. Additionally, the need for an early proof of concept for industrial applications was identified. This should be considered within the development approach.

### 5.3 Added Value and Lifecycle Perspective

The perspective on the added value differs with the lifecycle phase of the product and the DT services consumed. From the perspective of BoL, a potential reduced environmental impact could not be quantified yet, as the main DT service implemented aims at monitoring. The LCA showed that the material choice and corresponding supply chain and logistics have a high impact on the resulting LCA. The use phase is only in focus for use case 3, mainly due to the difficulties in gathering usage information from privately owned vehicles. The EoL phase could only be connected to the product development phase in use case 3. In the industrial environment, information on the environmental impact of the EoL phase is still not systematically integrated within the BoL.

### 5.4 Operation of the DT for LCA

The main feedback can be reduced to the high effort required for data collection, data management, and cross-IT system and cross-company data exchange. Data collection efforts are particularly high when monitoring systems are not in place during the digital shadow phase. During the operation of a DT for LCA, the IT infrastructure and respective processes need continuous assessment for efficiency. Therefore, defining trigger events and identifying an appropriate level of detail and data quality is crucial but often complex to execute in advance. For a holistic and accurate LCA with DTs, the provisioning of data across different companies is necessary, leading to the need for standardized data models and interfaces. Main challenges identified include interconnecting heterogeneous IT systems in an industrial environment and dealing with scattered data across different databases.

## 6 Conclusion

The presented concept and validation within this chapter show how DTs can be used for the asset-specific execution of an LCA and specifically its monitoring along the whole lifecycle. The presented results can be used as a blueprint for further industrial implementation—specifically the core functionalities and system elements that need to be considered to enable the establishment of DTs for LCA. The use cases from the automotive industry focus on the integration of production data. Here, especially the material choice, the energy, and resource consumption as well as the waste, i.e., defective parts, carry significant weight. The appropriate effort of establishing a DT for LCA needs to be weighed against the potential added value in terms of reduced environmental impact. On the other hand, the findings show that the current regulatory developments will demand an increased product-specific reporting in the future, justifying the effort of setting up the respective IT architecture.

The validation exhibits some relevant limitations. First, no long-term analysis of a DT for LCA has been conducted within the use cases yet. Therefore, it was not possible to actually measure the added value in terms of reduced environmental impact, but rather their monitoring. Additionally, for this chapter, the necessary organizational processes and roles for the development and operation in an industrial environment were considered out of scope.

The future research should focus on a rapidt identification of an appropriate level of primary and dynamic data, autonomous decision-making within DTs, further enabling IT systems, such as PLM systems, to fully and automatically integrate LCA as well as broadening the scope on different levels:

- Considering the impact of the digital infrastructure,
- Integrating the entire lifecycle in the LCA execution,
- Integrating Circular Economy strategies.

To facilitate the establishment across different industries, it is necessary to develop a standardized approach that enables the efficient and sustainable definition of the frequency and triggers of LCA updates based on primary data and ultimately enables cross-company data exchange.

**Acknowledgments** This chapter was written as part of the research and development project "Biological Transformation 4.0: Further development of Industry 4.0 by integrating biological principles (BioFusion 4.0)", which is funded by the German Federal Ministry of Education and Research (BMBF) and supervised by the Project Management Agency Karlsruhe (PTKA). The responsibility for the contents of this publication lies with the authors.

# References

1. European Parliament, The European Green Deal. [Online] Available: https://www.europarl.europa.eu/legislative-train/theme-a-european-green-deal/file-european-green-deal. Accessed 11 Jul 2023
2. European Parliament, *New circular economy action plan: In "A European Green Deal"* [Online] Available: https://www.europarl.europa.eu/legislative-train/theme-a-european-green-deal/file-new-circular-economy-action-plan. Accessed 11 Jul 2023
3. European Commission, Sustainable products initiative. [Online] Available: https://ec.europa.eu/info/law/better-regulation/have-your-say/initiatives/12567-Sustainable-products-initiative_en. Accessed 11 Jul 2023
4. European Commission, Ecodesign for Sustainable Products Regulation: proposal for a new Ecodesign for Sustainable Products Regulation (ESPR). [Online] Available: Ecodesign Directive 2009/125/EC. Accessed 11 Jul 2023
5. European Commission, Proposal for a REGULATION OF THE EUROPEAN PARLIAMENT AND OF THE COUNCIL concerning batteries and waste batteries, repealing Directive 2006/66/EC and amending Regulation (EU) No 2019/1020. [Online] Available: https://www.europarl.europa.eu/RegData/docs_autres_institutions/commission_europeenne/com/2020/0798/COM_COM(2020)0798_EN.pdf. Accessed 11 Jul 2023
6. European Parliament, Directive (EU) 2022/2464 of the European parliament and of the council: Corporate sustainability reporting. [Online] Available: https://eur-lex.europa.eu/legal-content/EN/TXT/?uri=CELEX%3A32022L2464&qid=1689947469993. Accessed 11 Jul 2023
7. Riedelsheimer T, Gogineni S, Stark R (2021) Methodology to develop Digital Twins for energy efficient customizable IoT-Products. Proc CIRP 98:258–263
8. Negri E, Fumagalli L, Macchi M (2017) A review of the roles of Digital Twin in CPS-based production systems. Proc Manuf 11:939–948
9. Seegrün A et al (2023) Sustainable product lifecycle management with Digital Twins: a systematic literature review. Proc CIRP 119:776–781
10. Grieves M (2014) Digital twin: Manufacturing excellence through virtual factory replication. White paper
11. Boschert S, Rosen R (2016) Digital Twin—the simulation aspect. In: Mechatronic futures: challenges and solutions for mechatronic systems and their designers. Springer International Publishing, Cham, pp 59–74
12. Riedelsheimer T, Lindow K, Stark R (2018) Feedback to design with Digital Lifecycle-Twins: literature review and concept presentation. In: Krause, Paetzold et al. (Hg.) 2018—Design for X, pp 1–12
13. Tao F et al (2018) Digital twin-driven product design, manufacturing and service with big data. Int J Adv Manuf Technol 94(9):3563–3576
14. Plattform Industrie 4.0, Details on the Asset Administration Shell (AAS): Specification. [Online] Available: https://www.plattform-i40.de/IP/Redaktion/DE/Downloads/Publikation/Details_of_the_Asset_Administration_Shell_Part1_V2.pdf?__blob=publicationFile&v=6
15. Eclipse Foundation, BaSyx. [Online] Available: https://wiki.eclipse.org/BaSyx. Accessed 11 Jul 2023

16. Stark R, Damerau T (2019) Digital Twin. In: CIRP encyclopedia of production engineering. Springer Berlin Heidelberg, pp 1–8
17. The greenhouse gas protocol: A corporate accounting and reporting standard. World Business Council for Sustainable Development; World Resources Institute, Geneva/Washington, DC, 2004.
18. Riedelsheimer T, Lünnemann P, Wehking S, Dorfhuber L (2020) Digital Twin Readiness Assessment: A survey in the discrete manufacturing industry. Fraunhofer-Institut für Produktionsanlagen und Konstruktionstechnik, Berlin. [Online] Available: https://www.ipk.fraunhofer.de/de/medien/markt-trendstudien/digital-twin-readiness-assessment/download-studie-digital-twin-readiness-assessment.html. Accessed 11 Jul 2023
19. Riedelsheimer T, Neugebauer S, Lindow K (2021) Progress for life cycle sustainability assessment by means of Digital Lifecycle Twins: a taxonomy. In: EcoDesign and Sustainability II. Springer Singapore, Singapore, pp 245–329
20. Seegrün A, Mügge J, Riedelsheimer T, Lindow K (2023) Digital Twins for sustainability in the context of biological transformation. In: 18th Global Conference on Sustainable Manufacturing 2022. Proceedings, Manufacturing Driving Circular Economy. Springer International Publishing, Cham, pp 576–584
21. Hagedorn L, Riedelsheimer T, Stark R (2023) Project-based learning in engineering education: developing Digital Twins in a case study. Proc Des Soc 3:2975–2984
22. Barni A et al (Sep. 2018) Exploiting the Digital Twin in the assessment and optimization of sustainability performances. In: Theory, research and innovation in applications, Portugal, pp 706–713
23. Rückert M, Merkelbach S, Alt R, Schmitz K (2018) Online life cycle assessment for fluid power manufacturing systems–challenges and opportunities. In: IFIP Advances in information and communication technology, vol. 536, Smart manufacturing for industry 4.0. Springer, Cham, pp 128–135
24. Ghita M, Siham B, Hicham M, Griguer H (2021) Digital Twins based LCA and ISO 20140 for smart and sustainable manufacturing systems. In: Advances in sustainability science and technology, sustainable intelligent systems. Springer Singapore, Singapore, pp 101–145
25. Mügge J et al (2023) Empowering end-of-life vehicle decision making with cross-company data exchange and data sovereignty via Catena-X. Sustainability 15(9):7187
26. Mügge J et al (2023) End-of-life decision support to enable circular economy in the automotive industry based on digital twin data. Proc CIRP 119:1071–1077
27. Mügge J, Hahn R, Riedelsheimer T, Chatzis J Digital Twins for circular economy: enabling decision support for R-strategies. In: Industrie 4.0 management, pp 42–46

# Toward a Framework for Sustainable Cross-Border Electronic Commerce

**Jiahe Chen and Yi-Chen Lan**

**Abstract** Cross-border electronic commerce (CBEC) is an essential contributor to global economic development as it opens up a global digital market for both production and consumption ends. However, CBEC inevitably poses threats to environmental sustainability because of the intensified carbon footprints through the convoluted and prolonged logistics processes for exporting and importing parties. While various CBEC logistics fulfillment operations (e.g., direct shipment or consolidated shipping; by sea or air freight; via free trade zone or not) have caused different environmental impacts, little is known about the environmental impacts of CBEC businesses, especially those impacts under different modes of CBEC logistics operations. This study contributes to this gap by developing a conceptual framework that dissects the processes of the CBEC logistics operations and assesses their carbon footprints throughout these processes. This chapter reports on an in-depth literature review that informs the conceptual framework that not only evaluates the environmental impacts of CBEC logistics but also enables comparisons of carbon footprints under different modes of logistics operations through scenario analysis. The proposed conceptual framework also illustrates the overall distribution of environmental impacts across CBEC operations. In general, this study is one of the earliest studies that focus on sustainability in CBEC. Theoretically, this study advances the knowledge of optimally managing the carbon footprints of various logistics operations modes in the CBEC context. Practically, this information can help CBEC business leaders to set the organization's environmental strategic directions. The proposed framework also provides CBEC practitioners (e.g., suppliers, third-party logistics service providers, and platforms) with a comprehensive guideline to improve production and operation patterns and move toward greener strategies of business and operations management. Furthermore, and from a consumer perspective, the

J. Chen
School of Business, Western Sydney University, Sydney, NSW, Australia
e-mail: Joe.Chen@westernsydney.edu.au

Y.-C. Lan (✉)
College of Management, Fu Jen Catholic University, New Taipei City, Taiwan
e-mail: Francis.Lan@mail.fju.edu.tw



S. Fukushige et al. (eds.), *EcoDesign for Circular Value Creation: Volume I*,
https://doi.org/10.1007/978-981-97-9068-5_29

findings may inform future research on behavioral science to explore consumers' willingness to comprise efficiency for environmentally-friendly alternatives. In these ways, this study on sustainability in CBEC logistics aims to contribute to achieving Sustainable Development Goal 12 (Responsible Consumption and Production).

**Keywords** Cross-Border E-Commerce · Sustainability · Responsible Consumption and Production · Sustainable Development Goal · Carbon Footprint · Logistics Operations

## 1 Introduction

Globalization boosts the vigorous development of cross-border electronic commerce (CBEC). CBEC refers to "online transactions between sellers and buyers in different countries or customs territories that are conducted through e-commerce platforms and fulfilled by both domestic and cross-border logistics and that entail online cross-border financial transfers" [1, 2]. Due to its convenient and useful business model, CBEC is playing an increasingly important role in both e-commerce and international trade sectors. CBEC is predicted to account for 17% of e-commerce by 2023 and is regarded as an indispensable component of international trade [3].

However, the continuous growth of CBEC has been accompanied by serious environmental challenges—the intensified carbon footprint spreading over CBEC logistics operations being a prominent example. Large geospatial distances in CBEC have substantial impacts on the environment [4], as do the reverse logistics of CBEC. Moreover, the unprecedented international trade has led to between 10% and 70% environmental and social impacts worldwide [5], notably 20% to 33% as to greenhouse gas emissions. Improving CBEC practices and processes to be more environmentally sustainable is thus an important issue and addressing it can contribute to the United Nations Sustainable Development Goal 12 (UNSDG 12 Responsible Consumption and Production).

Despite the relevance of improving CBEC logistics to promoting environmental sustainability, to the best of our knowledge, few studies have explored this topic in the existing literature. This is despite the importance of incorporating carbon footprint management into business operations and decisions being long advocated [6]. In CBEC, optimizing logistics modes and freight transports may shape the carbon footprint of CBEC operations. To raise awareness of the necessity of enhancing the environmental sustainability of CBEC business, this study provides a sustainable framework for CBEC logistics to inform carbon-efficient logistics operations and acquire greener environmental advantages. In particular, this study addresses the following research questions:

- RQ1: What elements should be captured to develop a framework to promote sustainable CBEC, especially for logistics?
- RQ2: How would the environmental and socioeconomic performance in carbon footprint assessment be balanced in this framework?

From a theoretical perspective, this study advances the knowledge of managing and optimizing the carbon footprints of various logistics operations modes in the CBEC context. From a practical perspective, the proposed framework provides CBEC practitioners (e.g., suppliers, third-party logistics service providers, and platforms) with a comprehensive guideline to improve production and operation patterns and to help them move toward greener business strategies and operations management.

The remainder of this chapter is organized as follows. Section 2 introduces the UN SDGs and SDG12 in particular. Section 3 reviews the literature on CBEC logistics and identifies key logistics processes and modes. Section 4 details the literature on carbon footprint assessment in logistics and identifies elements that would be useful to capture to assess carbon emissions in CBEC logistics operations. Section 5 details the sustainable CBEC framework. A discussion and future work are presented in Sects. 6 and 7 concludes the chapter.

## 2 Sustainable Development Goals

The SDGs were formulated in 2015 by the United Nations. They consist of 17 goals and accompanying targets to provide "a shared blueprint for peace and prosperity for people and the planet, now and into the future" [7]. Achieving these goals requires the joint efforts of all sectors and individuals in society. SDG12 aims at promoting sustainable and responsible consumption and production patterns [8]. This is relevant to CBEC and consumers' sustainable behaviors (e.g., carbon-neutral product purchases and low-carbon delivery modes) warrant exploration [9, 10]. Moreover, identifying carbon-intensive operations in the CBEC supply chain and addressing them can facilitate green transformation on the CBEC production end. This study focuses on the production end and aims to propose a framework that evaluates and compares the carbon footprint derived from various CBEC logistics activities.

## 3 Cross-Border e-Commerce Logistics

### *3.1 CBEC Logistics Characteristics and Processes*

CBEC logistics consists of forward and reverse logistics. CBEC forward logistics shares characteristics (e.g., package pick-up, tracking, and delivery) with home delivery services in traditional local e-commerce. They further entail cross-border transportation, and customs declaration and clearance [11, 12]. Through employing online text mining on CBEC logistics-related content, Y-H Hsiao, M-C Chen and W-C Liao [13] suggest a framework that generalizes CBEC logistics into four stages, namely, goods collection in the territory of the exporting party, international transportation, goods distribution and delivery in the territory of the importing party, and other value-added services, along with further detailed elements and attributes under each stage. Similar evidence can also be found in the work of M Giuffrida, R Mangiaracina, A Perego and A Tumino [11], X Wang, J Xie and Z-P Fan [14], and L Cheah and Q Huang [15]. Thus, this framework provides a fundamental basis which this study extends to incorporate a carbon footprint analysis/assessment across these stages in CBEC logistics. In addition, non-carbon-related components in the original framework are removed and other relevant components are supplemented in this study.

### 3.2 Forward Logistics Modes

Two predominant operation modes are recognized in CBEC, namely "bonded warehouse" and "direct mail" [14, 16, 17]. The "bonded warehouse" mode is an operation strategy by which products are shipped in bulk and in advance and then stored in the tariff-free or free trade zones in the destination countries before being delivered to end consumers. In contrast, the "direct mail" mode ships the ordered products directly from overseas warehouses to end customers via cross-border logistics services, such as international express post or special logistics lines. Figure 1 illustrates the process of these two CBEC operation modes.

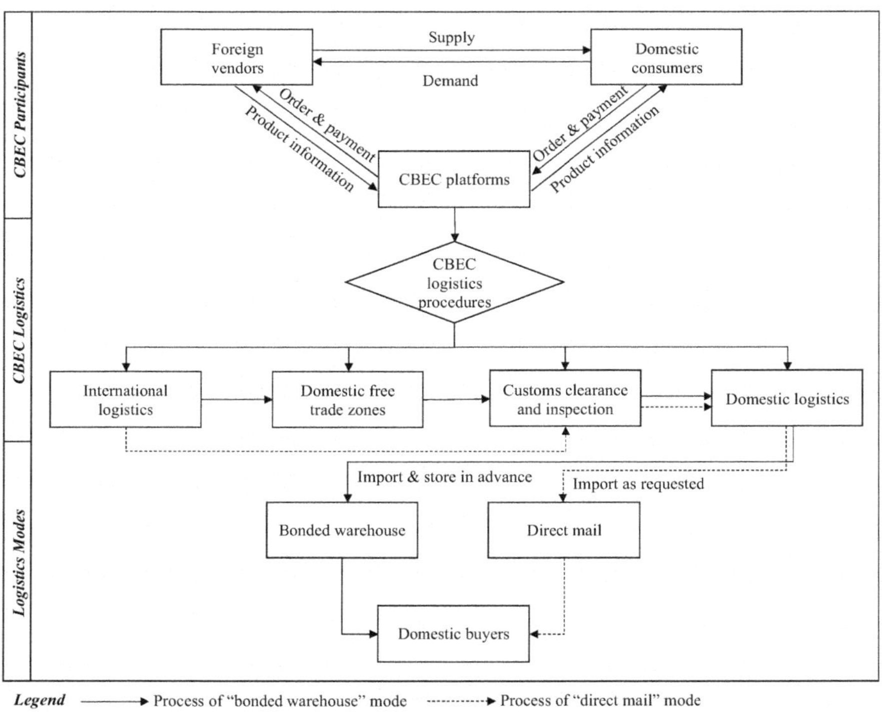


**Fig. 1** CBEC logistics process

### 3.3 Reverse Logistics Modes

Given the aforementioned two forward logistics modes, X Wang, J Xie and Z-P Fan [14] identify two CBEC reverse logistics practices: "return-to-domestic" mode and "return-to-overseas" mode. In the return-to-domestic mode, returned products are delivered to the bonded warehouse in the consumers' countries, while the return-to-overseas mode requires customers to return the products to the sellers' overseas warehouse. The two modes lead to significant distinctions in terms of time efficiency and return costs. The return-to-overseas mode involves longer reverse logistics processes and higher costs than the return-to-domestic mode.

### 3.4 Freight Transport Modes

Freight operations in CBEC logistics mainly include road, railway, air, and sea transport [18], with billions of tons of cargo being moved worldwide by trucks, trains, planes, and ships every year [19]. Most of these multimode transportation burn fossil fuels and thereby generate massive carbon footprints.

At present, few studies have incorporated carbon footprint as a primary indicator of environmental impacts to assess the sustainability of CBEC logistics. Thus, this study further reviews the literature on carbon footprints in international trade logistics to inform the development of a framework to effectively assess the environmental impacts of CBEC logistics.

## 4 Carbon Footprint in Logistics

Studies have explored carbon efficiency in the logistics or e-commerce domain. For example, with emphasis on last-mile logistics, J Edwards, A McKinnon and S Cullinane [20] compared carbon intensity between home delivery and traditional shopping trips. They found that the former outperformed the latter if the number of items purchased was relatively small. EYC Wong, AH Tai and E Zhou [21] considered carbon emission along with other trade-off factors including cargo size, destination, and transportation duration, to optimize truckload in third-party logistics. Hence, these studies confirm logistics management as an important and effective measures to mitigate a company's carbon footprint [22].

Despite the fact that CBEC can easily be carbon-intensive due to its wide-ranging logistics, very little is known about how CBEC can move toward greener and more sustainable logistics management and practices. A review of the literature on carbon footprint analysis of logistics identifies aspects it would be useful to capture in the framework for sustainable CBEC, as detailed below.

## 4.1 Forward Logistics

Logistics modes have diverse impacts on carbon emissions. For instance, among direct mail freight forwarder networks in a distribution, the many-to-many network, which avoids long distances to the consolidation points, has been proven to be the least carbon-intensive [23]. Aiming to optimize the carbon footprint of end-to-end maritime supply chains, P Rigot-Muller, C Lalwani, J Mangan, O Gregory and D Gibbs [24] found that carbon emissions varied substantially in different transportation modes (e.g., direct delivery versus transhipment). In CBEC, the direct mail mode generally fulfills order delivery one by one; however, the bonded warehouse mode enables products to be shipped in bulk and in advance. Hence, the carbon footprints of bonded warehouse and direct mail modes should presumably vary.

## 4.2 Reverse Logistics

Furthermore, reverse logistics intensifies the carbon footprint of CBEC operations [25, 26]. As such, reverse logistics is integral to the entire carbon footprint assessment [6]. Most studies on carbon footprint in logistics have been conducted under the qualifications of no product returns [20]; thus, offering an incomplete understanding of carbon footprint in logistics operations. Thus, reverse logistics modes, the return rate of products, and carbon emissions associated with product returns in CBEC should be captured [6].

## 4.3 Transport Modes

The carbon footprint of freight varies dramatically in different transport modes. The latest report has revealed that in 2020, road transport moved 26,807 million of ton-kilometers of freight; this figure for rail, air, and sea transport was 10,842, 303, and 101,486, respectively [27]. However, road freight emitted 2230 million of tons of carbon dioxide, and this figure for rail, air, and sea freight was 170, 155, and 657 [27]. This implies that sea transport, drawing on its ability to ship great amounts of cargo at once, prevails in terms of carbon effectiveness, followed by rail, road, and air transport. Air transport emits far more greenhouse gases (GHGs) than other modes, and air cargo leaves, therefore, a larger carbon footprint. On the other hand, air transport is the fastest channel for delivering products to consumers, whereas sea transport is considered the slowest delivery method.

## 4.4 Vehicle Types

The different types of energy that power vehicles for freight movement are crucial in determining the carbon footprint of CBEC logistics, for example, electric vehicles are far greener than fossil fuel vehicles.

## 4.5 Warehouse Operations

Logistic-related carbon emissions can also originate from warehousing. The review of JM Ries, EH Grosse and J Fichtinger [28] shows how much warehouse operations rely on electricity consumption for lighting, heating, cooling, air conditioning as well as material handling.

## 4.6 Last-Mile Delivery Modes

Consumption-based accounting is another consideration for carbon footprint assessment [5]. For example, as rightly argued by L Cheah and Q Huang [15], consumers' passenger trips to a collection point, if any, should also be accounted for in assessing the carbon footprint of CBEC logistics. Similarly, studies have suggested that last-mile delivery accounts for a considerable part of carbon emissions of online retailing [29]. The last-mile delivery options include walking, public transports, bicycles, and cars, which have different environmental impacts. Furthermore, it has been found that consumers are willing to adopt last-mile delivery alternatives that are less time-efficient but reduce carbon emissions [10].

## 4.7 Order Consolidation

Order consolidation is also a vital aspect that should not be overlooked. It has been indicated that a single order delivered to a home yields more carbon emissions compared to consolidating multiple orders given the other transport modes constant [15].

In short, the elements that would be useful to capture in assessing the carbon footprint of CBEC logistics are summarized in Table 1. From the perspective of the logistics process, carbon footprint assessment in CBEC logistics should capture processes: (1) domestic delivery in the exporting country, (2) international transport, (3) warehouse operations, (4) domestic delivery in the destination country, and (5) consumers' passenger trips to a collection point (if any). Further, other crucial elements that should be considered and distinguished contain forward logistics

**Table 1** Carbon footprint assessment elements in CBEC logistics

| Logistics process | Forward logistics mode | Reverse logistics mode | Transport mode | Vehicle type | Order consolidation |
|---|---|---|---|---|---|
| Domestic delivery in the exporting country (DDEC) | Bonded warehouse<br>Direct mail | Return-to-overseas<br>Return-to-domestic | Road transport<br>Rail transport<br>Air transport<br>Sea transport | Fossil fuel vehicle<br>Electricity vehicle<br>Hybrid vehicle | Single order<br>Consolidated order |
| International transport (IT) | | | | | |
| Warehouse operations (WO) | | | | | |
| Domestic delivery in the destination country (DDDC) | | | | | |
| Consumers' passenger trip to a collection point (if any; CCCP) | | | | | |

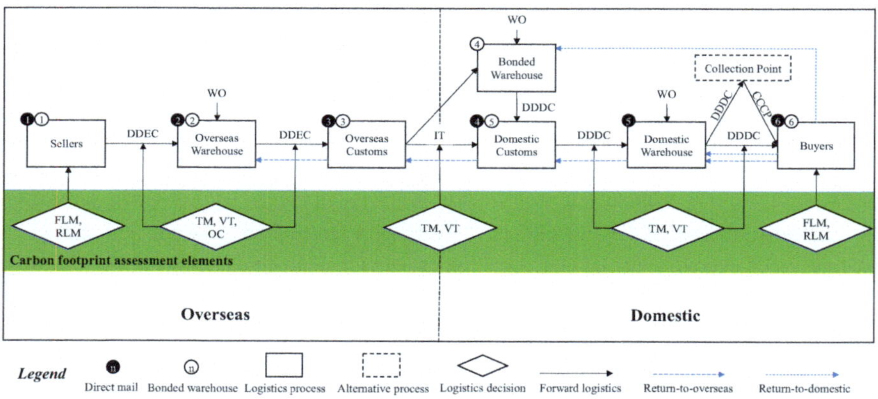


**Fig. 2** Carbon footprint assessment elements in CBEC logistics. Notes: DDEC: domestic delivery in the exporting country; IT: international transport; DDDC: domestic delivery in the destination country; WO: warehouse operations; CCCP: consumers' passenger trip to a collection point; FLM: forward logistics mode; RLM: reverse logistics mode; TM: transport mode; VT: vehicle type; OC: order consolidation

modes, reverse logistics modes, freight transport modes, vehicle types, and order consolidation. Figure 2 illustrates the abovementioned carbon footprint elements (e.g., decisions and operations) alongside the CBEC logistics.

## 5 Framework for Carbon Footprint Analysis

Approaches to carbon management strategies may involve trade-offs among environmental, social, and economic aspects [5]. According to a review, such trade-off analysis models represent the overwhelming majority (88%) in the studies addressing sustainable development issues, compared with the others involving a single aspect [30]. For example, in the logistics context, JMC Martí, J-S Tancrez and RW Seifert [31] characterize responsiveness (socioeconomic aspect) and carbon footprint (environmental aspect) as two aspects of trade-off analysis in optimizing a supply chain network.

This study extends existing approaches and proposes a framework that integrates the socioeconomic aspect comprised of delivery cost and delivery time to assess carbon emissions as a crucial indicator of the environmental aspect (Table 2). Delivery cost and time are identified as two essential determinants of logistics service designs from logistics service providers' perspectives [14, 15] and logistics service satisfaction from customers' perspectives [13]. By jointly considering delivery time and cost, researchers and practitioners can effectively assess if carbon management would compromise delivery cost and time for lower carbon emissions. Thus, the integration of environmental and socioeconomic variables enables our framework to provide a more balanced and comprehensive understanding of trade-offs stemming from different logistics processes in CBEC logistics.

Furthermore, the proposed framework also incorporates scenario analysis by combining five elements of carbon footprint assessment as identified above. "Scenario analysis has emerged as a method that is particularly well-suited to the task of taking a long-term view and attempting to harmonize socioeconomic and environmental goals" [30]. Hence, this is an important and useful analytical approach to carbon footprint assessment [28]. By analyzing scenarios comprising those elements in aspects of socioeconomic and environmental performance, this framework enables a comparative understanding of carbon footprint analysis of CBEC logistics.

## 6 Discussion and Future Work

CBEC businesses have caused undeniable environmental impacts, but little is known about the extent of these impacts, especially those impacts under different modes of logistics operations. This study proposes a conceptual framework that identifies the processes of the CBEC logistics operations and assesses their carbon footprint. The framework involves trade-offs between the environmental aspect (i.e., carbon emissions) and the socioeconomic aspects (i.e., delivery cost and time). Five stages of CBEC logistics and five types of elements are captured for scenario analyses. Our approach aims to enable a comprehensive long-term assessment of the carbon footprint of the industry so it can be used to improve practices to reduce carbon emissions.

**Table 2** Conceptual framework for scenario analysis

| | | Element combination for scenario analysis | | | | | Carbon footprint analysis | | |
|---|---|---|---|---|---|---|---|---|---|
| CBEC scenario | Logistics process | Forward logistics mode | Reverse logistics mode | Transport mode | Vehicle type | Order consolidation | Delivery cost | Delivery time | Carbon emissions |
| S1 | DDEC<br>IT<br>WO<br>DDDC<br>CCCP | Bonded ware-house<br>Direct mail | Return-to-over-seas<br>Return-to-domestic | Road<br>Rail<br>Air<br>Sea | Fossil fuel<br>Electricity<br>Hybrid | Single<br>Consolidated | | | |
| S2 | | | | | | | | | |
| S3 | | | | | | | | | |
| … | | | | | | | | | |
| Sn | | | | | | | | | |

Notes: *DDEC*: domestic delivery in the exporting country; *IT*: international transport; *DDDC*: domestic delivery in the destination country; *WO*: warehouse operations; *CCCP*: consumers' passenger trip to a collection point.

### 6.1 Anticipated Implications

From a theoretical perspective, this study advances the knowledge of optimally managing the carbon footprints of various logistics operations modes in the CBEC context. The conceptual framework facilitates the identification and improvement of carbon-intensive operation patterns in CBEC logistics and helps CBEC logistics providers and CBEC practitioners move toward effective carbon management strategies and greener business and operations practices. From the perspectives of the midpoints and endpoints of life cycle impact assessment [34], the research outcomes aim to help mitigate global warming by reducing carbon emissions in CBEC logistics operations, ultimately reducing damage to human health and ecosystem. In general, this study is one of the earliest studies that focus on sustainable production in CBEC; thus, contributing to SDG12 (Responsible Consumption and Production).

### 6.2 Future Work and Research Directions

In the next stage, the case study method will be employed to collect empirical data from CBEC logistics service providers to validate the conceptual framework. Relevant guidelines for calculating carbon emissions across CBEC logistics procedures will be used, such as guidelines for freight transport operations [18, 32] or electricity and fossil fuel consumption [33].

Future researchers can also adopt this framework to assess carbon footprint in other international trade or cross-border logistics contexts. Moreover, the actual effect of green logistics operations is also dependent on end consumers' willingness to adopt. Scenarios yielded from the proposed framework can be fed into experiment studies to explore consumer behavior in support of sustainable and greener CBEC logistics practices on the consumption end.

## 7 Conclusion

The increasing development of CBEC has significant environmental impacts. Carbon management strategies such as assessing multiple aspects of CBEC through the framework we propose help facilitate low-carbon or carbon-effective CBEC logistics operations, which ontributes to more sustainable and greener CBEC. Only by compiling metrics on environmental friendliness and socioeconomic effectiveness can carbon emission minimization be pursued with the joint goal of maximizing social and economic benefits, while moving toward more sustainable CBEC practices and UNSDG12.

## References


1. Chen J, Lan Y-C, Chang Y-W, Samaranayake P, Chen K-C (2022) Buyers' psychological situations in cross-border electronic commerce. In: 33th Australasian Conference on Information Systems proceedings, University of Melbourne, Melbourne, 4–7 December 2022
2. Chen J, Lan Y-C, Chang Y-W (2023) Consumer behaviour in cross-border e-commerce: systematic literature review and future research agenda. Int J Consum Stud 47:1–61
3. Forrester (2019) Forrester analytics: Online cross-border retail forecast, 2018 To 2023 (Global). [Cited 1 June 2023]; Available from: https://www.forrester.com/report/Forrester-Analytics-Online-CrossBorder-Retail-Forecast-2018-To-2023-Global/RES153276
4. Yu Y, Feng K, Hubacek K (2013) Tele-connecting local consumption to global land use. Glob Environ Chang 23(5):1178–1186
5. Wiedmann T, Lenzen M (2018) Environmental and social footprints of international trade. Nat Geosci 11(5):314–321
6. Choudhary A, Sarkar S, Settur S, Tiwari MK (2015) A carbon market sensitive optimization model for integrated forward–reverse logistics. Int J Prod Econ 164:433–444
7. UN (2013) THE 17 GOALS. [Cited 4 June 2023]; Available from: https://sdgs.un.org/goals#history
8. UN (2023) Ensure sustainable consumption and production patterns. [Cited 4 June 2023]; Available from: https://sdgs.un.org/goals/goal12
9. Liu Y, Yang D, Xu H (2017) Factors influencing consumer willingness to pay for low-carbon products: a simulation study in China. Bus Strateg Environ 26(7):972–984
10. Caspersen E, Navrud S (2021) The sharing economy and consumer preferences for environmentally sustainable last mile deliveries. Transp Res Part D: Transp Environ 95:102863
11. Giuffrida M, Mangiaracina R, Perego A, Tumino A (2017) Cross-border B2C e-commerce to greater China and the role of logistics: a literature review. Int J Phys Distrib Logist Manage 47(9):772–795
12. Gomez-Herrera E, Martens B, Turlea G (2014) The drivers and impediments for cross-border e-commerce in the EU. Inf Econ Policy 28:83–96
13. Hsiao Y-H, Chen M-C, Liao W-C (2017) Logistics service design for cross-border e-commerce using Kansei engineering with text-mining-based online content analysis. Telematics Inform 34(4):284–302
14. Wang X, Xie J, Fan Z-P (2021) B2C cross-border E-commerce logistics mode selection considering product returns. Int J Prod Res 59(13):3841–3860
15. Cheah L, Huang Q (2022) Comparative carbon footprint assessment of cross-border e-commerce shipping options. Transp Res Rec 2676(1):584–595
16. Zha X, Zhang X, Liu Y, Dan B (2022) Bonded-warehouse or direct-mail? Logistics mode choice in a cross-border e-commerce supply chain with platform information sharing. Electron Commer Res Appl 54:101181
17. Shao B, Cheng Z, Wan L, Yue J (2021) The impact of cross border E-tailer's return policy on consumer's purchase intention. J Retail Consum Serv 59:102367
18. Martínez LM, Kauppila J, Castaing M (2015) International freight and related carbon dioxide emissions by 2050: new modeling tool. Transp Res Rec 2477(1):58–67
19. Greene S (2023) Freight transportation. [Cited 4 June 2023]; Available from: https://climate.mit.edu/explainers/freight-transportation
20. Edwards J, McKinnon A, Cullinane S (2009) Carbon auditing the 'last mile': modelling the environmental impacts of conventional and online non-food shopping, Green Logistics Report. Heriot-Watt University
21. Wong EYC, Tai AH, Zhou E (2018) Optimising truckload operations in third-party logistics: a carbon footprint perspective in volatile supply chain. Transp Res Part D: Transp Environ 63: 649–661
22. Parashar S, Sood G, Agrawal N (2020) Modelling the enablers of food supply chain for reduction in carbon footprint. J Clean Prod 275:122932

23. Kellner F, Igl J (2015) Greenhouse gas reduction in transport: analyzing the carbon dioxide performance of different freight forwarder networks. J Clean Prod 99:177–191
24. Rigot-Muller P, Lalwani C, Mangan J, Gregory O, Gibbs D (2013) Optimising end-to-end maritime supply chains: a carbon footprint perspective. Int J Logist Manage 24(3):407–425
25. Kannan D, Diabat A, Alrefaei M, Govindan K, Yong G (2012) A carbon footprint based reverse logistics network design model. Resour Conserv Recycl 67:75–79
26. Sun Q (2017) Research on the influencing factors of reverse logistics carbon footprint under sustainable development. Environ Sci Pollut Res 24(29):22790–22798
27. ITF (2021) International transport forum transport outlook. [Cited 31 May 2023]; Available from: https://www.itf-oecd.org/itf-transport-outlook-2021
28. Ries JM, Grosse EH, Fichtinger J (2017) Environmental impact of warehousing: a scenario analysis for the United States. Int J Prod Res 55(21):6485–6499
29. van Loon P, McKinnon AC, Deketele L, Dewaele J (2014) The growth of online retailing: a review of its carbon impacts. Carbon Manage 5(3):285–292
30. Allen C, Metternicht G, Wiedmann T (2016) National pathways to the Sustainable Development Goals (SDGs): a comparative review of scenario modelling tools. Environ Sci Pol 66:199–207
31. Martí JMC, Tancrez J-S, Seifert RW (2015) Carbon footprint and responsiveness trade-offs in supply chain network design. Int J Prod Econ 166:129–142
32. ECTA (2011) Guidelines for Measuring and managing CO2 emission from freight transport operations. [Cited 5 June 2023]; Available from: https://www.ecta.com/wp-content/uploads/2021/03/ECTA-CEFIC-GUIDELINE-FOR-MEASURING-AND-MANAGING-CO2-ISSUE-1.pdf
33. EPA (2023) Greenhouse Gases Equivalencies Calculator - Calculations and References. [Cited 28 May 2023]; Available from: https://www.epa.gov/energy/greenhouse-gases-equivalencies-calculator-calculations-and-references#:~:text=Carbon%20dioxide%20emissions%20per%20therm,one%20therm%20(EIA%202021
34. Huijbregts MAJ, Steinmann ZJN, Elshout PMF, Stam G, Verones F, Vieira M, Zijp M, Hollander A, van Zelm R (2017) ReCiPe2016: a harmonised life cycle impact assessment method at midpoint and endpoint level. Int J Life Cycle Assess 22(2):138–147

# How Should AI Technique Support Wastewater Treatment and Sludge Management? A Case Study from Mongolia

**Richao Cong, Tumurtogtokh Oyunchimeg, Atsushi Fujiyama, and Toru Matsumoto**

**Abstract** In the current years, the artificial intelligence (AI) technique has been utilized for waste management due to its higher accuracy. Understanding the future disposal demand of wastewater (WW) and sludge generation accurately is essential for WW treatment and sludge management. This study proposed an AI-based approach to predict the daily demand for WW and generation amounts of sludge, demonstrated for Ulaanbaatar, Mongolia. A machine learning approach was developed to predict the daily disposal amounts of WW and generation amounts of sludge based on the daily records from a WW treatment plant. In the validation process, the monthly-mean accuracy of the predicted disposal amounts of WW for the last month of 2020 was about 98.0% (predicted by the Medium Gaussian SVM model). Meanwhile, the monthly-mean prediction accuracy of the sludge generation in the same period was increased from about 83.1% to 84.7% by using the proposed approach. The proposals on future applications were made for WW treatment, e.g., optimizing the personnel arrangement, procurement of chemicals, and prediction of the sludge generation. For sludge management, our proposals were made from future treatment plans and recovery of resources and energy aspects. This study demonstrated that the AI technique is helpful for waste management.

**Keywords** AI · Future prediction · Machine learning · Sludge generation · Wastewater treatment

R. Cong (✉) · T. Matsumoto
Institute of Environmental Science and Technology, University of Kitakyushu, Kitakyushu, Japan
e-mail: r-cong01@kitakyu-u.ac.jp

T. Oyunchimeg · A. Fujiyama
Faculty of Environmental Engineering, University of Kitakyushu, Kitakyushu, Japan

 
S. Fukushige et al. (eds.), *EcoDesign for Circular Value Creation: Volume I*,
https://doi.org/10.1007/978-981-97-9068-5_30

# 1 Introduction

In recent years, artificial intelligence (AI) techniques for analyzing "big data" have increasingly been utilized for such problems owing to their higher prediction accuracy in many fields [1–3].

To support waste management, an improved support vector machine (SVM) model was developed to forecast the weekly generation of municipal solid waste (MSW) in the city of Mashhad, Iran [4]. To optimize the cost of waste collection and transportation, an artificial neural network (ANN)–genetic algorithm model was combined with the response surface method to predict MSW generation in Langkawi Island, Malaysia [5]. ANN and decision tree algorithms were applied to predict MSW in Ontario, Canada, using socioeconomic and demographic parameters [6]. In Japan, Cong et al. developed a machine learning (ML) approach to predict daily industrial plastic waste collection demands in Fukuoka [7], explored the details for improving the prediction accuracy from multiple facilities [8], and made applications in plastic waste collection system integration [9]. For wastewater (WW) treatment, previous AI-related studies mainly focused on the prediction of plant performances, e.g., water quality [10–12], chemical oxygen demand (COD) [13, 14] and biochemical oxygen demand (BOD) [15]. For sludge management, only a few AI-related studies predicted the generation amount of sewage sludge from daily to annual scales [16, 17]. However, the future predictions by utilization of one response variable for another between WW disposal amount and sludge generation amount (especially at the daily level) to improve the prediction accuracy have not been discussed in these studies.

The Ulaanbaatar central wastewater treatment plant (CWWTP) operated in 1986 with basic mechanical and biological treatment, renewed for new treatment technologies in the past decades, and started a cogeneration combined heat and power project (to be finished by 2025). To build a novel WW treatment and sludge management system for the future, an AI-based system utilizing the accumulated disposal records (big data) needed to be designed.

In this context, the aim of this study is to answer the following questions: (1) How to make future predictions by ML approach for WW disposal demand and sludge generation amount? (2) Is it possible to improve the prediction accuracy by utilizing one response variable for another? (3) What kinds of potential applications are available from such an AI-based waste management system?

# 2 Methodology

## *2.1 Study Area and Workflow*

The target city is Ulaanbaatar (Fig. 1), which is the capital city of Mongolia. It is located in north central Mongolia at an elevation of about 1300 meters. Its total area is about 4704 km$^2$ and its population is about 1.61 million in 2021. As shown in

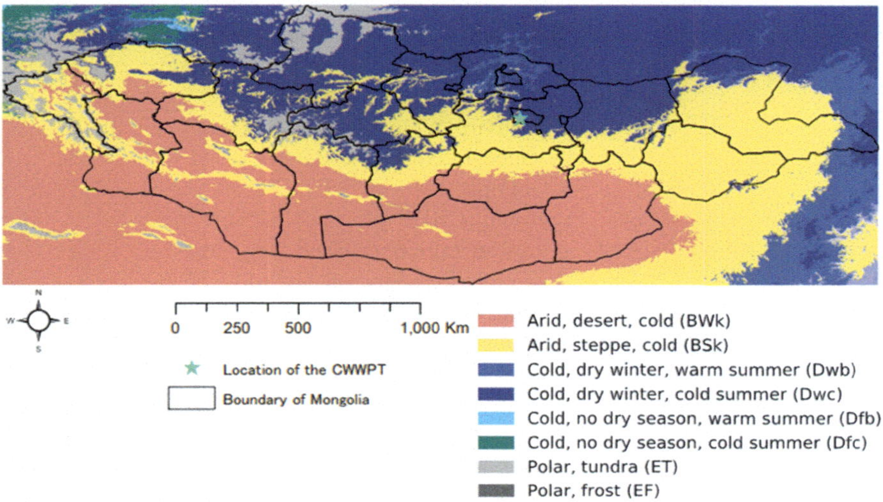


**Fig. 1** Location of the CWWTP in Mongolia and its climate zone

Fig. 1, this city belongs to the cold climate zone with dry winter and warm summer conditions [18]. It has been regarded as the coldest capital city in the world. Our target facility, the CWWTP is located in the southwest of Ulaanbaatar city in the Songinokhairkhan district, was commissioned in 1969 and built in several stages until its completion in 1986. As the country's largest wastewater treatment plant, CWWTP occupies an area of 176,271 $m^2$ and currently has the capacity to treat 170,000 $m^3$ of wastewater per day [19].

The workflow on the future prediction by AI techniques is described in Fig. 2. At first, the daily records on two types of activity data (WW disposal amount and generation amount of sludge) are combined with climate data in the data preparation process (prior processing). The daily amount of each activity is made as the response variable and other independent variables are also prepared. Then, model fitting experiments by all models are conducted to determine the best fitting model evaluated by using the indicator of root mean square error (RMSE). For prediction accuracy validation, the observed and predicted activity amount in a designed period is accessed to calculate the mean absolute percentage error (MAPE) for each prediction. To improve the prediction accuracy, WW disposal amount is also used as an independent variable in predictions for sludge generation amount. Finally, future predictions are made for a new data period by approaches with higher accuracy, and applications for optimal planning on WW treatment and sludge utilization are proposed.

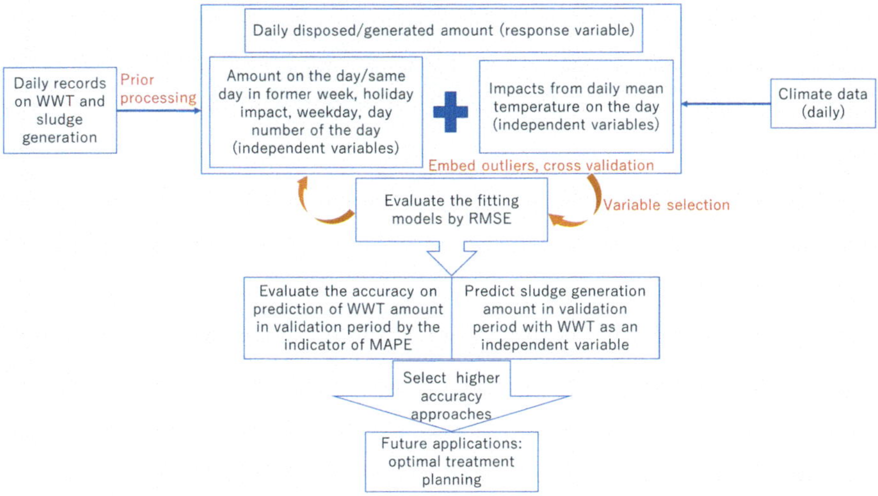


**Fig. 2** Workflow on predicting the WW treatment disposal demands from the CWWTP by using AI-based approach

## 2.2 *Data Preparation*

For analysis, this study used daily disposal amounts of WW and sludge generation amounts recorded in the CWWTP, Mongolia (collected from January 1, 2018 to December 31, 2020). Figure 3a shows the Monthly total amounts of WW disposed and sludge generated (refer to the right-y axis) in CWWTP in this period. The peak of monthly WW disposal amounts is always met in the summer seasons (August to October). For example, one peak occurred in October 2018, one occurred in August 2019, and the other occurred in September 2020. This trend has not changed even during the COVID-19 pandemic period. From this figure, we could see that the trend on monthly sludge generation is a little different from that on disposed amounts of WW in these three years. Figure 3b shows the monthly trends in mean temperature in Ulaanbaatar city from 2018 to 2020. We could see that there is a similar trend in temperature variation each year.

Table 1 shows the description of the independent variables used in this study. We included day number, weekday, climatic impact on daily-mean temperature [20], holiday impact on disposed WW, and disposal records in the previous day or week. For sludge generation prediction, the disposal amounts of WW were also made as a variable focusing on improving the prediction accuracy.

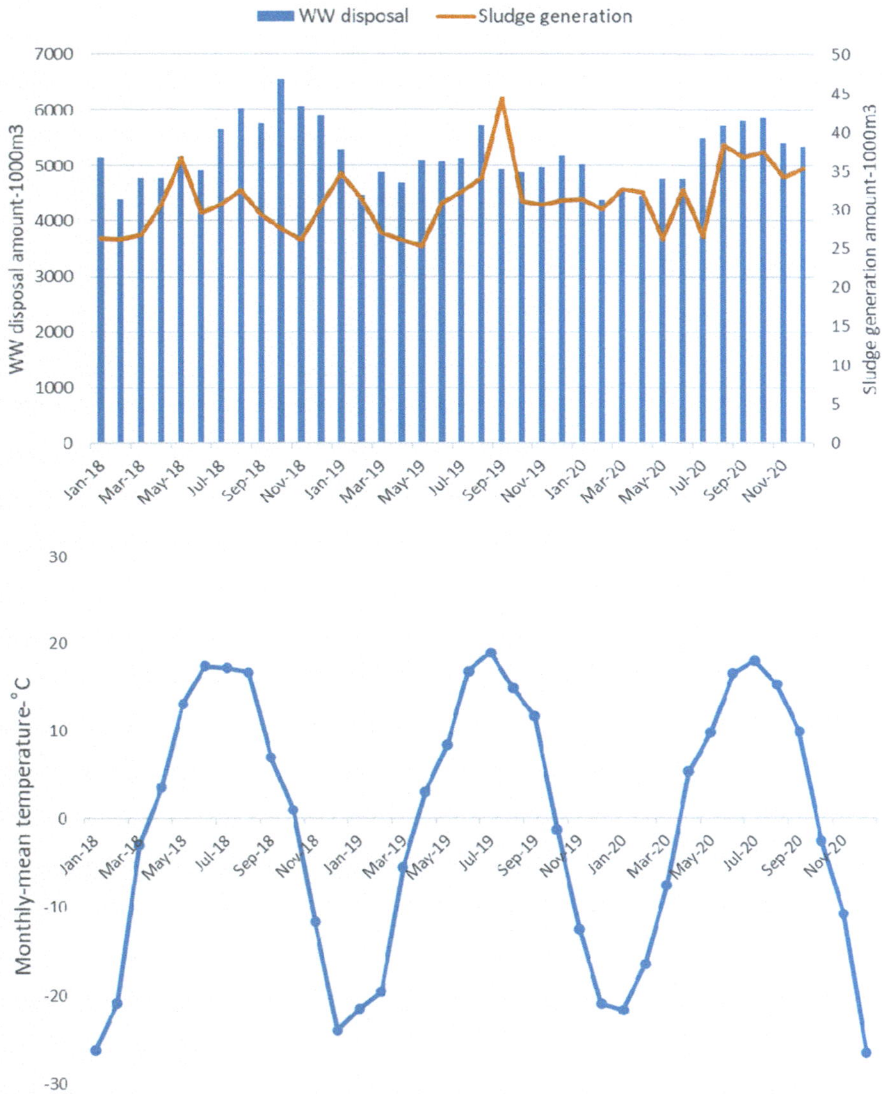


**Fig. 3** Monthly trends from 2018 to 2020 on **a** (above) total amounts of WW disposed and sludge generated (refer to the right-y axis) in CWWTP, unit: 1000 $m^3$; **b** (below) the monthly mean temperature in Ulaanbaatar

**Table 1** Description of the independent variables used in this study

| Category | Variable | Description |
|---|---|---|
| Common use | Day_num | Day number from 2018.1.1 to 2020.12.31 |
| | Weekday | What day was it? |
| | Climate_d | Category type on climate impact on daily-mean temperature |
| | Day_wk | Category type whether the day belongs to holiday or not |
| Wastewater disposal demand prediction use | Num_fd_ww | Disposed amount of WW on former day |
| | Num_fw_ww | Disposal amount on the same day of the former week |
| Sludge generation prediction use | Num_ww | Disposed amount of WW on the day |
| | Num_fd_ww | Disposed amount of WW on former day |
| | Num_fw_ww | Disposal amount of WW on the same day of the former week |
| | Num_fd_sd | Generated amount of sludge on former day |
| | Num_ fw_sd | Generated amount of the sludge on the same day of the former week |

Note: the category of common use means the relevant variables were used in predictions of two targets

# 3 ML on WW Disposal Demand and Sludge Generation Amount

In this study, ML-based experiments were made by using MATLAB_2021b version. There were 6 types total of 28 models (stored in MATLAB) used in model fitting by a five-cross-validation method. The whole data in these three years were split for training use (daily records before December 2020; 97.2% of all) and validation/test use (31 days in December 2020; 2.8%).

## 3.1 Model Fitting on WW Disposal Demand

In model fitting processes, the indicators of RMSE, coefficient of determination ($R^2$), and mean absolute error (MAE) were calculated for each model to show their fitting performances. Figure 4a shows the fitting performance of the Medium Gaussian SVM model (RMSE: 8210; $R^2$: 0.81; MAE: 5090). Blue dots are the observed values (disposal records on WW) and yellow dots refer to the machine-learned values (the fitted ones by model). X-axis is the day number in this period, Y-axis shows the daily disposal amount in a unit of a cubic meter ($m^3$). Figure 4b shows the importance of variables on the prediction results of WW disposal amount. It implies that the variable of dm_previousDay affects the prediction result most and that of

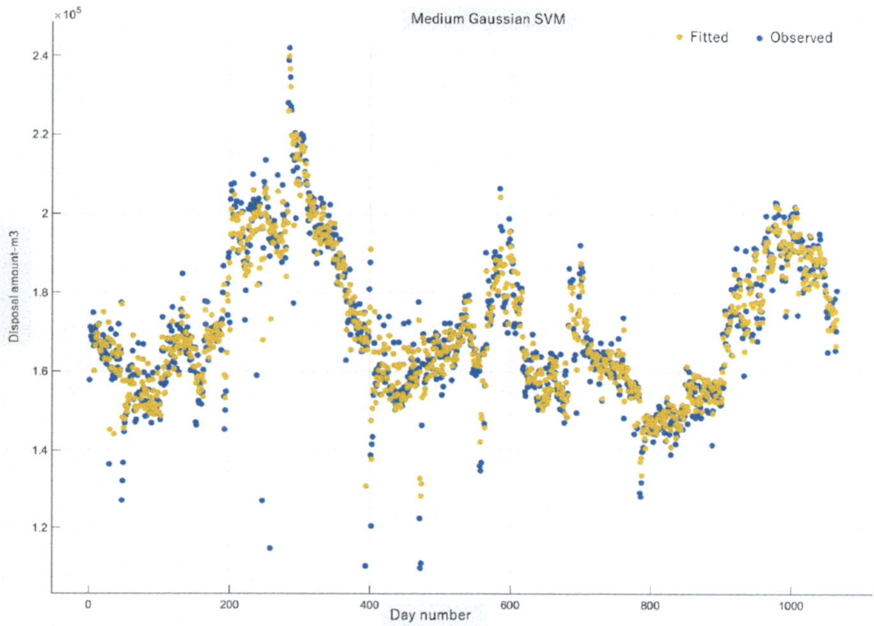


A

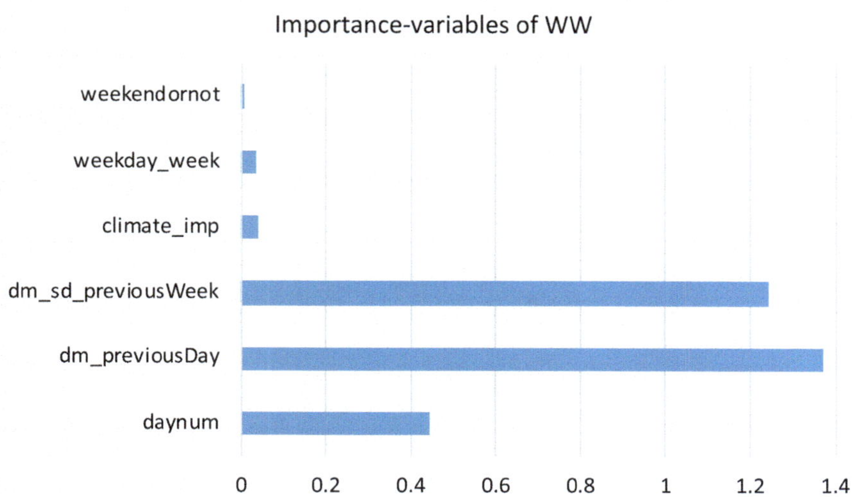


B

**Fig. 4** The fitting performance (a) and the importance of variables (b) on WW disposed amounts by the Medium Gaussian SVM model (RMSE: 8210; $R^2$: 0.81; MAE: 5090), unit: $m^3$

weekendornot affects the result least. We could see that most of the yellow dots fit well with the blue ones. Based on these indicators, the best-fitting model with the lowest RMSE was selected for future prediction.

### 3.2 Model Fitting on Sludge Generation Amount

Similar to the fitting processes for WW disposal demand, the best-fitting model was selected during the learning processes of sludge generation amount. Figure 5a shows the fitting performance of the Gaussian Process Regression (GPR) Rational Quadratic model (RMSE: 176; $R^2$: 0.43; MAE: 104). Compared with the performance of WW disposal demand, the best-fitting model for sludge generation showed a weaker performance in learning its daily trend which would cause a lower accuracy in future prediction. Figure 5b shows the importance of variables on the prediction results of WW disposal amount. It implies that the variable of dm_previousDay affects the prediction result most and that of weekendornot affects the result least.

### 3.3 Model Validation by the Selected Best-Fitting Models

Table 2 shows the fitting results for two targets by all models from which the best fitting models for them are selected. Then, new predictions for two targets in the validation period (31 days in December 2020) were made by using the best-fitted models. Subsequently, the accuracies of the predictions for them were evaluated as the indicator of MAPE based on the ML predicted and observed values. The RMSE used for measuring the absolute errors in model fitting process and MAPE used for measuring the prediction errors in percentage were calculated according to Eqs. 1 and 2, respectively,

$$\mathrm{RMSE} = \sqrt{\frac{1}{n}\sum_{i=1}^{n}(f_i - y_i)^2} \tag{1}$$

$$\mathrm{MAPE} = \frac{100}{n}\sum_{i=1}^{n}\left|\frac{f_i - y_i}{y_i}\right|, \tag{2}$$

where

$n$: the number of data
$y_i$: the observed value of data $i$
$f_i$: the predicted value of data $i$

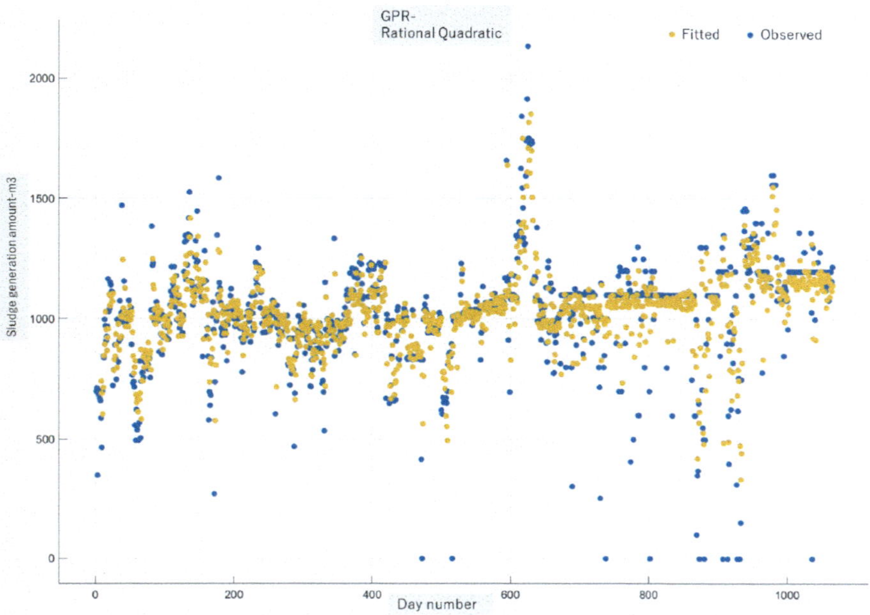


A

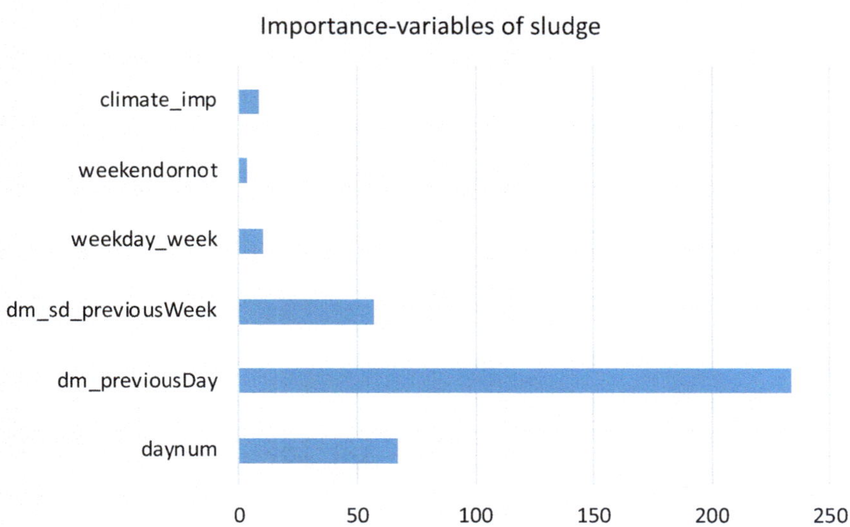


B

**Fig. 5** The fitting performance (**a**) and the importance of variables (**b**) on sludge generation amounts by the GPR Rational Quadratic model (RMSE: 176; $R^2$: 0.43; MAE: 104), unit: $m^3$

**Table 2** The indicators of RMSE for two targets by all models explored by model fitting

| Category | Model | RMSE for WW | RMSE for sludge |
|---|---|---|---|
| Linear regression | Linear regression | 8624 | 182 |
| | Interactions linear | 8682 | 189 |
| | Robust linear | 8566 | 190 |
| | Stepwise linear | 8652 | 184 |
| Decision tree | Fine tree | 9773 | 203 |
| | Medium tree | 9266 | 189 |
| | Coarse tree | 9320 | 186 |
| SVM | Linear SVM | 8567 | 184 |
| | Quadratic SVM | 8386 | 188 |
| | Cubic SVM | 8680 | 184 |
| | Fine Gaussian SVM | 11277 | 214 |
| | Medium Gaussian SVM | 8210 | 181 |
| | Coarse Gaussian SVM | 8344 | 180 |
| Ensemble | Boosted Trees | 11490 | 189 |
| | Bagged Trees | 8724 | 182 |
| Gaussian process regression (GPR) | Rational Quadratic | 8586 | 176 |
| | Squared Exponential | 8613 | 178 |
| | Matern 5/2 | 8556 | 179 |
| | Exponential | 8660 | 185 |
| Neural Network | Narrow Neural Network | 8351 | 178 |
| | Medium Neural Network | 8345 | 187 |
| | Wide Neural Network | 8640 | 274 |
| | Bilayered Neural Network | 8403 | 194 |
| | Trilayered Neural Network | 8626 | 199 |
| Optimal models | Optimal tree | 9095 | 184 |
| | Optimal SVM | 18722 | 235 |
| | Optimal GPR | 25717 | 193 |
| | Optimal ensemble | 10105 | 183 |

# 4 Results and Discussion

## *4.1 Future Prediction of Daily WW Disposal Amount by ML Approach*

As our evaluation, the MAPE in the validation period was calculated for the WW disposal amount as 2.0%. Thus, the accuracy of our prediction reached approximately 98.0% (Fig. 6). On the contrary, when we used the linear regression tool of SPSS to make predictions using the same data, the accuracy of the predictions was

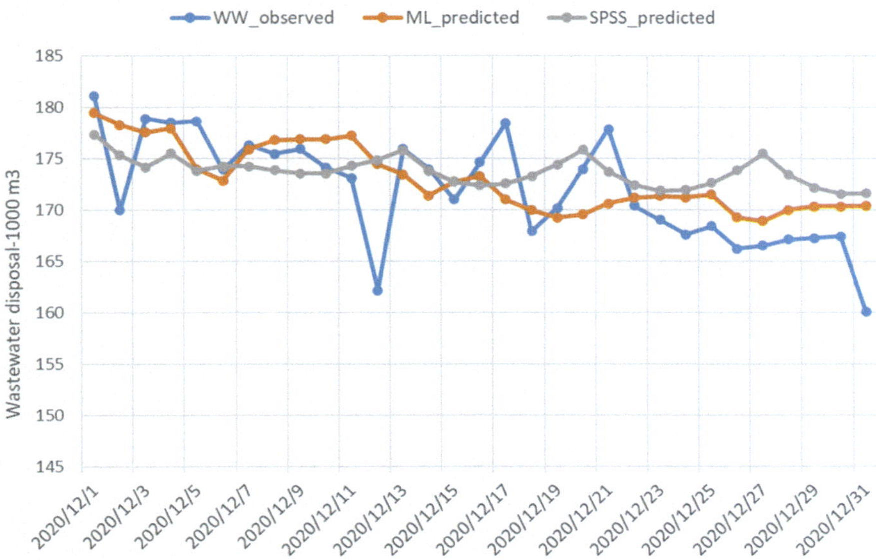


**Fig. 6** The comparisons between the prediction results by ML (98.0%) and SPSS (97.6%) with the observed disposal records from 2020.12.1 ~ 31

97.6%. We concluded that in this case, the ML approach provided higher prediction accuracy than that of the statistical approach. The data and model processing processes proposed could be used to improve the accuracy of future predictions.

## *4.2 Future Prediction of Daily Sludge Generation Amount by ML Approach*

As our evaluation, the MAPE in the validation period for the sludge generation amount was calculated as 16.9%. Thus, the prediction accuracy of it reached approximately 83.1% (Fig. 7). Notably, the MAPE decreased to 15.3% (the prediction accuracy: 84.7%) when predictions were made by the same model with disposed WW amounts as an independent variable; this implies that utilization of WW disposal demand improves the accuracy of prediction of sludge generation amounts.

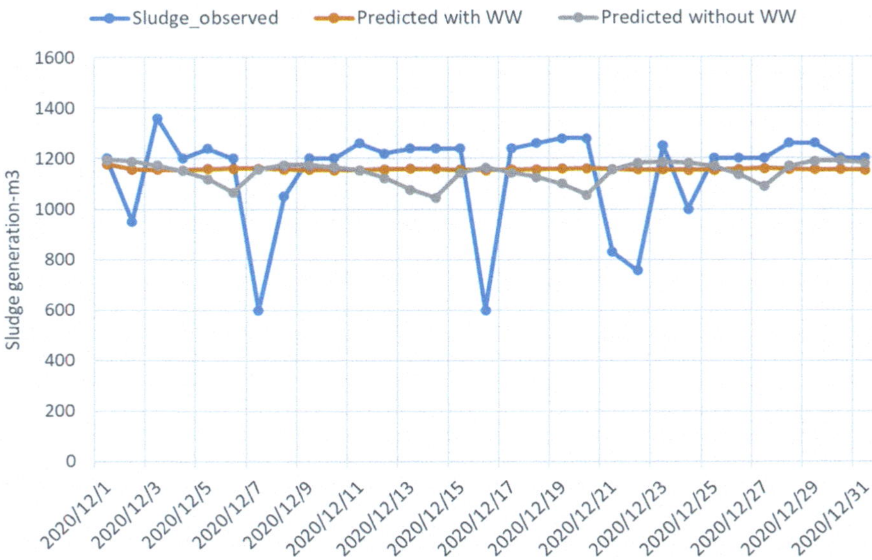


**Fig. 7** The comparisons between the prediction results by ML with WW (84.7%) and without WW (83.1%) with the observed disposal records from 2020.12.1 ~ 31

## 4.3 *Proposals on Future Applications by Using AI Techniques*

Figure 8 shows an example that predicting the daily amount of WW disposal demand and sludge generation amount (refer to right y-axis) to be occurred in a future week in January 2021 by using our proposed ML approach. Once combing with the other data such as methane generation intensity, caloric values of sludge for incineration, etc., daily amounts of methane generation or collected heating could be accurately predicted for the same future period. Based on findings from this work, our proposals on future applications for WW treatment and sludge management by using AI techniques are as follows:

1. Make advanced treatment plans on essential plant size design, relevant utilities (chemical materials and energy [21]) procurement, disposal schedule, and personnel arrangements for WW treatment.
2. Make advanced treatment plans on essential dumping sites, vehicle deployment, and personnel arrangements for sewage sludge management.
3. Clarify the potential for recovery of resources, e.g., nutrients [22], biogas gasification [23], heating, and power generation [24] from sewage sludge utilization in time.

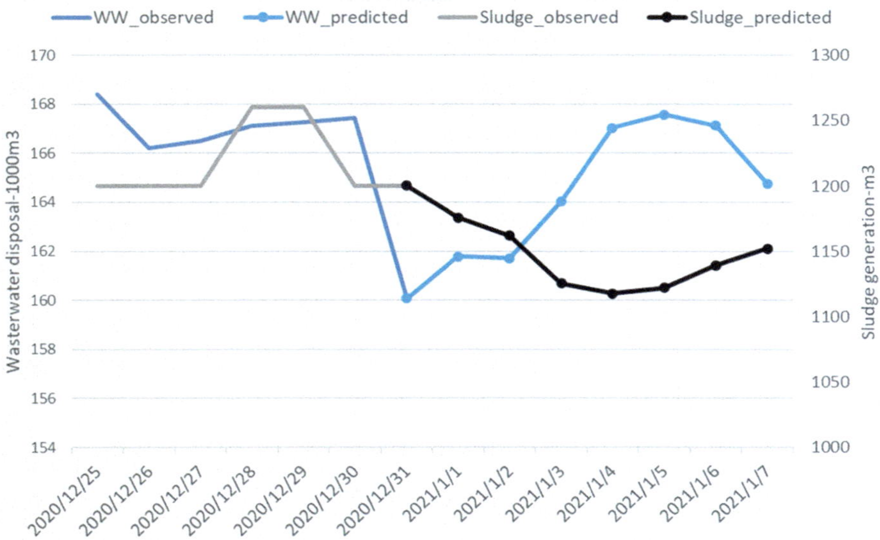


**Fig. 8** Future prediction of daily amount of WW disposal demand (1000 $m^3$) and sludge generation amount (refer to the right y-axis, $m^3$) to be occurred in a future week in January 2021 by using ML approach

# 5 Conclusion

This study proposed an AI-based approach to predict the daily demand for WW and generation amounts of sludge, demonstrated for Ulaanbaatar, Mongolia. In the validation process, the monthly-mean accuracy of the predicted disposal amounts of WW for the last month of 2020 was about 98.0% (predicted by the Medium Gaussian SVM model). Meanwhile, the monthly-mean prediction accuracy of the sludge generation amounts in the same period (predicted by the GPR Rational Quadratic model) was increased from about 83.1% to 84.7% by using the proposed approach. It implies that utilization of WW disposal demand improves the accuracy of the prediction of sludge generation amounts.

The predicted results are potentially used in future WW treatment, e.g., optimizing the personnel arrangement, procurement of chemicals, and prediction of the sludge generation. For sludge management, our proposals were made from future treatment plans and recovery of resources and energy aspects. This study demonstrated that the AI technique is helpful for waste management.

Future works will be made focusing on improving the prediction accuracy of the model, utilizing the proposed approach in other fields, and devising applications of this work. Once the relevant daily parameters (BOD [19], COD, other water quality indexes, and energy consumption) are collected, the proposed approach could be used to predict the performance, operate it in a greener manner, and manage this treatment plant intelligently in future periods.

## References

1. Hatton CM, Paton LW, McMillan D, Cussens J, Gilbody S, Tiffin PA (2019) Predicting persistent depressive symptoms in older adults: a machine learning approach to personalised mental healthcare. J Affect Disord 246:857–860. https://doi.org/10.1016/j.jad.2018.12.095
2. Alizamir M, Kisi O, Ahmed AN, Mert C, Fai CM, Kim S, Kim NW, El-Shafie A (2020) Advanced machine learning model for better prediction accuracy of soil temperature at different depths. PLoS One 15(4):e0231055. https://doi.org/10.1371/journal.pone.0231055
3. Oliveira V, Sousa V, Dias-Ferreira C (2019) Artificial neural network modelling of the amount of separately-collected household packaging waste. J Clean Prod 210:401–409. https://doi.org/10.1016/j.jclepro.2018.11.063
4. Noori R, Abdoli MA, Ghasrodashti AA, Jalili Ghazizade M (2009) Prediction of municipal solid waste generation with combination of support vector machine and principal component analysis: a case study of Mashhad. Environ Progress Sustain Energy 28(2):249–258. https://doi.org/10.1002/ep.10317
5. Shamshiry E, Mokhtar M, Abdulai AM, Komoo I, Yahaya N (2014) Combining artificial neural network-genetic algorithm and response surface method to predict waste generation and optimize cost of solid waste collection and transportation process in Langkawi Island, Malaysia. Malaysian J Sci 33(2):118–140. https://doi.org/10.22452/mjs.vol33no2.1
6. Kannangara M, Dua R, Ahmadi L, Bensebaa F (2018) Modeling and prediction of regional municipal solid waste generation and diversion in Canada using machine learning approaches. Waste Manag 74:3–15. https://doi.org/10.1016/j.wasman.2017.11.057
7. Cong R, Fujiyama A, Matsumoto T (2022) Collection of industrial plastic waste during the COVID-19 pandemic: a case study of the wholesale and retail trade sector in Fukuoka prefecture, Japan. J Human Environ Symbios 38(1):46–55. https://doi.org/10.32313/jahes.38.1_46
8. Cong R, Fujiyama A, Matsumoto T (2022) AI techniques aid for predicting the collection demands of industrial plastic waste from multiple facilities. Environ Sci 35(5):237–247. https://doi.org/10.11353/sesj.35.247
9. Cong R, Fujiyama A, Matsumoto T (2022) AI techniques aid for optimizing the collection system of industrial plastic waste. Environ Sci 35(5):248–257. https://doi.org/10.11353/sesj.35.237
10. Hamed MM, Khalafallah MG, Hassanien EA (2004) Prediction of wastewater treatment plant performance using artificial neural networks. Environ Model Softw 19(10):919–928. https://doi.org/10.1016/j.envsoft.2003.10.005
11. Alsulaili A, Refaie A (2021) Artificial neural network modeling approach for the prediction of five-day biological oxygen demand and wastewater treatment plant performance. Water Supply 21(5):1861–1877. https://doi.org/10.2166/ws.2020.199
12. Lotfi K, Bonakdari H, Ebtehaj I, Mjalli FS, Zeynoddin M, Delatolla R, Gharabaghi B (2019) Predicting wastewater treatment plant quality parameters using a novel hybrid linear-nonlinear methodology. J Environ Manag 240:463–474. https://doi.org/10.1016/j.jenvman.2019.03.137
13. Pai TY, Tsai YP, Lo HM, Tsai CH, Lin CY (2007) Grey and neural network prediction of suspended solids and chemical oxygen demand in hospital wastewater treatment plant effluent. Comput Chem Eng 31(10):1272–1281. https://doi.org/10.1016/j.compchemeng.2006.10.012
14. Matheri AN, Ntuli F, Ngila JC, Seodigeng T, Zvinowanda C (2021) Performance prediction of trace metals and cod in wastewater treatment using artificial neural network. Comput Chem Eng 149:107308. https://doi.org/10.1016/j.compchemeng.2021.107308
15. Dogan E, Ates A, Yilmaz EC, Eren B (2008) Application of artificial neural networks to estimate wastewater treatment plant inlet biochemical oxygen demand. Environ Prog 27(4): 439–446. https://doi.org/10.1002/ep.10295
16. Zeinolabedini M, Najafzadeh M (2019) Comparative study of different wavelet-based neural network models to predict sewage sludge quantity in wastewater treatment plant. Environ Monit Assess 191(3):163. https://doi.org/10.1007/s10661-019-7196-7

17. Wongburi P, Park JK (2022) Prediction of sludge volume index in a wastewater treatment plant using recurrent neural network. Sustain For 14(10):6276. https://doi.org/10.3390/su14106276
18. Beck HE, Zimmermann NE, McVicar TR, Vergopolan N, Berg A, Wood EF (2018) Present and future Köppen-Geiger climate classification maps at 1-km resolution. Sci Data 5(1):1–12. https://doi.org/10.1038/sdata.2018.214
19. Nyamjav B, Badarch A, Sangi C, Khurelbaatar G (2023) Improvement of the treatment performance of the conventional wastewater treatment plant: a case study of the central wastewater treatment plant in Ulaanbaatar, Mongolia. Sustainability 15(5):4528. https://doi.org/10.3390/su15054528
20. Meteorological and Environmental Analysis Department, Past Weather in Ulaanbaatar, Mongolia, http://w.tsag-agaar.gov.mn/observation/environment-monitoring (in Mongolian, accessed 16 May 2022)
21. Yang X, Wei J, Ye G, Zhao Y, Li Z, Qiu G et al (2020) The correlations among wastewater internal energy, energy consumption and energy recovery/production potentials in wastewater treatment plant: an assessment of the energy balance. Sci Total Environ 714:136655. https://doi.org/10.1016/j.scitotenv.2020.136655
22. Simoes F, Colston R, Rosa-Fernandes C, Vale P, Stephenson T, Soares A (2020) Predicting the potential of sludge dewatering liquors to recover nutrients as struvite biominerals. Environ Sci Ecotechnol 3:100052. https://doi.org/10.1016/j.ese.2020.100052
23. Khan MNA, Haq ZU, Ullah H, Naqvi SR, Ahmed U, Zaman M, Amin NAS (2023) Prediction of hydrogen yield from supercritical gasification process of sewage sludge using machine learning and particle swarm hybrid strategy. Int J Hydrog Energy 54:512. https://doi.org/10.1016/j.ijhydene.2023.01.033
24. Đurđević D, Blecich P, Jurić Ž (2019) Energy recovery from sewage sludge: the case study of Croatia. Energies 12(10):1927. https://doi.org/10.3390/en12101927

# Development of an Ideal Training Dataset for Visual Analysis-Based Waste Sorting Robot: An Experiment with Mixed-Construction Waste

**Tianhao Cheng, Hao Hu, Andante Hadi Pandyaswargo, and Hiroshi Onoda**

**Abstract** To cope with challenges influenced by population aging and workforce insufficiency, artificial intelligence (AI) and robotics have been considered to improve productivity in the recycling industry. To understand the challenges in recycling better, we analyzed five mixed constructional waste recycling factories in Tokyo areas and nine kinds of sorting robots implemented worldwide. Research showed that sorting waste by materials is the hardest part of automated recycling; this activity still relies on the worker's ability in Japan. Also, today's waste sorting technology is still insufficient to replace manual sorting due to sorting accuracy problems and the inability to sort waste under unclear conditions. This research tries to figure out the ideal characteristic of waste recognition AI's training data to improve the efficiency and effectiveness of the AI training process. Three characteristics tested in this research are "quantity of the targeted object," "presence of shadows," and "whole/partial appearance."

**Keywords** Mixed-construction waste · Automation · Sorting robot · AI · Training data

## 1 Introduction

Along with the efforts to develop a sustainable society and to cope with the increasing demand for recycling, automation technologies have been considered in the recycling industry. It has been considered a potential solution to replace the lack of workforce and increase the recycling rate [1]. Indeed, automation technologies

T. Cheng (✉) · H. Onoda
Graduate School of Environment and Energy Engineering, Waseda University, Tokyo, Japan
e-mail: tianhao.cheng@fuji.waseda.jp; onoda@waseda.jp

H. Hu
Environmental Intelligence and Innovation, Inc., Tokyo, Japan

A. H. Pandyaswargo
Environmental Research Institute, Waseda University, Tokyo, Japan



S. Fukushige et al. (eds.), *EcoDesign for Circular Value Creation: Volume I*,
https://doi.org/10.1007/978-981-97-9068-5_31

have successfully been implanted in varied fields such as the manufacturing industry [2], medical diagnosis [3], and fire alert [4]; however, automizing recycling procedures has different sets of challenges due to the great amount of variation of the waste condition. This study specifies the mixed constructional waste [5] recycling industry as the target to analyze the technical difficulties of a sorting robot. We experimented with a mixed-construction waste recycling facility in the Tokyo area. This study also performs: (1) a literature review of existing sorting robots, (2) participatory observation, and (3) analyze the AI training process for recycling.

In this chapter, the recycling procedure of mixed constructional waste will be treated as an example to introduce the work that robots are expected to replace humans in the recycling factory. The recycling industry's challenges will also be introduced to analyze the expected development target for automation technologies.

The second part of this chapter focuses on the challenges of waste sorting robots in the operation stage and artificial intelligence (AI) training stage. In total nine kinds of waste sorting robots with implementation over the world are taken as samples to analyze the challenges of robots in the operation stage. Sorting robot is mainly composed of computer vision-based AI and robotic grippers. The challenges of the robots' AI training stage are analyzed through the participation of annotating training data by the authors and the research team.

This chapter identifies the characteristics of the waste target to construct an ideal training data set for visual analysis-based waste sorting robotics. Three characteristics, "quantity of the targeted object," "presence of shadows," and "whole/partial appearance," are tested in this research. The sample data sets are selected from a common training data set of concrete identification AI to eliminate the bias from data collection. Each set of data is used to train an AI model, and the effectiveness of training data is evaluated by comparing the precision and recall rates of each model. Finally, the data sets of each character will be combined by two sets in a ratio of 50: 50 to test whether the separation improved the AI training result.

## 2 Mixed Constructional Waste Recycling Facilities

Constructional waste is released from newly established, renewed, or demolished buildings. While newly established and renewed buildings usually emit well-sorted wastes, demolition waste is mostly released in a mixed state. Demolition waste usually comes from constructions built without a recycling concept. Moreover, the building material varies by year and by building type. Therefore, releases of mixed constructional waste are unavoidable [6]. It is also not practical to educate the demolition workers due to the huge number of subcontractors, unclear definition of disintegration, and the lack of rules and regulations that created the assumption that inadequate demolition work is not illegal. Sorting the waste by materials is the hardest and the most important procedure to recycle mixed constructional waste.

Five mixed constructional waste recycling facilities in the Tokyo area are analyzed to understand the procedures and challenges of waste sorting. Through the field research, the recycling process in recycling facilities mainly contains five stages: field sorting, first sorting, crashing, second sorting, and recycling, as shown in Fig. 1. With the five stages, field sorting and the first sorting take about 80% of labors in the factory shows the highest needs of automation technology.

Field sorting eliminates large waste that cannot be handled on belt conveyors. However, the existence of laborers, carriers, and heavy machinery working in the same field makes the task dangerous. The only protection for human laborers in the area is a helmet. The image of field sorting is shown in Fig. 2.

The first sorting shown in Fig. 3 separates waste by genre. Since the process is manually done on a belt conveyor by laborers, it is also called manual sorting. Even the stage is called first sorting; the process would be repeated until the waste is

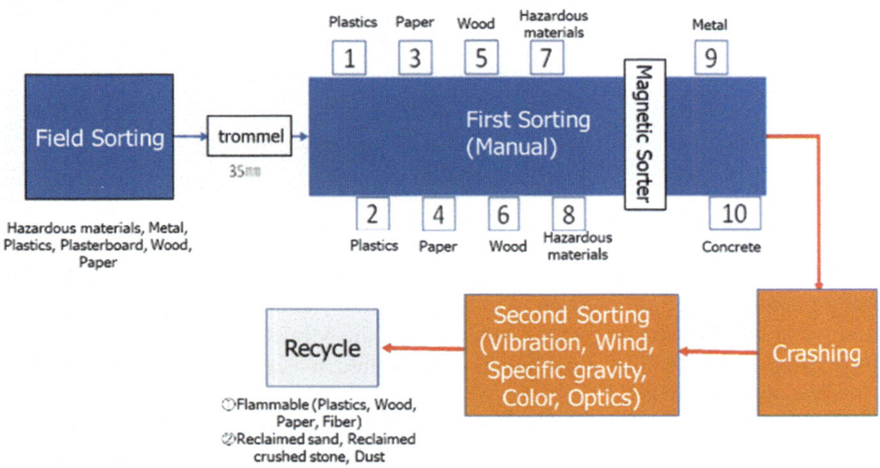


**Fig. 1** Recycling process in recycling facilities

**Fig. 2** Field sorting

**Fig. 3** Manual sorting

separated. Skilled workers have shown high efficiency in sorting wastes, but the aging population and insufficient workforce make the skill hard to pass down to the next generation. Based on the research done by the Ministry of the Environment in 2021, there were 46% of recycling facilities hired workers who were older than 70 years old in 2020 [7].

The rest of the recycling facility activity stages require minimum technological improvements. The second sorting stage sorts the waste that passes through the first sorting into uniform materials. For example, the metal genre among the waste in the first sorting is separated into iron, copper, aluminum, and other metals. A machine does this sorting activity based on the waste's physical characteristics. Since each machine could only use single physical properties, the sorting quality of the first sorting would determine the efficiency of the second sorting. Moreover, a mechanical failure may occur if the mixed waste is sorted by machines directly without the preceding stages. Based on our observation of the overall stages, robotics with improved accuracy is particularly needed in the first sorting stage.

## 3 Challenges of Waste Sorting Robot

### *3.1 Waste Sorting Robots with Implementation*

To identify the characteristics of waste sorting robots in the operation stage, this study takes nine kinds of constructional waste sorting robots from the literature that have already been implemented in factories worldwide as examples to analyze. The case study robots are listed in Table 1 and are evaluated by Technology Readiness Level (TRL) [8]. Based on the literature review, all robots apply visual identification methods to recognize sorting targets and represent impressive abilities to save the labor force, perform 24-hour operations, improve sorting accuracy, increase the recycling rate, and lower the safety risk. However, it has also been determined that introducing these sorting robots requires the factory to be prepared at the construction stage. In other words, the robot size is too large for factories with limited space. Also, today's robot sorting system still requires workers to stand by to eliminate hazardous materials and protect the robot from failures.

Although sorting robots have improved factories' capacity, it is hard to say that the robot's ability has reached its optimal potential. Based on our research, it is also determined that AI's accuracy in identifying sorting objects is insufficient to achieve automation over the whole recycling facilities. Especially for the congested scene shown in Fig. 4, robots cannot identify and pick the target precisely. Take mixed constructional waste as an example, the robots found it hard to identify the target that is covered by other materials. Moreover, when other materials are in contact with targets, the robot would make inaccurate picking. The gripper must also be developed to pick different targets from the sorting line. For example, using a robot hand to pick up plastic bags or heavy concrete is challenging for present auto-sorting robots.

**Table 1** Case study robot

| Robot | Producing country | TRL |
|---|---|---|
| Zenrobotics | Finland | 9 |
| AIBenkei | Japan & US | 9 |
| AIMusashi | | 9 |
| URANOS | | 9 |
| V-PICKER | Japan | 7 |
| Waste Sorting Robot | | 7 |
| MAX-AIAQC | US & Spain | 8 |
| Samur AI | Canada | 9 |
| waste robot | US | 8 |

**Fig. 4** Congested scene on sorting line

## *3.2 Annotation for Visual Analysis-Based Waste Sorting AI's Training Set*

To understand the challenges of the data collection stage, we interviewed the operating technicians. Based on the interviews, we collect original data in the form of photographs or videos with the laboratory or sorting site as background. We annotate and produce a visual-based waste sorting AI training set using the collected data. The strengths and weaknesses of the collected data types are:

- Photographic data effectively identifies objects because only the necessary screens are taken. However, the labeling process takes a long time since only one picture can be operated simultaneously. Thus, training data collection with photographic methodology requires much workforce.
- Video data acquire continuous image data at a given location, so a semi-automatic labeling method can be used. Thus, less workforce is needed; however, since all objects on the screen are captured, foreign objects mixed on the screen are captured simultaneously. This situation made it difficult to identify the target.
- Image data acquired in the laboratory have few foreign objects in the background. Furthermore, the characteristics of the objects can be grasped to some extent and easily processed into training data. However, data collected in the laboratory has significantly different backgrounds with a crowded sorting site, where multiple objects appear simultaneously. Thus, there is a risk that the AI identification rate in the experimental stage will decline if only this type of data is used.
- The training data taken at the sorting site is suitable for practical use because it records the actual state of the work as it is. However, some issues may arise depending on the labeling operator's ability to identify the object on the screen because the appearance state and shadow of the object on the screen are mixed up. The problem is that the annotator must be able to identify the object on the screen.

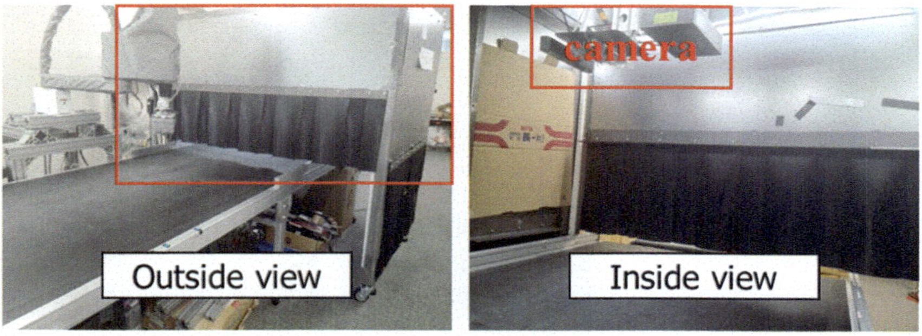


**Fig. 5** Training data collecting box

Since image identification AI processes image information as numerical values, the brightness at the time of the shooting and the resolution of the image must be unified. Otherwise, the AI's training will be affected, and the image cannot be processed into training data. Even though many photos of objects to be sorted have been taken in the waste resource recycling field, only a limited number of objects have been photographed in a uniform environment. Furthermore, only a small fraction of them can be used for AI training. Because of the limitations, developers must acquire training data independently. Thus, there is a need to explore more efficient methods of acquiring training data and to find initiatives that fit the development of visual analysis AI. Figure 5 represents a training data collecting box used by developers.

Through the participation of annotating concrete as a target to create training data sets for robots, two challenges in the labeling stage are found as follows:

- Labeling relies on annotators' skills, and human error cannot be avoided.

The object labeling work currently performed by manufacturers of automatic sorting robots is a simple task. However, identifying the objects to be labeled is still based on the worker's judgment. Unlike in production plants, labeling in waste recycling involves many kinds of waste and different varieties. The similar screening information is projected on the same screen, making human error inevitable. Figure 6 represents three kinds of common errors: mislabeling by annotators, errors caused by lighting changes, and errors caused by shadows.

- Labeling task has low efficiency.

Labeling tasks must be performed on photos and videos. We observed the labeling task performed for three months, where ten annotators worked 418 hours. Twenty-seven thousand frames of training data were employed in the task. A unit of frame/min*person determines the labeling efficiency. The result shows that the average speed of labeling work on photos is 1.2 frame/(min*person), and the speed of labeling work on videos using the tool is 13.2 frame/(min*person). The results are influenced by the fact that photo labeling is a manual task of marking the objects in each photo. The speed of the process is greatly affected by the number of

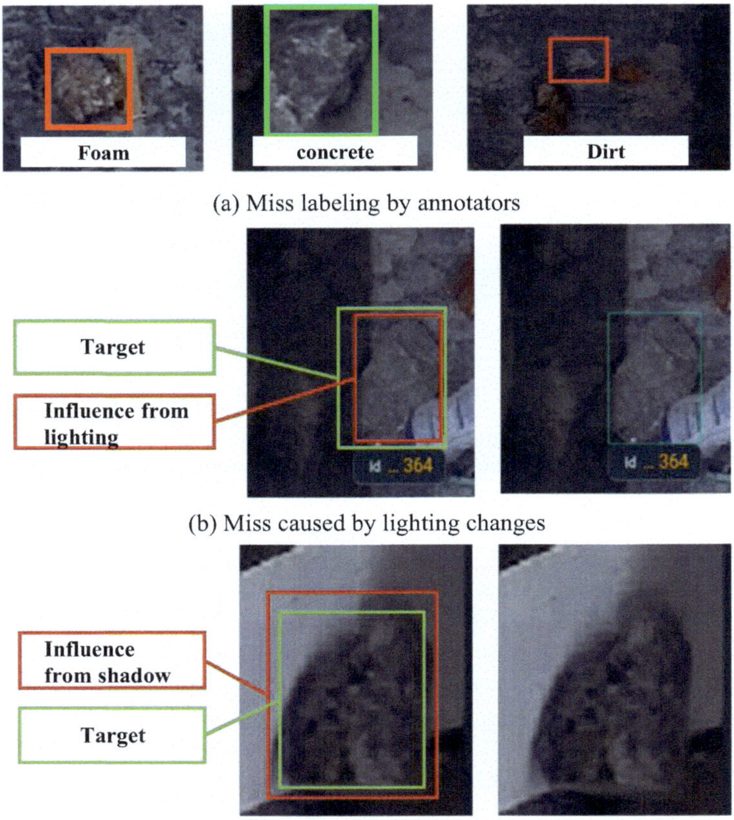


(a) Miss labeling by annotators

(b) Miss caused by lighting changes

(c) Miss caused by shadow

**Fig. 6** Example of challenges on annotation of waste sorting AI's training data. (**a**) Miss labeling by annotators. (**b**) Miss caused by lighting changes

objects in the photo and the mass of the screen. On the other hand, video labeling is a continuous representation of the image; the work consists of manually marking objects as they appear and automatically labeling them afterward. Since video labeling is primarily a tool operation, the work speed is greatly affected by the operating conditions of the tool.

Figure 7 shows the speed change in the labeling operation. The number of days is limited because only data from days when workers are labeling simultaneously are extracted. Each worker's work speed does not vary much with experience. However, it varies greatly depending on the quality of the images they are responsible for (photos) and the operational status of the labeling tools (videos). The final speed of creating training data declines because some frames that cannot be used as training data are mixed in due to errors by the workers. The only way to remove unusable frames from the data set is for the technician to set the criteria and check each frame

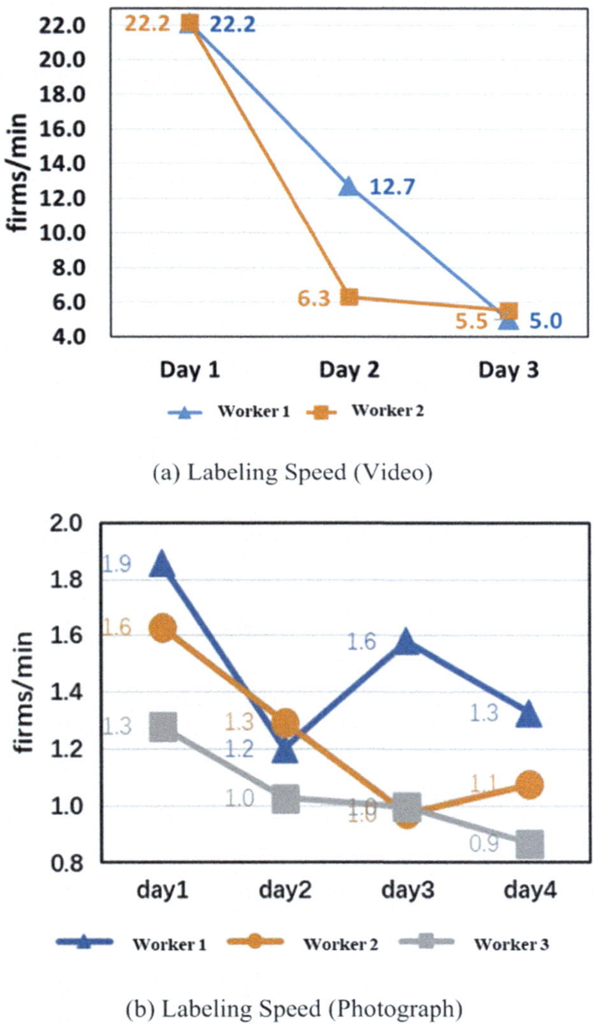


**Fig. 7** Speed of labeling operation. (**a**) Labeling Speed (Video). (**b**) Labeling Speed (Photograph)

individually. Since the AI training model always applies deep learning methods, in which AI makes all decisions, no one knows which characteristics positively influence improving AI's accuracy. Technicians can only adjust training data by experience, which leads to the inefficiency of AI training.

# 4 Ideal Training Data

Using the training data from Fig. 5, a concrete identification model with 55% precision and 63% recall is trained. Improving an AI's accuracy can be achieved by increasing the amount or improving training data quality. Increasing the numbers of training data is well-known and proven effective, but the quality of training data is insufficiently studied in waste sorting AI. To improve the efficiency of data collection and test the quality of training data, an experiment is designed to train an AI model with the sample data chosen from the previous data set by characteristics.

## 4.1 *Overview of Experiment*

This study summarized the questions faced while creating the concrete training data and based on the data obtained in Sect. 3.2. The training data were classified according to the following characteristics: "quantity of the targeted object," "presence of shadow," and "whole/partial appearance." With each feature as a criterion, data samples were extracted 1000 times based on those criteria. At the same time, the original composition of the training data was ascertained. Each sample was used as training data to train an image identification AI model for each classification. Table 2 represents the classification standard for each characteristic, and the ratio identifies the persistence of each classification within the characteristics.

Figure 8 represents the classification example of "Quantity of the targeted object." The classification only depends on the number of targets.

Figure 9 represents the classification example of "presence of shadow." If one target is under shadow while others are not, the frame will be classified into the "with shadow" class.

Figure 10 represents the classification example of "whole/partial appearance." If one target shows partially while the other targets show in a whole, the frame will be classified into the "shows a part of the target" class. We repeated the classification task until the class with the minimum ratio reached 1000. After we collected the sample, we calculated the ratio of each classification. After the calculation, each class's sample size is adjusted to 1000 by eliminating similar frames in each standard.

**Table 2** Classification of characteristics

| Characteristic | Class | Standard | Ratio |
|---|---|---|---|
| Quantity of the targeted object | I | 1 target | 0.25 |
| | II | 2 ~ 5 targets | 0.57 |
| | III | 6 or more targets | 0.18 |
| Presence of shadow | (1) | With shadow | 0.46 |
| | (2) | Without shadow | 0.54 |
| Whole/partial appearance | a | Whole appearance | 0.6 |
| | b | Partial appearance | 0.4 |

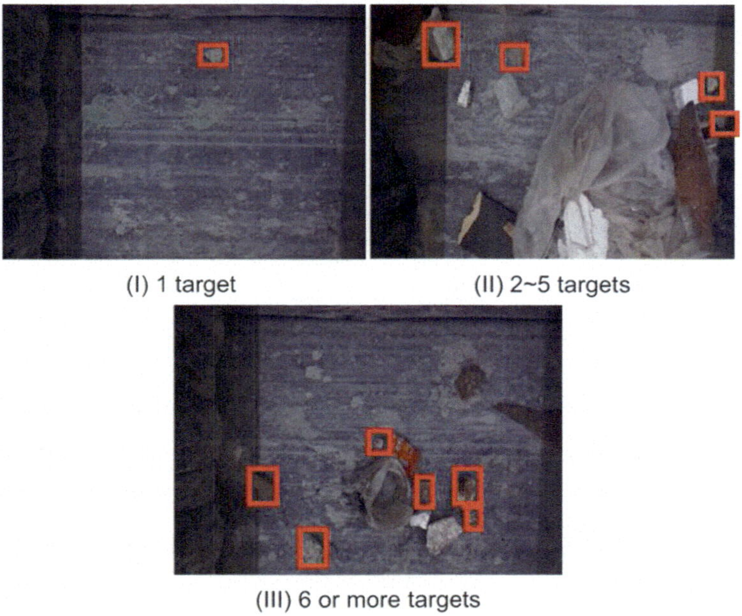


**Fig. 8** Quantity of the targeted object. (**I**) 1 target. (**II**) 2 ~ 5 targets. (**III**) 6 or more targets

**Fig. 9** Presence of shadow. (**1**) With shadow. (**2**) Without shadow

Each data set is trained in a common AI model as, YOLO7, to eliminate bias from the training method. Moreover, each model is trained from scratch.

To eliminate the bias from sample data, all frames are chosen from the 27 thousand concrete training data, and each frame can only show once in each characteristic. Within the characteristic, the standard is the only variable component; each characteristic is independent. The combination of classified data sets only performs within the characteristic, and the composition of the combined data set is evenly distributed among each class.

(a) Whole appearance (b) Partial appearance

**Fig. 10** Whole/partial appearance. (**a**) Whole appearance. (**b**) Partial appearance

To eliminate the bias from testing data, all samples are tested by a common test set selected randomly from the original data set without classification. A sample size of 1000 for AI training is too small to obtain sufficient accuracy for each class to work in the real world, but it is enough to analyze the tendency that each classification could lead to the training result.

The loss model and MAP (mean average precision) are checked before comparison to ensure that each standard is trained with a converge result shown on the graphic result.

## *4.2 Result and Analysis of the Experiment*

Figure 11 shows each model's precision and recall rate results by "quantity of the targeted object" from the experiments.

Further descriptions of each figure are as follows:

I: AI models trained on training data with only one object on the screen showed an average of 35% precision and 40% recall. II: AI models trained on training data with 2–5 objects on the screen showed an average of 40% precision and 47% recall. III: AI models trained on training data with six or more objects on the screen showed, on average, a 39% precision rate and a 37% recall rate.

The AI models trained with training data having 2–5 objects on the screen are the most stable and have the highest precision and recall rates.

Figure 12 represents the training result of the "Presence of shadow." classification.

1. The AI model trained by the "with shadow" class showed an average precision rate of 41% and a recall rate of 48%.
2. AI models trained on the "without shadow" class showed, on average, a 42% precision rate and a 42% recall rate.

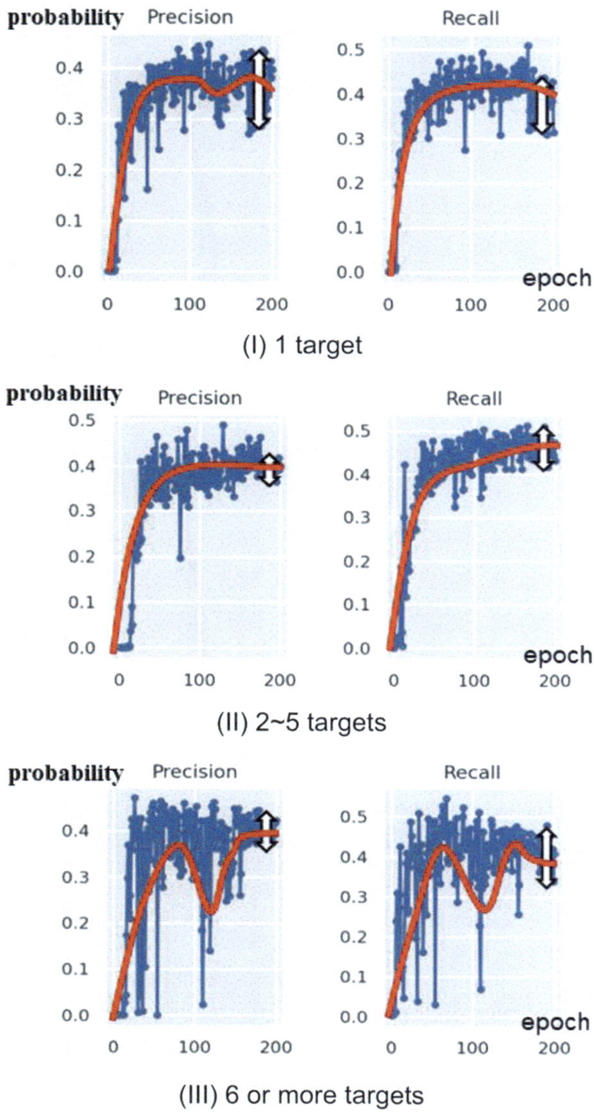


**Fig. 11** Training result of 'Quantity of the targeted object". **(I)** 1 target. **(II)** 2 ~ 5 targets. **(III)** 6 or more targets

Although the presence of shadows had little effect on the stability of the training effect, a higher recall rate was obtained by using training data with shadows.

Figure 13 represents the training result of the "Whole/Partial Appearance" classification.

(a) The AI model trained on training data showing the entire object on the screen averaged a 42% precision and 49% recall rate.

**Fig. 12** Training result of 'Presence of shadow'. (**1**) With shadow. (**2**) Without shadow

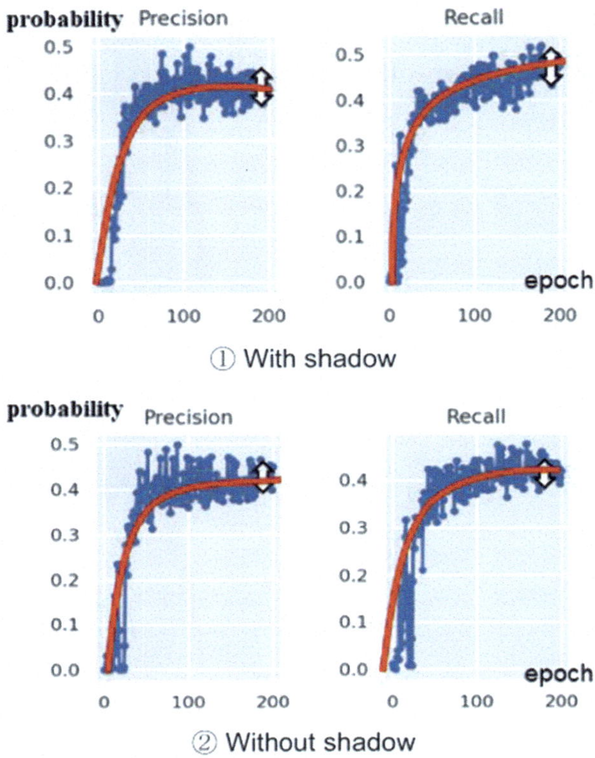


(b) The AI model trained on the training data with a partial view of the object on the screen showed, on average, a 33% precision rate and a 40% recall rate.

The training effect of the training data showing the entire object was more effective than if only parts of the object is seen, in obtaining a higher precision and recall rate.

To determine the impact of the combination of various data sets on the training effectiveness of the AI model. The training data is reconstructed with a 50:50 ratio of the above data sets and verified the AI training effectiveness by each feature. The combinations are "I + II," "I + III," "II + III," "(1) + (2)," and "a + b." Figure 14 represents the training result of the precision and recall rate combinations.

The effectiveness is evaluated by the compare the training result in precision and recall rate of the original data set, which took values of (1)+(2) class since the combination has the same ratio of components.

Our experiment confirmed that the preferred characteristics of training data for the sorting line in the waste recycling field are "with shadow" and "Whole appearance." The best training results for the number of objects on the screen were "two ~ five targets." However, the comparison with the "1 target" classification was based on the same number of frames in the training data. Since the number of objects in the

**Fig. 13** Training result of 'Whole/Partial Appearance'. (**a**) Whole appearance. (**b**) Partial appearance

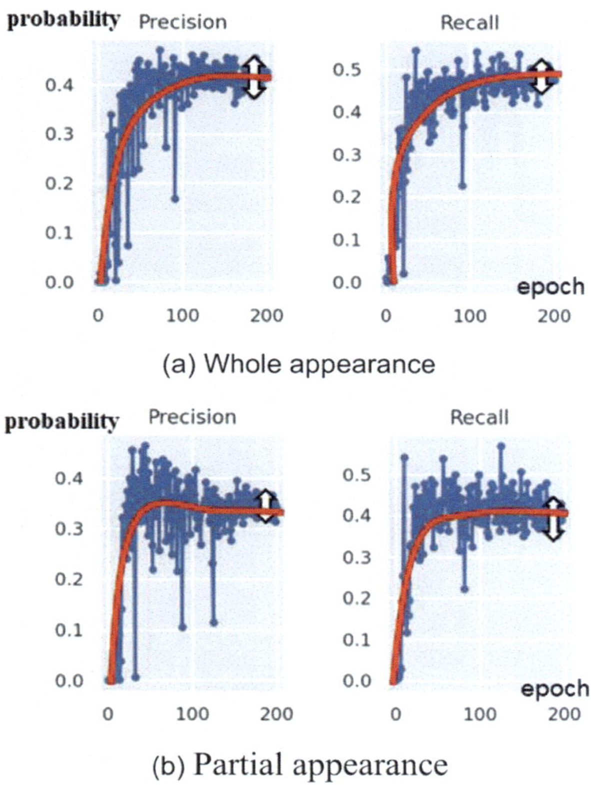


training data was unknown, further study is needed. However, we were able to ascertain that data representing "6 or more objects" on the same screen are inefficient as training data.

Training data with six or more objects generally appears within a more complex background of the sorting site which increases the probability of human error that occurs in the annotation stage. Moreover, the targets are more likely to be covered by other waste resulting in a higher ratio of "partial appearance targets" which lowers the recognition accuracy of AI. The situation also made the targets more likely to be in contact with each other, which increased the difficulty of edge identification by targets. Figure 15 represents an example of six or more identified objects. The yellow squares identify styrofoam waste, which could incorrectly be considered concrete waste in the annotation stage. In addition, our AI model is training for robotic implantation, which recognition should be made in a limited time. However, having more targets on a frame means a longer time is needed for calculation; AI may lower the calculation accuracy to provide a result in a limited time.

The specific training result is listed in Table 3. The evaluation results in the table column are calculated by adding the precision and recall of VS. original data. The

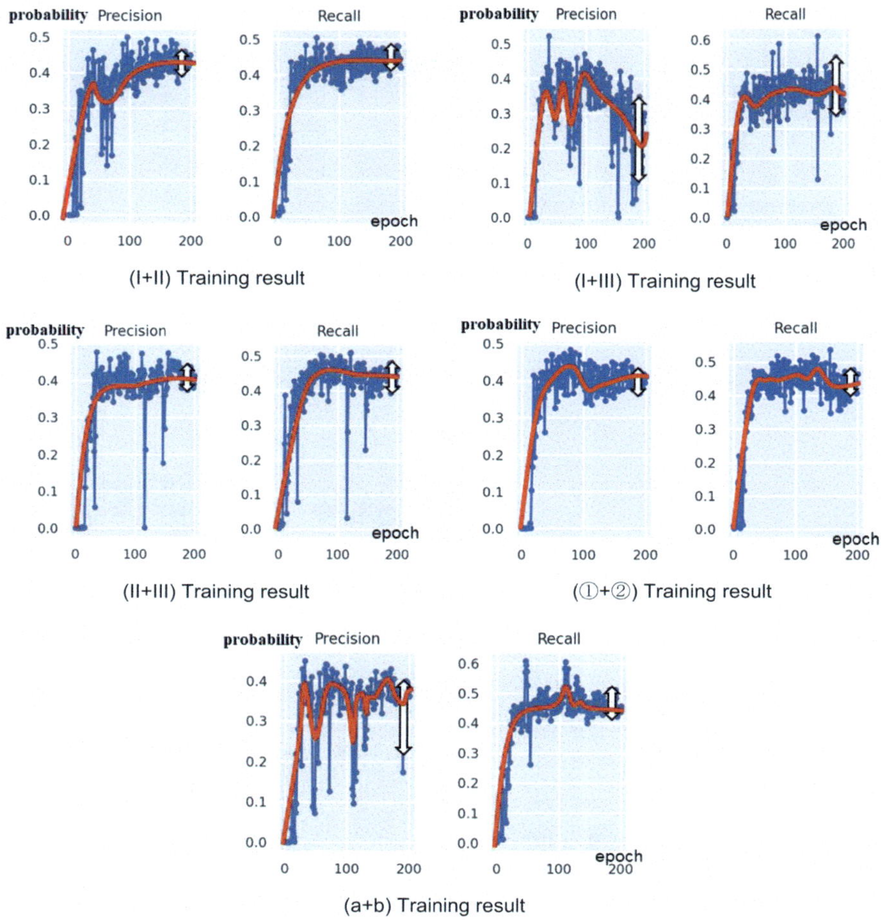


**Fig. 14** Training result of the combinations

green highlight represents the favorable class and the evaluation result for each characteristic.

## 5 Discussion

This research has studied the three distinct categories of training data. However, it is insufficient to say that we have produced ideal training data for waste sorting AI. Other characteristics such as "the complexity of background" or "presence of stains" should also be tested. To eliminate the bias from test set, additional test with

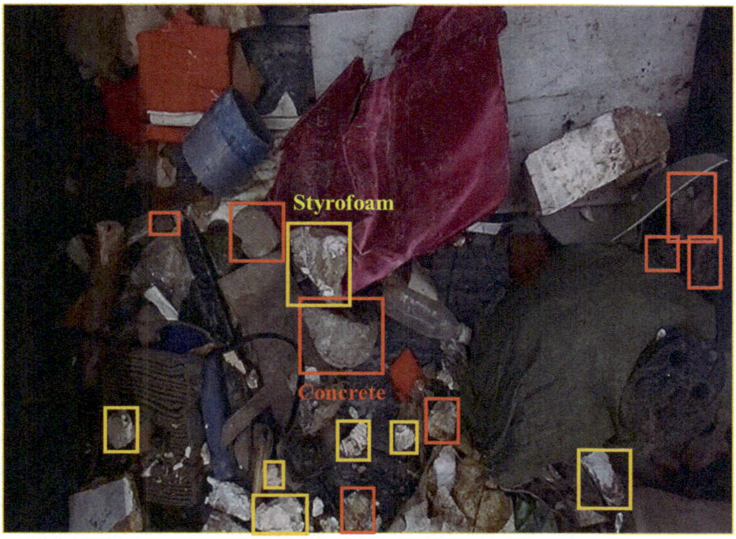


**Fig. 15** An example of six or more identified objects

**Table 3** The training result of training data experiment

| Class | Standard | Precision % | Recall % | Precision VS. Original data % | Recall VS. Original data % | evaluation |
|---|---|---|---|---|---|---|
| Original | ①+② | 40 | 44 | 0 | 0 | 0 |
| I | 1 target | 35 | 40 | -5 | -4 | -9 |
| II | 2～5 targets | 40 | 47 | 0 | 3 | 3 |
| III | 6 or more targets | 39 | 37 | -1 | -7 | -8 |
| ① | With shadow | 41 | 48 | 1 | 4 | 5 |
| ② | Without shadow | 42 | 42 | 2 | -2 | 0 |
| a | Whole appearance | 42 | 49 | 2 | 5 | 7 |
| b | Partial appearance | 33 | 40 | -7 | -4 | -11 |
| I+II | (50:50) | 43 | 44 | 3 | 0 | 3 |
| I+III | (50:50) | 25 | 40 | -15 | -4 | -19 |
| II+III | (50:50) | 40 | 44 | 0 | 0 | 0 |
| a + b | (50:50) | 38 | 43 | -2 | -1 | -3 |
| ①+② | (50:50) | 40 | 44 | 0 | 0 | 0 |

evenly distributed standards should be done. The number of targets discussed in this study considered the training data uniformed by frames, while taking the numbers of targets within the data set should also be tested as a unique characteristic.

To minimize human errors in the annotation stage, transferring and incorporating the knowledge from the recycling facilities to the robots should be considered. To achieve this, research should be conducted to test the difference between labeling with data collected on the sorting line and with data collected after the manual sorting process.

Creating a standard that uniforms the lighting, resolution, number of targets, etc., could also be helpful to improve the data collection efficiency. The standard would allow the simultaneous collection of usable data from different robot makers.

## 6 Conclusion

Based on our research results at the mixed constructional waste recycling facilities, we found that the automation technology most needed in the field is sorting robots and hazardous material removers. This study also found that the challenge of waste sorting robots is to minimize the robot size, improve robot performance, and improve the accuracy of implementation. Recycling facilities aim to improve efficiency and minimize human error during training data collection. Through experiments conducted in this study, we understand that the "2 ~ 5 targets," "with shadow," and "Whole appearance" are the favorable characteristics for visual analysis-based waste sorting AIs' training data. To improve the efficiency of data collection and the accuracy of AI recognition, further studies are needed for the other characteristics.

While the effectiveness of visual analysis-based sorting robots has been clarified, researchers and developers aiming to adopt the technology into the recycling field must remember the difference between production and recycling. Adding on the training set could improve sorting accuracy; however, eliminating the bias within the training data could affect the AI training and robot sorting efficiency.

The experiments conducted in this study contribute to the development of an ideal dataset to train AI-based sorting robots in construction waste. However, further experiments with different waste characteristics are still required to develop a failproof training dataset.

## References

1. Onoda H (2023) Current status and prospects of digital technology contributing to Circular Economy (CE). J Life Cycle Assess, Japan 19(2):79–84
2. Dauth W, Findeisen S, Südekum J, Woessner N (2017) German robots-the impact of industrial robots on workers
3. Chen K, Zhang J, Beeraka NM, Sinelnikov MY, Zhang X, Cao Y, Lu P (2022) Robot-assisted minimally invasive breast surgery: recent evidence with comparative clinical outcomes. J Clin Med 11(7):1827
4. Cheng T, Hu H, Kobayashi H, Onoda H (2022) Visual identification-based spark recognition system. Int J Autom Technol 16(6):766–772

5. Cheng T, Kobayashi H, Hu H, Cheng T, Onoda H (2022) Study on introduction of AI and robotics to improve the efficiency of construction mixed waste disposal. In: Proceedings of the Symposium on Environmental Engineering (pp 2205–2208). The Japan Society of Mechanical Engineers
6. Ministry of the Environment (2021) FY 2020 Construction Waste Recycling Study Report
7. Waste Regulation Division, Environmental Reclamation and Resource Recycling Bureau, Ministry of the Environment (2021) Outline of Results of Survey on Securing Diverse Human Resources in the Industrial Waste Management Industry in FY 2020
8. Ihara I, Pandyaswargo AH, Pang D, Onoda H (2019) Technology readiness assessment of biomass energy projects using J-TRA method: Application on Southeast Asian countries. In: Proceedings of the international workshop on environmental engineering

# Exploring the Current Applications and Potential of Extended Reality for Environmental Sustainability in Manufacturing

**Huizhong Cao, Henrik Söderlund, Mélanie Despeisse, and Björn Johansson**

**Abstract** In response to the transformation towards Industry 5.0, there is a growing call for manufacturing systems that prioritize environmental sustainability, alongside the emerging application of digital tools. Extended Reality (XR) — including Virtual Reality (VR), Augmented Reality (AR) and Mixed Reality (MR)—is one of the technologies identified as an enabler for Industry 5.0. XR could potentially also be a driver for more sustainable manufacturing: however, its potential environmental benefits have received limited attention. This paper aims to explore the current manufacturing applications and research within the field of XR technology connected to the environmental sustainability principle. The objectives of this paper are two-fold: (1) Identify the currently explored use cases of XR technology in literature and research, addressing environmental sustainability in manufacturing; (2) Provide guidance and references for industry and companies to use cases, toolboxes, methodologies, and workflows for implementing XR in environmental sustainable manufacturing practices. Based on the categorization of sustainability indicators, developed by the National Institute of Standards and Technology (NIST), the authors analyzed and mapped the current literature, with criteria of pragmatic XR use cases for manufacturing. The exploration resulted in a mapping of the current applications and use cases of XR technology within manufacturing that has the potential to drive environmental sustainability. The results are presented as stated use-cases with reference to the literature, contributing as guidance and inspiration for future researchers or implementations in industry, using XR as a driver for environmental sustainability. Furthermore, the authors open up the discussion for future work and research to increase the attention of XR as a driver for environmental sustainability.

**Keywords** Extended Reality · Virtual Reality · Augmented Reality · Mixed Reality · Sustainable Manufacturing · Sustainability · Industry

H. Cao (✉) · H. Söderlund · M. Despeisse · B. Johansson
Chalmers University of Technology, Gothenburg, Sweden
e-mail: huizhong@chalmers.se



S. Fukushige et al. (eds.), *EcoDesign for Circular Value Creation: Volume I*,
https://doi.org/10.1007/978-981-97-9068-5_32

# 1 Introduction

To stay competitive in a market with a growing demand for sustainable products and services, manufacturing companies are embracing smart and sustainable manufacturing practices as the core drivers for their Industry 5.0 transformational shift [1]. Extended reality (XR) as an enabler for this industry shift has previously been explored and discussed by multiple researchers, defining its applications in the manufacturing areas of the future. Previous research has proven a strong relationship between XR technologies in manufacturing together with economical and social sustainability, as XR provides great decision support and enables human-centric solutions. However, the effects that XR technologies potentially have on the environmental aspects of sustainability are more seldomly discussed and to some extent still uncertain [2] as the potential benefit to environmental sustainability that XR brings is usually only indirectly proven.

## *1.1 Extended Reality*

Extended Reality (XR) is the umbrella term for technologies that act as interfaces between the real and virtual worlds. The term includes Virtual Reality (VR), Mixed Reality (MR), and Augmented Reality (AR), which all offer different levels of immersion and levels of interactions with digital objects between the real and virtual world [3].

VR applications completely immerse the user in a virtual world, usually using a head-mounted display (HMD) or projections that encapsulate the users with a full visual experience of a virtual world. By tracking the motions and position of the user, the motions and position can be mimicked in the virtual world, giving the perception of full immersion [3]. For applications where full virtual immersion is not needed, MR technology can help us to mix and overlay 3D elements from both the real and virtual worlds in the same experience and environment. Usually, this is done via an HMD in combination with visual inputs from the real world using cameras or AR headsets with a transparent display [3]. Finally, AR technology allows us to remain in the real world while overlaying digital information and augmented visuals on top of our perception of the real world. Examples of such technologies could include AR headsets, smart glasses, smartphone applications, or projections [3].

The use of XR technology is often associated with industries such as gaming or entertainment. However, XR technology has also been widely implemented in manufacturing and has been identified as one of the enabling technologies for Industry 5.0 [4].

## 1.2 Environmental Sustainability in Manufacturing

Sustainable manufacturing has been defined in literature as a manufacturers' ability to continue to assure demand and improve the quality of human life while satisfying economical, societal, and environmental objectives, not compromising future generations to meet their needs [5]. However, as a manufacturer, it might not be clear how to actually measure and therefore achieve, sustainable manufacturing [6]. One possible way to measure the sustainability impact of a manufacturing system is by using the sustainability indicators developed by the US National Institute of Standards (NIST). NIST has identified a total of 213 indicators, that can be used to measure the sustainability in a manufacturing process, categorized under these three pillars of sustainability, environmental-, social- and economic sustainability [6, 7]. Out of these, 77 indicators have been identified under the pillar of environmental sustainability and primarily focused on minimizing emissions, resource consumption, and waste as a result of the manufacturing processes [6]. In Fig. 1, an illustration of the areas covered by the 77 environmental indicators proposed by NIST can be seen.

By considering the sustainability indicators, identified and categorized by NIST, manufacturers, and researchers are able to measure certain actions and activities in their manufacturing system to understand how they affect environmental sustainability and furthermore how they can improve their sustainable manufacturing practices [7].

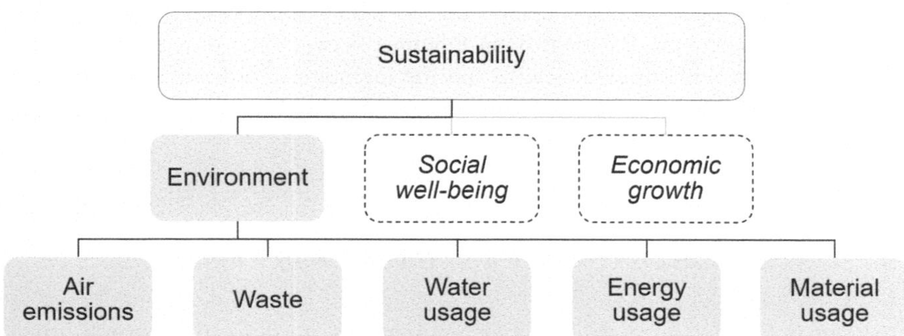


**Fig. 1** Illustration of areas covered by the environmental indicators defined by NIST, adapted from [8]

### 1.3 Research Questions

In recent years, we have seen an increase in literature exploring XR technologies in manufacturing as well as actual real-case implementations in industry. At the same time, some efforts have been put toward documenting the relationship between various XR tools and sustainable manufacturing [21]. However, there is still a gap in the knowledge that merges the broader concepts of XR, its systemization, and use cases in manufacturing and environmental sustainability.

This chapter aims to provide a better understanding of the direct or indirect contributions that XR technologies potentially can have on environmental sustainability in various manufacturing use cases. By mapping existing manufacturing use cases towards proven environmental impacts from existing literature, the authors hope to provide directions for future XR development for industry and researchers. By providing clear examples and closing this gap, acceleration towards sustainable manufacturing and Industry 5.0 using XR technology is aspired. The research questions for the paper are:

RQ1: How has sustainability been addressed in research efforts on XR for manufacturing applications?
RQ2: What environmental benefits have been obtained in use cases of XR in manufacturing?

## 2 Method

The method developed for this chapter categorizes the XR use cases according to the sustainability indicators. By identifying relations between the two, such as direct/indirect and with/without evidence, to sort the literature in order to find the correlations and research gap in strengthening the relationship between XR and environmental sustainability.

### 2.1 Literature Search, Scope and Inclusion of Papers

As shown in Fig. 2, to meet the objective requirements, we devised the search strings in the database Scopus, with the keywords "sustaina*," "XR" and its extension, "Production" OR "Manufac*" to look for the nonbiased trend in the XR development for sustainability. The first selection was based on the criteria of the year of publication, subject area, and language. After the first selection, the use cases derived from the selected papers were further screened by their conclusions, and 30 papers with clear use cases in sustainable manufacturing were processed and fully studied within the literature review.

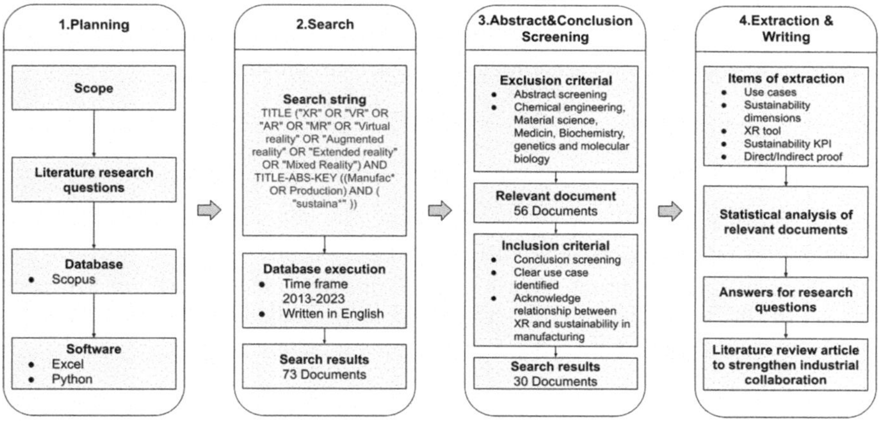


**Fig. 2** Illustration of the methodology used in the literature study

## *2.2 Analysis*

In order for the authors to conduct the analysis and to answer the posed research questions from Sect. 1, a spreadsheet was created to capture data from the examined papers used for the analysis. The spreadsheet recorded generic meta data of each paper such as year of publication, country of origin, etc., but also specific data points used in the analysis of use cases. Detailed information about the paper, including the hardware, software used, sustainability dimensions, evidence, limitations, and reflection, was collected and examined. The criteria recorded as part of the analysis were the following: (i) The application and use of XR technology proposed by the author, (ii) The corresponding sustainability pillar(s) impacted by the use case, (iii) Direct/Indirect impact on environmental sustainability.

To correctly identify the corresponding sustainability pillar addressed in the papers and use cases examined, the authors referred to the sustainability indicators defined by NIST [6, 7] under each sustainability pillar as described in Sect. 1. If the examined paper mentioned one or more indicators, tied to a specific sustainability pillar, as the result or impact of an XR implementation or use case, this paper would be categorized under corresponding pillar. This meant that the same paper could be categorized under more than one pillar and at the same time, papers not specifically mentioning a metric or outcome that could be tied to any of the indicators as suggested by NIST, would not be mapped towards any of the sustainability pillar. By mapping the recorded sustainability indicators and sustainability pillars towards the described application of XR proposed by the examined papers, clear examples and use cases could be identified that propose environmental impact and sustainability as the result of XR technology within manufacturing.

Furthermore, to determine the focus and main aim of the papers, the distinction between the directly affected indicator of the proposed use case and indirect affected indicator as the result of a use case was made.

- *Direct impact*: This is based on quantitative evidence directly cited by the paper.
- *Indirect impact*: This is derived from qualitative claims suggesting long-term benefits of applying XR technology to enhance sustainability.

# 3 Results

By analyzing the 30 papers selected for this study, a total of 39 use cases of XR technology in manufacturing were identified. In the following section a deeper analysis of the 39 use cases is described and their relationship to environmental sustainability is mapped out.

## *3.1 Sustainability Perspectives Analysis of Literature*

When categorizing the papers according to the NIST framework, a clear division in the attention between the three sustainability pillars given by the examined papers can be easily identified, see Fig. 3.

As shown on Fig. 3, a total of 36 use cases in the studied literature has been identified by its authors as having a direct or indirect impact on *economic* sustainability. For instance, via reduction of processing time or via increased production efficiency with the help of XR technology to support decision-makers or operators to act more quickly [9, 10]. Second, 15 of the examined use cases are described by their authors as having a positive impact on *social* sustainability. Often as a result of

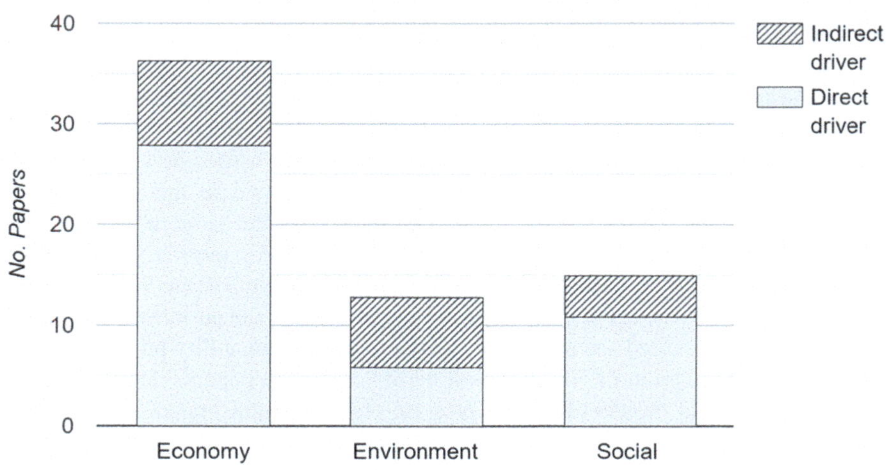


**Fig. 3** Graph showing the number of papers that identifies one or more of the NIST indicators of any of the three sustainability pillars

improved workers' health and safety or inclusion in the workforce. The examined use cases showed great examples of XR technology as a tool to support human-centric and ergonomic workplace design or as an aid for hearing-impaired workers to receive visual instructions and alerts [11, 12]. *Environmental* sustainability was the least referenced sustainability pillar in the examined literature and only 13 use cases were described as having a positive impact on environmental sustainability by its authors. Some examples of which included reduction of waste as a result of improved quality assurance, resource efficiency by virtual prototypes and design reviews, or by reduced travel via virtual shop floor meetings or optimized path planning using AR [13–17].

Furthermore, a clear distinction can be seen between the three sustainability pillars when it comes to the direct and indirect impact on the sustainability indicators. More than half of the use cases marked under the environmental sustainability pillar (7 out of 13) mention environmental indicators as indirectly impacted by the XR technology and not as a main focus or driver of the XR use case and publication. For instance, as described in the paper [17], AR technology was adopted with the aim to decrease the setup and maintenance time of machinery via remote support from a technician, however, the author also highlighted the reduced need for traveling and transportation as a positive consequence of the use case. Meanwhile, a clear majority of the use cases found under the social and economic sustainability pillar acknowledge its indicators as the direct consequence of the XR technology and the focus and driver of the publication, see Fig. 3.

## *3.2 Use Case Categorization*

The use cases described in the literature were categorized with a focus on the use and application of the XR technology as described by its authors using a deductive categorization approach. The categorization suggests cross-industry and generic use cases of XR that can be found in various areas of manufacturing, such as logistics, maintenance, or operations, etc. A total of seven unique categories of XR use cases were defined and are further presented as shown in Fig. 4 as well as detailed in the following paragraphs.

### Virtual Prototyping

*Virtual prototyping* uses XR as a co-creation testbed to involve customers and engineers to reduce real material for tests and prototypes. The paper [18] works on business-to-business interaction through VR for co-production, and the author addresses the potential of VR to include more parameters like energy consumption and carbon emissions with its high calculative power for sophisticated graphics. Customer involvement requires the VR demo to meet their unconscious requirement

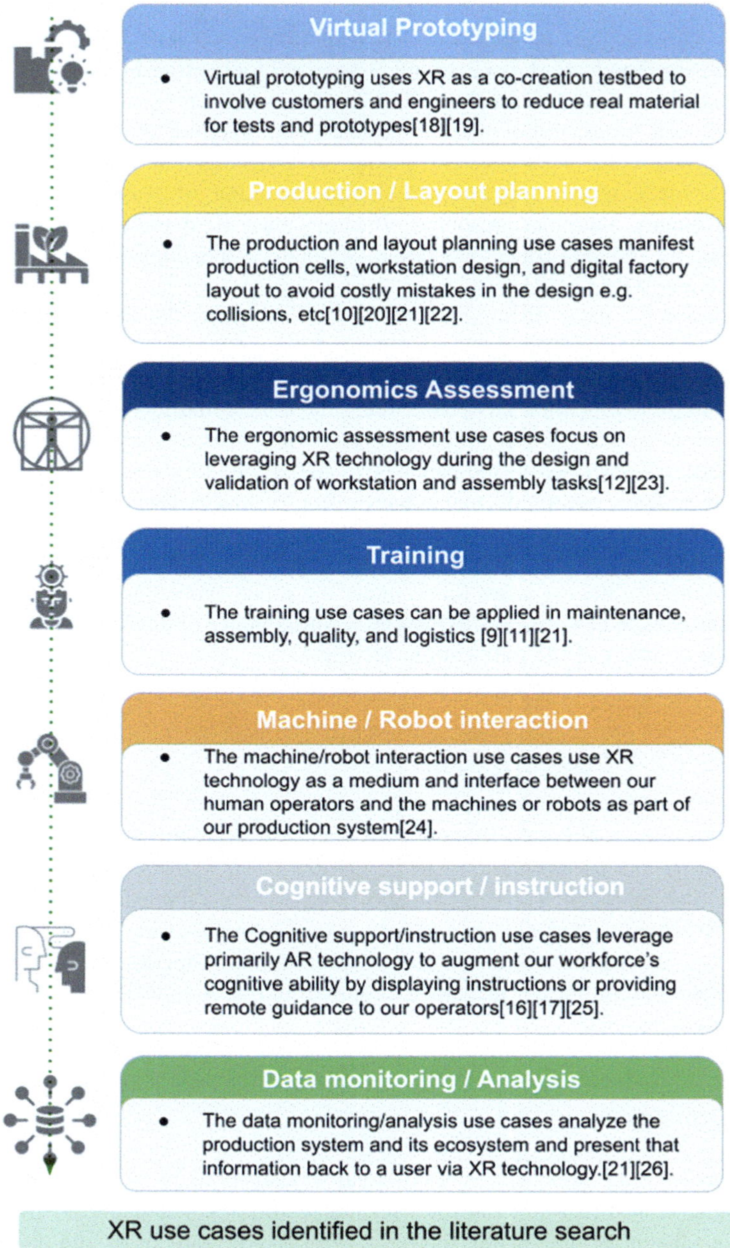


**Fig. 4** Summary of the seven XR use case categories identified in literature studies

like facial expressions to capture customers' quick attitudes which are mentioned in the paper [19], it shows the virtual prototype design strategy should be respondent to customer's needs, focusing on avoiding waste and optimizing the resources.

**Production and Layout Planning**

The *production and layout planning* use cases manifest production cells, workstation design, and digital factory layout to avoid costly mistakes in the design, e.g., collisions, etc. In the paper [10], the author clarifies by using Visual Component to visualize a production cell in VR would be able to verify the layout planning of the cell. Paper [20] tests the concept of using AR based on CAD software e to visualize production cells and workstations on top of a physical layout drawing. Paper [21] gives a clear design strategy on energy and material flow dynamic visualization, and paper [22] mentions measuring and disclosing accounting for virtual assets used in R&D activities could be another design strategy.

**Ergonomic Assessment**

The *ergonomic assessment* use cases focus on leveraging XR technology during the design and validation of workstation and assembly tasks. By virtually recreating a manual operation or workstation, engineers and researchers can leverage VR and AR technology to physically replicate the task in the virtual world to assess its ergonomic impact and grading. This is often with the aim of finding, and redesigning, tasks and stations that potentially pose a health and safety risk for our workforce. Often this is done in combination with a motion capture system, as seen in paper [12], or in combination with digital human modeling software and a digital manikin, such as Siemens Jack as presented in paper [23].

**Training**

The *training* use cases can be applied in maintenance, assembly, quality, and logistics [9]. Paper [9] shows the evidence that in their use case to compare training media through VR/AR and manual, which demonstrates the lowest operation speed with VR, followed by AR, which is better than manual instruction. Other publications, such as [11], utilize XR training to aid hearing-impaired workers in an assembly procedure. The training design strategy should be tested and iterated for inclusiveness, especially the elderly [9] for equal virtual environment tolerance. Another design strategy mentioned in the paper [21] points out the concern about raising environmentally related issues in training.

### Machine/Robot Interaction

The *machine/robot interaction* use cases use XR technology as a medium and interface between our human operators and the machines or robots as part of our production system. These use cases promote human–machine interaction, often by the projection of machine commands or interfaces toward the machine operator using an AR headset. As seen in paper [24] AR technology is used to visualize relevant diagnostic information and trajectories of machines and robots to its operator, thus acting as decision support with the aim to increase efficiency and workers' safety.

### Cognitive Support/Instruction

The *cognitive support/instruction* use cases leverage primarily AR technology to augment our workforce's cognitive ability by displaying instructions or providing remote guidance to our operators. These use cases often promote increased quality assurance, workforce inclusiveness and ramp-up time for new operators by providing on-the-job guidance and work instructions via an AR headset, for instance as described in paper [16, 25]. Furthermore, AR technology and cognitive support functions could be used to minimize maintenance time and travel by the possibility of remote assistance from a technician or manager via an AR application, as described in the paper [17].

### Data Monitoring/Analysis

The *data monitoring/analysis* use cases analyze the production system and its ecosystem and present that information back to a user via XR technology. Information on the well-being of the production system such as key measurements and KPI's can be displayed to an operator using and AR headset to support his/hers decision-making. As presented in the paper [21], digital twins of facilities and production systems, together with an AR headset, can be used to visualize energy flows and emissions of production facilities or systems. Another use case is presented in the paper [26], which showcases how a thermal camera mounted on an AR headset helps operators avoid potential hot and harmful surfaces to improve workers' safety, furthermore, this technology could also be used to identify heat or energy loses in the system as well as wear and tear on machines and equipment.

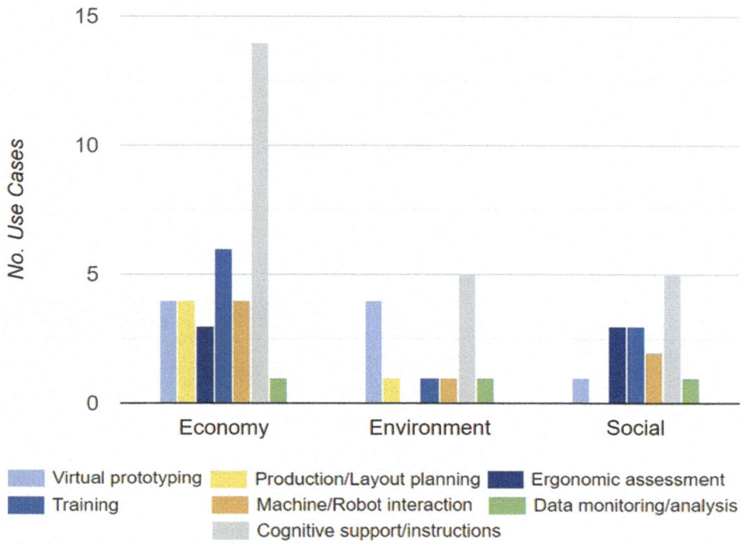


**Fig. 5** Graph showing the number of particular use cases addressed by the literature related to each sustainability dimension

## 3.3 Sustainability Coverage by Use Case

By mapping the use cases from the literature according to the categorization described above and the previously made categorization per sustainability pillar, clear trends can be seen on how the different categories of use cases impact different pillars of sustainability. Certain categories of use cases can be directly pinpointed and linked to having a potential positive environmental impact, which provides an answer to RQ1 as well as a first direction for RQ2 as posed in this chapter.

Figure 5 showcases that five use cases from the literature of Cognitive support/ instructions, as well as four use cases of Virtual prototyping, have reported a potentially positive impact on environmental sustainability.

## 3.4 Environmental Impact by Use Case

To better understand how the papers and use cases identified under the environmental sustainability pillar propose a potential positive contribution to the environment, an overview of the use cases and proposed environmental impacts was created. In Fig. 6, a summary of potential environmental impact per use case category can be seen with reference to relevant literature.

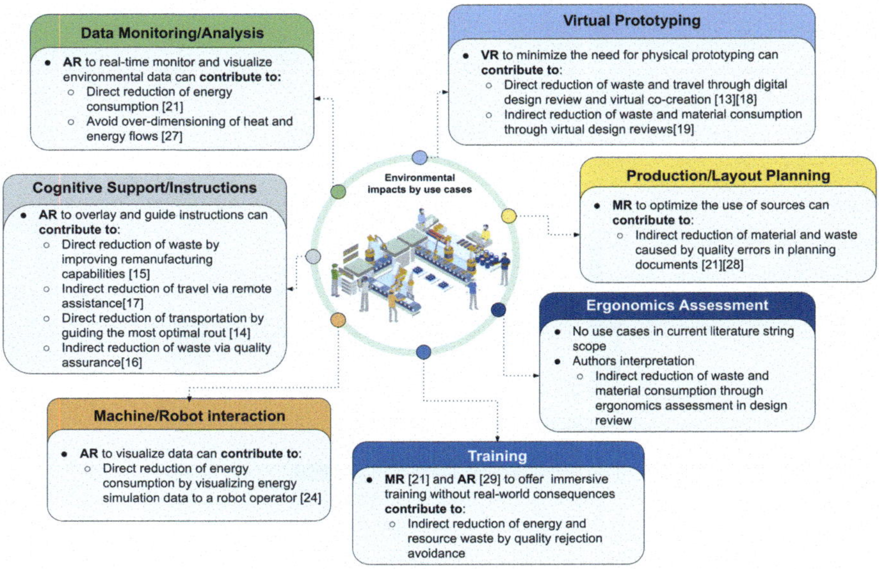


**Fig. 6** Illustration showing summary of environmental impacts collected from the use cases listed in Fig. 5

## 4 Discussion

**How has sustainability been addressed in research efforts on XR for manufacturing applications?**

The results of the literature study show that there are strong relationships between sustainability and XR technology in manufacturing that are being addressed in the literature. The identified XR use cases from Sect. 3 provide an overview of potential areas and applications that have documented benefits for sustainability, thus providing insights and answers to *RQ1* posed by this paper. However, this study shows that environmental sustainability is more seldomly the driver or direct indicator behind an implementation or adaptation of XR technology in manufacturing, when compared to the other sustainability pillars. Environmental sustainability is to a larger extent acknowledged as an indirect consequence of the implementation (if at all) rather than a driver. Meanwhile, the results show that the majority of examined literature addresses the relationship between economic sustainability and XR in Manufacturing. Indicators and KPIs reporting on financial targets and economic values have traditionally steered the manufacturing sector [6], thus the findings in Sect. 3 are to be expected. However, by promoting clear use cases and publications addressing the environmental impact of XR technology within manufacturing, the authors of this paper aim to challenge the view of financial KPIs as the driver of XR technology. We have shown that XR technology can have a positive impact on environmental sustainability and could be seen as a valid tool or aid to lower the environmental impact in a manufacturing system. By using referenced publications and use cases as inspiration, companies and researchers could find ways to keep developing and leveraging XR technology for their specific use case toward more sustainable manufacturing practices.

**What environmental benefits have been obtained in use cases of XR in manufacturing?**

As shown in Fig. 6, use cases underscore XR's invaluable role in intertwining technological advancements with environmental sustainability goals.

*Virtual Prototyping:*

To minimize resource consumption, waste, and unnecessary travel from physical prototyping through VR cases [13, 18, 19].

*Production and Layout Planning:*

To optimize, ensure efficient use of resources, and reduce environmental footprints through MR cases [21, 28]. This also reduces costly errors that might have environmental repercussions.

*Ergonomic Assessment:*

To assess ergonomic factors, ensuring healthy and productive environments through MR cases [30] and AR cases [23]. A sustainable workspace isn't just about eco-friendliness; it's about the well-being of workers.

*Training:*

To offer immersive training without real-world consequences. This reduces resource use and waste, aligning with sustainable and pedagogical practices while emphasizing eco-friendly methodologies through MR [21] and AR [29].

*Machine/Robot Interaction:*

To facilitate harmonious human-machine interactions by visualizing energy simulation data, optimizing energy and material usage through AR [24].

*Cognitive Support/Instruction:*

To overlay and guide instructions to ensure tasks are executed efficiently and correctly the first time to minimize waste via AR [14–17].

*Data Monitoring/Analysis:*

To real-time monitor and analyze environmental data and enable timely decisions that prioritize the planet's well-being, helping industries pivot towards greener processes via AR [21, 27].

In summary, XR acts as a potential bridge to balance between human endeavors and environmental preservation. However, as interpreted from Fig. 6, AR and MR seem to be more impactful than VR for environmental sustainability from the use cases identified in the research. AR and MR can guide on-site and real-time data that could more directly raise awareness of the environmental impact by overlaying critical digital data onto the actual production environment. From the use cases identified by the research string, AR and MR show more potential to take environmental sustainability as a main driver and contribute to environmental sustainability with both quantitative and qualitative impact.

## *4.1 Limitations*

This study addressed papers clearly acknowledging indicators categorized under the environmental sustainability pillar, as defined by NIST [6], use cases and papers with indirect effects on environmental sustainability that were not acknowledged by its original author have not been take into consideration. Furthermore, the delimitation of the search string and criteria used for this exploration also affects the number of publications and use cases addressing environmental sustainability found by the authors. The search string used for this study emphasized the use of XR terminology in the title of the examined literature. By expanding this search to also include papers mentioning XR terminology in the abstract, the numbers of papers would increase to 422 before exclusion and abstract screening. However, this would also introduce a majority of papers with non-concrete XR use cases and applications; hence, the decision was made to keep the XR terminology in the title as a requirement for the search. By looking at other industries and sectors such as e.g., agriculture, construction, or mining, other use cases might have been found that could have been relevant for cross-sectoral learnings for manufacturing companies. Likewise, expanding the search string to include more synonyms of environmental sustainability could have presented us with other use cases of relevance. Therefore, the referenced publications should be considered as samples of use cases that have documented a positive impact on environmental sustainability.

## *4.2 Future Research Directions*

The future research based on the study could be the qualitative or quantitative assessment of the impact of each use case and thus provide suggestions for the companies to prioritize the use cases for impact on sustainable manufacturing. Another parameter interesting to extract from the literature is the technical readiness level (TRL), which has a feasibility trend mapping for different use cases.

For the theoretical part, the limitation analysis of XR and different use cases are needed to see the research gap in the literature study on environmental sustainability issues, and it will help us to answer the important question: why is environmental sustainability receiving less attention in the development of XR from manufacturing? Thus, we could list the following research questions that would be potential for further research on this topic:

*RQ1*: How can XR support various indicators of environmental sustainability such as energy flow, material flow, noise and water pollution, etc.?
*RQ2*: What are the current gaps in the research between XR technology and environmentally sustainable manufacturing?
*RQ3*: What are the potential negative impacts that XR brings to environmental sustainability?

# 5 Conclusion

This chapter shows that XR technology has the potential to drive a positive impact on environmental sustainability in a manufacturing system. Thus, the objective of the paper to highlight and eliminate the gap between XR and the direct/indirect relationship with environmental sustainability has been met. To highlight the potential XR could bring to environmental sustainability and to act as inspiration for industry and researchers to continue to work together for developing XR applications for a sustainable future, this chapter presents a summary and examines seven categories of potential use case to adopt, from the identified use cases of the literature study.

## *5.1 Contribution to Theory*

The seven XR use case categories are concluded from the review to provide a visual representation of the XR development for sustainable manufacturing. It also details pointing out the direct/indirect impact of XR use cases on environmental sustainability, and thus a potential research direction to raise awareness of environmental sustainability with XR is identified.

## *5.2 Contribution to Practice*

The cross-analysis of the use cases to sustainability and seven categories could provide companies with toolboxes, cases and design strategies when a new case study is identified. The research also claims that the XR use cases should take sustainability principles into consideration, and corresponding KPI and evaluation should be followed to direct XR for sustainable manufacturing.

**Acknowledgments** The authors would like to thank the Swedish innovation agency Vinnova for their funding of the PLENUMproject, grant number: 2022–01704. The work was carried out within Chalmers' Area of Advance Production. The support is gratefully acknowledged.

## References

1. Ren S, Zhang Y, Liu Y, Sakao T, Huisingh D, Almeida CM (2019) A comprehensive review of big data analytics throughout product lifecycle to support sustainable smart manufacturing: a framework, challenges and future research directions. J Clean Prod 210:1343
2. Beier G, Fritzsche K, Kunkel S, Matthess M, Niehoff S, Reißig M, van Zyl-Bulitta V (2020) A green digitalized economy? Challenges and opportunities for sustainability. IASS Fact Sheet 1
3. Fast-Berglund Å, Gong L, Li D (2018) Testing and validating extended reality (XR) technologies in manufacturing. In: Proceedings of the 8th Swedish Production Symposium
4. Leng J, Sha W, Wang B, Zheng P, Zhuang C, Liu Q, Wuest T, Mourtzis D, Wang L (2022) Industry 5.0: Prospect and retrospect. J Manuf Syst 65:279
5. E. Commission (2022) What is industry 5.0? Available at https://research-and-innovation.ec.europa.eu/research-area/industrial-research-and-innovation/industry-50_en. Accessed 26 Apr 2023
6. Joung CB, Carrell J, Sarkar P, Feng SC (2013) Categorization of indicators for sustainable manufacturing. Ecol Indic 24:148
7. NIST (2021) Sustainable manufacturing indicator repository (SMIR), Available at https://www.nist.gov/services-resources/software/sustainable-manufacturing-indicator-repository-smir. Accessed 26 Apr 2023
8. Kim DB, Leong S, Chen C-S (2012) An overview of sustainability indicators and metrics for discrete part manufacturing, vol 2
9. Gabajová G, Furmannová B, Medvecká I, Grzor P, Krajcovic M, Furmann R (2019) Virtual training application by use of augmented and virtual reality under university technology enhanced learning in Slovakia. Sustain For 11
10. Törökováa M, Török J (2021) Simulation of a production workplace using virtual reality. TEM J:1390
11. Araujo T, Oliveira F, Costa C (2019) Creating job opportunities in computer assembly line for people with disabilities through augmented reality. In: IEEE Frontiers in Education Conference
12. Dreesbach T, Mertens A, Hellig T, Pretzlaff M, Nitsch V, Brandl C (2020) Mixed mock-up meets ergocam: feasibility study for prospective ergonomic evaluation of manual assembly processes in real-time using augmented reality and marker less posture analysis. Lect Notes Comput Sci
13. Chen X, Gong L, Berce A, Johansson B, Despeisse M (2021) Implications of virtual reality on environmental sustainability in manufacturing industry: a case study. In: 54th CIRP Conference on Manufacturing Systems
14. Sidiropoulos V, Bechtsis D, Vlachos D (2021) An augmented reality symbiosis software tool for sustainable logistics activities. Sustain For 13
15. Quesada Diaz R, Syberfeldt A (2022) Optimised shopfloor operator support on an integrated AR and PLM framework for remanufacturing. In: Swedish Production Symposium
16. Alves J, Marques B, Dias P, Santos B (2021) Using augmented reality for industrial quality assurance: a shop floor user study. Int J Adv Manuf Technol 115:105
17. Lund G, Ortova M (2022) A spontaneous adoption of AR technology a manufacturing industrial context. IFIP Advances in Information and Communication Technology

18. Royo-Vela M, Leszczyński G, Velasquez Serrano M (2022) Sustainable value co-production and co-creation in virtual reality: an exploratory research on business-to-business interactions. Sustainability 14:7754
19. Violante M, Marcolin F, Vezzetti E, Nonis F, Moos S (2019) Emotional design and virtual reality in product lifecycle management (PLM). In: Smart Innovation, Systems and Technologies
20. Husár J, Knapciková L, Hrehova S, Balog M (2022) The design of workplaces with augmented reality in engineering education. In: Lecture Notes in Mechanical Engineering
21. Thiede S, Damgrave R, Lutters E (2022) Mixed reality towards environmentally sustainable manufacturing - overview, barriers and design recommendations. In: 29th CIRP Life Cycle Engineering Conference
22. Abdelkhalik R, Elngar A (2020) Costs management of research and development in the factories of the future using virtual reality. In: International Conference on Advances in Computing and Communication Engineering
23. Hovanec M, Pacaiová H, Hrozek F, Varga M (2014) Proactive ergonomics based on digitalization using 3d scanning and workplace modeling in texnomatix jack with augmented reality. Nase More
24. Schmidt B, Sánchez De Ocãna Torroba A, Grahn G, Karlsson I, Ng AH (2022) Augmented reality approach for a user interface in a robotic production system. In: Swedish Production Symposium
25. Kontovourkis O, Georgiou C, Stroumpoulis A, Kounnis C, Dionyses C, Bagdati S (2019) Implementing augmented reality for the holographic assembly of a modular shading device. In: Education and Research in Computer Aided Architectural Design in Europe and XXIII Iberoamerican Society of Digital Graphics Conference
26. Damiani L, Revetria R, Morra E (2020) Safety in industry 4.0: the multi-purpose applications of augmented reality in digital factories. In: Advances in Science, Technology and Engineering Systems
27. Fukuda T, Yokoi K, Yabuki N, Motamedi A (2019) An indoor thermal environment design system for renovation using augmented reality. J Comput Des Eng 6:179
28. Muñoz A, Mahiques X, Solanes JE, Martí A, Gracia L, Tornero J (2019) Mixed reality-based user interface for quality control inspection of car body surfaces. J Manuf Syst 53:75
29. Palmarini R, Erkoyuncu JA, Roy R, Torabmostaedi H (2018) A systematic review of augmented reality applications in maintenance. Robot Comput Integr Manuf 49:215
30. Peruzzini M, Grandi F, Pellicciari M, Campanella CE (2018) A mixed-reality digital set-up to support design for serviceability. Proc Manuf 17:499

# Part VI
# Sustainable Manufacturing

# Exploring Factory Data for Resource Efficiency Assessment: A Case Study at a Truck Manufacturing Company

**Qi Fang, Mélanie Despeisse, Ebru Turanoglu Bekar, Lena Moestam, Helena Söderberg, Dennis Andersson, and Björn Johansson**

**Abstract** In the transition to circular economy, resource efficiency (RE) is at the intersection of environmental competitiveness and economic growth. However, studies found that industrial companies are struggling to use RE methods in daily operations as part of performance management beyond demonstrations in research projects. Both academia and industry need a better understanding about RE solutions in terms of their usability: production systems must be able to supply and manage good quality data. To close the knowledge gap, this chapter presents a feasibility study with a truck manufacturer focusing on processes to manufacture metal frame beam vehicle chassis. The feasibility study started both qualitatively and quantitatively with process mapping, material flow modelling, and energy efficiency analysis. The chapter presents findings from exploring and discussing data availability and quality to meet essential requirements for RE assessments. Besides proposing improvement actions to the case company, the chapter also recommends manufacturers perform a factory data evaluation before gathering additional data and working on specific RE indicators. Future collaborative studies with more manufacturers will help develop generalizable guidelines for factory data evaluation to support resource efficiency studies and sustainable manufacturing.

**Keywords** Resource efficiency assessment · Factory data · Sustainable production systems · Sustainable manufacturing

Q. Fang (✉) · M. Despeisse · E. T. Bekar · B. Johansson
Department of Industrial and Materials Science, Chalmers University of Technology, Gothenburg, Sweden
e-mail: qifa@chalmers.se

L. Moestam · H. Söderberg · D. Andersson
Volvo Group Trucks Operations, Gothenburg, Sweden

 
S. Fukushige et al. (eds.), *EcoDesign for Circular Value Creation: Volume I*,
https://doi.org/10.1007/978-981-97-9068-5_33

# 1 Introduction

The concept of resource efficiency (RE) originates in eco-efficiency and acts as an indicator of the circular economy (CE) at the micro level [1, 2]. It also becomes an intersection of environmental and economic competence in achieving CE [3]. Therefore, many manufacturers are focusing on minimising resource consumption while maintaining high quality and productivity levels.

A systematic review of methods and technologies regarding RE, such as environmental impact assessment, monitoring, and improvement methods, highlighted the potential of digital simulation techniques [4]. This requires the production systems to provide quality data for different layers, from the manufacturing process, multi-machine system, and facility to the supply chain.

However, supporting RE assessment with quality data can be challenging for manufacturing companies running complicated production systems. In a study about challenges in data life cycle management, M. Despeisse and E. T. Bekar summarised that the common risk is being trapped by inconsistencies and incompatibilities without standardised system boundaries and integrated feedback loops in the planning stage [5]. They also concluded in the same paper that though RE assessment methodologies have been developed and demonstrated over the years, industrial practitioners rarely use these methodologies in the context of daily operations.

Hence, both academia and industry need to better understand the gaps between methodologies and applications in practice. Between 2022 and 2023, a case study was conducted together with a truck manufacturer in Sweden that is actively working on RE improvement. The truck frame beam manufacturing area was selected as a feasibility study in this chapter.

The primary objective of this feasibility study is to identify the data requirements for conducting RE assessments in a manufacturing company. By addressing this research objective, the study aims to identify hidden gaps that may arise during the implementation of basic RE assessment methods in the case company. The study also seeks to develop initial directions for future scale-up implications.

# 2 Methodology

The method (Fig. 1) of the feasibility study was developed from a factory-level prototype methodology for identifying improvements developed by Despeisse et al. [6] in 2012. This methodological framework served as a foundation and was adapted to suit the specific requirements of the feasibility study.

**Step1: Manufacturing System Understanding**

Throughout the feasibility study, a series of Gemba walks were conducted, wherein specialists from the production line provided guidance and explanations. These walks aimed to provide researchers with insights into the processes, resource utilisation, and waste management at each station.

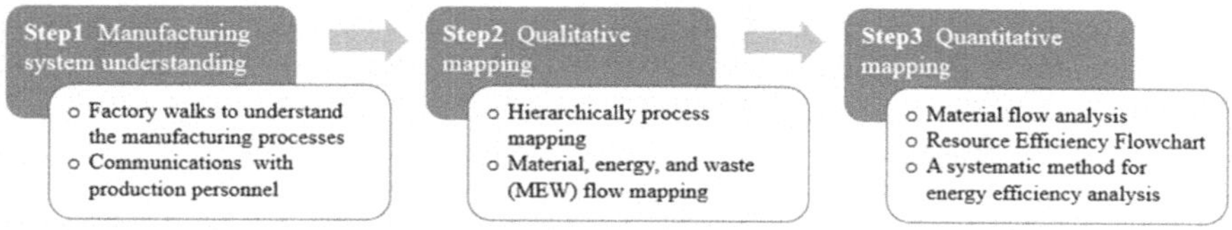


**Fig. 1** Three steps developed for the feasibility study

**Table 1** Classification approach of data in input and output elements

| | | |
|---|---|---|
| Category A | Collected | Data in this category are already collected for assessments in the operation |
| Category B | Available but not collected | Data are available from equipment or other sources but require extra efforts to be collected for assessments |
| Category C | Not available | Data are not available in the current production system; Category C requires further investigation or proper estimation in assessments |

The participation of individuals from diverse hierarchies within the factory, including line leaders, area managers, and factory sustainability managers, was integral to this step. By involving personnel at various levels, a comprehensive understanding of the operational intricacies could be attained.

Furthermore, maintaining regular communication and updates with personnel across the organisation was deemed imperative to ensure the ongoing accuracy and relevance of the data and information inputs throughout the subsequent stages.

**Step2: Qualitative Mapping**

Shifting to qualitative studies, two process mapping tools were chosen to help get an in-depth understanding of the existing factory data and to refine the selection of feasible RE assessment methods.

The IDEF0 methodology has been widely used in designing and analysing structured systems for its effectiveness since the last century [7]. In this study, the IDEF0 framework and its functional units modelling principle were chosen to illustrate hierarchies of current manufacturing processes. On top of a basic decomposition view, associated data for each input and output element was classified into three categories, as outlined in Table 1. This classification approach drew inspiration from a method for simulation data categorisation developed by Robinson and V. Bhatia [8]. Adaption of a proper classification approach will enable a better understanding of data availability and collection, serving as the cornerstone of production system performance assessments [9]. Furthermore, the integration of distinct colours within the process mapping aided in clear visualisation.

The Spaghetti Diagram, a well-known tool from Kaizen Solutions, is commonly used during the initial stages of optimisation planning [10]. In this feasibility study, the Spaghetti Chart was utilised to visualise the material, energy, water, chemicals, and waste flows (thereafter, resource flows), incorporating relevant production layout information.

The visual representation allowed for a comprehensive understanding of the existing resource flows and provided a foundation for identifying areas of improvement and optimisation.

**Step3: Quantitative Mapping**

To conduct a quantitative study on manufacturing processes, Material Flow Analysis (MFA) was adapted to model a balanced material flow, considering inputs, outputs, and inventory [11]. MFA can be applied for each layer, ranging from multi-factory to individual process analysis. In the case study, the MFA approach was used to examine the resource flows within the beam manufacturing process.

Another method selected in the feasibility study is the Resource Efficiency Flowchart (REF) developed by Thomas Lundholm et al. [12]. The REF method provides a cost-based tool that enhances transparency in RE losses analysis. In addition to resource flows, the REF tool incorporates losses associated with human resources and equipment efficiency, contributing to a comprehensive cost calculation.

Considering the increasing interest in energy consumption within the manufacturing sector, a systematic guided method was selected for energy efficiency analysis in the feasibility study [13]. The method involved the development of an energy portfolio tool, enabling the classification of energy consumers and subsequent prioritisation of optimisation efforts. Energy consumption and work sequence are two essential variables required in modelling the load profile of production equipment. The sampling rate [14] of energy consumption plays a key role in determining the resolution of the measurement. A low resolution of energy consumption measurement can limit the identification of improvement potentials.

Although the three quantitative methods chosen for the feasibility study are fundamental, they provide a straightforward starting point for assessing the RE performance of manufacturing processes in the case company.

# 3 Results

## *3.1 Results of Qualitative Process Mapping*

**Hierarchical Process Mapping of the Truck Manufacturing Company**

Figures 2 and 3 depict a hierarchical process mapping of the truck manufacturing company, showcasing three distinct layers: factory, production area, and beam manufacturing process. Within each functional unit, activities, constraints, and mechanisms are illustrated to highlight their essential functions. The colour-coded input and output arrows featured in the figures correspond to the data categories outlined in Table 1.

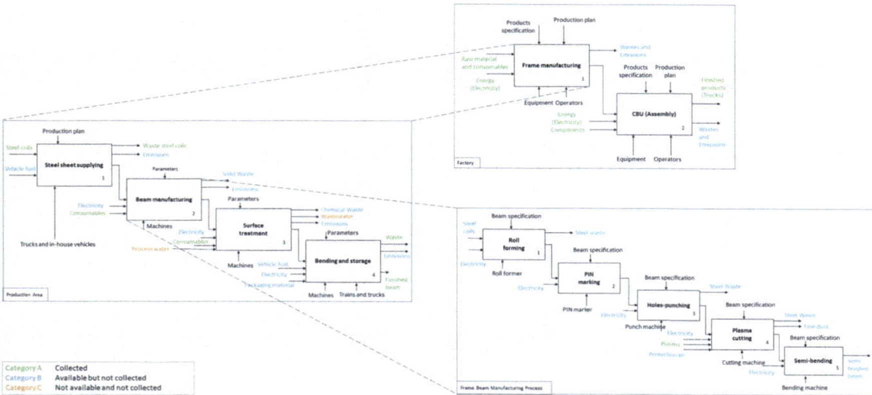


**Fig. 2** Hierarchical process mapping of factory, production area, and beam manufacturing process

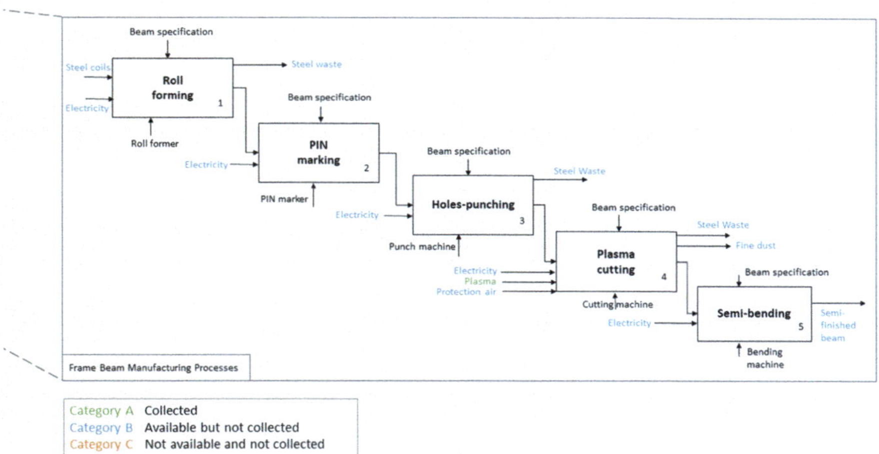


**Fig. 3** A detailed process mapping of frame beam manufacturing process

The hierarchical structure allows for a clear understanding of the relationships between different units and their respective functions.

**Resource Flow Mapping of Beam Manufacturing Area**

Figure 4 shows a Spaghetti diagram to map the resource flows specifically in the frame beam production area. The focus is on the flow of metal material, which serves as the main product. Additionally, other materials are observed to join the process at different stations, while waste is generated at the same time. The mapping also includes the depiction of electricity supply and ventilation systems based on the actual layout of production facilities.

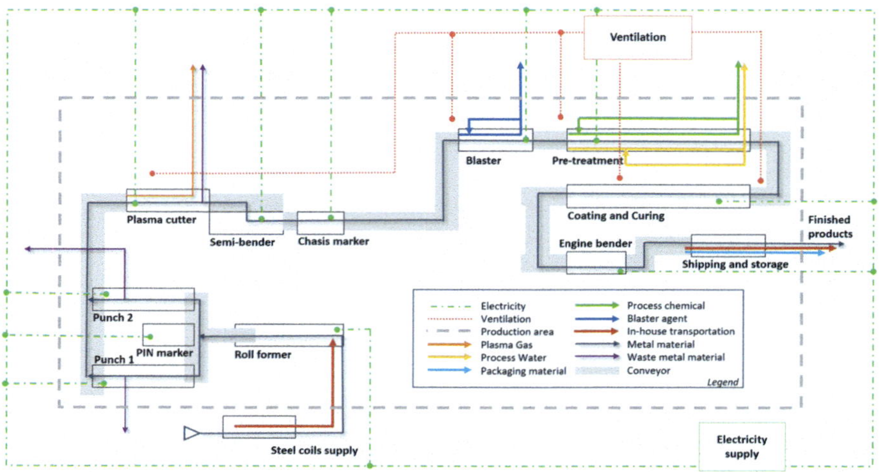


**Fig. 4** Resources flow mapping of frame beam manufacturing area

**Table 2** Overview of data measurement in the frame beam manufacturing process

| Resource Flow / Sampling Rate | Material | | | Energy | Waste | | |
|---|---|---|---|---|---|---|---|
| | Metal consumption Category A | Consumable consumption Category A | Water consumption Category C | Electricity Category A | Solid Waste Category B | Consumable waste Category A | Wastewater Category C |
| Every 10s | | | | X | | | |
| Daily | X | | | O | | | |
| Weekly | O | | | O | | | |
| Monthly | O | X | | O | | X | |
| Annual | O | O | | O | | O | |
| X – data are sampling at the corresponding rate | | | | | | | |
| O – data are available by accumulation | | | | | | | |

The resource flow mapping provides a comprehensive overview of the materials involved, waste generation points, and the infrastructure supporting the production processes.

## 3.2 *Results of Quantitative Mapping*

Table 2 provides an overview of the required data, including information related to resource flow data, sampling rate, and data availability. The table outlines the data required by MFA, REF, and energy analysis methods. Each column represents a specific resource flow data, and rows indicate the corresponding sampling rate.

For instance, focusing on metal material, data is collected daily to track metal consumption for the frame beam manufacturing process. Furthermore, the accumulated metal consumption is available for weekly, monthly, and annual calculations, providing a broader perspective on metal consumption over longer time periods.

The implementation of the MFA and REF methods was not fully realised in the feasibility study due to incomplete data regarding solid waste (in category B). For the REF method specifically, it requires the resource flow data to align with production to analyse losses, necessitating a consistent and reliable sampling rate [12].

Regarding the analysis of electricity consumption, the case company had measuring equipment (switchgears) in place for the production area with a sampling rate of every 10 s. This allowed for the allocation of electricity consumption to multi-equipment groups. However, despite having a good measurement resolution, the establishment of a load profile and energy portfolio was limited by incomplete historical electricity consumption records (attributable to breakdowns in measuring switchgears) and unsynchronised time stamps among individual equipment.

These limitations in resource flow data impacted the implementation of MFA, REF, and electricity consumption analysis in this feasibility study. Further discussion and interpretation will underscore the importance of having accurate and adequate data in category A for conducting quantitative assessments.

# 4 Discussion and Interpretation

In this case study conducted at the truck manufacturing company, both qualitative and quantitative study methods have revealed that the available and collected data (in categories A and B) within the factory are not adequately prepared for implementing RE assessments across different operational levels. The chapter delves into a deeper analysis of the identified limitations and offers interpretations regarding future improvements.

## *4.1 Challenges in Data Completeness and Reliability*

A conceptual data quality framework, developed by Levitin and Redman in 1995, provided an approach for analysing data quality. The framework suggests five dimensions that are crucial in data quality analysis: contents, scope, level of detail, composition, and consistency [15]. Additionally, in the context of production data management, two common constraints in system modelling and assessment are data scarcity and reliability issues (Data scarcity refers to the limited availability of data and reliability issues encompass the trustworthiness and accuracy of the data itself) [5].

**Data Scarcity Hinders Full-Scale RE Assessments**

Based on the mapping in Figs. 2 and 3, an observation is the distribution of data categories across different hierarchies. Figure 5 illustrates that category A (available) at the highest hierarchy holds the largest share. This indicated that at the factory level, the relatively higher availability of data has enabled the implementation of basic RE calculations.

As the analysis delves into more detailed layers in the production site, the percentage of data in Category B (available but collected) and Category C (not available) increases. This poses challenges for performing RE assessments at these hierarchies.

This pattern highlights the need for improving data collection practices, particularly at the production area and manufacturing process levels. However, to ensure the effectiveness of RE assessments, it is necessary to align the data collection with the purpose of the RE assessment methods during the initial stage. Manufacturing companies can prioritise the acquisition of data that is most relevant to the specific assessment being conducted.

**Data Reliability Hinders RE Assessments**

Whilst gaps in data completeness were found at a very early stage by qualitative study, gaps in data reliability required closer interpretations of collected resource flow data. Table 3 shows some issues encountered during the quantitative study, specifically in the areas of electricity consumption and waste management data.

**Unstandardised Data Collection Templates Hinder RE Assessments**

In addition to the defects found in data characteristics such as timeliness, granularity, and correctness, standardised data acquisition also poses challenges in the feasibility study. Particularly in this feasibility study, manual obtaining data on operation activities and material flows has limited the standardisation of data collection. It makes it difficult to compare and combine data from different resource flows, with various units, or across different time periods.

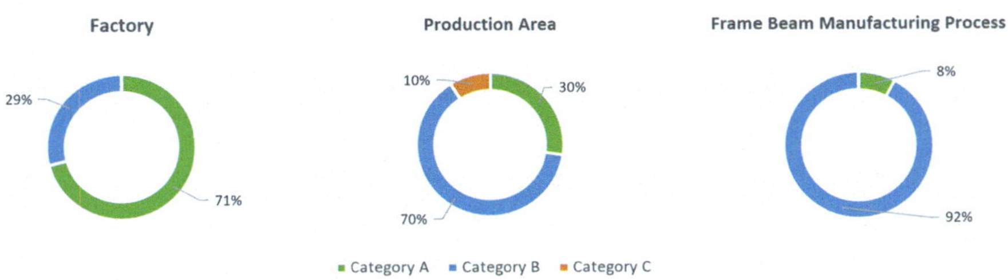


**Fig. 5** Distribution of three data categories at factory, production area, and process levels

**Table 3** Examples of limitations in resource flow data reliability

| Limitations found in in resource flow data from quantitative study |
|---|
| Waste management: Lagging update from external waste management service supplier |
| Waste management: Generating location information has not been included in waste data yet |
| Electricity energy: Timestamps on different equipment were not synchronized to be consistent |

Addressing the standardisation of data templates is crucial in improving the efficiency of quantitative analysis in real-life operations [16, 17]. Implementing standardised data templates can streamline data management processes, enhance data quality, and enable more robust and reliable quantitative analysis.

## 4.2 *Improvement in Effectiveness and Efficiency of RE Assessments*

**Action Plan for the Case Company**

The qualitative and qualitative studies have provided insights and direction for the next steps in improving data quality in the truck manufacturing company. To ensure practicality and effectiveness, an action plan has been proposed to address issues identified in factory data in a systematic and sustainable manner.

The action plan recognises the limitations of addressing all potentials simultaneously in a business environment. Therefore, it outlines both immediate actions and long-term suggestions to gradually improve the existing factory data.

For instance, to address the limitations in electricity consumption data, the immediate action plan suggested calibrating the timestamps among manufacturing equipment and maintaining the proper working condition of measurement gears. In the long run, the manufacturing company planned to work on further allocations of electricity consumption to the machine level, allowing for a more precise classification of energy consumers. For water and waste data, water meters were installed to collect data manually for the entire production area. Additionally, a Kaizen team is also working on breaking down waste data to the process level.

Those immediate and long-term actions demonstrate the industrial partner's commitment to efficiently complete the next iteration of RE assessments in a few months with minimum resources. Furthermore, the case study company recognises the continuous improvement for the potential introduction of new RE assessment methods.

**Research Plan for the Next Steps**

The research project continues with future research activities that incorporate insights from both researchers and industrial partners based on the findings of the feasibility study.

In terms of methodology, the hierarchical process mapping will be selectively used based on the availability of data patterns derived from Gemba walks. Integrating data sampling points into the resource flow mapping will provide more comprehensive information. Digital technologies will serve as a crucial enabler in improving data from production systems for quantitative studies. The quantitative mapping tools used in the third step will be extended to cover more advanced RE assessment tools in future implementations.

Furthermore, future implications are planned across four stages: feasibility study, pilot study, scale-up study, and transplanting. Building upon the immediate actions proposed in this feasibility study, the same case study company will be able to complete a pilot study of RE assessments for the beam manufacturing area and gradually scale up to other production areas. To facilitate future transplanting to other manufacturers, it will be necessary to integrate and standardise the work steps developed from this feasibility study. Collaboration among companies from different manufacturing sectors will be encouraged to explore and leverage factory data for RE assessments.

## 5 Conclusion

The feasibility study conducted in the truck manufacturing company offered valuable insights into the challenges faced by manufacturers when systematically assessing RE and maximising the utilisation of available data. The study highlighted data scarcity and reliability issues as key areas for improvement. By focusing on the data required for selected RE assessment methodologies, the manufacturer can progress towards RE performance management, leading to enhanced environmental and economic competencies.

The manufacturing area chosen for the feasibility study, which was frame beam manufacturing, served as a representative case of a resource-intensive process. Regardless of the specific manufacturing type, ensuring adequate data quality is crucial for embedding RE assessments into production management routines.

The study also recommends that manufacturers conduct similar feasibility studies to evaluate their factory data before collecting additional data and selecting specific RE indicators. Deliverables of the feasibility study serve as a foundation for iterative implementations in the case company and in other industrial practitioners.

Collaborative studies involving diverse manufacturers will further contribute to the development of generalisable guidelines for factory data management. By following the strategic (feasibility study, pilot study, scale-up study, and transplanting) roadmap, the research aims to advance RE assessment methods and foster collaboration among manufacturers for the adoption of sustainable practices.

**Acknowledgements** This work was supported by the Swedish innovation agency Vinnova and the strategic innovation programme Produktion2030 projects FREED under grant no. 2022-02460, and VIVACE under grant no. 2023-00868. The work was conducted within Chalmers' Area of Advance Production. The support is gratefully acknowledged.

## References


1. Ghisellini P, Cialani C, Ulgiati S (2016) A review on circular economy: the expected transition to a balanced interplay of environmental and economic systems. J Clean Prod 114:11–32
2. Kristensen HS, Mosgaard MA (2020) A review of micro level indicators for a circular economy—moving away from the three dimensions of sustainability? J Clean Prod 243. Elsevier Ltd
3. Reike D, Vermeulen WJV, Witjes S (2018) The circular economy: new or refurbished as CE 3.0? — exploring controversies in the conceptualization of the circular economy through a focus on history and resource value retention options. Resour Conserv Recycl 135:246–264
4. Duflou JR et al (2012) Towards energy and resource efficient manufacturing: a processes and systems approach. CIRP Ann Manuf Technol 61(2):587–609
5. Despeisse M, Bekar ET (2020) Challenges in data life cycle management for sustainable cyber-physical production systems. In: IFIP advances in information and communication technology. Springer, pp 57–65
6. Despeisse M, Ball PD, Evans S, Levers A (2012) Industrial ecology at factory level: a prototype methodology. Proc Inst Mech Eng B J Eng Manuf 226(10):1648–1664
7. Presley A, Liles DH (1998) The use of IDEF0 for the design and specification of methodologies
8. Robinson S, Bhatia V (1995) Secrets of successful simulation projects. SIMTEGR8-Simulation to evaluate healthcare View project Linking simulation and AI View project
9. Institute of Electrical and Electronics Engineers, Simulation Conference, 2008
10. Hys K, Domagała A (2018) Application of spaghetti chart for production process streamlining. Case study
11. Daniels PL (2002) Approaches for quantifying the metabolism of physical economies: a comparative survey. Part II: review of individual approaches. J Ind Ecol 6(1):65–88
12. Lundholm T, Lieder M, Rumpel G (2012) Resource efficiency assessment system
13. Thiede S, Bogdanski G, Herrmann C (2012) A systematic method for increasing the energy and resource efficiency in manufacturing companies. In: Procedia CIRP, vol 2. Elsevier B.V, pp 28–33
14. Herrmann C, Kara S (n.d.) Sustainable production, life cycle engineering and management series editors
15. Levitin A, Redman T (1995) Quality dimensions of a conceptual view
16. Ćwikła G (2014) Methods of manufacturing data acquisition for production management—a review. Adv Mater Res 837:618–623
17. Institute of Electrical and Electronics Engineers, 2018 Winter Simulation Conference (WSC)

# An Analysis Method Considering Manufacturing Skills and Technologies Toward Slow Manufacturing

**Yuta Irino, Hidenori Murata, Hideki Kobayashi, and Giampaolo Campana**

**Abstract** Continuous productivity improvement is an essential target in the manufacturing industry, and two key approaches to improving labor productivity are reducing costs and increasing value added. This study focuses on increasing value added by improving labor productivity, specifically through craftsmanship. Craftsmanship is based on slow manufacturing and contributes to the well-being of workers and sustainability. The study proposes a method for analyzing the pattern of importance of customer requirements by considering a combination of manufacturing skills and automation technologies. The method consists of three steps: introducing the Skill–Technology Index, making a quality function deployment matrix, and calculating and analyzing the importance pattern of customer requirements. The proposed method was applied to a glass manufacturer, and the calculated result was not significantly different from the pattern of the current importance of customer requirements set by the company. In future work, the proposed method will be improved to find the appropriate balance of craftsmanship and automation technology.

**Keywords** Value-added process · Craftsmanship · Sustainable manufacturing · Slow manufacturing

## 1 Introduction

In the manufacturing industry, continuous productivity improvement is an essential target. Both reducing costs and increasing value-added are key to improving labor productivity. Cost reductions can be achieved by applying the "design for manufacturing" concept and digitizing and automating manufacturing processes.

Y. Irino (✉) · H. Murata · H. Kobayashi
Osaka University, Osaka, Japan
e-mail: irino@ssd.mech.eng.osaka-u.ac.jp

G. Campana
University of Bologna, Bologna, Italy

 
S. Fukushige et al. (eds.), *EcoDesign for Circular Value Creation: Volume I*,
https://doi.org/10.1007/978-981-97-9068-5_34

Since the advent of Industry 4.0, a German national project that began in 2011, research and practical experiments have been conducted on smart manufacturing using digital twin and cyber-physical production systems [1, 2]. Meanwhile, various efforts have been made to increase the added value. In this study, we refer to craftsmanship as "manufacturing skills."

In the manufacturing industry, craftsmanship is a key competency for building the overall manufacturing process. For example, several Italian entrepreneurs operating in specific industrial districts (mainly luxury brands) rely on skilled artisans whose expertise contributes to manufacturing attractive high-quality products, even though automated machines may be implemented for a portion of the production cycle [3]. Mazda is working on incorporating craftsmanship into mass production systems through "mass craftsmanship" by analyzing the mechanisms of craftsmanship [4]. One of the authors has proposed the slow manufacturing concept as a paradigm of manufacturing that is not necessarily restricted by efficiency issues and with craftsmanship at its core [5]. Slow manufacturing contributes to making long-life products that continue to provide value and improve the well-being of workers in terms of sustainability.

It is worth mentioning that, once lost, manufacturing skills are difficult to recreate owing to their implicit aspects. Therefore, manufacturing companies need to make efforts to identify and retain their crucial manufacturing skills. To achieve this, it is necessary to sooner understand the user demand patterns for those products that require manufacturing skills. Indeed, understanding user demand patterns leads to an increase in added value and productivity.

The concept of Industry 5.0 has been presented as a green and sustainable next step in human-centered digital manufacturing systems that not only respond to new markets and global social trends but also do not exclude workers from the process [6, 7]. Collaborative robots are important for improving productivity, and Industry 5.0 concludes that the use of robots will create job opportunities for humans [8]. Thus, in this study, we assume the coexistence of manufacturing skills and digitalization.

This study proposes a method for analyzing the appropriate combination of manufacturing skills and automation technology for the slow manufacturing paradigm.

## 2 Related Works

### *2.1 Studies Focused on Craftsmanship*

Efforts have focused on improving the value added to industrial products through craftsmanship for a long time. A recent example is the Italian "slow factory." Based on the concept of "slowness," it aims to brand and systematize the "made in Italy" approach that emphasizes craftsmanship [3].

The slowness concept originated from the "slow food" movement. Proponents seek efforts against the "fast life thinking" that emphasizes efficiency in modern society. While the slow factory is an Italy-specific concept, slow manufacturing was proposed as a more general concept. This focuses on increasing value added in industrial products through the manufacturing process by craftsmanship. Regional factors such as cultural tradition are important. Also, key to slow manufacturing is manufacturing management, such as corporate decision-making [5]. One of the authors has previously proposed the "Craftsmanship Index," which is used to evaluate quantitatively the contribution of craftsmanship in the manufacturing process. It consists of four components: craft skills; creativity skills; culture, history, and tradition; and territory vocation [9, 10].

A second example of efforts to improve value added is "mass craftsmanship" at Mazda. Mazda analyzes the mechanism of craftsmanship and incorporates it into mass production systems [4]. To this end, the company is working on visualizing craftsmanship by using motion capture to measure the movements of craftsmen who have cultivated their skills over many years and analyzing the combined movements of their eyes and musculoskeletal system [11]. This is only one example of using digital technology to address craftsman knowledge that is difficult to formalize. The "digital triplet" has been proposed as a structured way to describe an engineer's intentions and judgments by mapping them to operations in a cyber-physical production system [12]. This method is expected to represent the engineer's problem-solving process as reusable knowledge.

## 2.2 Industries 4.0 and 5.0

Industry 4.0 was first proposed by the German government in 2011 as a strategy to promote advanced manufacturing. Since then, many countries and companies have launched initiatives and projects related to Industry 4.0. The automation of the manufacturing industry using modern smart technologies, large-scale machine-to-machine communication, and the Internet of Things has led to an ongoing fundamental change in the way global production and supply networks operate.

Thus, while Industry 4.0 is a concept that describes the transformation of industries through digitization and data, it is limited to technological and economic perspectives. To address this, Industry 5.0 has been proposed as a concept that complements the technological and economic characteristics of Industry 4.0 and adds environmental and social elements. A document published by the EU in January 2021 defines Industry 5.0 as follows: "Industry 5.0 recognizes the power of industry to achieve societal goals beyond jobs and growth to become a resilient provider of prosperity by making production respect the boundaries of our planet and placing the wellbeing of the industry worker at the center of the production process" [7]. Previous studies on Industry 5.0 often discuss human–robot collaboration. In fact, hybrid manufacturing, in which craftsmen use digital tools such as CAM to apply their knowledge, has already been implemented [13].

# 3 Proposed Method

This study developed a method for analyzing appropriate patterns of customer demand importance by considering the combination of manufacturing skills and automation technology. The method consists of three steps, as shown in Fig. 1.

*Step 1: Calculation of Skill–Technology Index*

The Skill–Technology Index ($STI_k$) is an indicator that represents the degree of manufacturing skills and automation technology used in each process. A mix of craftsmanship and automation technology is involved in the manufacturing process. The level of craftsmanship and automation technology used in each process is evaluated separately and expressed as a single number considering the percentage of each in the process. The degree of craftsmanship and automation technology in a production process $k$ is defined as the Skill Index ($SI_k$) and the Technology Index ($TI_k$), respectively, using the evaluation criteria in Table 1. The contribution of automation technology in a production process is represented as the mixing ratio ($mix_k$), where $mix_k = \{0, 0.1, 0.3, 0.6, 1\}$. $STI_k$ is defined as follows:

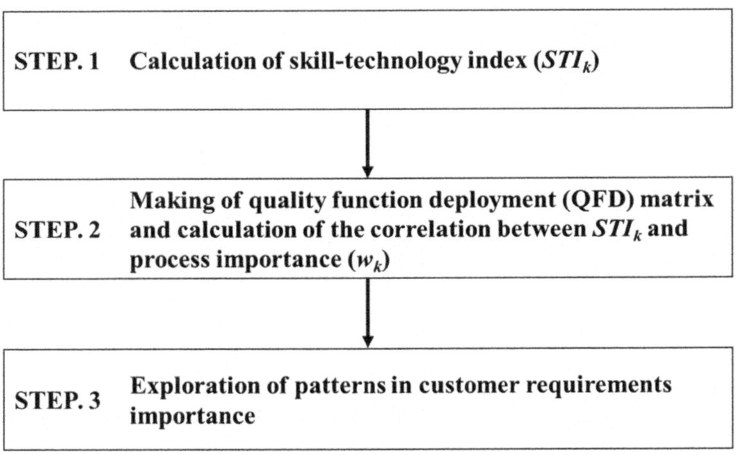


**Fig. 1** Flow of the proposed method

**Table 1** Evaluation criteria for the Skill Index and Technology Index

| Index | Evaluation subject | Evaluation value | | | | |
|---|---|---|---|---|---|---|
| | | 0 | 1 | 2 | 3 | 4 |
| Skill Index $SI_k$ | Worker skill | No skill needed | Elementary skill level | Intermediate skill level | Expert skill level | Only human can |
| Technology Index $TI_k$ | Automation technology | | | | | Only machine can |

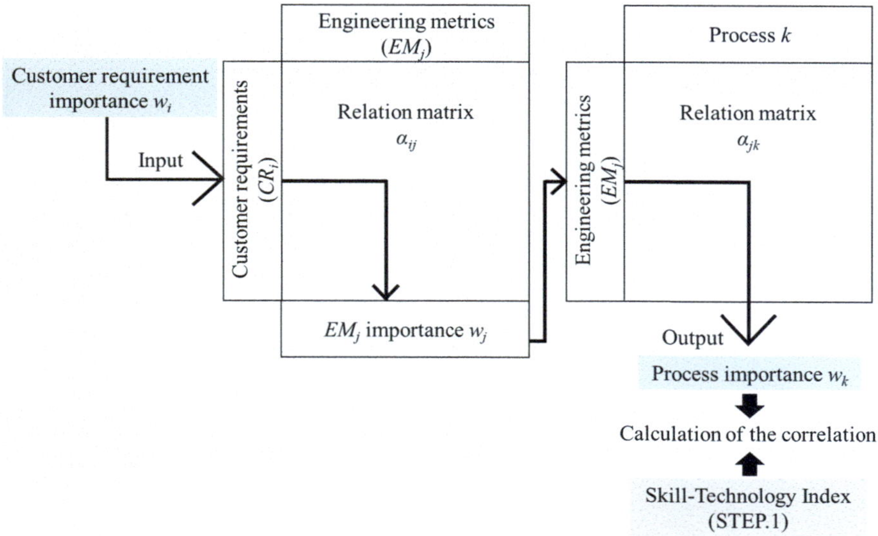


**Fig. 2** Overview of Step 2

$$STI_k = (1 - \text{mix}_k) \cdot SI_k + \text{mix}_k \cdot TI_k. \tag{1}$$

*Step 2: Creation of a quality function deployment matrix and calculation of the correlation between $STI_k$ and process importance ($w_k$)*

Exploration of a quality function deployment (QFD) matrix is made from customer requirements ($CR_i$) to engineering metrics ($EM_j$) and processes. Figure 2 shows an overview of the calculations, which can be expressed as follows:

$$w_j = \sum_i^I w_i \cdot \alpha_{ij} \ (1 \le i \le I), \tag{2}$$

$$w_k = \sum_j^J w_j \cdot \alpha_{jk} \ (1 \le j \le J). \tag{3}$$

Generally, the importance of customer requirements is set by an analyst using market research. The importance of customer requirements is denoted as $w_i$, the importance of engineering metrics is denoted as $w_j$, process importance is denoted as $w_k$, and the degree of association is denoted as $\alpha$.

Next, the correlation between $w_k$ and $STI_k$ is calculated as follows:

$$r = \frac{\frac{1}{K}\sum_k^K (w_k - \overline{w})(STI_k - \overline{STI})}{\sqrt{\frac{1}{K}\sum_k^K (w_k - \overline{w})^2}\sqrt{\frac{1}{K}\sum_k^K (STI_k - \overline{STI})^2}}, \tag{4}$$

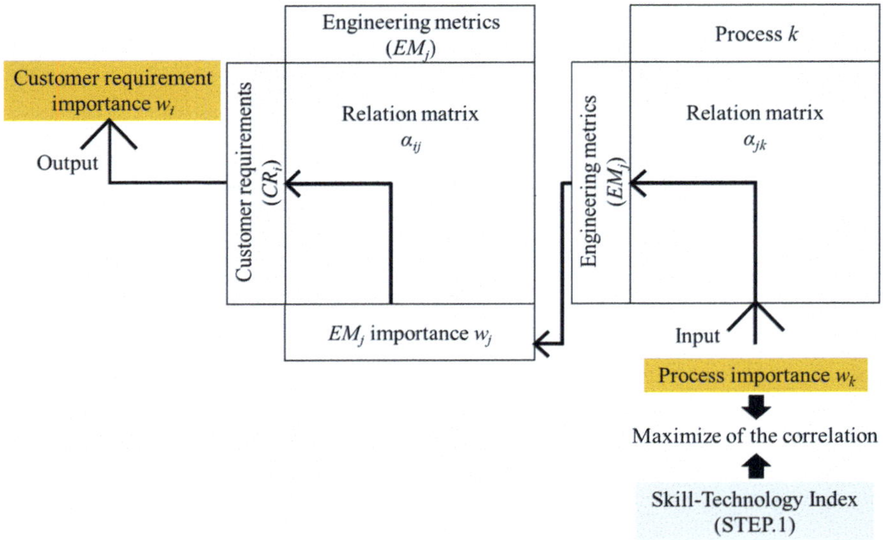


**Fig. 3** Overview of Step 3

where $\overline{w}$ is the average of $w_k$, $\overline{STI}$ is the average of $STI_k$, and $K$ is the number of processes.

As the correlation increases, so too does the level of skills and technologies used in the process important to the customer, thus improving compatibility.

*Step 3: Exploration of patterns in customer requirements importance*

The pattern of the importance of customer requirements $w_i$ that maximizes the correlation between the process importance $w_k$ and $STI_k$ is calculated (Fig. 3).

## 4 Case Study

A case study on glass bottle manufacturing was carried out. The manufacturer satisfies customer demands with high-quality manufacturing by leveraging its craftsmanship and know-how to produce glass bottles.

*Step 1: Calculation of the Skill–Technology Index*

We created an overview of the manufacturing process from the process analysis for the target product (glass bottles; Fig. 4), which is produced continuously. The flow of the main operation is shown in the upper row of the figure. The preparatory operations that are necessary for manufacturing but do not directly change the material and have unique processes are shown in the lower row of the figure.

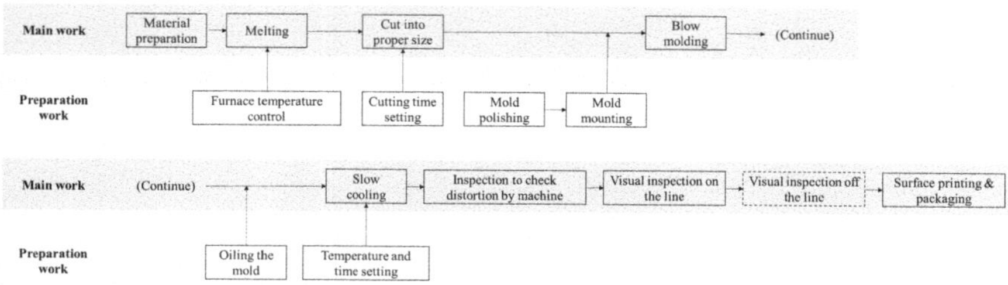


**Fig. 4** Process flow for manufacturing glass bottles

**Table 2** Skill–Technology Index of glass bottle manufacturing processes

| *K* | Production process | Main work | Preparation work | *TIk* | *SIk* | *mixk* | *STIk* |
|---|---|---|---|---|---|---|---|
| 1 | Material preparation | ✓ | | 2 | 2 | 0.6 | 2 |
| 2 | Melting (furnace temperature control) | ✓ | | 1 | 2 | 0.6 | 1.4 |
| 3 | Cut into proper size | ✓ | | 0 | 0 | 1 | 0 |
| 4 | Line-checking | | ✓ | 0 | 0 | 0 | 0 |
| 5 | Mold selection/manufacturing | | ✓ | 2 | 3 | 0.3 | 2.7 |
| 6 | Mold polishing | | ✓ | 0 | 0 | 0 | 0 |
| 7 | Mold mounting | | ✓ | 0 | 1 | 0 | 1 |
| 8 | Blow molding | ✓ | | 1 | 1 | 0.6 | 1 |
| 9 | Oiling the mold | | ✓ | 0 | 1 | 0 | 1 |
| 10 | Slow cooling | ✓ | | 1 | 0 | 1 | 1 |
| 11 | Inspection to check distortion by machine | ✓ | | 4 | 0 | 1 | 4 |
| 12 | Visual inspection on the line | ✓ | | 0 | 4 | 0 | 4 |
| 13 | Visual inspection off the line | ✓ | | 0 | 4 | 0 | 4 |
| 14 | Surface printing & packaging | ✓ | | 0 | 1 | 0 | 1 |

Furthermore, if the defect rate in the online visual inspection exceeds a certain value, the product is re-inspected offline. (This process is not always performed, so it is indicated by a dotted border.) From there, the manufacturing process was classified into 14 steps, and $STI_k$ was calculated according to the criteria listed in Table 1 and Eq. 1 (Table 2). The values of $STI_k$ are high for the three inspection steps. Machine inspection is used to check for distortions and cracks in bottles that are not visible to the human eye, while human inspection is used subjective judgment of quality.

*Step 2: Creation of a QFD matrix and calculation of the correlation between* $STI_k$ *and* $w_k$

Table 3 is the QFD matrix deployed from customer requirements ($CR_i$) to engineering metrics ($EM_j$). The manufacturer produces bottles according to the required specifications of the customer. In addition, because the manufacturer's

**Table 3** QFD matrix deployed from customer requirement to engineering metrics

| Customer requirements ($CR_i$) | Customer requirement importance (*Wi*) | Engineering metrics ($EM_j$) | | | | | | | | | | | |
|---|---|---|---|---|---|---|---|---|---|---|---|---|---|
| | | Transparency | Number of bottle cap types | Number of bottle mouth sizes | Capacity flexibility | Strict inspection standards | Surface roughness | Shape flexibility | Stability of bottle material | Airtightness | Lead time | Strength and heat resistance | Yield rate |
| High quality | 9 | | | | | 9 | | | 3 | | | | 9 |
| Appearance that matches the product image | 9 | 9 | 3 | 3 | 3 | 1 | 1 | 9 | | | | | |
| Prevent deterioration of content | 9 | | 1 | 1 | 1 | 1 | | 1 | 9 | 9 | | 9 | 3 |
| Suitable for small lot production | 3 | | 1 | | 3 | | | | | | 9 | 1 | 1 |
| Convenient shape | 3 | | | 1 | 3 | | | 9 | | | | | |
| Good touch | 1 | | | | | | 9 | 1 | | | | | |
| Engineering metrics importance ($w_j$) | | 81 | 39 | 39 | 54 | 99 | 18 | 118 | 108 | 81 | 27 | 84 | 111 |

client base includes cosmetics companies, it is important to prevent deterioration of the contents of the bottles. Therefore, the important customer requirements for "high quality," "appearance that matches the product image," and "prevent deterioration of contents" are set as high values. Table 4 is a QFD matrix deployed from engineering metrics to process importance $w_k$.

Figure 5 shows an analysis chart of production process $k$ by normalizing $STI_k$ and process importance $w_k$ using the current $CR_i$ importance. The correlation coefficient between $w_k$ and $STI_k$ was calculated as 0.729 using Eq. 4.

*Step 3: Exploration of patterns in the importance of customer requirements*

Figure 6 shows an analysis chart of production process $k$ by normalizing $STI_k$ and process importance $w_k$ using the optimized $CR_i$ importance. It indicates that the correlation is even stronger than that in Fig. 5, while deviations are smaller. In fact, the correlation coefficient was calculated as 0.916 using Eq. 4.

Figure 7 compares the importance pattern of $CR_i$ assumed by the analyst (blue dotted line) with the optimal solution as determined by our method (orange line). According to the analyst's assumption, "quality," "appearance that matches the product image," and "preventing deterioration of contents" were all considered the most important factors. By seeking the optimal solution, it was found that both skill and technology are highly effective in ensuring "quality." Conversely, the analyst had also set the importance level of "preventing deterioration of contents," but it was found that this metric cannot be achieved solely through production skills and technology.

## 5 Discussion

The glass manufacturer's inspection standards for "high quality" are very strict, and inspections often find defects that might pass inspection in other companies. As a result, the bottles are manufactured with higher quality and reliability, meaning the company has a competitive advantage for customers who value "high quality." To "prevent deterioration of contents," glass bottles are generally very airtight and chemically stable, and they do not undergo chemical reactions with their contents. Thus, it is believed that "prevent deterioration of contents" applies to glass bottles in general and does not lead to an advantage through skills and techniques.

In this method, the degree of craftsmanship and automation technology is considered by $mix_k$, and by changing $mix_k$, it is possible to analyze how the importance pattern for customer requirements changes when technology replaces skill. Figure 8 shows the influence of changes in the $CR_i$ importance when the $mix_k$ of "mold selection/production," which is $k = 5$ in Table 2, changes from 0.3 to 1. Figure 8 indicates that only the importance of "appearance that matches the product image" has been changed. This means that the manufacturer cannot produce a bottle with a good appearance without craftsmanship in the mold fabrication process.

**Table 4** QFD matrix deployed from engineering metrics to process importance

| Engineering metrics ($EM_j$) | Engineering metrics importance ($w_j$) | Engineering metrics ($EM_j$) Material preparation | Melting (furnace temperature control) | Cut into proper size | Line-checking | Mold selection/ manufacturing | Mold polishing | Mold mounting | Blow molding material | Oiling the mold | Slow cooling | Inspection to check distortion by machine | Visual inspection on the line | Visual inspection off the line | Surface printing & packaging |
|---|---|---|---|---|---|---|---|---|---|---|---|---|---|---|---|
| Transparency | 81 | 9 | 1 | | | | 3 | | | 1 | | | 1 | 3 | |
| Number of bottle cap types | 39 | 9 | 3 | | | | | | | | | | | | 3 |
| Number of bottle mouth sizes | 39 | | | | | 9 | | | 3 | | | | 1 | 1 | |
| Capacity flexibility | 54 | 3 | 3 | 9 | | 3 | | | 3 | | | | | | 3 |
| Strict inspec-tion standards | 99 | | 3 | | | | | 1 | | 3 | | 9 | 9 | 9 | |
| Surface roughness | 18 | | | | | 3 | 9 | 1 | | 3 | | | | 1 | |
| Shape flexibility | 118 | 3 | 9 | | | 9 | | 1 | 3 | | | | 1 | 1 | 3 |
| Stability of bottle material | 108 | 9 | 3 | | | | | | | | 1 | | | | |

| | | | | | | | | | | | | | | | |
|---|---|---|---|---|---|---|---|---|---|---|---|---|---|---|---|
| Airtightness | 81 | | | | | | | | 1 | | | | | | |
| Lead time | 27 | | 9 | | 1 | | | | 1 | | 9 | 1 | 3 | 3 | 3 |
| Strength and heat resistance | 84 | 9 | 9 | | | | | | | | 9 | 9 | | 3 | |
| Yield rate | 111 | | | | 3 | | | | 1 | 3 | | 9 | 9 | 9 | |
| Process importance ($w_k$) | | 3324 | 30421 | 486 | 360 | 1629 | 405 | 235 | 852 | 765 | 1107 | 2673 | 2209 | 2641 | 714 |

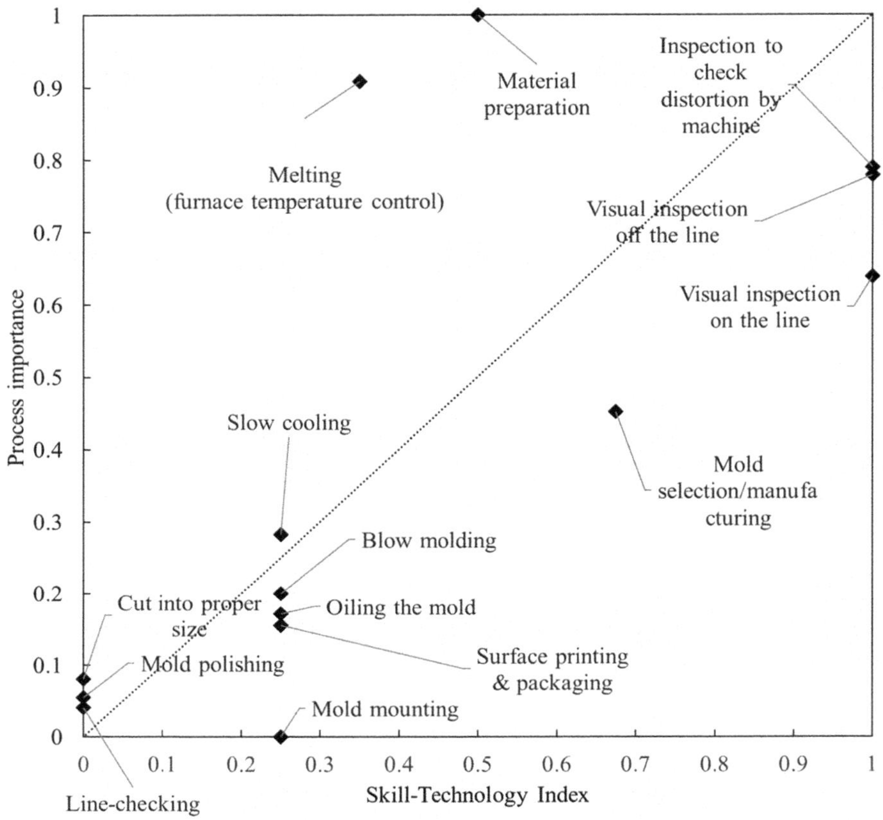


**Fig. 5** Normalized Skill–Technology Index and process importance $w_k$

This suggests that the process should not be fully automated. Thus, this method could be applied to manufacturing management.

## 6 Conclusion

In this study, we developed a three-step method for analyzing the importance of customer requirements for industrial products in combination with specific issues related to the manufacturing environment, the quality of the manufacturing skills, and the level of automation technology. It is based on a QFD matrix and the newly proposed SKI.

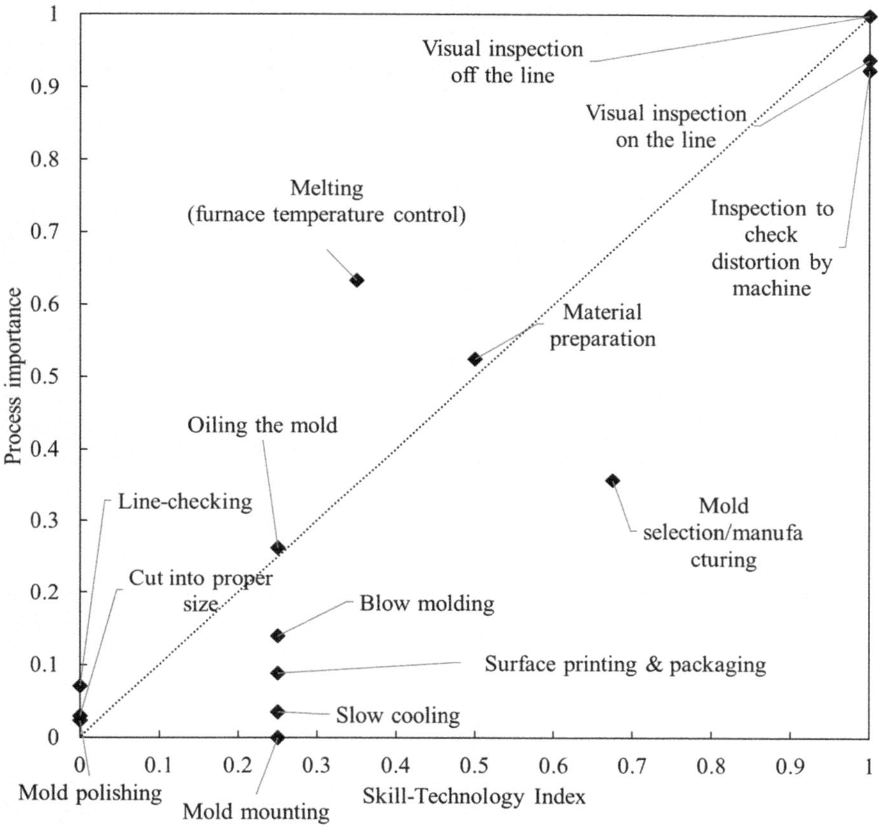


**Fig. 6** Analysis chart of production process $k$ by normalizing $STI_k$ and $w_k$ using the optimized $CR_i$ importance

As a final result, a graphical form was devised to compare the importance of customer requirements with the analyses through the QFD matrix.

The usefulness of the proposed method was shown through the implementation of a case study. It was suggested that manufacturing skills and technology alone might not necessarily meet the customer requirements that were considered significant. Also, it was indicated that this method could support manufacturing management by giving measurements or balancing manufacturing skills and automation technology.

Further studies investigating an optimized and appropriate balance of craftsmanship and automation technology within industrial manufacturing environments are needed for future work.

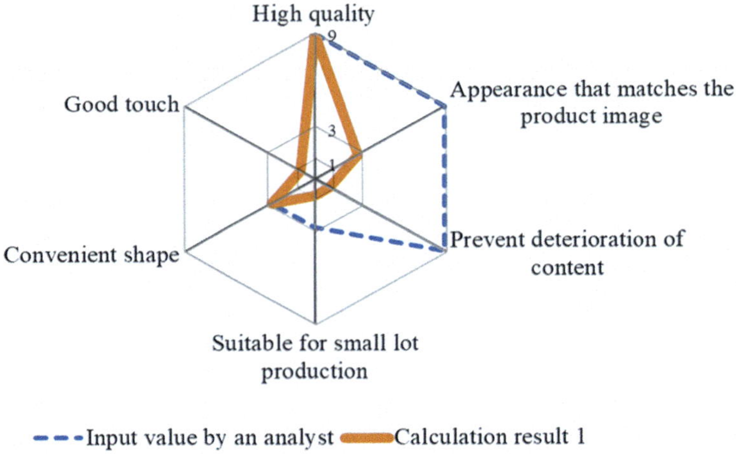


**Fig. 7** Comparison of importance of $CR_i$ between QFD input and the result of STEP 3

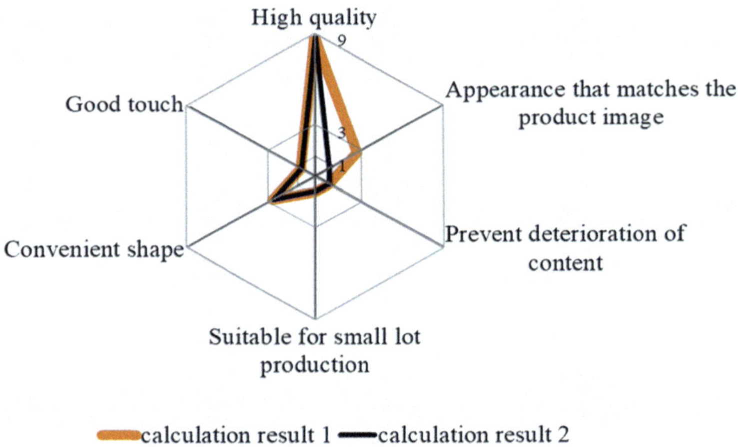


**Fig. 8** Comparison of importance of $CR_i$ using different $mix_k$ values

**Acknowledgments** This work was supported by the Japan Society for the Promotion of Science (JSPS) Bilateral Joint Research Project No. JPJSBP 120229925. We thank Japan Seiko Glass Co., Ltd. for their cooperation in the case study. We also thank Momo Kojima, who was involved in the preliminary research for this study at the School of Engineering, Osaka University.

## References

1. Monostori L et al (2016) Cyber-physical systems in manufacturing. CIRP Ann 65:621–641
2. Schleich B et al (2017) Shaping the digital twin for design and production engineering. CIRP Ann 66:141–144
3. Campana G, Cimatti B (2013) The slow factory: a new paradigm for manufacturing. 11th Global Conference on Sustainable Manufacturing, pp 273–277
4. Adachi N (2019) Development of mass craftsmanship production system (in Japanese). Mazda Gihou 36:154–158
5. Kobayashi H (2022) A concept proposition of slow-manufacturing (in Japanese). JSME, MSD 2022:136–138
6. Trstenjak M et al (2023) Key enablers of industry 5.0- transition from 4.0 to the new digital and sustainable system. In: GCSM 2022, LNME, pp 614–621
7. European Commission (2021) Industry 5.0-towards a sustainable. Human-Centric and Resilient European Industry
8. Nahavandi S (2019) Industry 5.0-a human-centric solution. In: Sustainability 2019, 11, 4371
9. Campana G et al (2016) A proposal for the evaluation of craftsmanship in industry. Procedia CIRP 40:668–673
10. Campana G et al (2022) Evaluation of competencies for a sustainable industrial environment. In: GCSM 2022, MDCE, pp 836–844
11. Kubo Y et al (2019) Nurturing of craftsmen to support "KODO design" (in Japanese). Mazda Gihou 36:165–170
12. Umeda Y et al (2022) Digital triplet and its implementation on learning factory, vol 55. IFAC Papers OnLine, pp 55–62
13. Zoran A (2015) Hybrid craft: showcase of physical and digital integration of design and craft skills. Leonardo 48(4):384–398

# Reconfigurable Tooling for Circular Manufacturing: A Review

**Cherukupally Shivaprasad, S. Suryakumar, Carla Susana A. Assuad, Kristian Martinsen, Sverre Gulbrandsen-Dahl, and N. Venkata Reddy**

**Abstract** Flexible manufacturing technologies are gaining importance in recent years to produce a wide variety of geometries cost-effectively and eco-friendly. Reconfigurable tooling is one such technology that has the potential to replace costly dedicated tools with discrete pins that can be configured to the desired part shape by adjusting them locally. In the present work, a comprehensive review of reconfigurable tooling used in various manufacturing processes is presented, and challenges that need to be addressed to utilize it in the circular manufacturing context are emphasized. In addition, the beneficial role of component-specific cushion design on the tooling cost and maintenance is discussed.

**Keywords** Reconfigurable tooling · Sheet metal forming · Circular manufacturing · Component specific elastic cushion

## 1 Introduction

Considering the global push for sustainability and the increasing demand for customized components, industries are moving toward flexible manufacturing processes to produce a wide variety of geometries cost-effectively and eco-friendly [1–4]. Traditional manufacturing processes, such as injection molding, stamping, stretch forming, etc., usually employ component-specific tooling/molds/dies, and these processes are most suitable for mass production as the tooling cost is distributed over a number of components being produced. However, factors such as costly component-specific dies/molds, intensive energy consumption in fabrication, the

C. Shivaprasad · S. Suryakumar · N. V. Reddy (✉)
Indian Institute of Technology Hyderabad, Kandi, Telangana, India
e-mail: nvr@mae.iith.ac.in

C. S. A. Assuad
Norwegian University of Science and Technology, Gjøvik, Norway

K. Martinsen · S. Gulbrandsen-Dahl
SINTEF Manufacturing AS, Raufoss, Norway



S. Fukushige et al. (eds.), *EcoDesign for Circular Value Creation: Volume I*,
https://doi.org/10.1007/978-981-97-9068-5_35

longer product development cycle, and additional costs incurred in their storage make traditional component-specific tooling/die/mold-based processes unsustainable for low volumes. Furthermore, the fabrication of dies or molds for new part designs entails significant material consumption and generates substantial waste, which contributes to environmental degradation and resource depletion. To address the above-mentioned challenges, many attempts [1, 2, 5, 6] are made to develop flexible and rapid response processes/technologies to produce customized components at a reasonable cost with minimal adverse environmental effects.

Reshaping and reuse of components in the end-of-life (EoL) products is gaining importance in recent years to increase circularity and reduce the possibility of using energy-intensive conventional recycling (i.e., re-melting) and associated environmental effects. Several researchers [4, 7–11] mentioned that the most effective way to improve circularity in manufacturing is through "Reuse" or "Remanufacturing" [4], the latter involving "upcycling," where used components/products are raw materials for new products with contemporary function, quality, and value for the customer. Ali et al. [8] have described a recovering strategy for sheet metal scrap from the car-body manufacturing process. Falsafi et al. [7] suggest a reformability index addressing possible deterioration effects from flattening sheets with complex shapes. Abduah et al. [9] made a sustainability assessment on the benefits of remanufacturing end-of-life (EOL) passenger car waste sheet steel into a mesh sheet [9]. Takano et al. [12] used single-point incremental forming to deform the flattened sheet (which was obtained from the EoL bent component) into conical geometry. Reddy et al. [10] emphasized that the combination of incremental sheet forming (ISF) and reverse engineering can be used to remanufacture sheet metals to remove dents and/or reshape to new geometry (within the formability limits). Ingarao et al. [11] reshaped the undeformed flat base on the deep-drawn square cup into the conical/pyramidal components using SPIF without affecting the already deformed regions. Despite using the SPIF process for reshaping, exploring other flexible/hybrid manufacturing processes is essential to reduce the cost and improve process flexibility by eliminating the necessary component-specific tooling during reshaping.

Reconfigurable tooling is one of the flexible manufacturing technologies that has the potential to replace costly dedicated tools with discrete pins that can be configured to the desired part shape by adjusting them locally. Figure 1 illustrates the difference between the component-specific die and reconfigurable tooling. It can be clearly seen from Fig. 1 that there is a need for intensive energy (that is associated with a negative environmental impact) consumption and extensive material utilization (material is removed from the cast block) to fabricate the mono-block tool for each geometry during the product development cycle and it will lead to environmental degradation and resource depletion. Whereas, in the case of reconfigurable technology, having the flexibility to configure the pins into desired component geometry provides an opportunity to save the resources and environment for future generations.

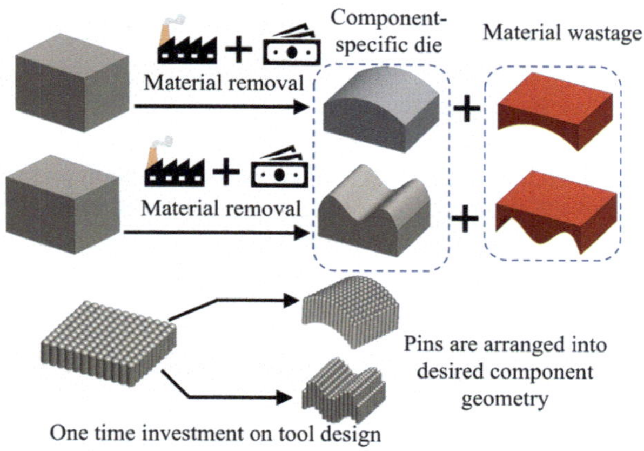


**Fig. 1** Demonstration of the need for reconfigurable tooling in the digital era to meet the circular manufacturing goals (representative schematic only)

The reconfigurable tooling is not a nascent technology. Cochran [13] was the first to patent the idea of reconfigurable tooling in 1863. Nakajima [14] brought it to practice (at the University of Tokyo, Japan) in 1960. The substantial advancements in reconfigurable tools are carried out by Hardt, Walczyk, and their coworkers [15–19] on mechanical design of reconfigurable tools [15, 16] and development of a shape control algorithm [17–19] to produce the components with reasonable accuracy. With the progress of time, reconfigurable tooling technology has been embraced in various manufacturing processes, namely, injection molding [20, 21], stamping (sheet metal forming) [15, 22, 23], stretch forming [24, 25], roll forming [26], sheet/tube hydroforming [27, 28], rubber pad forming [29], thermoforming [30, 31], and incremental forming [32], to enhance geometric flexibility and shorten the lead time. Thus, reconfigurable tooling technology is one of the promising approaches for mass-customization/on-demand manufacturing as it offers geometrical freedom and promotes circular manufacturing/economy practices in the digital era.

To the best of the author's knowledge, there is no comprehensive review that collates the application of reconfigurable tooling in various manufacturing processes, and the same is attempted in the present work. Emphasizing the need for reconfigurable tooling in the context of circular manufacturing is presented in this work. The main objective of the present paper is to conduct a comprehensive review of reconfigurable tooling exploited in various manufacturing processes and bring out challenges that need to be addressed to be effective in circular manufacturing. Some of the works carried out by our team is also be presented, for instance: the beneficial role of component-specific cushion design on the tooling cost and maintenance as the number of pins required in reconfigurable tooling is significantly less with better accuracy in case of SF. In addition, the potential of reconfigurable tooling in repairing parts or producing components from recycled material or the deformation of newly developed materials will be emphasized.

# 2 Reconfigurbale Tooling Technology in Different Shaping Processes

In the present work, the application of reconfigurable tooling in shaping processes is classified into two types based on the solidification and deformation process. Figure 2 shows the detailed taxonomy of reconfigurable tooling in several shaping processes. In this section, the working principle and summary of each process are explained.

## 2.1 *Stretch Forming of Sheet Metals*

Stretch forming (SF) process is extensively used to produce three-dimensional gentle curvature sheet metal components in the aerospace and automotive industries [25]. In the stretch-forming process, opposite sides of the rectangular sheet are clamped (Fig. 3) and is wrapped (i.e., bending) around the die, followed by stretching to produce a component. Figure 3 illustrates the difference between conventional and reconfigurable tooling-based stretch-forming (called multipoint stretch forming) process. An elastic cushion (Fig. 3) is placed between the reconfigurable pin head and sheet metal surfaces to avoid the direct contact, thereby, reduce the height of dimples on the component. Despite the flexibility in reconfigurable tooling, producing the component with reasonable accuracy is difficult, and it depends on the various process parameters, namely elastic cushion

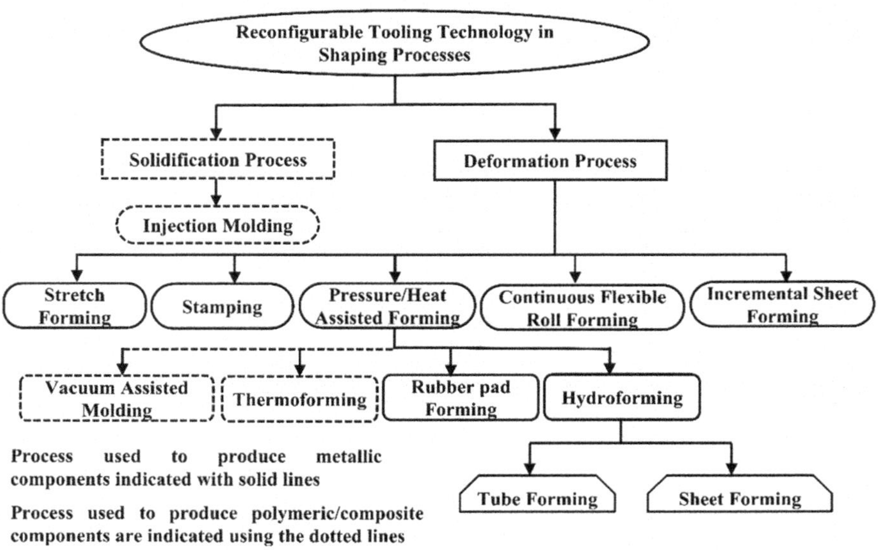


**Fig. 2** Classification of shaping processes that used reconfigurable tooling

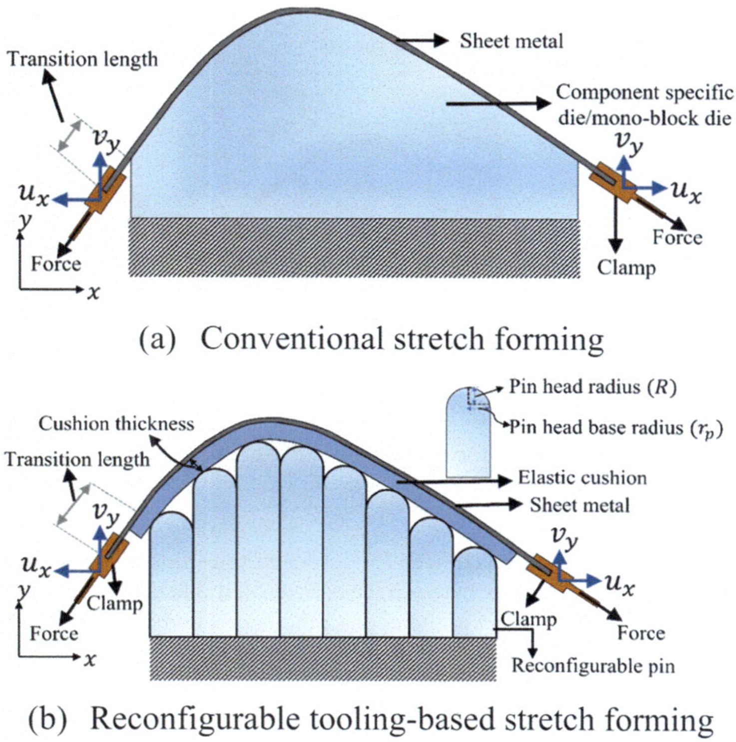


**Fig. 3** Difference between conventional and reconfigurable tooling-based stretch forming process. (Adopted from shivaprasad et al. [34] and is redrawn with suitable modifications)

thickness, pin radius, number of pins in the forming area, and transition length (Fig. 3).

Various attempts [24, 33–36] have been made to study the effect of process parameters on global/local shape error (global shape error is defined as deviation between ideal and formed geometries after spring back; local shape error is the maximum dimple height) and deformation behavior of sheet metal and the following conclusions are drawn:

- More uniform stress and strain distribution in the sheet metal is attained with an increase in elastic cushion thickness (due to its high bending stiffness) [24, 34, 36], pin head radius (due to the increased contact area between the cushion and pinhead surface for a given pin base radius; Fig. 1), and a number of pins within the same forming area [33, 37] or applying the stretch to the cushion (due to induced tensile stresses in the elastic cushion) [34]; Thus, the local shape error is reduced as the dimples height is decreased due to more uniform deformation in the sheet metal components.
- Global shape error of components decreased either with the decrease in the cushion thickness [24, 34] or an increase in the pin head radius [33]. However,

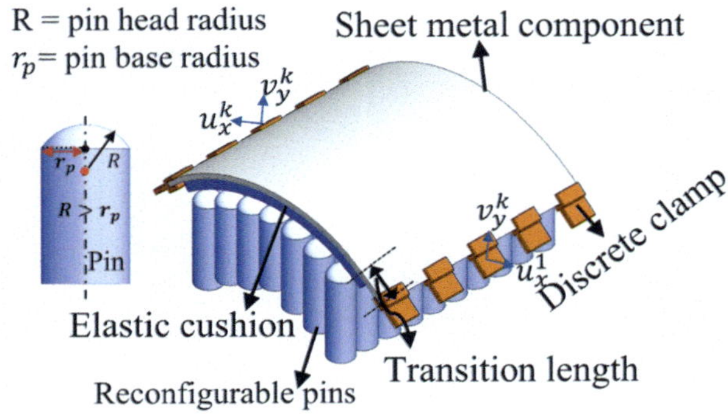


**Fig. 4** Usage of discrete clamps to improve the flexibility in sheet metal clamping (Concept taken from Wang et al. [42] and is redrawn with suitable modifications)

the usage of a larger pin head radius confines the range of curvatures of the components to be formed.

- The possibility of wrinkle formation is high with the longer transition length [36]. In addition, more material wastage can be generated with the increase in transition length.

It is well known that there is a need for sufficient transition length to clamp the sheet and ensure conformability between the cushion and sheet surfaces. In addition, there is a possibility of thinning at the edge of the compound curvature geometries during the forming with the single clamp. Hence, discrete clamps (Fig. 4) are used, in which each clamp can be controlled (displacement [38–40] or force [41, 42]) independently to improve the flexibility to clamp the sheet, reduce material wastage by decreasing the transition length, and form large curvature components without significant thinning. It is reported that more uniform stress and strain distribution is achieved in the component formed using the multiple clamps (with displacement or force-controlled loading paths), and conformability between a cushion and sheet is enhanced with lesser transition length [38–40, 42]. Force-controlled loading paths are preferable to form the large curvature geometries without failure as lower equivalent strain can be attained [41, 42]. The configuration of pins (i.e., arrangement of pins in pin matrix) will also affect the contact area between the cushion and pin head. Among three configurations (aligned, declining staggered (i.e., pins of rectangular cross section are rotated by 45 degrees), parallel staggered), the component formed using a parallel staggered pins configuration resulted in more uniform stress and strain distribution as well as enhancement in accuracy due to the large contact area [43, 44].

## 2.2 *Stamping of Sheet Metals*

Stamping is a sheet metal forming process that employs component-specific tools, namely punch and die, to deform the sheet blank into the desired geometry. The stamping process is extensively employed in the automotive and aerospace

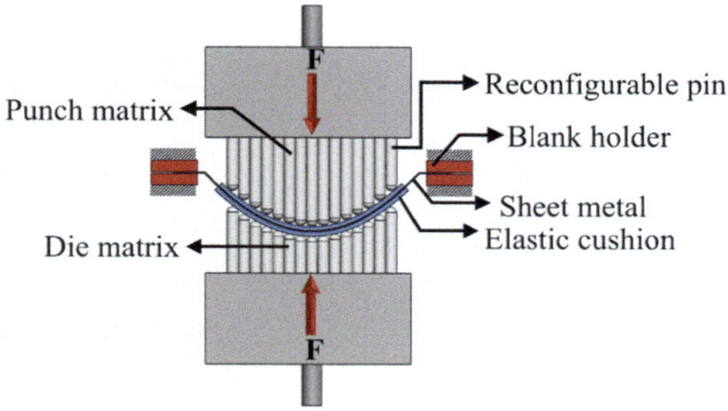


**Fig. 5** Schematic representation of the multipoint forming (Concept taken from Nakajima et al. [14] and is redrawn with suitable modifications)

industries to produce components. During stamping, the punch moves to a predetermined position to draw a blank (which is held with a blank holder with appropriate force) into the die cavity to attain the desired shape. To enhance the process flexibility, one of the alternatives to conventional stamping is multipoint forming which uses the reconfigurable tooling-based punch and die (shown in Fig. 5). Multipoint forming has four variants based on the pin adjustment method [45], namely, multipoint tool (pins in the punch and die matrix are configured to desired geometry before the start of forming, i.e., ex-situ configured multipoint tool), in-situ configured multipoint tools (both pins in the punch and die matrix are configured to desired geometry incrementally during deformation based on intermediate geometry), multipoint half die/punch (punch/die matrix pins configured to desired shape before forming, whereas in die/punch matrix configured incrementally during the deformation).

Two essential features of multipoint forming make it different from conventional stamping. One is that adjusting the pin positions (based on the intermediate geometry) during the forming processes can vary the deformation path to form the component (in-situ configured multipoint tools variant, Fig. 6), which is impossible using geometry-specific tooling in conventional stamping [22, 45–47]. Particularly, it will be more beneficial for the sheet materials that are sensitive to deformation path dependence. Advantages of the varying deformation path are enhancement in formability, better component surface quality, and suppression of both wrinkles as well as dimples on the component [22, 45–47]. Another feature is section-by-section forming (i.e., sectional forming, refer to Fig. 7), in which pins are configured for each section of the total geometry [22, 45–48]. Sectional forming can produce large parts in small multipoint forming presses. Li and their co-workers [49] have made significant progress in multipoint forming, including machine development [49], predicting the optimum forming path and intermediate geometry [22, 46, 47], developing discrete blank holders [49, 50], process design for sectional forming [48, 51], and integrating CAD, CAM, and FEA to create the digital die-forming system [49, 52, 53]. A comprehensive review of multipoint forming is not presented in this work as it is already discussed by Li et al. [49] recently.

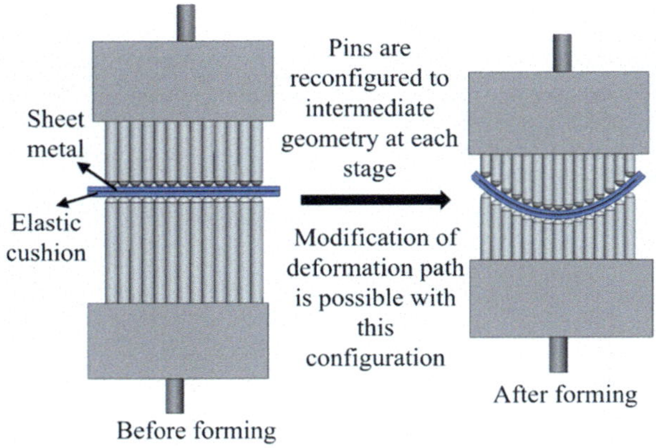


**Fig. 6** Multipoint press variant in which deformation path can be varied during forming (Concept taken from Li et al. [49] and is redrawn with suitable modifications)

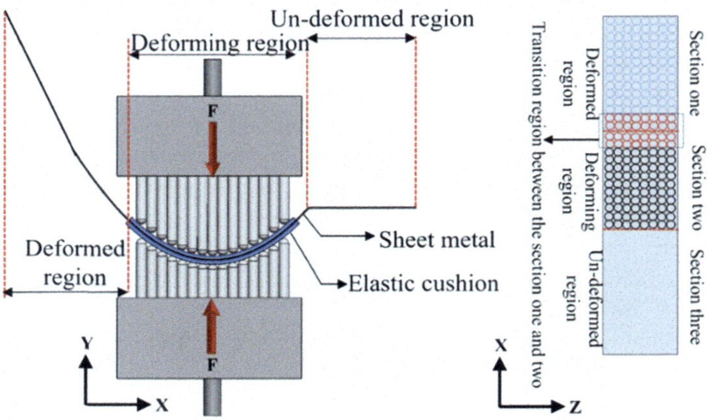


**Fig. 7** Representation of section-by-section forming to produce the large parts in a small press (Concept taken from Li et al. [49] and is redrawn with suitable modifications)

## *2.3 Pressure/Heat-Assisted Forming of Metals and Polymers*

***Sheet/Tube Hydro-Multipoint Forming*** High-pressure hydraulic fluid is used to deform the sheet/tube into a die cavity to get the desired component at room temperature [54]. It has the potential to produce complex geometries (with a high strength-to-weight ratio) in a single step, which is difficult using conventional/ flexible forming processes such as stamping or SF [54]. Figure 8 shows the schematic view of the sheet hydro-multipoint forming process [27, 55]. In sheet/tube hydro-multipoint forming, there is a high possibility of wrinkles and dimples on the formed component due to high forming pressure and less stiffness of elastic cushion in addition to the discrete die surface of reconfigurable tooling. Hence, a thick metallic sheet [27, 55] or elastic cushion [28] will be used as an interpolator between the die surface and sheet metal to produce the components as it reduces the height of

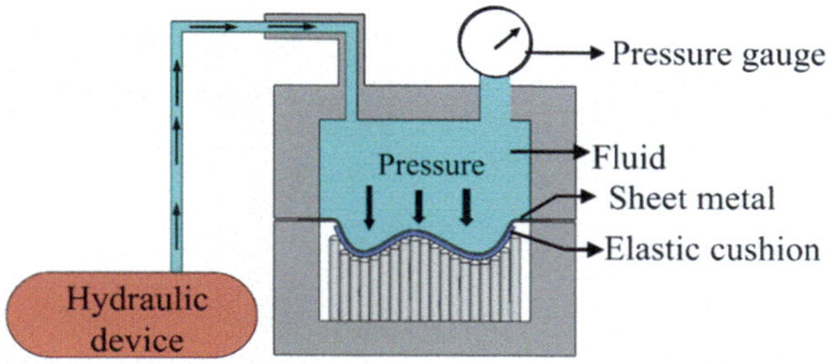


**Fig. 8** Schematic of sheet hydro-multipoint forming. (Concept taken from Panuoiu et al. [27] and is redrawn with suitable modifications)

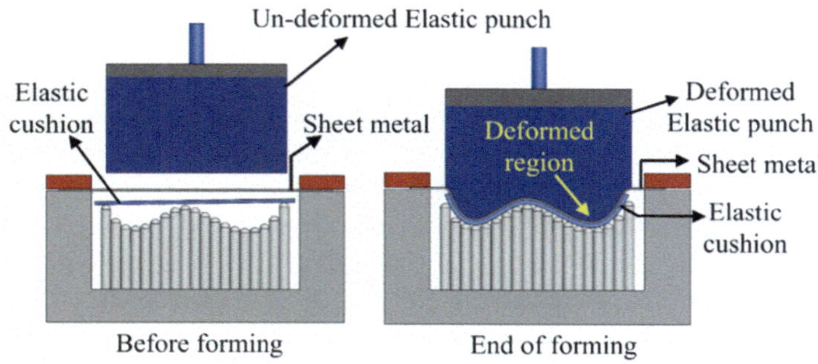


**Fig. 9** Illustration of rubber pad multipoint forming. (Concept taken from Boers et al. [56] and is redrawn with suitable modifications)

wrinkles and dimples (or local shape error) on the component due to its increased stiffness, consequently reduction in the shape error [27, 55]. However, there is a possibility of an increase in the forming forces when the thick metallic interpolator is used, leading to higher energy consumption by the machine tool. Also, a material used for the interpolator will be sacrificed as a wastage after forming the components. In tube hydro-multipoint forming, there is an attempt [28] to analyze the number of pins on local and global shape errors of tube components. It is reported that local shape error is reduced significantly with increasing the number of pins (having smaller diameters), whereas there is an insignificant effect on the global shape error [28].

***Rubber-Pad Multipoint Forming*** It is similar to hydroforming; the main difference is that an elastomeric punch is used to deform the sheet into a die cavity instead of the high-pressure fluid medium. In this process (Fig. 9), elastomeric punch starts deforming as soon as it comes in contact with sheet metal, and forming force is increased gradually with the deformation of both punch and sheet; finally, both punch and sheet metal takes the shape of die cavity [29, 56, 57]. It is reported that the wrinkle height on the formed component using the rubber pad multipoint forming (solid elastic punch is used) is less (due to more uniform stress and strain distribution) than the one formed using multipoint forming [29]. In addition, it is suggested by Tolipov et al. [57] that the elastic punch with structured holes is preferred over a solid punch to further reduce the wrinkle height in formed components.

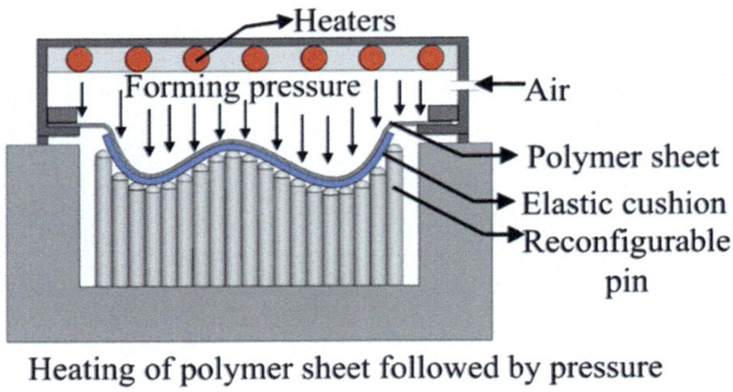


**Fig. 10** Depiction of multipoint thermoforming. (Concept taken from Cao et al. [59] and is redrawn with suitable modifications)

***Multipoint Thermoforming*** Thermoforming is used to produce thermoplastic polymer components, which have extensive applications in aerospace, automobile, architecture, and medical instruments [30, 31, 58, 59]. In multipoint thermoforming, a peripherally clamped thermoplastic polymer sheet is heated above its glass transition temperature before placing it over the elastic cushion, and pressure is applied on a heated sheet to deform it to the desired shape (Fig. 10). After cooling, pressure is removed, and obtain the part geometry [30, 31, 58, 59]. The critical challenge in multipoint thermoforming is producing parts with good accuracy because the possibility of wrinkles formation on the polymer sheet is higher as the deformation occurs at elevated temperatures. Several attempts [30, 31, 58, 59] have been made to analyze the process parameters, namely polymer sheet thickness, the glass-transition temperature, curvature of geometry to be formed, thickness and hardness of elastic cushion, friction between sheet and cushion interface, on the quality of the components. It is observed that the height of wrinkles is increased with a decrease in the radius of curvature (due to large deformation), the thickness of the polymer sheet being formed (due to less stiffness), and increasing the forming temperature (due to more softening) [30, 58]. Also, global shape error is increased, and local shape error (i.e., maximum dimple height) is decreased with an increase in elastic cushion thickness, hardness, and friction (between the cushion and polymer) [59].

***Vacuum-Assisted Multipoint Molding*** In this process, carbon fiber reinforced plastic (CFRP) curved panels are produced, which have extensive applications in the aerospace and automotive industries due to their lightweight, high strength, and stiffness. Vacuum-assisted multipoint molding [60] is an alternative to CFRP laminating with a component-specific die process. In this process, a reconfigurable tooling system is placed in the vacuum chamber; pins are arranged to the desired part geometry, and an elastic cushion is placed on the adjusted reconfigurable pins; after that, the elastic cushion conforms to the reconfigurable die surface as soon as a vacuum is created in the chamber; finally; CFRP sheets are laminated over the elastic cushion followed by curing under heat and pressure. An attempt made by Wimmer et al. [60] on the influence of elastic cushion thickness on dimple height reported that increased cushion thickness reduced the dimple height.

## 2.4 Roll Forming of Sheet Metals

The continuous and flexible roll forming (CFRF) process can produce doubly curvature parts at much lower costs than the one obtained using conventional stamping or multipoint forming [26, 61, 62] as the tooling cost is reduced significantly. Despite the flexibility in multipoint forming, the work envelope in this process is constrained as it depends on the number of pins available in rows and columns as well as pin size. Working/Forming area can be increased by including additional pins if the component dimensions exceed the existing work envelope. Otherwise, the joining of two small components can produce large-size components. However, the additional sheet material is needed to hold the sheet (that will be wasted after trimming), and including the extra pins to increase the working area makes the process costly. Also, the strength of the large component will be different if the two sheet metal components are joined. To overcome the above-mentioned challenges, conventional roll bending is combined with reconfigurable tooling, which is called continuous flexible roll forming [26, 61, 62]. In this process, the one upper and two lower flexible rolls are employed as a forming tool to generate curvatures simultaneously in the longitudinal and transverse directions [26, 61, 62]. The shape of the flexible roll is adjusted based on the desired geometry by moving the pins vertically. Flexible rolls that are rotated using motors will deform the sheet continuously and moves the sheet forward using frictional forces at the sheet and roll interface. The component curvature in the transverse direction is generated by configuring the upper and lower flexible rolls into the desired shape, and vertical movement of the upper roll is employed to generate the longitudinal curvature. Li and co-workers [26, 61, 62] demonstrated the potential of this process in producing the double curvature component, and they also developed a digitally integrated system that includes the process mechanics to control the CFRF process and produce the accurate components.

## 2.5 Incremental Sheet Forming of Metals

Incremental Sheet Forming (ISF) is a die-less forming process that has the potential to form complex three-dimensional components without using component-specific tools [5, 63]. Advantages of the ISF process include higher formability, less lead time, and suitability for low-volume production [64]. Limitations of the process are less productivity, sheet thinning, and forming the high wall angle features components (multistage DSIF can be employed, but energy consumption and lead time to produce the component increases). A combination of multipoint forming and ISF will overcome the limitations of the ISF process. For example, Zhang et al. [32] demonstrated that components with high wall angle features can be produced by combining multipoint forming and single-point incremental sheet forming (SPIF) in shorter lead times. In this hybrid process, preform geometries are formed using multipoint forming; then, preforms are mounted on the ISF setup to form the final geometry. Another variant of this hybrid process is the usage of reconfigurable pins as pattern support, and this hybridization will not enhance productivity but improves the accuracy of the component (if the proper elastic cushion thickness is used) as it provides the support [65, 66]. However, the same component can be formed using the double-sided incremental sheet forming (in which two independently controlled tools on either side of the sheet, in which one tool is employed to deform the sheet, whereas the other provides local support) [5, 63].

### *2.6 Injection Molding of Polymers*

In injection molding or casting, the design and fabrication of mold is a decisive phase of product manufacturing. For each component, there is a need to design and manufacture geometry specific molds. There are attempts [21, 67, 68] to use reconfigurable pin-based molds in the injection molding to enhance the process flexibility and nearly eliminate the tooling cost for each component. In addition, there is a possibility to eliminate the ejection system in injection molding if reconfigurable pins are used (because pins can be adjusted after molding to remove the molded parts) [67]. Note that there will be a considerable difference in the reconfigurable tooling design for sheet metal forming and the molding process, as the thermal and mechanical conditions (i.e., high pressure and temperature of molten material) need to be considered [21]. One of the important aspects of tooling design in injection molding is that there should not be any gap between the pins in the contact [21]. Even small gaps will lead to leakage of molten material and result in cracks or flashes similar to conventional molding, eventually, failure of the reconfigurable mold tool [21]. Koc and Thangaswamy [21] analyzed the design and fabrication of the reconfigurable tooling for molding and successfully demonstrated the feasibility of the usage of reconfigurable pins by manufacturing parts.

## 3 Role of Elastic Cushion in Reconfigurable Tooling

Discontinuous forming surface area in reconfigurable tooling is a major source of inaccuracy and forming defects (wrinkles and dimples) on the surface of the formed geometry. An elastic cushion is placed between the sheet metal and pinhead surfaces to achieve continuous pressure distribution on the sheet metal and to prevent the possibility of dimples on the component. However, incomplete conformability of cushion surface (related cushion of constant thickness) with pinhead and less stiffness of cushion will not reduce the height of dimples effectively as it is compressed between the sheet metal and pinhead. Consequently, it is tended to flow between two successive pins due to compressive load. Therefore, the elastic cushion becomes thin at the contact region (pinhead and sheet) and thick at the other regions, and it leads to non-uniform deformation of the cushion and the same will be

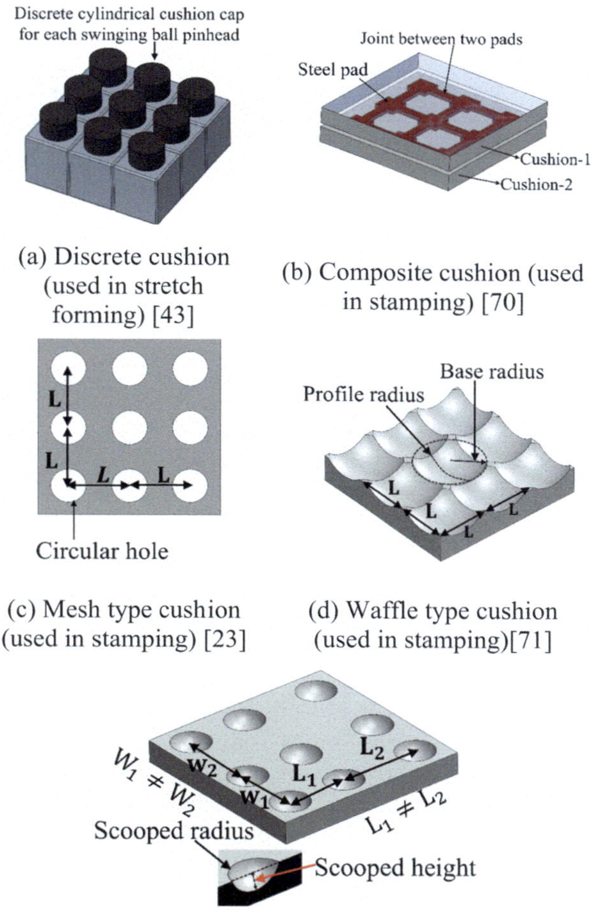


(a) Discrete cushion (used in stretch forming) [43]

(b) Composite cushion (used in stamping) [70]

(c) Mesh type cushion (used in stamping) [23]

(d) Waffle type cushion (used in stamping)[71]

(e) Component specific cushion (used in stretch forming) [25]

**Fig. 11** Elastic cushion types in reconfigurable tooling (Adopted from various resources [23, 25, 43, 70, 71] and are redrawn with suitable modifications)

replicated in the sheet metal component [25]. In this subsection, the main focus is on the different types of cushions used in the reconfigurable tooling is presented (the effect of cushion thickness and hardness on accuracy is already discussed in earlier sections). Various attempts [23, 25, 69–71] have been made to use different elastic cushion types, namely discrete [69], composite [70], mesh [23], waffle [71], and component-specific elastic cushions [25] (shown in Fig. 11), in reconfigurable tooling and enhanced the accuracy, reduced the height of wrinkles as well as dimples on the formed components.

It is well known that local and global shape errors reduce with an increase in the number of pins (i.e., pin size is reduced) in the given forming area. However, the complexity of the design and assembly of a reconfigurable tooling system will be

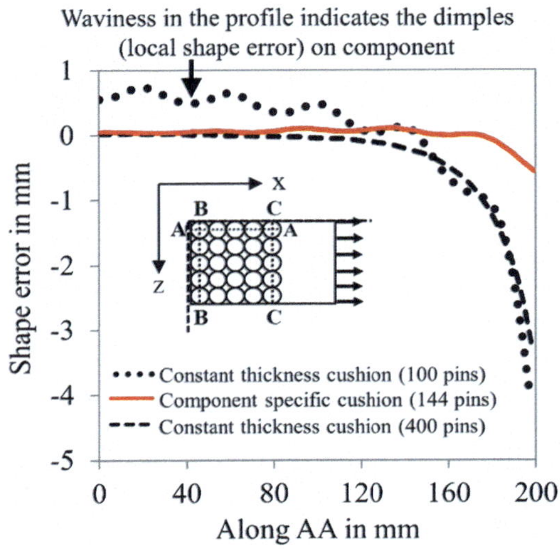


**Fig. 12** Effect of component-specific elastic cushion on shape error and number of pins used in the forming area (Details are available in Shivaprasad et al. [25])

increased. In addition, cost incurred with maintenance of pins will be higher with more number of pins. There is also a possibility of inaccuracies due to the smaller dimensions of the pin, as it can be deflected during the forming due to low flexural rigidity. Recently, an attempt was made by Shivaprasad et al. [25], which demonstrated that the shape error/surface quality of components produced from a lesser pin matrix size (12 × 12; 144 pins only) with component-specific elastic cushion is nearly the same as one formed using the 20 × 20 pin matrix (400 pins) with constant cushion thickness (shown in Fig. 12). In addition, the maximum shape errors obtained with a cushion of constant thickness are 5.2 mm (100 pins) and 3.25 mm (400 pins), whereas it is reduced to 0.7 mm with a component-specific cushion with less number of pins (144 pins) [25]. It can be clearly realized from the above discussion that the amount of energy and material consumption needed for 256 pins (=400–144) and associated mechanical and electrical elements can be avoided with usage of component-specific cushion in reconfigurable tooling. Hence, design of component-specific cushion can be one of the promising approaches in producing accurate components with less energy and material consumption in tool design.

## 4 Reconfigurable Tooling in Circular Manufacturing

Based on the above comprehensive review, it can be realized that reconfigurable tooling is one of the candidates that meet the goals of circular manufacturing/economy practices in the digital era due to its flexibility. It is worth noting that the widespread usage of reconfigurable tooling is limited due to achievable accuracies (without reducing the pin size), automation effort needed, and determination of pin positions by predicting the cushion deformation. This paper emphasizes the challenges that need to be addressed to use reconfigurable tooling concept in circular manufacturing/economy practices, which are discussed below:

- Recycling (cold/hot, solid/liquid) is beneficial economically and ecologically due to the significant reduction in energy consumption and carbon footprints compared to the use of virgin material. Solid-state recycling is always advantageous (if that can be achieved) as it avoids melting. However, the material properties of recycled ones usually differ from the virgin material. Hence, tool design needs to be modified to produce the components with the desired accuracy, but it is cost intensive. In those situations, the use of reconfigurable tooling reduces the cost and time. In addition, testing the deformation behavior of lightweight materials can be possible with reconfigurable tooling.
- Remanufacturing of sheet metal products can be performed by reshaping, i.e., taking used sheet components and reforming (cold recycling) them into new components for new products. Reshaping strategies are meant to directly use a metal panel avoiding shredding and compaction and providing new shape or directly performing new stamping operations on it. This method is often used for automotive EoL parts, where it is possible to salvage the metal from old cars and repurpose it into new parts. Reshaping sheets with reconfigurable/flexible tooling can thus be an excellent way to reduce waste and preserve natural resources.
- The ability to locally adjustment of pins before forming or during the process provides an opportunity to repair dented sheet metal components. One can scan the dented component and transform the point data into a computer-aided design model, which will be used to determine the pin positions.
- One can analyze the design of component-specific elastic cushions for different reconfigurable tooling processes and develop the necessary analytical model to design with and without stretching.

# 5 Summary

Reconfigurable tooling is suitable for circular manufacturing/economy practices to produce a wide variety of geometries cost-effectively and eco-friendly. A comprehensive review on the use of reconfigurable tooling in various manufacturing processes, and challenges to be addressed are presented in this chapter. In addition, the role of elastic cushion design in reconfigurable tooling is discussed. One can conclude that reconfigurable tooling technology has great potential to form the components without using the component-specific tooling quickly with obvious energy and material savings, and it can be considered a sustainable tool in the deformation process if it produces the components with the desired accuracy and quality.

**Acknowledgments** This paper is written as part of CIRMAN-INPART project on Circular Manufacturing, which is supported by the Norwegian Research Council (#32275).

## References


1. Yang DY, Bambach M, Cao J, Duflou JR, Groche P, Kuboki T, Sterzing A, Tekkaya AE, Lee CW (2018) Flexibility in metal forming. CIRP Ann 67:743–765
2. Schuh G, Bergweiler G, Fiedler F, Bickendorf P, Colag C (2019) A review on flexible forming of sheet metal parts. IEEE Int Conf Ind Eng Eng Manag:1221–1225
3. Ingarao G (2017) Manufacturing strategies for efficiency in energy and resources use: the role of metal shaping processes. J Clean Prod 142:2872–2886
4. Cooper DR, Gutowski TG (2017) The environmental impacts of reuse: a review. J Ind Ecol 21: 38–56
5. Reddy V, Lingam R, Cao J (2015) Incremental metal forming processes in manufacturing. Hand Book of Manufacturing Engineering and Technology, pp 411–452
6. Trzepieciński T (2020) Recent developments and trends in sheet metal forming. Metals 10:1–53
7. Falsafi J, Demirci E, Silberschmidt VV (2016) Computational assessment of residual formability in sheet metal forming processes for sustainable recycling. Int J Mech Sci 119:187–196
8. Ali AK, Wang Y, Alvarado JL (2019) Facilitating industrial symbiosis to achieve circular economy using value-added by design: a case study in transforming the automobile industry sheet metal waste-flow into Voronoi facade systems. J Clean Prod 234:1033–1044
9. Abdullah ZT (2021) Assessment of end-of-life vehicle recycling: remanufacturing waste sheet steel into mesh sheet. PLoS One 16:1–17
10. Reddy NV, Suryakumar S, Matsumoto M, Martinsen K, Guldbrandsen S (2019) Incremental sheet metal forming: a candidate process for circular manufacturing. Presented at the EcoDesign 2019 International Symposium
11. Ingarao G, Zaheer O, Fratini L (2021) Manufacturing processes as material and energy efficiency strategies enablers: the case of single point incremental forming to reshape end-of-life metal components. CIRP J Manuf Sci Technol 32:145–153
12. Takano H, Kitazawa K, Goto T (2008) Incremental forming of nonuniform sheet metal: possibility of cold recycling process of sheet metal waste. Int J Mach Tools Manuf 48:477–482
13. Cochrane J (1863) Improvement in presses for bending metallic plates. U.S. Patent No. 39,886, issued September 15
14. Nakajima N (1969) A newly developed technique to fabricate complicated dies and electrodes with wires. Bull JSME 12:1546–1554
15. Walczyk DF, Hardt DE (1998) Discrete dies for sheet metal forming. J Manuf Syst 17:436–454
16. Papazian JM (2002) Reconfigurable forming dies raise the efficiency of small-lot production. Mechnaical Eng 129:52–55
17. Anagnostou EL, Papazian JM (2004) Optimized tooling design algorithm for sheet metal forming over reconfigurable compliant tooling. Proc AIP Conference 712:741–748
18. Socrate S, Boyce MC (2001) A finite element based die design algorithm for sheet-metal forming on reconfigurable tools. J Eng Mater Technol 123:489–495
19. Hardt DE, Webb RD, Suh NP (1982) Sheet metal die forming using closed-loop shape control. CIRP Ann Manuf Technol 31:165–169
20. Galizia FG, ElMaraghy W, ElMaraghy H, Bortolini M, Mora C (2019) The evolution of molds in manufacturing: from rigid to flexible. Procedia Manuf 33:319–326
21. Koc B, Thangaswamy S (2011) Design and analysis of a reconfigurable discrete pin tooling system for molding of three-dimensional free-form objects. Robot Comput Integr Manuf 27: 335–348
22. Li MZ, Cai ZY, Sui Z, Yan QG (2002) Multi-point forming technology for sheet metal. J Mater Process Technol 129:333–338
23. Tolipov A, Elghawail A, Abosaf M, Pham D, Hassanin H, Essa K (2019) Multipoint forming using mesh-type elastic cushion: modelling and experimentation. Int J Adv Manuf Technol 103: 2079–2090
24. Cai ZY, Wang SH, Xu XD, Li MZ (2009) Numerical simulation for the multi-point stretch forming process of sheet metal. J Mater Process Technol 209:396–407

25. Shivaprasad C, Praveen K, Venkata Reddy N (2024) Component specific elastic cushion design to enhance the accuracy with the usage of reconfigurable tools in stretch forming. Proc Inst Mech Eng Part B J Eng Manuf. 238:406-418
26. Cai ZY, Sui Z, Cai FX, Liu L (2013) Continuous flexible roll forming for three-dimensional surface part and the forming process control. Int J Adv Manuf Technol 66:393–400
27. Panuoiu V, Boazu D (2017) Hydro-multipoint forming, a challenge in sheet metal forming. Lect Notes Mech Eng:79–94
28. Alavizadeh SM, Shahbazi Karami J (2019) Experimental and numerical investigation on metal tubes forming with a novel reconfigurable hydroforming die based on multi-point forming. Prod Eng 13:489–500
29. Elghawail A, Essa K, Abosaf M, Tolipov A, Su S, Pham D (2019) Low-cost metal-forming process using an elastic punch and a reconfigurable multi-pin die. Int J Mater Form 12:391–401
30. Peng H, Chen X, Liu H (2019) Numerical simulation and experiment of wrinkle for multipoint thermoforming of resin sheet. IOP Conf Ser Mater Sci Eng 631:0220025
31. Simon D, Kern L, Wagner J, Reinhart G (2014) A reconfigurable tooling system for producing plastic shields. Procedia CIRP 17:853–858
32. Zhang H, Lu B, Chen J, Feng S, Li Z, Long H (2017) Thickness control in a new flexible hybrid incremental sheet forming process. Proc Inst Mech Eng Part B J Eng Manuf 231:779–791
33. Wang S, Cai Z, Li M (2010) Numerical investigation of the influence of punch element in multi-point stretch forming process. Int J Adv Manuf Technol 49:475–483
34. Cherukupally S, Konka P, Venkata Reddy N (2022) Enhancement of accuracy in multi-point stretch forming: cushion stretching. Manuf Lett 33:205–213
35. Wang S, Cai Z, Li M, Lan Y (2012) Numerical simulation on the local stress and local deformation in multi-point stretch forming process. Int J Adv Manuf Technol 60:901–911
36. Wang SH, Cai ZY, Li MZ, Lan YW (2011) Numerical analysis of the multi-point stretch forming process of aircraft outer skin part. Adv Mater Res 154–155:1068–1072
37. Wang SH, Cai ZY, Li MZ (2010) FE simulation of shape accuracy using the multi-point stretch-forming process. Int J Mater Prod Technol 38:223–236
38. Cai ZY, Yang Z, Che CJ, Li MZ (2016) Minimum deformation path sheet metal stretch-forming based on loading at discrete points. Int J Adv Manuf Technol 87:2683–2692
39. Yang Z, Cai ZY, Che CJ, Li MZ (2016) Numerical simulation research on the loading trajectory in stretch forming process based on distributed displacement loading. Int J Adv Manuf Technol 82:1353–1362
40. Wang Y, Li MZ (2014) Research on three-dimensional surface parts in multi-gripper flexible stretch forming. Int J Adv Manuf Technol 71:1701–1707
41. Cai ZY, Liang X bo, Yang Z, Li XJ (2017) Force-controlled sheet metal stretch-forming process based on loading at discrete points. Int J Adv Manuf Technol 93:1781–1789
42. Wang Y, Li M, Wang D, Wang A (2014) Modeling and numerical simulation of multi-gripper flexible stretch forming process. Int J Adv Manuf Technol 73:279–288
43. Xing J, Cheng YY, Yi Z (2019) The effect of swinging ball heads with different arrangements in multi-point stretch-forming process. Materials 12:1–10
44. Xing J, Li M zhe, Cheng Y yan, Wang B long, Yang Z, Wang Y (2016) Effect of the arrangement of the punch units in multi-point stretch forming process. Int J Adv Manuf Technol 86:2309–2317
45. Li M, Liu Y, Su S, Li G (1999) Multi-point forming: a flexible manufacturing method for a 3-d surface sheet. J Mater Process Technol 87:277–280
46. Cai ZY, Li MZ (2002) Multi-point forming of three-dimensional sheet metal and the control of the forming process. Int J Press Vessel Pip 79:289–296
47. Zhongyi C, Mingzhe L (2001) Optimum path forming technique for sheet metal and its realization in multi-point forming. J Mater Process Technol 110:136–141
48. Liu W, Jia SS, Zhang CM, Li MZ (2007) Generation and suppression of local severe plastic deformation in sectional multi-point forming. Int J Adv Manuf Technol 32:705–710

49. Li MZ, Han QG, Cai ZY, Liu CG, Su SZ (2014) Multipoint forming. In: Comprehensive materials processing (eds). Elsevier, pp 107–147
50. Peng L, Lai X, Li M (2006) Transition surface design for blank holder in multi-point forming. Int J Mach Tools Manuf 46:1336–1342
51. Qian ZR, Li MZ, Tan FX (2007) The analyse on the process of multi-point forming for dish head. J Mater Process Technol 187–188:471–475
52. Li MZ, Cai ZY, Liu CG (2007) Flexible manufacturing of sheet metal parts based on digitized-die. Robot Comput Integr Manuf 23:107–115
53. Cai ZY, Li MZ, Di CX (2006) Digitized die forming system for sheet metal and springback minimizing technique. Int J Adv Manuf Technol 28:1089–1096
54. Ramezani M, Ripin ZM (2012) Rubber-pad forming processes: technology and applications. Elsevier
55. Selmi N, Belhadjsalah H (2013) Finite element and experimental investigation of the multipoint flexible hydroforming. Key Eng Mater 554–557:1290–1297
56. Boers SHA, Schreurs PJG, Geers MGD (2007) Optimum forming strategies with a 3D reconfigurable die. Poster session presented at Mate Poster Award 2006: 11th Annual Poster Contest
57. Tolipov A, Hassanin H, El-Sayed MA, Eldessouky HM, Alsaleh NA, Alfozan AK, Essa K, Ahmadein M (2022) Multipoint forming using hole-type rubber punch. Metals 12:491
58. Peng H, Li M, Liu C, Cao J (2013) Study of multi-point forming for polycarbonate sheet. Int J Adv Manuf Technol 67:2811–2817
59. Cao J, Fu W, Li M, Peng H (2015) Research on the elastic cushion conditions in multi-point thermoforming for polymer sheet. Proc Inst Mech Eng Part E J Process Mech Eng 229:149–161
60. Wimmer MSJ, Lušić M, Maurer C (2016) Vacuum assisted multipoint moulding—a reconfigurable tooling technology for producing spatially curved single-item CFRP panels. Procedia CIRP 57:368–373
61. Hu ZQ, Li MZ, Cai ZY, Gong XP (2009) Continuous flexible forming of three-dimensional surface parts using bendable rollers. Mater Sci Eng A 499:234–237
62. Cai ZY, Li MZ, Lan YW (2012) Three-dimensional sheet metal continuous forming process based on flexible roll bending: principle and experiments. J Mater Process Technol 212:120–127
63. Peng W, Ou H, Becker A (2019) Double-sided incremental forming: a review. J Manuf Sci Eng Trans ASME 141:1–12
64. Praveen K, Ch S, Venkata Reddy N (2022) Effect of support force on quality during double-sided incremental forming: an experimental and numerical study. Int J Adv Manuf Technol 122: 4275–4292
65. Li XJ, Li MZ, Liu CG, Cai ZY (2009) Principle and simulation study on multi point-single point incremental combined forming for sheet metal. In: Materials Science Forum Trans Tech Publications Ltd 626:273–278
66. Boudhaouia S, Gahbiche MA, Giraud E, Ben Salem W, Dal Santo P (2016) Investigation of hybrid forming process. Int J Mech Mechatronics Eng 10:1762–1766
67. Kelkar A, Nagi R, Koc B (2005) Geometric algorithms for rapidly reconfigurable mold manufacturing of free-form objects. CAD Comput Aided Des 37:1–16
68. Zäh MF, Hagemann F, Teufelhart S (2009) Form-flexible tools for injection molding: approach for the economic application of injection molding for small lot sizes. Prod Eng 3:281–285
69. Xing J, Cheng Y yan, Li M zhe (2017) Research on swinging unit multi-point die with discrete elastic cushion in flexible stretch forming process. Int J Adv Manuf Technol 91:237–245
70. Erhu Q, Mingzhe L, Rui L, Liang Z, Zhuo Y (2018) Inhibitory effects of a flexible steel pad on wrinkling in multi-point die forming. Int J Adv Manuf Technol 95:2413–2420
71. Moheen M, Abdelwahab A, Hassanin H, Essa K (2020) Reconfigurable multipoint forming using waffle-type elastic cushion and variable loading profile. Metals 13:1–21

# Sustainable Disjoining of Aluminium Multi-material Joints

**Siri Marthe Arbo, Sverre Gulbrandsen-Dahl, Kristian Martinsen, Mitsutaka Matsumoto, and S. Surya Kumar**

**Abstract** Multi-material joints consisting of aluminium joined with a higher strength metal such as steel, titanium or copper enable the production of hybrid products utilizing the strength of both metals, at the same time achieving a lower weight and design flexibility. However, upon recycling, the separation of the two metals into the original base metal constituents is a major challenge since cross-contamination between different metals can degrade the recycling possibilities. This chapter discusses possible solutions and methods for the automated disjoining of multi-material joints by the application of fracture strain, utilizing the differences in mechanical properties between the base materials and the nature of the intermetallic phases to separate the materials. The chapter exemplifies the method for steel–Al and Cu–Al welds, including results from laboratory experiments showing how this method can ensure a separation with minimum iron contamination, and suggest how the disjoining using fracture strain can be automated.

**Keywords** Disjoining · Aluminium · Steel · Copper · Welded joints · Fracture straining

## 1 Introduction

### *1.1 Joining of Dissimilar Materials*

The durability of products is important for sustainability, and the design of products with the ability to sustain extensive use in several circular loops of reuse and

---

S. M. Arbo · S. Gulbrandsen-Dahl · K. Martinsen (✉)
SINTEF Manufacturing AS, Raufoss, Norway
e-mail: kristian.martinsen@sintef.n

M. Matsumoto
National Institute of Advanced Industrial Science and Technology, Tsukuba, Japan

S. S. Kumar
Indian Institute of Technology, Hyderabad, India



S. Fukushige et al. (eds.), *EcoDesign for Circular Value Creation: Volume I*,
https://doi.org/10.1007/978-981-97-9068-5_36

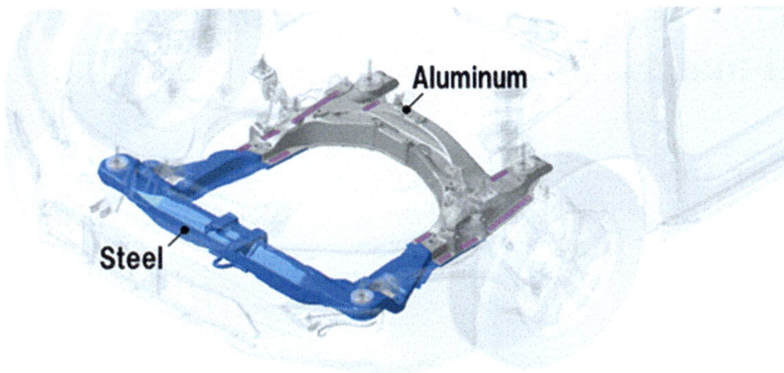


**Fig. 1** Honda Accord sub-frame FSW Al-steel joint [2]

remanufacturing is a key factor. On the other hand, the reduction of mass and lighter products is important for energy efficiency in transport and mobility. One way to achieve both durability and lightweight is to combine dissimilar materials in hybrid structures utilizing a mix of different material properties [1], for example, by combining aluminium and steel components. Combinations of steel and aluminium have been used extensively in the automotive sector, and in many applications, the two dissimilar materials are welded together.

An industrial case of Friction Stir Welding (FSW) of aluminium to steel is the Honda Motor Co., Ltd case of the front sub-frame used in the Honda Accord from 2012 [2], see Fig. 1. FSW is a solid-state process where a rotating pin is stirring the materials into a joint, resulting in a small Heat-affected zone (HAZ). The FSW allows for the introduction of more aluminium into the sub-frame structure resulting in 25% weight reduction, 50% less welding electricity consumption, 20% increased rigidity of the mounting point and welding strength similar to arc fusion welding. When such products reach a final End-of-Life (EoL) and remelting of the materials are the preferred option, a clean and effective separation of the different materials is needed. Shredding into smaller fragments followed by automated sorting is currently a widely used method. Novel sensing technologies such as LIBS (Laser Induced Breakdown Spectroscopy) [3] are being implemented to sort the Al scrap not only from other elements, but also separate between alloys with different alloying elements. Still, this will not ensure separation without contamination and for high-end products, the use of recycled Al is still very limited.

Cross contaminations such as Fe or Cu contamination of Al, complicates the recycling process and reduces the recyclability of aluminium scrap. If additional cleaning/dilution steps are necessary in the remelting process, this will consume additional energy, time and resources, increase the overall cost of recycling and lower the efficiency of aluminium recovery. This is especially the case for high-strength wrought aluminium alloys. One approach is to purify or dilute the material and/or improve the quality when remelting. Separation by disassembly or disjoining the components and keeping the material streams separated after disjoining is another option where purification is less needed. For a separation strategy to be a sustainable solution, automation might be needed for socially (avoiding HES hazards), economically as well environmentally reasons.

## 1.2 Metallurgical Considerations During Multi-material Disjoining and Subsequent Recycling

Metals can be mixed into different alloys (such as Al-Cu, Cu-Al-Fe, etc.), to provide improved mechanical properties for specific purposes. There are, however, limitations established for the maximum amount of each alloying element the alloys should contain without risking a reduction in performance. Al alloys are especially sensitive to cross-contamination by too high addition of, for instance, Fe and Cu [4]. It can lead to reduced strength, ductility, corrosion resistance and fatigue resistance of the recycled aluminium [5, 6]. For example, the increased presence of Fe can form large intermetallic phases, such as $FeAl_3$, which are brittle and reduce alloy performance [7]. Hence, a too high Fe content can reduce the lifespan and performance of the recycled aluminium components.

By minimizing excessive Fe and Cu contamination above critical alloying limitations in aluminium scrap during disassembly and implementing effective purification techniques, the quality and recyclability of the recycled aluminium can be improved, contributing to a more sustainable and efficient circular economy. Often, the metal components are shredded before recycling. In recent years, advances have been made in sorting to obtain as pure metal fractions as possible. An extensive introduction to advances in scrap sorting is found in Raabe et al. [4] for Al alloys. However, a lot of the potential lies in optimizing the disjoining strategy, especially for multi-material products.

On the other hand, Al contamination of steel or copper is not as critical. For over 100 years, aluminium has been actively added to steel to remove oxygen and nitrogen [8], and Cu–Al is a well-known aluminium–bronze alloy. Al is also a majority alloying element in the commonly used Ti-alloys such as TiAl6-V4. Hence, during the recycling of copper, titanium and steels, Al contamination can be accepted in the production of high-quality alloys of equal quality as alloys produced from primary sources [8]. In today's steel industry, secondary material sources are widely used.

## 2 Multi-material Joining

One of the key challenges with multi-material welding/joining of metals is the formation and growth of intermetallic phases at the joint interface at elevated temperatures. The intermetallic phases form due to the limited solubility between the metals. The intermetallic phase growth ensures a solid metallurgical joint between the metals. However, these phases have been reported to be brittle, and if the growth is not controlled, the layer will deteriorate the mechanical properties of the joint [9, 10]. Commonly reported intermetallic phases and reported hardness values are listed in Table 1 for Steel–Al, Cu–Al, and Ti–Al joints. The presence of intermetallic phases reduces the possible applications and performance of the joints. The thickness of the intermetallic phase layer is strongly influenced by the temperatures reached during joining or during post-welding heat treatments, i.e., thicker layers are formed for fusion welding techniques compared to solid-state welding techniques.

Several thermal joining techniques have been demonstrated to make robust multi-material joints, such as: resistance spot welding [15], laser welding [16], friction welding [17], FSW [18], roll bonding [19] and hybrid metal extrusion & bonding [20]. All these techniques produce solid-state connections between aluminium and steel based on the formation of intermetallic phases in the joining zone. For some techniques, such as resistance spot welding and the friction-based welding techniques, a heat-affected zone on the aluminium side of the joint is also formed due to the heat input during the welding process. Studies of the design of aluminium to steel connections cover a wide variety of designs: butt welding [21], overlap welding [22], composite laminates [23], welding based on transition elements [24], and overlap joints [25]. All these designs have in common that the structural weak link is either on the aluminium side or at the joint interface.

**Table 1** Commonly reported intermetallic phases in multi-material joints

| Metal combination | Common intermetallic phases | Reported hardness values |
|---|---|---|
| Steel–Al | $FeAl_3/Fe_4Al_{13}$<br>$Fe_2Al_5$ | ~ 800–900 HV [11]<br>~ 1000 HV0.5 [11, 12] |
| Cu–Al | $Al_2Cu$<br>CuAl<br>$Al_3Cu_4$<br>$Al_4Cu_9$ | 600 HV10 [13]<br>900 HV10 [13]<br>850–900 HV10 [13]<br>600–700 HV10 [13] |
| Ti–Al | $TiAl_3$ | 600–700 HV [14] |

## 3 Disjoning of Multi-material Joints by Fracture Strain

Results and evaluations of disjoining of multi-material adhesively joined can be found in the literature [26]. However, little can be found on disjoining of multi-materials produced through joining by welding. An efficient method that could enable a desired disjoining of welded multi-material joints could be to utilize the nature that lies within a multi-material joint consisting of dissimilar metals, i.e., the differences in mechanical strength between the metals and/or the brittle intermetallic phase layer along the joint interface. Straining the joint parts until fracture (by pulling or stretching) involves applying force to separate the dissimilar base metals, breaking the joint by exceeding the mechanical strength of the metals and or weld interface, as performed during mechanical testing such as traditional tensile tests.

Depending on the nature of the joint, three scenarios can occur, illustrated in Fig. 2:

1. *Fracture in the joint due to lack of metallic bonding in the joint.* If the joint strength is poor, due to lack of sufficient bonding, the fracture will occur at the original interface. This will usually leave a clean interface between the dissimilar metals.
2. *Fracture in the weaker material, beyond intermetallic layers (usually thin intermetallic layer, end strong interface).* If a strong joint with a minimal intermetallic phase layer is produced, the forces can be strategically applied in a way that the fracture occurs on the weakest material side. This requires careful design and control to ensure fracture at the desired location. When the fracture occurs on the weak material side, the risk of contamination of the weaker material during disjoining is significantly reduced, and enables selective recycling without significant mixing, allowing recycled material to maintain the integrity, purity, and desired properties.
3. *Fracture in brittle intermetallic layer (usually thick intermetallic layer).* If the joint strength is poor due to the formation of a thick intermetallic phase layer, the fracture might occur in the brittle intermetallic layer. This will most probably leave residue on both parts after fracture strain disjoining.

The disjoining methods have been observed for both fusion and solid-state welding techniques such as by TIG welding [27] and Cold Metal Transfer [28] where Al-steel joints are failing in the intermetallic layer. Another example shows FSW failing at the joint interface due to insufficient bonding in Al-steel joints [29]. Fracture in the base material or intermetallic layer depending on thickness is also reported for Al-Ti joints [30]. Finally, fracture in the Al has been observed for Electromagnetic pulse welds between Al-Cu as shown by Wang et al. [31].

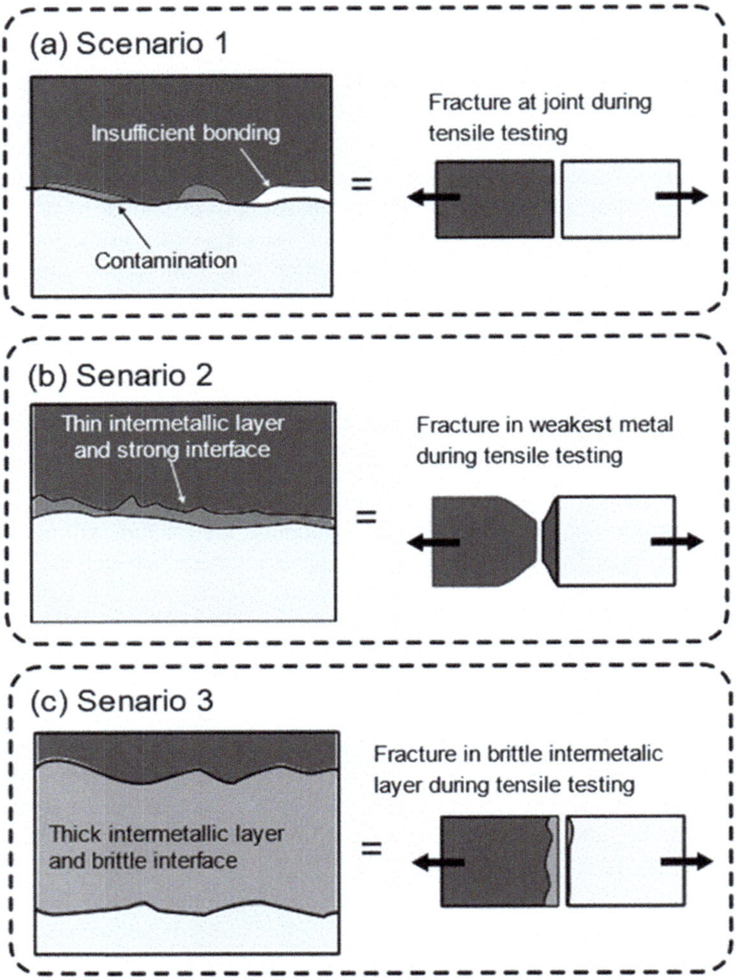


**Fig. 2** Illustration of the three scenarios of fracture in a multi-material joint between a material of different strength, the dark material is the weakest. (**a**) Scenario 1, fracture in a weak joint, (**b**) Scenario 2. fracture in the weakest material and (**c**) Scenario 3, fracture along the weak interface due to thick and brittle intermetallic layer

## 4 Results From Laboratory Tests Disjoining Multi-material Al Joints

The concept of disjoining by fracture strain was tested for three different thermal joining methods:

1. Roll bonding of steel–Al
2. Hybrid Metal Extrusion and Bonding (HYB) of steel–Al, Ti–Al and Cu–Al
3. FSW of Cu–Al

After joining, the joints were strained until fracture, either in the form of a peel-test or tensile testing of machined dog-bone specimens 90 degrees across the joints. The fracture surfaces were examined to assess possible cross-contamination.

## 4.1 Roll Bonding

Roll Bonding (RB) is a solid-state welding technique used to manufacture composite materials consisting of two or more dissimilar metals. In RB, metal sheets are stacked together and sent through a rolling mill achieving a certain reduction in thickness. A metallic bond is formed between the stacked metal layers as a result of the high normal pressure. According to the film theory, the surface layer of the metals will crack during rolling because of the surface expansion [32, 33]. The high normal pressure ensures that fresh metal can be extruded locally through the surface cracks. A metallurgical bond is then formed where fresh metals from each opposing metal surface meet [32, 33]. See illustration in Fig. 3. The strength of the bond is a function of the achieved plastic deformation as well as the properties of the metal (s) to be joined [33]. The bond strength is also dependent on other process parameters such as surface preparation, rolling temperature and post- or pre-rolling heat treatments [34].

The process allows for a high level of design freedom regarding material combinations for the production of sheets, foils or plates. A thorough introduction to the technique and its variants and future perspectives is found in Khan et al. [19]. Several articles have shown the possibilities of joining dissimilar metals by RB [35–39].

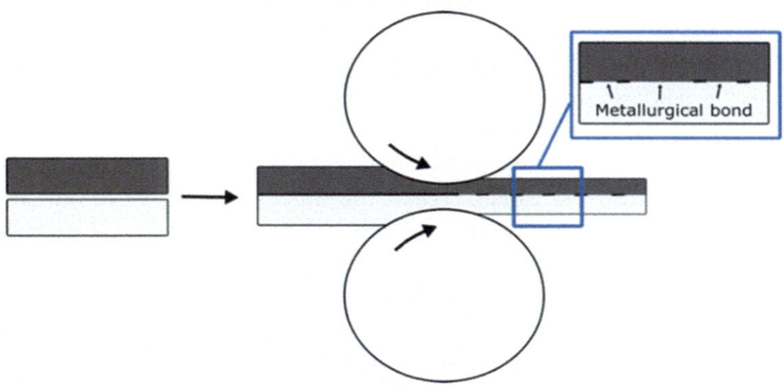


**Fig. 3** Illustration of the principles of roll bonding between metal plates/sheets

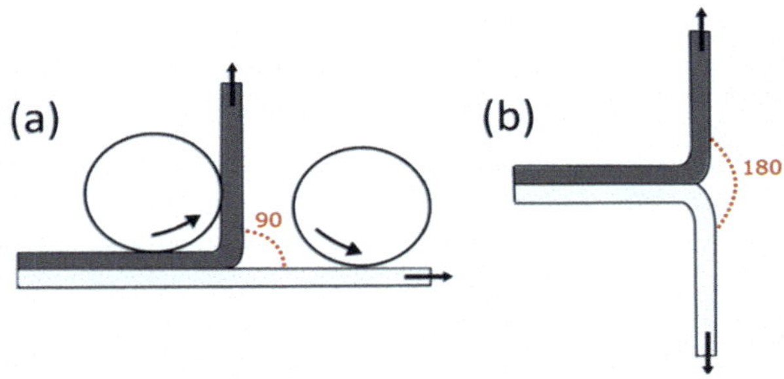


**Fig. 4** Illustration of the setup for straining of multi-material components produced with RB, (**a**) illustrates the rolling and (**b**) the peeling

RB components differ from other joining techniques by producing sheets or plates where the joint is a part of the entire component. Also, the joined sheets/ plates are in some cases very thin and can fracture easily upon strain. However, published literature has shown that by performing peel-testing, the dissimilar metals can be separated with a simple setup as illustrated in Fig. 4. The chosen setup will depend on the thickness of the material and its strength. For instance, the 90-degree peel test setup only requires bending and stretching of one of the metals in the composite and thus could be more suited in multi-material composites where one alloy is very weak or thin.

**Example of Disjoining of Al–Steel Joint Produced by Roll Bonding**

Figure 5 show the results from peel testing of a roll bonded joint consisting of a low alloyed steel (IF Steel) and a AA6082 aluminium alloy. The production and characterization of the joints are found in Arbo et al. [35]. Additional images of the cross-section after the fraction are added in this study to by the author.

The images show the initial joint interface for two different joints, (a.1) as rolled and (b.1) after post-rolling T6 heat treatments where an intermetallic layer consisting of α-AlFeSi, $Fe_2Al_5$ and $Fe_4Al_{13}$ has formed [35]. The steel and Al fracture surfaces after peeling obtained by scanning electron microscopy (SEM) and SEM combined with Energy Dispersive Spectroscopy (EDS) are shown in (a.2-3) and (b.2-5) for the two joints respectively. In the specimen without post-aging there is no sign of intermetallic phases at the interface (see Fig. 5 a.1). This implies the fracture has occurred in the Al base material during straining. A ductile fracture has occurred in the Al, leaving Al residue on the steel fracture surface. Some steel fragments are also found on the Al fracture surface after disjoining. In this case, the amount of surface Fe contamination was very small. When the amount of Fe is within the alloy tolerance, the contamination will not significantly hamper the recycled Al properties.

When an intermetallic layer is present in substantial thickness (max of 0.3 μm), the fracture has occurred in the intermetallic layer, leaving residue on both fracture surfaces. Depending on the thickness of the layer and the respective amount of residue on each fracture surface, the quality of the subsequent recycled material could suffer.

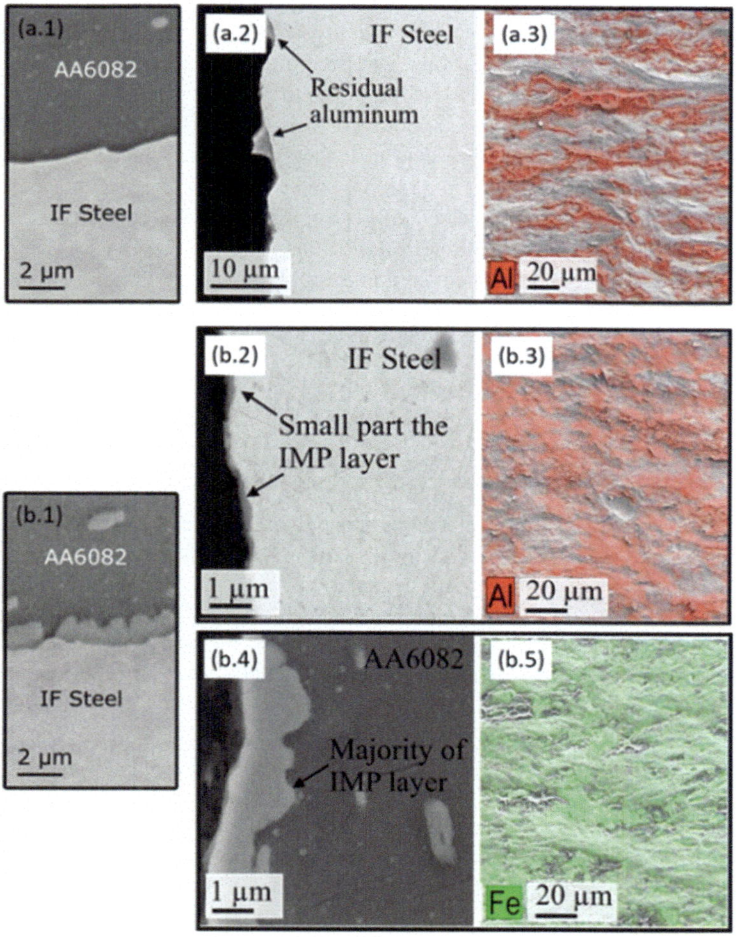


**Fig. 5** Characteristics of the interface between AA6082 and IF steel; (**a**.1) as rolled, (**b**.1) after post-rolling T6 heat treatment where an intermetallic phase layer has formed along the interface. (**a**.2-3) and (**b**-2-5) show the fracture surfaces after disjoining by peel testing

## *4.2 Friction Stir Welding*

Friction Stir Welding (FSW) is a joining process without adding material, where a rotating FSW tool creates simultaneous rotation and translational motion, which leads to asymmetry between the adjoining sides [40]. FSW, Friction Stir Processing, Friction Stir Casting (also known as FTMP), Friction Stir Micro forming, Friction Stir Powder Processing, Friction Stir Channelling, and Friction Stir Spot Welding (FSSW) are all different branches of the Friction Stir process family [1].

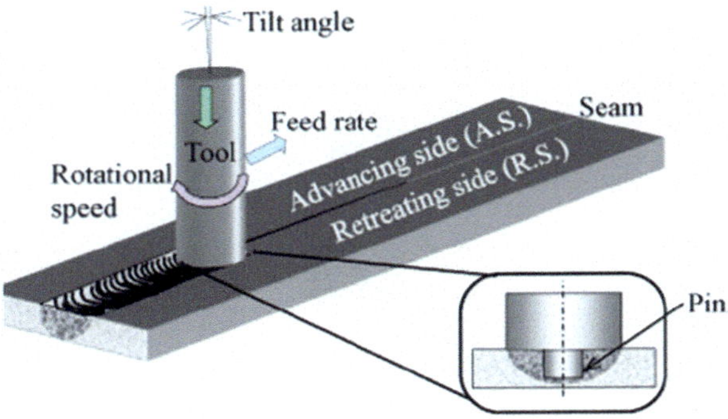


**Fig. 6** FSW process for joining workpieces from [47]

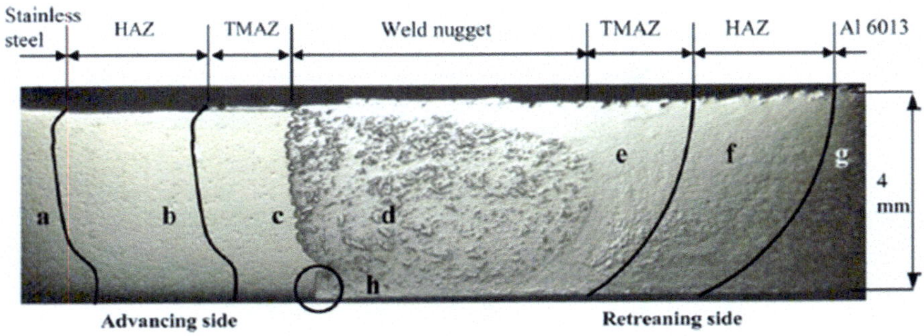


**Fig. 7** Cross section of friction stir welded Al 6013-T4 alloy to X5CrNi18–10 stainless steel [43]

The main parameters which influence weld quality are type of material(s), angle of tool (usually 90°), traversing tool speed, rotation speed, pressure of the tool, pin length and diameter, pin geometry and shoulder diameter [41, 42]. An illustration of the setup is shown in Fig. 6. An FSW weld microstructure is divided into four regions [43] (see Fig. 7). Unaffected material (a, g), Heat affected zone (HAZ) (b, f), thermo-mechanically affected zone (TMAZ) in (c, e) and the weld nugget (d). In the TMAZ, there might be recrystallization producing different grain structures.

FSW minimizes brittle intermetallic phases, metallurgical incompatibility, differences in melting points and thermal mismatch [44]. Mukuna & Akinlabi [45] give a general overview of the friction stir welding of dissimilar materials, including aluminium, steel, magnesium and titanium. There are several reports from successful joining of dissimilar metals by FSW for instance Al–steel [43, 44, 46, 47], Al–Cu [45, 48–50] and Al–Ti [51].

**Example of Disjoining of Al–Cu Joint Produced by FSW**

Figure 8 shows the characteristics of a Cu-Al joint produced by FSW, consisting of pure Cu and 6101 Al alloy in T7 condition. The weld nugget consists of fine Cu particles stirred into the Al as shown in Fig. 8a. The hardness curve obtained across the weld interface is shown in Fig. 8b. High hardness Cu particles make the nugget

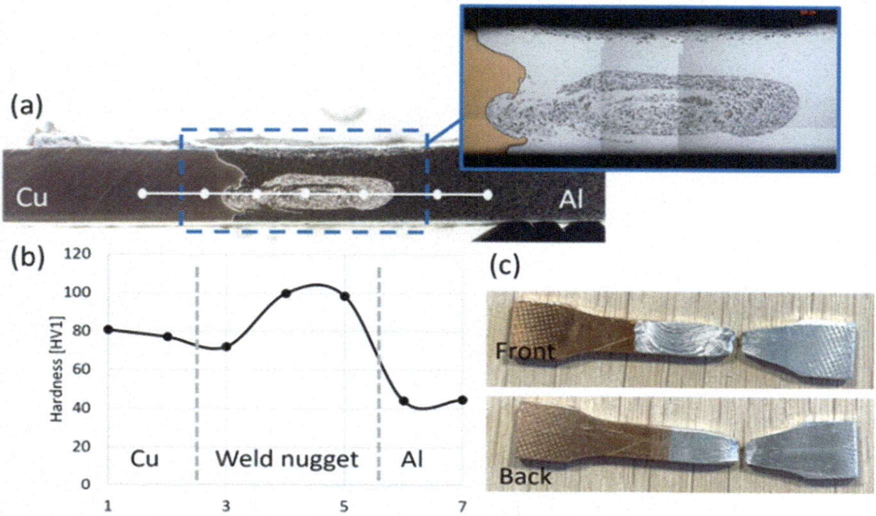


**Fig. 8** (**a**) Interface characteristics of a FSW between Cu and Al. (**b**) Hardness measurements across the interface, position of indents shown in (**a**). (**c**) Images of the fractured specimens after tensile strain

zone hardness higher than the Cu base material. During tensile strain, the fracture naturally occurs in the Al base material, as it is the weakest part of the joint. For an FSW multi-material joint, where the weld nugget could achieve higher hardness than the weakest metal, in this case, Al, a large amount of residue will be left on the strong metal side after disjoining. However, this is not detrimental to subsequent recycling, as previously introduced. Similar fracture characteristics have been reported also by Cao et al. [50] for FSW. However, for laser welded Cu–Al as in the work by Zuo et al. [52], the fracture is observed to occur in the intermetallic phase layer, leaving residue on both surfaces.

## 4.3 *Hybrid Metal Extrusion and Bonding*

Hybrid Metal extrusion and Bonding (HYB) technology was developed by HyBond as a solid-state welding technique utilizing continuous extrusion to squeeze a filler wire in between two metals to be joined under high pressure thus achieving a joint. The principle of the technique is illustrated in Fig. 9. The potential of technology has been reported for instance in the work by Bergh et al. [53], Sandnes et al. [54] and Bergh et al. [55] for multi-material joining. Overall, the technique has demonstrated that only a very thin intermetallic phase layer is formed after joining, resulting in good mechanical properties [53, 54].

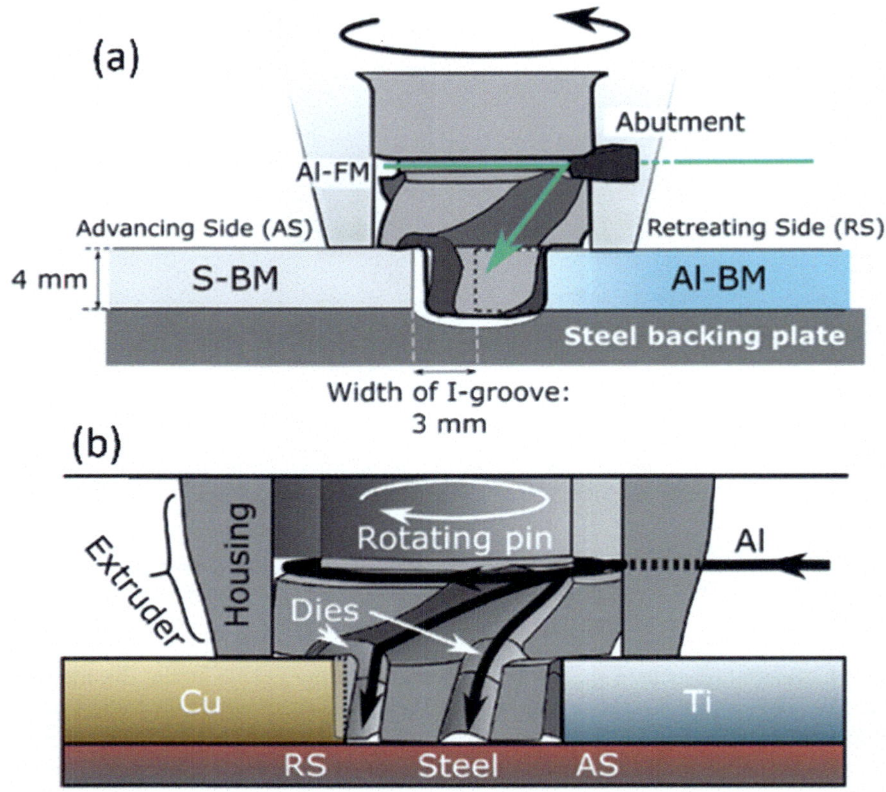


**Fig. 9** Illustration of the HYB technology and setup during (**a**) butt welding of two dissimilar metals, from [54] and (**b**) multi-material welding from [55].

**Example of Disjoining of Al-Ti, Al-Cu and Al-Steel Joints Produced by HYB**
The HYB technology was used to demonstrate the potential for multi-material joining by producing a joint between AL–steel, Al–Cu and Al–Ti simultaneously, as illustrated in Fig. 9b. The Al alloy was an AA6082 wire extruded into the gap between the three dissimilar metals as illustrated in the figure. From the joint, small specimens used for testing of tensile strain were produced as illustrated in Fig. 10a. The cross-section of the produced specimen is shown in (d).

During straining, the following was observed for the three dissimilar metal combinations, extracted from Bergh et al. [55]. SEM images of the fracture surfaces are shown in Fig. 11, using backscatter mode, i.e., Al appears dark in the images while Cu, Ti and steel are bright. For Al–Cu, the interface experienced a lack of bonding in some areas and poor joint strength, i.e., the specimens fractured at the interface, with only small regions with Al residue, see Fig. 11 (a.1–2). For Al-Ti joints, the fracture occurred in the Al base material as a ductile fracture, see Fig. 11

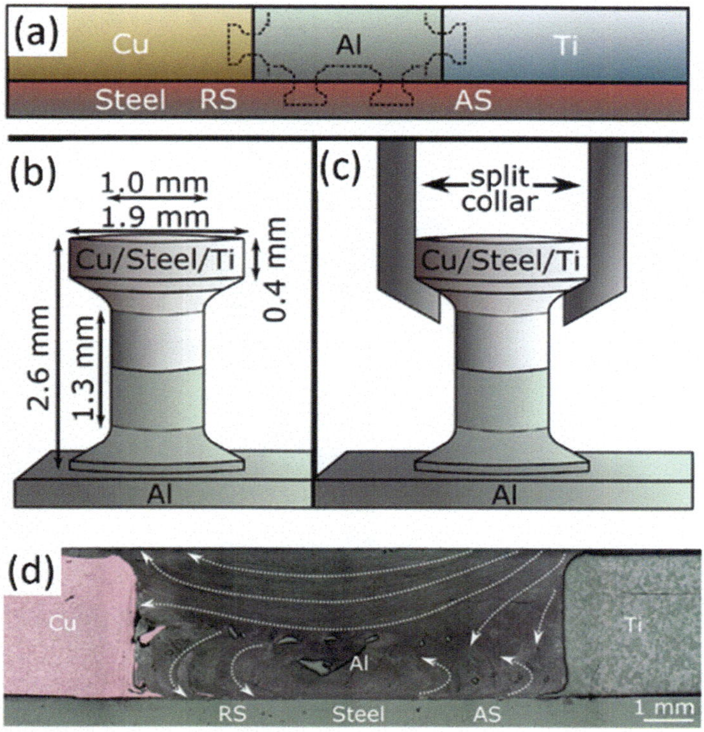


**Fig. 10** (**a**) Illustration of the position of the specimens used for tensile strain investigations. (**b**) and (**c**) show the size and setup during straining. (**d**) show the interface characteristics of the joined cross section. The size of the specimens and the tensile strain setup are shown in (**b**) and (**c**). Further description of the production of the specimens, characterization and testing is found in Bergh et al. [55]

(b.1–2). A thin, approximately 50 nm thick Al–Ti intermetallic layer was found along the interface, providing sufficient metallurgical bonding between the metals.

For the Al–steel joint, a 0.1–1 μm thin Fe–Al intermetallic layer was found after joining. During straining, the fracture occurred at the interface between the metals with a brittle characteristic, see Fig. 11 (c.1–2). Residue from the opposing metal is found on both the steel and Al after disjoining. Overall, the HYB specimen illustrates the three possible scenarios for disjoining that can occur in multi-material joints.

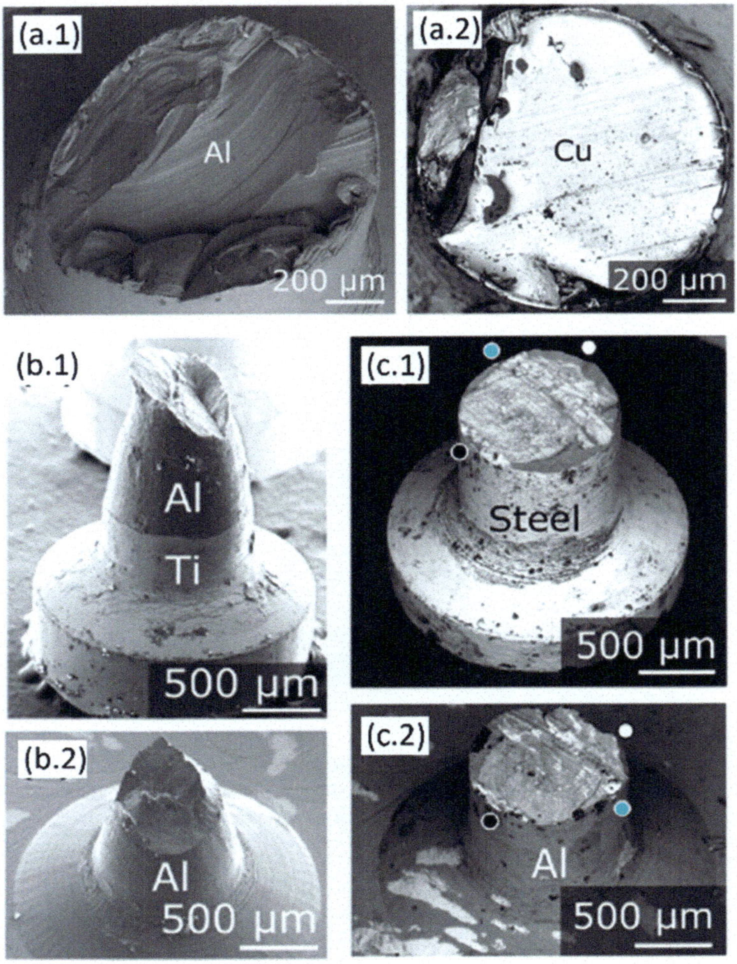


**Fig. 11** SEM images of the fracture surfaces of the HYB multi-material joints from Bergh et al. [55]. (**a**.1–2) Al–Cu joint. (**b**.1–2) Al–Ti joint and (**c**.1–2) Al–steel joint. (Al appears dark in the images)

## 5 Summary and Further Work

This chapter has described fracture strain as a possible process for disjoining dissimilar welded joints. The methodology entails utilizing the nature of the multi-material joints, where two metals with differences in strength are joined together with the risk of formation of intermetallic phases along the interface. The chapter highlights the three possible disjoining scenarios that can occur during straining and how this affects the subsequent recycling of the separated base materials. The chapter shows results from disjoining of roll bonded, HYB and FSW multi-material joints illustrating the described scenarios; (1) Fracture in the joint due to lack of metallic bonding, (2) Fracture in the weaker material due to usually thin intermetallic layer and strong interface and (3) Fracture in brittle intermetallic layer due to thickness.

The results show how contamination of the Al-part of the weld can be avoided or minimized through this method. This way, the Al parts can be recycled, remelted, and reused for high-end products without any loss of performance compared to contemporary virgin alloys. Further work will contain tests on automation of the fracture strain process.

**Acknowledgement** We would like to thank The Norwegian Research Council for supporting this work through the SFI Manufacturing (520460).

## References

1. Martinsen K, Hu SJ, Carlson BE (2015) Joining of dissimilar materials. CIRP Ann Manuf Technol *64*:679–699. https://doi.org/10.1016/j.cirp.2015.05.006
2. Honda Motor Co., L (2012) Honda develops new technology to weld together steel and aluminum and achieves world's first application to the frame of a mass-production vehicle—Hybrid-Structured Front Subframe Achieves Both Weight Reduction and Increased Rigidity
3. Werheit P, Fricke-Begemann C, Gesing M, Noll R (2011) Fast single piece identification with a 3D scanning LIBS for aluminium cast and wrought alloys recycling. J Anal At Spectrom *26*:2166–2174. https://doi.org/10.1039/C1JA10096C
4. Raabe D, Ponge D, Uggowitzer PJ, Roscher M, Paolantonio M, Liu C, Antrekowitsch H, Kozeschnik E, Seidmann D, Gault B et al (2022) Making sustainable aluminum by recycling scrap: the science of "dirty" alloys. Prog Mater Sci *128*:100947. https://doi.org/10.1016/j.pmatsci.2022.100947
5. Kumari S, Wenner S, Walmsley JC, Lunder O, Nisancioglu K (2019) Copper enriched by dealloying as external cathode in intergranular corrosion of aluminium alloy AA6005. Corros Sci *158*:108090. https://doi.org/10.1016/j.corsci.2019.07.014
6. Ji S, Yang W, Gao F, Watson D, Fan Z (2013) Effect of iron on the microstructure and mechanical property of Al–Mg–Si–Mn and Al–Mg–Si diecast alloys. Mater Sci Eng A *564*:130–139. https://doi.org/10.1016/j.msea.2012.11.095
7. Que Z, Wang Y, Mendis CL, Fang C, Xia J, Zhou X, Fan Z (2022) Understanding Fe-containing intermetallic compounds in Al alloys: an overview of recent advances from the LiME research hub. Metals *12*:1677. https://doi.org/10.3390/met12101677
8. Hsiao CC (1958) Fine aluminium nitride precipitates in steel. Nature *181*:1527–1528. https://doi.org/10.1038/1811527b0
9. Bozzi S, Helbert-Etter AL, Baudin T, Criqui B, Kerbiguet JG (2010) Intermetallic compounds in Al 6016/IF-steel friction stir spot welds. Mater Sci Eng A *527*:4505–4509. https://doi.org/10.1016/j.msea.2010.03.097
10. Chen N, Wang M, Wang H-P, Wan Z, Carlson BE (2018) Microstructural and mechanical evolution of Al/steel interface with Fe2Al5 growth in resistance spot welding of aluminum to steel. J Manuf Process *34*:424–434. https://doi.org/10.1016/j.jmapro.2018.06.024
11. Yasuyama M, Ogawa K, Taka T (1996) Spot welding of aluminium and steel sheet with an insert of aluminium clad steel sheet: dissimilar metal joining of aluminium and steel sheet (1st report). Weld Int *10*:963–970. https://doi.org/10.1080/09507119609549121

12. Kobayashi S, Yakou T (2002) Control of intermetallic compound layers at interface between steel and aluminum by diffusion-treatment. Mater Sci Eng A *338*:44–53. https://doi.org/10.1016/S0921-5093(02)00053-9
13. Chen C-Y, Hwang W-S (2007) Effect of annealing on the interfacial structure of aluminum-copper joints. Mater Trans *48*:1938–1947. https://doi.org/10.2320/matertrans.MER2006371
14. Ma Z, Wang C, Yu H, Yan J, Shen H (2013) The microstructure and mechanical properties of fluxless gas tungsten arc welding–brazing joints made between titanium and aluminum alloys. Mater Des *45*:72–79. https://doi.org/10.1016/j.matdes.2012.09.007
15. Miyamoto K, Nakagawa S, Sugi C, Sakurai H, Hirose A (2009) Dissimilar joining of aluminum alloy and steel by resistance spot welding. SAE Int J Mater Manf *2*:58–67. https://doi.org/10.4271/2009-01-0034
16. Sun Z, Ion JC (1995) Laser welding of dissimilar metal combinations. J Mater Sci *30*:4205–4214. https://doi.org/10.1007/BF00361499
17. Fratini L, Buffa G, Cammalleri M, Campanella D (2013) On the linear friction welding process of aluminum alloys: experimental insights through process monitoring. CIRP Ann *62*:295–298. https://doi.org/10.1016/j.cirp.2013.03.056
18. Mishra RS, Ma ZY (2005) Friction stir welding and processing. Mater Sci Eng: R: Reports *50*:1–78. https://doi.org/10.1016/j.mser.2005.07.001
19. Khan HA, Asim K, Akram F, Hameed A, Khan A, Mansoor B (2021) Roll bonding processes: state-of-the-art and future perspectives. Metals *11*:1344. https://doi.org/10.3390/met11091344
20. Berto F, Sandnes L, Abbatinali F, Grong Ø, Ferro P (2018) Using the Hybrid Metal Extrusion & Bonding (HYB) process for dissimilar joining of AA6082-T6 and S355. Procedia Struct Integr *13*:249–254. https://doi.org/10.1016/j.prostr.2018.12.042
21. Beygi R, Zarezadeh Mehrizi M, Akhavan-Safar A, Safaei S, Loureiro A, Da Silva LFM (2021) Design of friction stir welding for butt joining of aluminum to steel of dissimilar thickness: heat treatment and fracture behavior. Int J Adv Manuf Technol *112*:1951–1964. https://doi.org/10.1007/s00170-020-06406-3
22. Meschut G, Janzen V, Olfermann T (2014) Innovative and highly productive joining technologies for multi-material lightweight car body structures. J of Materi Eng and Perform *23*:1515–1523. https://doi.org/10.1007/s11665-014-0962-3
23. Zhang T, Wang W, Yan Z, Zhang J (2021) Interfacial morphology and bonding mechanism of explosive weld joints. Chin J Mech Eng *34*:8. https://doi.org/10.1186/s10033-020-00495-7
24. Meco S, Pardal G, Ganguly S, Williams S, McPherson N (2015) Application of laser in seam welding of dissimilar steel to aluminium joints for thick structural components. Opt Lasers Eng *67*:22–30. https://doi.org/10.1016/j.optlaseng.2014.10.006
25. Mathieu A, Shabadi R, Deschamps A, Suery M, Matteï S, Grevey D, Cicala E (2007) Dissimilar material joining using laser (aluminum to steel using zinc-based filler wire). Opt Laser Technol *39*:652–661. https://doi.org/10.1016/j.optlastec.2005.08.014
26. Banea MD, Rosioara M, Carbas RJC, Da Silva LFM (2018) Multi-material adhesive joints for automotive industry. Compos Part B *151*:71–77. https://doi.org/10.1016/j.compositesb.2018.06.009
27. Furuya HS, Sato YT, Sato YS, Kokawa H, Tatsumi Y (2018) Strength improvement through grain refinement of intermetallic compound at Al/Fe dissimilar joint Interface by the addition of alloying elements. Metall and Mat Trans A *49*:527–536. https://doi.org/10.1007/s11661-017-4442-x
28. Jácome LA, Weber S, Leitner A, Arenholz E, Bruckner J, Hackl H, Pyzalla AR (2009) Influence of filler composition on the microstructure and mechanical properties of steel—aluminum joints produced by metal arc joining. Adv Eng Mater *11*:350–358. https://doi.org/10.1002/adem.200800319
29. Tanaka T, Morishige T, Hirata T (2009) Comprehensive analysis of joint strength for dissimilar friction stir welds of mild steel to aluminum alloys. Scr Mater *61*:756–759. https://doi.org/10.1016/j.scriptamat.2009.06.022

30. Shehabeldeen TA, Yin Y, Ji X, Shen X, Zhang Z, Zhou J (2021) Investigation of the microstructure, mechanical properties and fracture mechanisms of dissimilar friction stir welded aluminium/titanium joints. J Mater Res Technol *11*:507–518. https://doi.org/10.1016/j.jmrt.2021.01.026
31. Wang P, Chen D, Ran Y, Yan Y, Peng H, Jiang X (2019) Fracture characteristics and analysis in dissimilar Cu-Al alloy joints formed via electromagnetic pulse welding. Materials *12*:3368. https://doi.org/10.3390/ma12203368
32. Campbell FC (2011) Joining: Understanding the basics
33. Bay N (2011) Cold welding 1 characteristics, bonding mechanisms, bond strength. Met Constr *18*:369–372
34. Li L, Nagai K, Yin F (2008) Progress in cold roll bonding of metals. Sci Technol Adv Mater *9*. https://doi.org/10.1088/1468-6996/9/2/023001
35. Arbo SM, Bergh T, Solhaug H, Westermann I, Holmedal B (2018) Influence of thermomechanical processing sequence on properties of AA6082-IF steel cold roll bonded composite sheet. Procedia Manufacturing *15*:152–160. https://doi.org/10.1016/j.promfg.2018.07.189
36. Arbo SM, Bergh T, Holmedal B, Vullum PE, Westermann I (2019) Relationship between Al-Ni intermetallic phases and bond strength in roll bonded steel-aluminum composites with nickel interlayers. Metals *9*:827
37. Bagheri A, Toroghinejad MR, Taherizadeh A (2018) Effect of roughness and surface hardening on the mechanical properties of three-layered brass/IF steel/Brass composite. Trans Indian Inst Metals *71*:2199–2210. https://doi.org/10.1007/s12666-018-1351-7
38. Manesh HD, Shahabi HS (2009) Effective parameters on bonding strength of roll bonded Al/St/Al multilayer strips. J Alloys Compd *476*:292–299. https://doi.org/10.1016/j.jallcom.2008.08.081
39. Naseri M, Reihanian M, Borhani E (2016) Bonding behavior during cold roll-cladding of tri-layered Al/brass/Al composite. J Manuf Process *24*:125–137. https://doi.org/10.1016/j.jmapro.2016.08.008
40. Shin H-S, Park J-S, Jung Y-C, Ahn J-H, Yokoyama Y, Inoue A (2009) Similar and dissimilar friction welding of Zr–Cu–Al bulk glassy alloys. J Alloys Compd *483*:182–185. https://doi.org/10.1016/j.jallcom.2008.07.231
41. Sahin M (2009) Joining of stainless-steel and aluminium materials by friction welding. Int J Adv Manuf Technol *41*:487–497. https://doi.org/10.1007/s00170-008-1492-7
42. Sahin M, Çıl E, Misirli C (2013) Characterization of properties in friction welded stainless steel and copper materials. J. of Materi Eng and Perform *22*:840–847. https://doi.org/10.1007/s11665-012-0310-4
43. Uzun H, Dalle Donne C, Argagnotto A, Ghidini T, Gambaro C (2005) Friction stir welding of dissimilar Al 6013-T4 To X5CrNi18-10 stainless steel. Mater Des *26*:41–46. https://doi.org/10.1016/j.matdes.2004.04.002
44. Watanabe T, Takayama H, Yanagisawa A (2006) Joining of aluminum alloy to steel by friction stir welding. J Mater Process Technol *178*:342–349. https://doi.org/10.1016/j.jmatprotec.2006.04.117
45. Mukuna Patrick M, Akinlabi E (2013) Friction stir welding of dissimilar materials: an overview. In International Scholarly and Scientific Research & Innovation
46. Taban E, Gould JE, Lippold JC (2010) Dissimilar friction welding of 6061-T6 aluminum and AISI 1018 steel: properties and microstructural characterization. Mater Des *31*:2305–2311. https://doi.org/10.1016/j.matdes.2009.12.010
47. Coelho RS, Kostka A, Dos Santos JF, Kaysser-Pyzalla A (2012) Friction-stir dissimilar welding of aluminium alloy to high strength steels: mechanical properties and their relation to microstructure. Mater Sci Eng A *556*:175–183. https://doi.org/10.1016/j.msea.2012.06.076
48. Sahu PK, Pal S, Pal SK (2017) Al/Cu dissimilar friction stir welding with Ni, Ti, and Zn foil as the interlayer for flow control, enhancing mechanical and metallurgical properties. Metall Mater Trans A *48*:3300–3317. https://doi.org/10.1007/s11661-017-4093-y

49. Sharma N (2017) Friction stir welding of aluminum to copper—an overview. Trans. Nonferrous Met. Soc, China, p 24
50. Cao F, Li J, Hou W, Shen Y, Ni R (2021) Microstructural evolution and mechanical properties of the friction stir welded Al Cu dissimilar joint enhanced by post-weld heat treatment. Mater Charact *174*:110998. https://doi.org/10.1016/j.matchar.2021.110998
51. Zhang Z, Li B, Feng X, Shen Y, Hu W (2014) Friction-stir welding of titanium/aluminum dissimilar alloys: joint configuration design, as-welded interface characteristics and tensile properties. Proc Inst Mech Eng B J Eng Manuf *228*:1469–1480. https://doi.org/10.1177/0954405413518514
52. Zuo D, Hu S, Shen J, Xue Z (2014) Intermediate layer characterization and fracture behavior of laser-welded copper/aluminum metal joints. Mater Des *58*:357–362. https://doi.org/10.1016/j.matdes.2014.02.004
53. Bergh T, Sandnes L, Johnstone DN, Grong Ø, Berto F, Holmestad R, Midgley PA, Vullum PE (2020) Microstructural and mechanical characterisation of a second generation hybrid metal extrusion & bonding aluminium-steel butt joint. Mater Charact. https://doi.org/10.1016/j.matchar.2020.110761
54. Sandnes L, Bergh T, Grong Ø, Holmestad R, Vullum PE, Berto F (2021) Interface microstructure and tensile properties of a third generation aluminium-steel butt weld produced using the Hybrid Metal Extrusion & Bonding (HYB) process. Mater Sci Eng A *809*:140975. https://doi.org/10.1016/j.msea.2021.140975
55. Bergh T, Fyhn H, Sandnes L, Blindheim J, Grong Ø, Holmestad R, Berto F, Vullum PE (2023) Multi-material joining of an aluminum alloy to copper, steel, and titanium by Hybrid Metal Extrusion & Bonding. Metall Mater Trans A *54*:2689–2702. https://doi.org/10.1007/s11661-023-07047-3

# Applying Principles of Biological Transformation for Product Development and Manufacturing

**Magda Berkhahn, Ina Peters, and Rainer Stark**

**Abstract** Biological Transformation (BT) in manufacturing, also referred to as biologicalization, describes the new paradigm to converge technical, digital, and biological systems to promote sustainability-optimized product and production systems and, thus, meet the current challenges regarding sustainability and resilience. The project BioFusion 4.0 strives to refine Industry 4.0 by integrating biological principles, materials, and organisms. Therefore, relevant principles of Biological Transformation have been identified and classified as a taxonomy. However, there is a strong need for guidance regarding how to apply those principles in the manufacturing industry. Thus, a method to systematically align BT principles with specific technological problems was developed and tested by conducting a workshop series with six technology experts from applied research and industry. The method is based on an iterative top-down approach and follows four basic steps: (1) Characterizing the technological problem functionally and precisely, yet in a universal non-specialist language (including its framework, requirements, and state of development), (2) finding analogies from biology which match the characteristics of the established technical problem without contradiction, (3) identifying suitable biological solution strategies for these specific cases, and (4) abstracting those strategies into technological modes of action by means of the predefined BT principles. This chapter describes the method in detail, gives instructions for applying it, and discusses the central findings of the workshop series on utilizing principles of Biological Transformation for innovating product development.

**Keywords** Biological transformation · Biologicalization · Product development · Requirement engineering

M. Berkhahn (✉) · I. Peters · R. Stark
Technische Universität Berlin, Berlin, Germany
e-mail: m.berkhahn@tu-berlin.de

 
S. Fukushige et al. (eds.), *EcoDesign for Circular Value Creation: Volume I*,
https://doi.org/10.1007/978-981-97-9068-5_37

# 1 Introduction

Employing natural materials and using natural phenomena as a source of inspiration for problem-solving is as old as humankind. However, in our highly technologized world with all the challenges that occur as side-effects of technical progress it pays off to remember those approaches. Evolution has been tremendously creative in finding solutions for the crudest of problems, making biology an almost inexhaustible source of ideas. When applied to complex modern technological problems, such biological strategies can be a true driver of innovation. Thus, systematically including biology in product creation and manufacturing is seen as a transformational development of Industry 4.0 [1]. This Biological Transformation is about aligning means of production and information technology with means of nature while at the same time ensuring ecological sustainability.

Industrial progression evolved around new (information) technologies fueled by engineering research. Its Biological Transformation, however, depends on the ability to merge knowledge sets from different fields, demanding interdisciplinary creation. Despite this challenge, new opportunities arise regarding resilience and reliability in production and supply, as well as achieving overarching societal goals such as equality and sustainability. The key driver of striving for a biologically transformed manufacturing industry is the presumption of positive effects, as can be observed in natural ecosystems, such as sustained availability of energy and resources, high viability through adaptivity, or the ability to exist in individual niches. The growing demand for such innovative solutions has changed the fuzzy process of searching for inspiration in nature to a scientific discipline of modern science, backed up by increasing research activities [1–4].

To borrow ideas from biology, engineers first need to know about biological systems and strategies to then gain a deeper understanding of them. Collaborating with experts from biological disciplines is the most promising approach when it comes to detailed questions and deep understanding. However, to know which expertise to look for, engineers must be enabled to identify general principles of Biological Transformation that might apply to their specific project. In this chapter, we demonstrate a method for identifying suitable biologically transformative principles for existing technological challenges as the first step towards biologically transformed solutions.

# 2 State of the Art

## *2.1 Biological Transformation in Industry*

The term *bionics* was defined by Jack E. Steele in 1960 [5] as the transfer of biological forms and functions to technical application fields. Following his argument, more and more researchers elaborated on this special field of manufacturing under multiple terms like *biomimicry, biomimetics* or *biologically inspired design*. As terms and approaches varied over the past decades, so did the definitions to describe the Biological Transformation (BT) in manufacturing. One widely used definition is that of Byrne et al. [1], which states a connection between the introduction of biological principles, materials, functions, structures and resources and an intelligent and sustainable manufacturing industry.

In the research project BIOTRAIN, Fraunhofer-Gesellschaft introduces a definition of Biological Transformation for the manufacturing industry [6]. In this context, BT is seen as optimizing a manufacturing system to meet its societal and business challenges as the ultimate goal of the convergence of the biosphere and technosphere. This chapter defines BT in the context of industrial manufacturing and engineering as defined within the research project BioFusion 4.0. The definition is consistent with the discussed literature while additionally addressing the superordinate goal of sustainability as well as the basic modes of action of BT [7]:

Biological Transformation in manufacturing is a holistic approach to change industrial value creation toward sustainable optimized product and production systems, by an accelerating convergence of technical, digital, and biological systems in the manufacturing environment. BT proceeds in three complementary developmental modes: (1) the integration of biological materials, structures, organisms, processes, and functionalities, (2) the inspiration by nature and the transfer to the design of products and manufacturing technologies as well as (3) the interaction of the bio- and technosphere by means of information technology.

As a systematic application of knowledge is key to achieving the aspired effects, principles of Biological Transformation (BT principles) in manufacturing have been captured and classified in a taxonomy in earlier works [7]. This taxonomy functions as a classification system to establish effective principles that lead to transformation of the manufacturing industry in the form of either inspiration from, integration of or interaction with natural systems. The BT principles describe mechanisms that can be applied in industrial manufacturing contexts. Natural organisms and ecosystems are following certain principles or strategies to survive just as technological systems rely on complementary principles for functionality and efficiency. This chapter aims at utilizing these principles for the early phases of product development.

## 2.2 *Biological Transformation of Product Development*

General procedures for biomimetic product development have been depicted by standards of the International Standard Organization (ISO) and the Association of German Engineers (VDI). The ISO standard 18458:2015 [8] describes a procedure according to which a biomimetic system can be developed for a technical problem. This procedure is also regarded in the VDI standards VDI 6220 and VDI 6226 [9, 10]. According to those standards, the basic steps of a biomimetic engineering process (BEP) are those depicted in Fig. 1 [11].

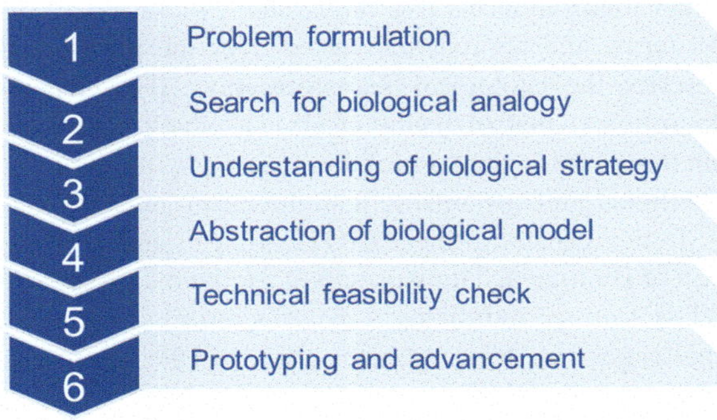


**Fig. 1** Workflow for biomimetic engineering process [11]. (© Plant Biomechanics Group BIOKON)

Each result of this procedure is assumed to be per se a bionic system and thus holds the promise of bionic products. The criteria for a bionic system following the ISO standard do not so much relate to outcome quality as to the fulfillment of three general steps along the procedure: "Functional analysis of the biological system," "Abstraction from the system to the model," and "Transfer and application without using the biological system." The same approach can be found in the *Biomimicry Design Process* of the *Biomimicry Institute*, which originated from Carl Hasttrich in 2005 [12, 13].

Product developers can find support when following this approach in tools and methods from research. Table 1 gives an overview of seven easily accessible tools, according to their characteristics, and their position in the process of biomimetic engineering process, referring to the ISO and VDI standards. Those tools are good examples of how, in the field of Biological Transformation, design support so far concentrates mostly on bio-*inspiration* and is less suited for its other two dimensions of *integration* of or *interaction* with biological systems for technical problem-solving.

Familiar biomimetic development processes are not aiming at fulfilling quality requirements regarding biomimetic properties. Applying them does not ensure that the target product systematically fits into the concept of Biological Transformation and can thus serve the overarching goal (i.e., the transformation of industry towards sustainability).

## 3 Linking Development Projects to Biological Transformation

The best time to integrate principles for a Biological Transformation of the manufacturing industry into the product development process is in an early stage i.e., during the requirements analysis up to conceptual. However, to do so, it is necessary to understand how the specific project relates to Biological Transformation in the first place. Therefore, we have developed a method that helps product

**Table 1** Biomimetic tools

| Name | Description | Position in the biomimetic engineering process (Fig. 1) |
|---|---|---|
| AskNature Database [14] | Online, freely available, searchable database of strategies for living nature | Search for bio. Analogy<br>Understanding bio. Strategy |
| Biomimicry Taxonomy [15] | Hierarchy of functions and attributes that provides the principal classification scheme for biological functions and strategies | Problem formulation<br>Understanding bio. Strategy |
| BioCards [16] | Formal descriptions of biological phenomena and their underlying functional principles to propose conceptual ideas for design solutions | Understanding bio. Strategy<br>Abstraction of biol. Model |
| BioTRIZ [17] | Identifying innovation principles inspired by nature through requirement contradictions to be solved | Search for bio. Analogy |
| DANE [18] | Database of biological structure-behavior-functions that can be transferred into the technical solution | Search for bio. Analogy<br>Understanding bio. Strategy |
| E2BMO [19] | A computer-aided user interface to access an ontology of biological systems (including functions) via semantic translation | Search for bio. Analogy<br>Understanding bio. Strategy<br>Abstraction of biol. Model |
| Idea inspire 3.0 [20] | Web-based tool to search database natural and artificial solution systems by their multi-modal representations | Understanding bio. Strategy<br>Abstraction of biol. Model |

managers and designers to define their projects in the context of Biological Transformation by allocating BT principles. The process follows current procedures of biomimetic development, can be supported by available tools, and creates a systematic comprehension of Biological Transformation to be integrated into the solution generation.

## 3.1 The BT-Driven Cycle for Requirement Engineering

The introduced approach guides participants in defining the solution space with a focus on solutions that contribute to the Biological Transformation of the industry, whereby it connects different tools of biomimetic design. As illustrated in Fig. 2, it comprises four main steps, with different activities assigned to them. Traversing the cycle in multiple loops with different teams and growing biological expertise is creating an iterative work process that can facilitate the evaluation and refinement of solution concepts. The introduced approach guides participants in defining the solution space with a focus on solutions that contribute to the Biological

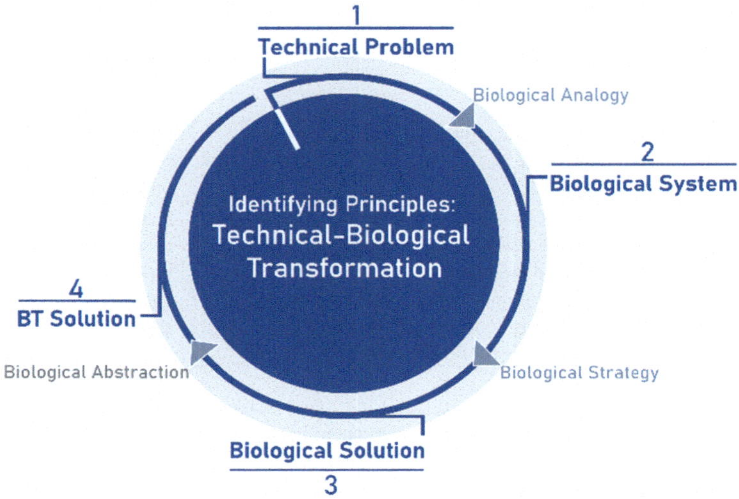


**Fig. 2** Illustration of the BT-driven cycle for requirement engineering

**Table 2** Characteristics of the technical problem in BT-driven cycle for requirement engineering

| Superordinate features of outcome | Technical obstacles | Structural organization | Physiological requirements | Framework conditions |
|---|---|---|---|---|
| Mandatory basic functions<br>Pursued benefits | Features to make current solutions inadequate | Division into sub-tasks<br>Building components | Dynamics<br>Color<br>Shape<br>Smell<br>Noise<br>Energetics<br>Size<br>Power/ strength | Surrounding media<br>Materials<br>Time period<br>Contact with humans or other living organisms<br>Natural non-living component<br>Required technologies |

Transformation of the industry, whereby it connects different tools of biomimetic design. Traversing the cycle in multiple loops with different teams and growing biological expertise is creating an iterative work process that can facilitate the evaluation and refinement of solution concepts. The single steps are described in more detail in the following sections.

**Technical Problem**

The BT-driven cycle for requirement engineering is started when a technical problem to solve has been identified. As an initial activity, the technical problem formulation in a general domain unspecific language is carried out. Information regarding various characteristics ought to be described, as outlined by Table 2. This results in a compiled problem description that displays all key characteristics and is comprehensible from a technical and biological perspective. This description facilitates the following interdisciplinary knowledge exchange.

**Biological System**

To accomplish the next step of identifying biological systems, analogies between the technical problem and biological phenomena are to be found. This is accomplished by brainstorming, researching, and describing systems where similar characteristics can be found in nature. Particularly systems that accumulate multiple characteristics are to be envisaged. In addition, it should be clarified whether a characteristic is fulfilled by a single organism, a group of the same species, or by a community of different species in interaction, as this will have an impact on the BT principles that will be considered later in the cycle.

The collected analogies are systematically examined following the items for characterization as displayed in Table 2. This activity is supported by further internet research and different tools that are assigned the step "2—Search for biological analogy" in the biomimetic product development workflow (Table 1).

As a result, a collection of analogous biological systems fitting several characteristics of the technical problem is available for the following process steps.

**Biological Solution**

The next step is the identification of solutions the biological systems adopt in analogous situations. The key activity to perform is to match biological strategies from identified organisms or natural systems in relation to the problem description. A biological strategy is an active or passive behavior of biological systems that serves a function to survive. Supportive tools are the ones that are assigned to "3 Understanding biological strategy" in the biomimetic product development workflow (Table 1).

**Biological Transformation Solution**

The central activity here is transferring biological solution strategies into BT principles. Leading questions with corresponding characteristics support the process step, as displayed in Table 3.

Each characteristic of Table 3 targets a specific Group of principles for Biological Transformation. By subsequently questioning the extent to which a technical solution fits a single principle of the taxonomy of Biological Transformation, it is possible to further condense the selection of principles that can be effectively applied.

The step results in the characterization of basic solution principles regarding the Biological Transformation. The conclusive result of the BT-driven cycle for requirement engineering is a deductive textual representation of the basic solution space and its effective principles to be elaborated and tested in interdisciplinary collaboration with domain-specific experts. Consequential steps in the overall product development process are the conceptualization of a solution, which can now address the Biological Transformation.

**Table 3** Leading questions to identify BT principles

| Bio inspiration | Bio integration | Bio interaction |
|---|---|---|
| Can the biological strategies be adopted to yield a technical solution through one of the following principles? Adopting physiological properties<br>Mimicking modes of communication and information processing<br>Emulating resource management<br>Mimicking to achieve self-empowering behavior<br>Maintaining function under changes in its environment | Can the biological system be implemented in the product or production to fulfill any of the following functions?<br>Material synthesis<br>Material substitution<br>Material removal/ decomposition<br>Energy supply and storage<br>Therapeutics<br>Modification of bio. material | Can the behavior be utilized to facilitate the interaction of biological and technical systems by means of information technology?<br>Status of biological system integrated technical functionality<br>Implementing biological components to process information<br>Transferring the biological actuators directly or synthetically<br>Reciprocal information exchange between natural and technical elements |

## 3.2 *Exemplary Implementation of the BT-Driven Cycle for Requirement Engineering*

The method has been applied in six online workshops of two hours each, in which technology experts (engineers and project manager) were led through the BT-driven cycle step by step to identify suitable BT principles for ongoing development processes in their fields of expertise.

**Workshop Series**

Each workshop dealt with a specific product development use case corresponding to the expertise of the participating engineers and project managers. Those use cases derive from the research project BioFusion 4.0 and purposefully cover different industrial product and manufacturing solutions, covering digital twins in bio-inspired product engineering [21], ecological intelligent services for production, intelligent recirculation of materials, biologically optimized process simulation of milling processes [22], bionic integration for networked production systems [23], additive manufacturing with biogenic and biodegradable polymers, and biointelligent assistant systems for workers. These were used to investigate the applicability of the development approach and its results.

The workshop was attended by two persons, constituting the development team. The workshop host took a passive role and led the workshop by guiding each team and providing answers on how to conduct the workshop. In addition, the host provided the tools as they were needed in each step (see Sect. 3.1) and explained their use. To support the participants, they were provided with a visual display of the cycle for requirement on a digital canvas to work on.

In the beginning, the procedure and the goal of the workshop were explained to the development team. Since this took place online, the procedure was explained with a visualization according to Fig. 2.

Participants then started with step 1 of the BT-driven cycle for requirement engineering and described the technical system using Table 2.

For identifying biological systems, participants were introduced to the AskNature database. For classifying biological solutions, they used the Biomimicry Taxonomy. For each process step, guiding questions were formulated to help acquire relevant information. Part of the canvas was also a synoptic table. There, the collected information for each formulated technical problem was sorted for better traceability of the deduction process.

**Workshop Results and Findings**

All participants were able to complete the task and generated a basic solution approach on what principles of BT are suitable to be integrated into their product development project. They were able to make a connection between the problem characteristics of technical products and BT principles to answer these problems.

Thus, for the case of the worker assistance system, it was recognized that a characteristic in the problem description is the ignorance of the workers about their own posture at work and therefore they cannot correct it into a healthy state. Through the BT-driven cycle, it was determined that the principles of self-awareness and self-optimization must be implemented by using feedback mechanisms inspired by nature. The integration of visual, haptic, or acoustic stimulus transmission was included in the development of the assistance system.

For the use case of additive manufacturing with biogenic polymer, it was identified that the materials should not only be biogenic but also biodegradable. In addition, the reason for using biogenic polymer was to create positive environmental attributes, namely low emissions and better resource efficiency. The result was that the principle of circularity: closed loop recycling shall be applied. Thus, the plan for the project was to produce the polymer from food waste, this can be composted, and is thereby returned to the crop production. Hence it should be avoided that the disposal of a 3D printed product takes place outside the natural carbon cycle, for example by incineration.

Although all participants were given the same tasks, tools, and guiding questions, they approached the process very differently. The biggest variation was apparent in the first task of defining technical problems. Since participants chose different abstraction levels, the defined problems ranged from open questions of how BT could affect product development in general to very concrete technological challenges such as collecting health data of workers to derive recommendations at the workplace (e.g., recreational breaks). Although the abstraction level of the technical problem influenced the abstraction level of the outcomes of the following process steps and despite the wide range of technical problems, the analogies, and biological systems identified showed remarkable overlaps. Out of the six experts, four identified analogous biological systems in organization and communication of ant colonies, three in beehives. Living creatures reacting to environmental conditions were also chosen three times. Two participants compared their technical problem to the interconnected quality of ecosystems and two concentrated on material cycles in nature. Also, two were reminded of human learning processes, just as two others found the mycelium in the forest connecting tree roots underground analogous. The accumulation of chosen biological systems might be due to restrictions on the AskNature database. However, it is striking that those analogies are rather well-known examples from public biological debates surrounding, i.e., swarm intelligence, sustainability, and the recent public realization that plants actually communicate. Thus, people seem to choose analogies that they can relate to. Those do not necessarily have to be the most suitable but the easiest to understand and work with.

However, all participants were consequently able to allocate BT principles to their technical problem and elaborately argue their case. They showed detailed understanding of the concept of Biological Transformation with respect to their technology in the end of the process whereas the initial problem definition indicated a rather blurry idea of how BT could affect their technology, often even stating this question explicitly.

Therefore, the method reached its goal to raise awareness for BT potential as well as to help engineers to locate their technical problems within the BT taxonomy for manufacturing.

## 4 Critical Review

Interdisciplinary collaboration between engineers, ICT teams, and biologists in the BT-driven cycle for requirement engineering (as every interdisciplinary collaboration) might be challenging due to different mind-sets, domain-specific language etc. While the described method has not been developed to be applied by interdisciplinary teams it can function as a tool to introduce experts from product development to biological principles, thus, help them build up basic understanding of biological phenomena and their classification, which is expected to be useful when entering interdisciplinary teamwork in later development stages.

However, looking at the workshop results and finsings in 3.2.2 it can be argued that even at this early stage a BT integration should be conducted by interdisciplinary teams of technical experts and biologists to ensure that also biological systems are considered that are not pivotal elements of public debate. For this stage of approaching BT, those biologists should rather be generalists. For further concretization of biologically transformed technical solutions, experts in specific fields might be needed.

Another option would be to implement the BT driven cycle for requirement engineering into more effective assisting tools, such as knowledge-based engineering systems (KBES) that offer detailed guidance through broad data sets to help identify the most suitable biological systems with respect to a concrete technical problem. The transfer of the BT-driven cycle into a digital information model for KBES would also offer great advantages in responding to known challenges in requirements engineering. The identified BT principles function as a solution requirement in the development process, which yield the systems requirements and subsequently the system architecture. For a continuous collaboration on and with requirements, bidirectional traceability of those requirements throughout the development process is important. Eliciting, tracing, and communicating changes to requirements helps collaboration in the development team, increases transparency, and helps to rapidly bring the product to market [24]. Since the variety and number of requirements are constantly increasing, they need to be managed with sufficient reliability. To keep up with fast evolving requirements a digital solution is to be preferred [24]. The BT-driven cycle for requirement engineering as a digital assistance tool can also provide interfaces to other development tools for better user experience, such as those shown in Table 1.

Another valid point is that the method does not consider the sustainability potentials of BT integration although this is part of the very definition of Biological Transformation. Still, the described method is not meant to be the final step in BT-driven product development but the first. Indisputably, further methods and tools need to be developed that together will form a comprehensive toolkit guiding interdisciplinary teams further on from where this method ends up to the BT product.

With such a toolkit it would also be possible to develop a design of experiment that follows a product idea from the first consideration up to the final product. This way the quality of results could be compared to product design processes that use the tool kit and those that do not.

## 5 Conclusion

Biological Transformation is said to revolutionize manufacturing. While biomimicry and bionic have long found their niche in product development, the more complex requirements of BT are not yet ready for broad integration into everyday manufacturing practices. By defining BT principles and developing processes, methods, and tools for their application, they are gradually becoming utilizable by practitioners. In this chapter, we have presented a method based on standards for biomimetic product development that enables engineers and ICT experts to make the link between their actual technological problems and BT principles as a first step towards developing biologically transformed or even biologically transforming products.

**Acknowledgments** This chapter was written as part of the research project "Biological Transformation 4.0: Further development of Industrie 4.0 by integrating biological principles (BioFusion 4.0)," funded by the German Federal Ministry of Education and Research (BMBF) and supervised by the Project Management Agency Karlsruhe (PTKA). The responsibility for the contents of this publication lies with the authors.

## References


1. Byrne G, Dimitrov D, Monostori L, Teti R, van Houten F, Wertheim R (2018) Biologicalisation: biological transformation in manufacturing. CIRP J Manuf Sci Technol 21: 1–32. https://doi.org/10.1016/j.cirpj.2018.03.003
2. van Houten F et al (2021) Bio-based design methodologies for products, processes, machine tools, and production systems. CIRP J Manuf Sci Technol 32:46–60. https://doi.org/10.1016/j.cirpj.2020.11.008
3. Bergs T et al (2020) Application cases of biological transformation in manufacturing technology. CIRP J Manuf Sci Technol 31:68–77. https://doi.org/10.1016/j.cirpj.2020.09.010
4. Miehe R et al (2018) The biological transformation of the manufacturing industry—envisioning biointelligent value adding. Procedia CIRP 72:739–743. https://doi.org/10.1016/j.procir.2018.04.085
5. Steele JE (1960) Living prototypes: the key to new technology. Dayton, United States
6. Bauernhansl T, Brecher C, Drossel W (2019) Biointelligenz: eine neue Perspektive für nachhaltige industrielle Wertschöpfung: Ergebnisse der Voruntersuchung zur biologischen Transformation der industriellen Wertschöpfung (Biotrain). 2019. [Online]. Available: http://publica.fraunhofer.de/eprints/urn_nbn_de_0011-n-5412052.pdf
7. Berkhahn M, Kremer G, Riedelsheimer T, Lindow K, Stark R Taxonomy for biological transformation principles in the manufacturing industry. in Proceedings of 18th Global Conference on Sustainable Manufacturing, 2022, to be Published
8. DIN ISO 18458:2015 Bionik—Terminologie, Konzepte und Methodik, DIN
9. Biomimetics—Fundamentals, conception, and strategy: part 1, VDI 6220 Part 1, VDI, Sep. 2021
10. Biomimetics—Architecture, civil engineering, industrial design—basic principles: part 1, VDI 6226 Part 1, VDI, Feb. 2015
11. Harst S, Früchtl M, Neugebauer R (2022) Biological transformation of manufacturing—from a vision to industrial transfer—interim evaluation from the perspective of applied research. Procedia CIRP 107:925–930. https://doi.org/10.1016/j.procir.2022.05.086
12. Biomimicry Institute, The power of the Biomimicry Design Spiral—Biomimicry Institute. [Online]. Available: https://biomimicry.org/biomimicry-design-spiral/. Accessed: 23 Jan 2023
13. Biomimicry Toolbox, The Biomimicry Process—Biomimicry Toolbox. [Online]. Available: https://toolbox.biomimicry.org/methods/process/. Accessed 23 Jan 2023
14. Search—Ask Nature. [Online]. Available: https://asknature.org/?s=&page=0&hFR%5Bpost_type_label%5D%5B0%5D=Biological%20Strategies&is_v=1. Accessed 23 Jan 2023
15. Hooker G, Smith E (2016) ASKNATURE and the biomimicry taxonomy. Insight 19(1):46–49. https://doi.org/10.1002/inst.12073
16. Lenau TA, Keshwani S, Chakrabarti A, Ahmed-Kristensen S (n.d.) Biocards and level of abstraction. Design Theory and Research Methodology, Design Processes
17. Bogatyrev N, Bogatyreva O (2014) BioTRIZ: a win-win methodology for eco-innovation. In: Springer Cham (ed) Eco-innovation and the development of business models: lessons from experience and new Frontiers in theory and practice. Springer International, Switzerland
18. Vattam S, Wiltgen B, Helms M, Goel AK, Yen J (2011) DANE: fostering creativity in and through biologically inspired design. In: Taura T, Nagai Y (eds) Design creativity 2010. Springer, London, pp 115–122
19. McInerney S et al (2018) E2BMO: facilitating user interaction with a BioMimetic ontology via semantic translation and Interface design. Designs 2(4):53. https://doi.org/10.3390/designs2040053
20. Chakrabarti A, Siddharth L, Dinakar M, Panda M, Palegar N, Keshwani S (2017) Idea inspire 3.0—a tool for analogical design. In: Chakrabarti A, Chakrabarti D (eds) Smart innovation, systems and technologies, research into Design for Communities, volume 2. Springer, Singapore, pp 475–485

21. Seegrün A, Mügge J, Riedelsheimer T, Lindow K Digital Twinsfor Sustainabilityin the Context of Biological Transformation. Proceedings of 18th Global Conference on Sustainable Manufacturing, 2022, to be published
22. König V, Berkhahn M, Riedelsheimer T, Plakhotnik D, Stautner M (2022) Biological transformation in process simulation for enhancing ecological sustainability indicators. Procedia CIRP 110:53–58. https://doi.org/10.1016/j.procir.2022.06.012
23. Lange A, Knothe T, Kohl H, Seliger G (2022) Biological transformation: principles to enhance holistic production systems. Procedia CIRP 110:293–298. https://doi.org/10.1016/j.procir.2022.06.053
24. Hull E, Jackson K, Dick J (2002) Requirements engineering. Springer, London. [Online]. Available: https://ebookcentral.proquest.com/lib/kxp/detail.action?docID=3074659

The manufacturer's authorised representative in the EU is Springer Nature Customer Service Centre GmbH, Europaplatz 3, 69115 Heidelberg, Germany. If you have any concerns regarding our products, please contact ProductSafety@springernature.com

Printed and bound by CPI Group (UK) Ltd, Croydon, CR0 4YY
30/09/2026
02225299-0013